Nutrition and Behavior

Human Nutrition
A COMPREHENSIVE TREATISE

General Editors:
Roslyn B. Alfin-Slater, University of California, Los Angeles
David Kritchevsky, The Wistar Institute, Philadelphia

A Continuation Order Plan is available for this series. A continuation order will bring delivery of each new volume immediately upon publication. Volumes are billed only upon actual shipment. For further information please contact the publisher.

Nutrition and Behavior

Edited by

Janina R. Galler

Boston University School of Medicine
Boston, Massachusetts

PLENUM PRESS · NEW YORK AND LONDON

Library of Congress Cataloging in Publication Data

Main entry under title:

Nutrition and behavior.

(Human nutrition; v. 5)
Bibliography: p.
Includes index.
1. Malnutrition — Complications and sequelae — Addresses, essays, lectures. 2.
Human behavior — Nutritional aspects — Addresses, essays, lectures. 3. Malnutrition in children — Complications and sequelae — Addresses, essays, lectures. 4.
Malnutrition — Complications and sequelae — Animal models — Addresses, essays, lectures. 5. Nutrition — Psychological aspects — Addresses, essays, lectures. 6. Food preferences — Addresses, essays, lectures. I. Galler, Janina R.,
1949-   . II. Series. [DNLM: 1. Behavior. 2. Diet therapy. 3. Diet — Adverse effects. 4. Nutrition disorders. QU 145 H9183 1979 v. 5]
QP141.H78 vol. 5   612.3′9   83-27035
ISBN 0-306-41435-X

© 1984 Plenum Press, New York
A Division of Plenum Publishing Corporation
233 Spring Street, New York, N.Y. 10013

Printed in the United States of America

*This volume is dedicated to
the late HERBERT G. BIRCH,
who inspired it,
and to HAMISH N. MUNRO,
who encouraged its fruition.*

Contributors

Arnold E. Andersen • Department of Psychiatry and Behavioral Sciences, The Johns Hopkins University School of Medicine, Baltimore, Maryland 21205

David E. Barrett • Children's Hospital Medical Center; Harvard Medical School, Boston, Massachusetts 02115

M. W. P. Carney • Northwick Park Hospital and Clinical Research Centre, Harrow, Middlesex HA1 3UJ, United Kingdom

Patricia Coleman • Department of Nutritional Sciences, Faculty of Medicine, University of Toronto, Toronto, Ontario M5S 1A3, Canada

C. Keith Conners • Children's Hospital National Medical Center, Washington, D.C. 20010

Mary A. Crawford • Department of Human Development and Family Studies, Cornell University, Ithaca, New York 14850

Susan F. Fleischer • Department of Psychology, Queens College, City University of New York, Flushing, New York 11367

Janina R. Galler • Division of Psychiatry, Boston University School of Medicine, Boston, Massachusetts 02118

Robin B. Kanarek • Department of Psychology, Tufts University, Medford, Massachusetts 02155

Magdalena Krondl • Department of Nutritional Sciences, Faculty of Medicine, University of Toronto, Toronto, Ontario M5S 1A3, Canada

L. Thomas Kucharski • Department of Child Psychiatry, Boston University School of Medicine, Boston, Massachusetts 02146

Daisy Lau • Brescia College, University of Western Ontario, London, Ontario N6G 1H2, Canada

Loy D. Lytle • Laboratory of Psychopharmacology, Department of Psychology, University of California, Santa Barbara, California 93106; Kellogg Co., Battle Creek, Michigan 49106

Robin Marks-Kaufman • Department of Psychology, Tufts University, Medford, Massachusetts 02155

Jean Mayer • President, Tufts University, Medford, Massachusetts 02155

Ellen Messer • Department of Anthropology, Yale University, New Haven, Connecticut 06520

Mark F. Nelson • Laboratory of Psychopharmacology, Department of Psychology, University of California, Santa Barbara, California 93106; Kellogg Co., Battle Creek, Michigan 49106

Nilla Orthen-Gambill • Department of Psychology, Tufts University, Medford, Massachusetts 02155

Henry N. Ricciuti • Department of Human Development and Family Studies, Cornell University, Ithaca, New York 14850

David Rush • Rose F. Kennedy Center, Bronx, New York 10461; Division of Pediatric and Perinatal Epidemiology, Departments of Pediatrics and of Obstetrics and Gynecology, Albert Einstein College of Medicine, Bronx, New York 10461

Gerald Turkewitz • Department of Psychology, Hunter College, City University of New York, New York, New York 10021; Departments of Psychiatry and Pediatrics, Albert Einstein College of Medicine, Bronx, New York 10461

Carla S. Whitacre • Laboratory of Psychopharmacology, Department of Psychology, University of California, Santa Barbara, California 93106; Kellogg Co., Battle Creek, Michigan 49106

Preface

After the appearance of the four-book series *Human Nutrition: A Comprehensive Treatise*, it became apparent to the editors that an important area of nutrition had been overlooked, namely, behavioral aspects of nutrition. There are two areas in which nutrition and behavior interact. On the one hand, malnutrition may play a major role in determining behavior; alternatively, often aspects of behavior influence the eating habits of populations and individuals and thus affect their nutritional status.

Volume 5 of this series speaks eloquently to both features of this important topic. Various aspects of the influence of behavior modification and nutrition have been explored by a number of qualified investigators. It is hoped that this volume will prove a valuable addition to the subjects covered in the other volumes.

Roslyn B. Alfin-Slater
David Kritchevsky

Los Angeles and Philadelphia

Contents

Chapter 3

The Behavioral Consequences of Malnutrition in Early Life
Janina R. Galler

Chapter 4

The Behavioral Consequences of Protein–Energy Deprivation and Supplementation in Early Life: An Epidemiological Perspective
David Rush

Chapter 5

Nutritional Therapy in Children

C. Keith Conners

Chapter 6

Vitamin Deficiencies and Excesses: Behavioral Consequences in Adults

M. W. P. Carney

Chapter 7

Mechanisms of Nutritient Action on Brain Function

Loy D. Lytle, Carla S. Whitacre, and Mark F. Nelson

Part II • Behavioral Determinants of Food Intake and Nutritional Status

Chapter 8

The Role of Mother–Infant Interaction in Nutritional Disorders

*Janina R. Galler, Henry N. Ricciuti, Mary A. Crawford, and
L. Thomas Kucharski*

Chapter 9

Anorexia Nervosa and Bulimia: Biological, Psychological, and Sociocultural Aspects

Arnold E. Andersen

Chapter 10

Obesity: Possible Psychological and Metabolic Determinants

Robin B. Kanarek, Nilla Orthen-Gambill, Robin Marks-Kaufman, and Jean Mayer

Chapter 11

Psychological Factors Affecting Food Selection

Daisy Lau, Magdalena Krondl, and Patricia Coleman

Chapter 12

Sociocultural Aspects of Nutrient Intake and Behavioral Responses to Nutrition

Ellen Messer

Introduction: The Challenge of Nutrition and Environment as Determinants of Behavioral Development

Janina R. Galler

In recent times, knowledge progressing along three different lines has converged to illuminate the complex interactions between nutrition and related environmental factors, on one hand, and behavior, on the other hand. These three advancing areas are the behavioral sciences, knowledge of the effects of specific nutrients on brain and behavioral function, and, finally, the study of gross malnutrition in underdeveloped regions of the world and its impact on behavior.

First, the study of human behavior is one of the major advances of the 20th century, beginning with Freud's discovery of the importance of early traumatic experiences in the genesis of neurotic disorders in adulthood. From Freud and his followers, we have learned about the complexity of human behavior, and also its underlying structure. In contrast to earlier psychoanalysts' emphasis on the self, later psychiatrists, notably Harry Stack Sullivan, whose work on the environmental correlates of schizophrenia remains a classic, added a new dimension to the understanding of mental illness by uncovering the role of environmental influences. Modern psychoanalysis and human psychology combine both these approaches and recognize the multiple determinants of human behavior ranging from genetic to environmental—that is, the interaction of nature and nurture (Alexander and Selesnick, 1966).

Second, at the same time as these developments were being made in the understanding of behavior, important discoveries were being made in the field of nutrition. Goldberger was the first to observe that a specific nutritional deficiency could cause marked behavioral abnormalities (Terris, 1964). Through

Janina R. Galler • Division of Psychiatry, Boston University School of Medicine, Boston, Massachusetts 02118.

his landmark epidemiologic studies, he discovered the causal relationship between nicotinic acid deficiency and pellagra, which is characterized by the clinical constellation of *diarrhea*, *dementia*, and *dermatitis*. It took a generation before routine treatment was made available for this nutritional deficiency. The foundation for understanding of the role of specific nutritional factors in behavioral function was laid down by these and similar early observations.

Third, although the behavioral effects of concurrent infantile malnutrition were recognized as early as the turn of the century in populations living in poverty, it is only in recent years that the long-term sequelae to early malnutrition have been identified. This has come about because in recent years a large percentage of children with severe malnutrition have survived owing to improved medical care and better living and housing conditions. As a result, there has been increasing interest in the outcome of such children. Since their illness most often coincided with the period of rapid brain growth, a special concern has been focused on mental development of these children. In the past fifteen years, there has been considerable controversy in this field of study concerning the conclusion that early malnutrition causes mental retardation. The more acceptable conclusion at present is that early malnutrition is responsible for long-term behavioral changes, many of which do limit a child's ability to adapt successfully, and that such changes are mediated by the presence or absence of a supportive home and socioeconomic environment.

The purpose of this book is to provide a comprehensive overview of the available evidence linking nutrition and behavior. Other texts can be consulted for details concerning the role of nutrition in biochemical and anatomical changes in the developing brain (Wurtman and Wurtman, 1977). This volume concentrates primarily on protein and energy deficits because these are the leading causes of malnutrition today. Other food constituents, such as vitamins (Chapter 6) and sugar (Chapter 5), are also considered in this volume, but certain micronutrients, such as EFA, zinc, and the trace metals, which have been shown to be important in animal studies of behavior, have not been included because these still require further study in humans. Iron deficiency and its effects on behavior have been reviewed in detail in an earlier volume of this series (Leibel *et al.*, 1979).

The volume is divided into two parts. Part I considers the effects of nutritional deficiencies or excesses on behavioral outcome. Chapter 1 presents a general overview of the methodologies and research designs in this field and their limitations and draws attention to the complexities of such designs.

Chapter 2 discusses the relevance of animal studies in elucidating the relationship between malnutrition and behavior. There is a very large body of data available from animal research that is often ignored in looking at the human condition. These authors conclude that animal studies have inherent limitations as "models" of the human condition, but, taken together with human studies, are complementary and provide useful information that could be gathered only under laboratory conditions. While the remaining chapters in the book focus on human nutrition, use is frequently made of the relevant animal literature.

Chapter 3 is a comprehensive review of studies of the behavioral consequences of severe early malnutrition from the time of recovery from the episode through childhood and adolescence. Although the most dramatic changes in behavior occur at the time of the acute episode, many behavioral deficits persist long after recovery from the malnutrition and remain at least through adolescence. In many cases, these deficits are cumulative and are the result of delays in earlier stages of development. The behavioral deficits are often modified by environmental conditions, which may exacerbate or diminish their manifestation. This chapter also reviews the author's current studies in Barbados, which illustrate the results of a comprehensive follow-up study of children with early malnutrition.

Chapter 4 reviews the various childhood intervention programs that have been used in communities where malnutrition is endemic. The experience from these programs has, in general, suggested that early interventions are important in modifying the deleterious sequelae of severe childhood malnutrition, but further careful evaluation is required.

Chapter 5 discusses the role of sugar and food additives in causing behavioral disorders in children and reviews the role of megavitamin use in severe behavioral disorders of childhood, such as infantile autism. The existing evidence is often inconclusive because of inadequacies in research design.

Chapter 6 is a comprehensive review of the effects of vitamin excesses and deficiencies on behavior. Selected vitamin deficiencies in neurological disease were reviewed in an earlier volume of this series (Dreyfus, 1979). Chapter 6 is much more specific in outlining the role of vitamins in psychiatric disease and includes the author's personal experience with hospitalized psychotic individuals, in whom nutritional deficiencies are widespread, but commonly overlooked. Many depressed patients, for example, have anorexia as a by-product of their illness—and their recovery may be delayed by unappreciated nutritional deficiency.

Chapter 7 reviews the mechanisms of nutrient actions on brain function, using serotonin deficiency as a specific example, since extensive laboratory studies have been conducted on this neurotransmitter. The way in which malnutrition affects behavior is not known, but it is believed to be mediated both by anatomical changes in the brain—especially in the case of childhood malnutrition—and by neurochemical changes in patterns of neurotransmitter release or uptake.

Part II of the book reviews the behavioral determinants of food intake and nutritional status. A wide range of normal and aberrant behaviors may actually lead to altered nutritional status—both malnutrition and overnutrition—and this relationship has hitherto not been adequately considered. An understanding of such behaviors may permit the identification of strategies for intervention. For instance, in Chapter 8, it is pointed out that a dysfunctional mother–child relationship may contribute to early childhood malnutrition. Consequently, identification and treatment of maternal depression in populations where malnutrition is endemic may prevent many cases of infantile malnutrition. Since feeding is one of the major ways in which the mother and child interact, adequate intake by the child most often parallels the adequacy of the mother–child interaction.

Chapter 9 reviews the etiology, clinical manifestations, and treatment of anorexia nervosa. Severe behavioral psychopathology leads to the partial or total rejection of food in adolescents with anorexia or, in the case of bulimia, to pathological eating patterns that may persist for years.

Chapter 10 comprehensively reviews the etiology and treatment of obesity in adults. While obesity is of great importance in the United States and other industrialized countries, its incidence is also increasing in disadvantaged populations that have recovered from malnutrition. This chapter should be read in conjunction with a chapter on obesity in an earlier volume of this series (Stern and Kane-Nussen, 1979) in which many diets currently being used to treat obesity and the limitations of each are listed.

Finally, Chapters 11 and 12 summarize the extensive information now available on psychological and cultural factors underlying food choice and their influence on the nutritional status of groups of individuals within different communities. These factors prescribe the conditions under which malnutrition will occur and also determine the limits within which interventions can occur.

From this overview of the contents of the volume, it is apparent that the study of nutrition and behavior is an established area. The challenge revealed by the chapters collectively is that there is a continuing interaction between, on one hand, nutrition and related environmental factors and, on the other hand, the behavioral development of man. This interaction is reciprocal insofar as the environment modifies the impact of behavioral changes generated by nutritional deficiencies or excesses and, conversely, behavioral abnormalities and cultural variations in food habits may determine nutritional status, e.g., obesity. We can confidently anticipate that closer cooperation between experts in the behavioral and nutritional sciences will amplify this picture. For example, areas that so far have escaped adequate attention, notably the adolescent and the elderly, should receive the benefit of advancing knowledge in these two fields. This will contribute toward completion of a total view of nutrition and behavior interacting over the entire life cycle, especially the late effects of antecedent malnutrition. Such cooperation will also result in a deeper understanding of the biological basis of these interactions, e.g., the role of nutrients as precursors of neuro-transmitters involved in behavior. Advances in the interaction between the nutritional and behavioral sciences will mean, for the clinician, more specific diagnostic indicators of clinical syndromes, their etiologies, and their treatment. For the public health expert, it will help to identify appropriate interventions that will mitigate the untoward conditions leading to poor nutrition and its consequences. Investments in determining and providing optimal nutrition especially in the early years of life will pay dividends in the terms of populations that will be better able to adapt to and benefit from the environment.

ACKNOWLEDGMENTS. I wish to thank Alina Martin and Eleanor LaBombard for their expert secretarial assistance and for their diligence in preparing multiple drafts of the text. I also wish to thank Barbara Gale for skillful editing of several of the manuscripts and Dorothy Thomas for the indexing. I especially wish to thank the Thrasher Research Fund and the Ford Foundation for support of my research and for their continued commitment to the field.

References

Alexander, F. G., and Selesnick, S. T., 1966, *The History of Psychiatry: An Evaluation of Psychiatric Thought and Practice from Prehistoric Times to the Present*, Harper and Row, New York.

Dreyfus, P. M., 1979, Nutritional disorders of the nervous system, in: *Human Nutrition: A Comprehensive Treatise*, Vol. 4, *Nutrition: Metabolic and Clinical Applications* (R. E. Hodges, ed.), pp. 53–81, Plenum Press, New York.

Leibel, R. L., Greenfield, D. B., and Pollitt, E., 1979, Iron deficiency: Behavior and brain biochemistry, in: *Human Nutrition: A Comprehensive Treatise*, Vol. 1, *Pre- and Postnatal Development* (M. Winick, ed.), pp. 383–439, Plenum Press, New York.

Stern, J. S., and Kane-Nussen, B., 1979, Obesity: Its assessment, risks, and treatments, in: *Human Nutrition: A Comprehensive Treatise*, Vol. 4, Nutrition: *Metabolic and Clinical Applications* (R. E. Hodges, ed.), pp. 347–407, Plenum Press, New York.

Terris, M., 1964, *Goldberger on Pellegra*, Louisiana State University Press, Baton Rouge.

Wurtman, R. J., and Wurtman, J. D., eds. 1977, *Nutrition and the Brain*, Raven Press, New York.

I

Nutritional Deficiencies or Excesses Modifying Behavioral Outcome

Methodological Requirements for Conceptually Valid Research Studies on the Behavioral Effects of Malnutrition

David E. Barrett

1. Introduction

A number of papers and chapters have been addressed to methodological issues in research on the effects of malnutrition on human behavior (Tuddenham, 1974; Warren, 1973; Elias, 1976; R. E. Klein, Irwin, Townsend, and Engle, unpublished manuscript). Most have stressed the importance of designing studies that permit the investigator to infer with a high degree of confidence that statistical relationships between nutritional and behavioral variables reflect the influence on behavior of nutrition *per se*, and not the effects of one or more confounding (i.e., nonnutritional) variables. For example, Warren's review criticized a number of studies that used nonexperimental designs on the grounds that the variables representing nutritional status may have been correlated with social and environmental factors that themselves could have accounted for the statistical relationships. Tuddenham reviewed five classes of research designs for studying the effects of malnutrition on behavior and described as adequate only those experimental designs that provided for random assignment to treatments and thus avoided the problem of selection bias. R. E. Klein and colleagues discussed several problems in the interpretation of retrospective studies of malnutrition and behavior, including possible confounding of onset and severity of malnutrition with duration of hospitalization, covariation of home environmental conditions with nutritional history, and difficulty in separating acute from chronic components of nutritional history. In short, the primary methodological concerns are summed up in the assertion by Elias (1976) that

David E. Barrett • Children's Hospital Medical Center; Harvard Medical School, Boston, Massachusetts 02115.

"the primary issue in the human studies is that of confounding of nutritional with non-nutritional factors" (p. 455).

It is interesting that the methodological critiques have focused on what Campbell and Stanley (1966) called the "internal validity" of the research designs. That is, the concern has generally been: has this study been designed in such a way that (1) variation in the outcome variable of interest can be unequivocally, although not necessarily uniquely, attributed to variation in the independent variable of interest (i.e., in the event that a statistically significant relationship is obtained) or (2) lack of relationship between independent and dependent variables is not an artifact of methodological or statistical features of the design or analysis that mask a true causal relationship? With respect to research on the effects of malnutrition on behavior, the internal validity of the design is the extent to which a significant relationship between malnutrition and behavior can be specifically interpreted as a nutrition effect and a lack of such relationship can be specifically interpreted as indicating no causal relationship between malnutrition and behavior.

However, internal validity is only one consideration in research design. There are several requirements that must be met if the research is to have heuristic value. A useful conceptual framework for evaluating the adequacy of a research design is Cook and Campbell's (1979) heirarchical classification of threats to the validity of a research design. There are four levels of evaluation.* The first is "statistical-conclusion validity," the extent to which the data permit the investigator to conclude that a statistical relationship does or does not exist—regardless of what the relationship may "mean." The second is internal validity, defined above. Next, there is external validity, the extent to which the results may be generalized to different samples, settings, and time periods. Finally, there is "construct validity of putative causes and effects," the degree to which our conceptualizations of the independent and dependent variables match our operationalizations (i.e., measurement operations) and therefore the extent to which the theoretical inferences we draw from the findings are empirically justified.

In this chapter, I will discuss specific problems of research design and methodology relevant to the issue of the effects of malnutrition on children's behavior. I will focus on those problems that derive from inadequate preliminary conceptualization of the independent and dependent variables and insufficient attention to the limitations and generalizability of the interpretations of significant relationships. I will then discuss the practical implications of these issues: the probable difficulties in formulating studies that adequately address all the methodological concerns, and suggested guidelines and priorities for designing studies that will have theoretical value. It should be emphasized that many of the problems I will discuss are relevant to research studies in a wide

---

* Cook and Campbell list the four classes of threats to the validity of a design in the order: statistical-conclusion validity, internal validity, construct validity of putative causes and effects, and external validity. I have reversed the order of the last two classes of validity for the purpose of clarity and logic of the presentation.

range of fields, and not solely to research on malnutrition and behavior. However, I will give particular attention to methodological issues that are particularly germane to this area of study and will cite examples to illustrate these concerns.

2. Statistical-Conclusion Validity

Threats to statistical-conclusion validity include such problems as insufficient numbers of subjects, low reliability of assessments of independent or dependent variables or both, high sampling error due to heterogeneity of sample or lack of uniformity in administration of treatments or tests, and use of inappropriate statistical tests (i.e., procedures that require mathematical assumptions that are not met by the data). In short, any factors that reduce the power of the statistical tests or bias the outcomes such that the probability of rejecting the null hypothesis differs from the chosen probability level may be considered threats to the statistical-conclusion validity of the study.

Two of these problems assume particular significance in the research on malnutrition and behavior. First, there is heterogeneity of scores on the dependent variable due to unintentional fluctuations in the application of treatment. In experimental studies of the effects of nutritional supplementation on selected behavior outcomes, it is often difficult to make the treatments uniform for subjects in the same treatment condition. When the investigator lacks control over how supplements are administered or consumed, variation in the use of supplement will result (assuming the supplement has some influence on the criterion variable) in increased variability of response within the treatment group and thus reduced power of statistical tests. Two studies that faced this potential problem are the Bogota, Colombia, study (Herrera *et al.*, 1980) and the New York City supplementation study (Rush *et al.*, 1980). In the Bogota study, food supplements consisting of dry skim milk, protein-enriched bread, and cooking oil were administered to families in which the children were at risk for malnutrition. For those groups receiving the supplements, supplements were provided weekly with no control for intrafamily distribution and no prescription regarding supplementation of or substitution for normal diet. Family use of supplementation was significantly affected by prestudy nutritional status. For mothers who had been consuming more than 1500 cal per day during pregnancy, supplementation did not significantly improve total daily caloric intake. Only for mothers who were consuming less than 1500 cal per day did the supplementation treatment significantly increase total energy intake. Thus, comparisons between supplementation and nonsupplementation conditions in this study must overcome an inflated error term within experimental conditions, at least insofar as individual differences in the criterion variables were sensitive to differences in nutrient intake. In the New York project, supplements were administered to women during pregnancy. Subjects were at risk for delivering a low-birth-weight infant. There were two types of supplement, one a beverage containing 322 cal and 40 g protein per can and one a beverage containing 322 cal and 6 g protein per can, provided daily. Although supplementation was intended to add to rather than substitute for diet calorie and protein intake, the supplement was used as a food substitute by both treatment groups, with 20% substitution of calories in the high-protein supplement group and 12% substitution of calories in the low-protein supplement group. Whether or not there was wide interindividual variability in the degree of substitution cannot be determined from the published data. To the extent that there was, the sensitivity of statistical tests may have been compromised.

The second type of problem involves reliability of assessments. The issue of reliability of indicators of nutritional status or nutritional history has been discussed in detail by Habicht *et al.* (1974). In their paper, the authors illustrate how the reliability of a nutritional assessment sets an upper limit on the correlation that that variable can have with any other variable. [If the criterion measure is perfectly reliable, the maximum value of the validity coefficient (i.e., the maximum correlation of the predictor variable with the criterion variable) is limited by the reliability of the predictor variable; specifically, the maximum validity coefficient is the square root of the reliability (see Magnussen, 1966, p. 149).] They show, for example, that when repeated measurements for length yield test–retest correlations of 0.97, the maximum correlation of length with another variable is 0.98, only 2% less than the perfect correlation. However, test–retest correlations for other nutritional assessments or indicators may be far lower. For example, in the data from the Guatemala Institute of Nutrition of Central America and Panama (INCAP) study that they report, test–retest reliability for home protein intake based on 3-month recall surveys was only 0.51, resulting in a maximum correlation coefficient of 0.71, almost a 30% reduction in sensitivity. The importance of the reliability of nutritional assessments is well understood by nutrition investigators, and the point need not be belabored. It should be noted, however, that a number of factors may influence the estimate of reliability—not only such factors as the type of statistic computed (test–retest correlation, coefficient of internal consistency, coefficient of equivalence), the heterogeneity of the sample (Magnussen, 1966), and the age at which measurements are made (particularly for test–retest reliability coefficients), but also the setting and conditions of assessment. A case in point is the study by Lechtig *et al.* (1976) on the usefulness of the urea/creatine (U/C) ratio as an indicator of protein ingestion. In a field study of Guatemalan preschool children, the test–retest correlation for U/C ratio was 0.96 when based on same-day same-urine-sample correlations, but only 0.50 when different urine samples were taken at 1-day intervals. In contrast, intraindividual variability of this magnitude is not found when measurements are obtained under the laboratory conditions of a metabolic ward. The results of the study of Lechtig and co-workers show the importance of collecting reliability data on nutritional measures under the same conditions that will characterize the research.

It is in assessing the reliability of behavioral variables that the researcher investigating the behavioral effects of malnutrition may confront particular problems. Very often, the researcher must study the behavior of persons in a culture that is different from that in which the assessment procedures were originally developed. However, for an assessment procedure to be reliable, to elicit consistent individual differences in response, it must be meaningful to the study population. If it is not, one will not obtain varied, codable, and consistent—that is, reliable—individual differences. I will illustrate this by discussing a behavior-assessment procedure that we used in our research in Guatemala (Barrett *et al.*, 1982; Barrett, in press). Since this procedure was developed in the United States for a non-Hispanic population, we had to modify it to make it appropriate for the Guatemala sample.

One of the hypotheses in our study was that differences in the adequacy of pre- and postnatal nutritional intake would be related to the child's ability at school age to control motor impulses. We used a Simon-Says type game to examine alertness and impulse control. When we first devised the procedures, we used the usual instructions: all behaviors were modeled, and the children were directed to imitate the model's movement if the model said "Simon Says do this" and not to imitate if the model said only "Do this." When we used the procedure with San Diego, California, schoolchildren (see Barrett *et al.*, 1981), there was, as expected, a wide range of responses. However, when we tried the game with village children in Guatemala, we found that *all* children imitated *all* modeled actions, regardless of the verbal directions. Thus, there was no variability of response. In thinking about this problem and how to rectify it, we considered two facts: First, the culture we were studying was more "traditional" than the population in which we had developed the procedures; in particular, there was a greater deference to authority figures and more rigid constraints on child and adult role behavior. Second, and related to the first point, the children at this age level—ages 6–8—were being trained in their school classrooms to imitate adults' actions. Much of the activity in the classrooms involved repetition of the teacher's words and actions (singing, marching, repeating the alphabet), and the children were probably not accustomed to having an adult demonstrate a behavior and direct them *not* to perform the modeled behavior. Therefore, we modified the procedures. Instead of providing visual cues for the subjects, we directed them to respond differentially to verbal cues only. We reasoned that these cues would exert less "pull" on the subjects. The new directions were: if the model said "One-two-three-yes," the child should clap; if the model said "One-two-three-no," the child should refrain from clapping. The new directions were very effective. There were wide individual differences in responding, and, as we have reported in other papers (Barrett, in press; Barrett and Yarrow, 1983), we found significant relationships between nutritional history and the child's speed and accuracy of responding in this task.

It should be emphasized that while the issue of reliability of behavioral assessments becomes particularly salient when one develops measures in one culture and then uses them in another, the problems of reliability of assessment must be faced by the researcher studying the effects of malnutrition on behavior regardless of the culture or setting. The interested reader is referred to Johnson and Bolstad (1973), Mitchell (1979), and Cairns and Green (1979) for in-depth treatments of issues in reliability and observer agreement of behavioral observations in naturalistic observational research.

3. Internal Validity

As indicated earlier, the major concern of methodologists evaluating the validity of research studies investigating the effects of malnutrition on behavior has been with the internal validity of the designs. Internal validity concerns the justifiability of causal inferences. In an internally valid design, we have a high degree of confidence in our conclusion that a given independent variable does or does not have a causal relationship with a given dependent variable. Our confidence in making such inferences is lowered when there are one or more alternative explanations for how the independent and dependent variables were statistically related or why they failed to relate.

Cook and Campbell (1979) list a number of threats to the internal validity of a design. Many are well known to researchers in this field and require no definition: selection effects, statistical regression, history, maturation, subject mortality, and instrumentation. All these effects are made less likely by the use of random assignment of subjects to treatments and the use of one or more appropriate control groups. Campbell and Stanley (1966) and Cook and Campbell (1979) describe a number of experimental and quasi-experimental designs that provide different amounts of protection against these various threats to internal validity.

There are four other types of occurrences that can reduce the internal validity of a design. Cook and Campbell have called these "diffusion or imitation of treatment," "compensatory equalization of treatment," "compensatory rivalry," and "resentful demoralization," and all are particularly likely when the experimental treatment is a large-scale intervention (for example, nutrition or educational intervention) that is widely perceived as bringing substantial benefits to the recipients. Diffusion or imitation means that the group that is not assigned the treatment that is perceived as the most beneficial receives the treatment informally—either by managing to share the treatment with the experimental group or by eliciting similar treatment from the investigators. Compensatory equalization means that the investigators depart from their planned administration of the treatments and nontreatments and attempt to give all groups similar experiences, so that no groups are deprived of a positive intervention. Compensatory rivalry occurs when a group receiving a less favorable treatment attempts to try harder—precisely because of their less advantageous condition—when tested on the dependent variable. Resentful demoralization would occur if the group receiving the less advantageous treatment tried *less* hard when tested on the behavioral measure than the group receiving the more favorable treatment. Diffusion or imitation of treatment, compensatory equalization, and compensatory rivalry reduce differences between experimental and control groups, while resentful demoralization increases them. All affect the internal validity of the design in that they reduce

one's confidence in interpreting the relationship or lack of relationship between independent and dependent variables in specifically causal terms.

All threats to the internal validity of a study are relevant to research on malnutrition and behavior. In the remainder of this section, I will illustrate the operation of three of these effects in studies on the behavioral effects of malnutrition: selection, statistical regression, and maturation–history.

As Warren (1973) showed in his review, a number of studies that have compared children with a natural history of undernutrition with matched controls who were not malnourished appear to have been compromised by selection effects. Two notable studies are those by Stoch and Smythe (1963, 1967) in South Africa and Chase and Martin (1970) in the United States. Stoch and Smythe compared school-age children who had been severely malnourished in infancy with controls matched for age and sex on a standardized intelligence test. Although the nonmalnourished controls scored significantly higher on the intelligence measure, they also scored significantly higher on measures of social, economic, and educational status that were likely predictive of school-age intellectual test performance. Chase and Martin compared 3- to 4-year-old children who had been severely malnourished in the first year of life with a sample of matched controls on a standard developmental assessment. The non-malnourished group scored significantly higher than the controls on the developmental assessment, but also scored higher on measures of the quality of the home environment. The authors themselves concluded (p. 938) that they could not separate the effects of malnutrition from those of "associated environmental influences."

The importance of statistical regression is illustrated by a study by Mora *et al.* (1974) that was a pilot study for the Bogata intervention study (Herrera *et al.*, 1980) cited earlier. In the pilot study, 218 children between the ages of 2 and 5 years were assigned to a nutritional intervention in which the children received, directly, food supplements containing 100% of the recommended daily allowances for protein and greater than two thirds of the recommended allowances for calories. Of these children, 102 had been classified as malnourished prior to the intervention and 116 as well nourished, based on weight and height for age. Seventy-six malnourished children and 84 well-nourished children were assigned to a nonintervention control group. Children were tested on entry into the study and at a 1-year follow-up on the Griffiths test of mental abilities. Both well-nourished groups showed slight decreases in Griffiths test scores, while the nonintervention malnourished group showed a slight increase. The results indicated regression-to-the-mean effects. It was by virtue of the significantly greater gain in the intervention malnourished group relative to the nonintervention malnourished group that the authors were able to conclude that nutritional supplementation had a significant and positive influence on cognitive development.

Maturation–history effects are changes in performance that are due to developmental and experiential changes in the subject that can be presumed to have occurred regardless of whether or not there was an intervention. A special case of maturation–history is the "cumulative deficit" effect, described

and documented by Jensen (1974, 1977). Cumulative deficit refers to a process whereby children from underprivileged populations—those characterized by poor educational history, impoverished physical environment, and poor health—show poorer functioning on tests of cognitive ability over time, relative to children from nonunderprivileged populations. According to the learning theory interpretation of Campbell and Frey (1970), such children are accumulating information and problem-solving skills at a lower rate than nondeprived children and thus are testing at lower and lower levels with repeated assessment. Since the child subjects of studies of malnutrition are frequently from populations in which formal learning opportunities are limited, such cumulative deficit effects are likely to occur. To the extent that they do, they present an obstacle that the environmental intervention must overcome.

Such a maturation–history effect is illustrated in a recent study by Zeskind and Ramey (1978, 1981). The investigators studied the effects of a cognitive and nutritional intervention program on the growth and development of prenatally undernourished infants. One group of subjects, one half of whom were assigned to the intervention group and one half to the control group, had been born with a low ponderal index (PI), the ratio of birth weight to birth length, which is an indicator of prenatal undernutrition. The other group, also randomly assigned to treatment and control conditions, was of normal PI. Children were assessed on the Bayley scales at 3 and 18 months and on the Stanford–Binet at 24 and 36 months. The pattern of results for the low-PI nonintervention children indicated a cumulative deficit with Bayley scores decreasing from 91 to 87 and Stanford–Binet scores from 78 to 71. In contrast, nonintervention normal-PI children showed a decline in Bayley performance from 3 to 18 months, but a slight increase in Stanford–Binet scores between 24 and 36 months. Intervention was associated with increases in the Bayley scores from 3 to 18 months and in the Stanford–Binet scores from 24 to 36 months for both groups. It appears that the effect of the intervention in the undernourished children was to help to reverse the cumulative effects of inadequate environmental stimulation and learning opportunities. The results suggest that in studies considering the effects of nutritional interventions on children's behavior and development, it may be important, in evaluating the improvements due to treatment, to consider whether one might expect decrements in performance over time without intervention. In short, the base line against which one evaluates the usefulness of the intervention may have to be adjusted to take into account the effects of cumulative deficit. It should be noted that the "direction of effects" for the maturation–history effect described above is opposite to that for regression to the mean. It is conceivable that both factors may be operative in a single study and to some degree cancel each other out.

4. External Validity

When an investigator has identified a relationship between an independent and a dependent variable and infers that the independent variable has a causal

influence on the dependent variable, he or she must also ask the question: to what extent can I expect this relationship to generalize to different populations, different settings (for treatment or testing or both), and different epochs? While the last question can be addressed directly by including different cohorts in the study (cf. McKay *et al.*, 1978; Herrera *et al.*, 1980), the issue of the generalizability across population groups or experimental conditions or both is often left as a matter of speculation.

In studies of the effects of malnutrition on behavioral development, the issue of generalizability across populations is extremely important. There are both theoretical grounds and empirical data to suggest that the relationship between individual differences in nutritional history and behavioral development will be highly dependent on the general level of nutritional status of the study population (as well as on variability in nutritional status) and on the adequacy of the environment, specifically, the opportunities for interaction with the social and nonsocial environment and thus for learning.

Lloyd-Still (1976) has argued that malnutrition may have its strongest effects on mental development when overall environmental stimulation—physical, linguistic, perceptual—is also limited. His argument is based on evidence that differences in cognitive development between children with a history of severe malnutrition in early life and those without such a history are consistently significant only when the children from both the index and the comparison group have experienced impoverished social environments. Lloyd-Still's position is consistent with the findings of Frankova (1974) and Levitsky and Barnes (1972), who found in their work with animals that provision of a highly stimulating environment could offset the behavioral effects of malnutrition. Further support for the proposed interaction of malnutrition and environmental deprivation is found in the neurophysiological literature showing effects of both malnutrition and environmental stimulation or deprivation on brain structure and biochemistry (Greenough, 1975; Im *et al.*, 1976; Rosenzweig and Bennett, 1972). With respect to human studies, the cognitive developmental theory of Piaget (1970) provides an interpretation of these interaction effects. According to Piaget's theory, changes in "cognitive structure" or the person's mental capacities occur as a result of the person's interaction with the environment, which forces him to accommodate to new experiences. Individual differences in rates of development are due to genetically determined individual differences in learning abilities (i.e., the ability to bring about changes in cognitive structure through interaction with the environment) and gross differences in the quantity and quality of individuals' interactions with the environment. From a cognitive–developmental view, one would expect that large changes in environmental stimulation (including stimulation mediated by nutritional improvements) might have their greatest effects when the prior level of environmental stimulation is lowest.

A number of studies on the effects of malnutrition on children's behavior are consistent with this interpretation and indicate important treatment–population interactions. R. E. Klein (1979) studied the effects of protein–calorie supplementation administered prenatally and in the first postnatal years on children's cognitive abilities at preschool age. Significant supplementation effects were identified only for subjects whose families scored lowest on a composite measure of social–environmental status. Barrett *et al.* (1982) examined the effects of pre- and postnatal caloric supplementation on the school-age behavioral functioning of Guatemalan children selected from the population of children studied by R. E. Klein. While nutritional supplementation showed significant relationships with measures of social behavior and emotional characteristics across villages, nutritional supplementation was related to cognitive functioning only in the village in which the absolute level of caloric intake was lowest. That is, only at the lowest level of energy intake were differences in caloric supplementation predictive of differences in intellectual functioning. In two studies cited by Lloyd-Still (1976), school-age children from middle-class populations who had been hospitalized for neonatal disorders and had been severely malnourished during their hospitalizations were compared with previously hospitalized but nonmalnourished controls on measures of intelligence (Valman, 1974; Lloyd-Still *et al.*, 1974). There were no differences between index and control children in either study. In contrast, in nine studies in which subjects were from lower socioeconomic classes, significant differences in cognitive development between children with a history of malnutrition and children without a history of malnutrition were identified (cf. Lloyd-Still, 1976, p. 119). These findings suggest that in studies investigating the effects of malnutrition on later behavior, the adequacy of both the nutritional and the nonnutritional environment may have a significant influence on the outcome of the research.

Generalizability of relationships across conditions of treatment or testing or both is a conceptually difficult issue. To the extent that a treatment has a different effect in different conditions of administration, we may assume that it does so because the setting or conditions of administration influences the very nature of the treatment (i.e., its impact on the recipient). But if this is the case, if we are talking about the conditions of administration *changing the nature* of the treatment, it seems more appropriate to view such influences not as threats to the external validity of the findings, but rather as threats to the construct validity of our interpretations. The same can be said for treatment–setting interactions with respect to conditions of assessment of criterion variables. It is frequently the case in behavioral research that measurement of a behavior variable obtained under one set of conditions is very different from measurement of the "same behavior" under other conditions. For example, frequencies of children's prosocial behaviors in one situation (e.g., a helping situation) may be very different from frequencies of prosocial behaviors in a different situation (e.g., a sharing situation), and the two sets of behaviors may have very different correlates (cf. Payne, 1980). Rather than view this problem as one of external validity or generalizability of findings, it seems more useful, heuristically, to treat the two sets of assessments as measurements of different behaviors, i.e., prosocial behavior in a helping situation vs. prosocial behavior in a sharing situation. For these reasons, when we consider the important issue of whether a nutrition–behavior relationship is replicable if we vary the conditions of administration of a treatment or the procedures for measuring our outcome variables, we may do so in the context of the broader question: are the relationships we have identified between our measures of malnutrition and behavior telling us what we wish to know about the relationship between malnutrition and behavior as we have conceptually defined these constructs? It is to this issue—the construct validity of our interpretations of relationships between nutrition and behavior—that we now turn.

5. Construct Validity of Putative Causes and Effects

The question of construct validity is really a question of the conceptual validity of our interpretations of findings. We are asking, here, both about the adequacy of our operationalizations of variables, that is, whether they correspond to our constructs, and about the appropriateness of our inferences about the relationships between variables, whether the processes that have brought about an empirical relationship are the same as those that we have hypothesized to account for the relationship.

Cook and Campbell (1979) identified a number of specific threats to the construct validity of a research study. These may be grouped into four general categories. The first problem involves inadequate conceptualization of the independent or dependent variables, leading to too narrow, broad, or otherwise inappropriate operationalizations of the variables. Treated as separate problems but actually relating to the same general issue are what Cook and Campbell call monooperation bias and monomethod bias. Monooperation bias occurs when the investigator measures only one aspect of the dependent variable or operationalizes the independent variable in a restricted way. An example of monooperation bias is examining effects of malnutrition on general cognitive development, but including as a dependent measure only a test of verbal ability. The risk, of course, is that the investigator will draw an inference about the effects of malnutrition on intellectual development solely on the basis of findings about verbal abilities. Monomethod bias occurs when an investigator examines the relationship between independent and dependent variables, but studies the dependent variable using only one "form" of assessment, for example, responses to specific structured testing situations, but not behavior in natural situations. If we consider this problem as a failure to consider the target behavior in its entirety—that is, a failure to comprehensively operationalize the construct—we have essentially another type of monooperation bias.

A second class of problem involves the interaction of treatment with prior testing and the interaction of different treatments. Both of these are threats to construct validity in that we may interpret a relationship as resulting solely from the influence of an independent variable, whereas in reality, the power of the independent variable to influence the dependent variable was affected by a prior measurement or a prior intervention. The first problem is an issue whenever the behavior being considered is psychological test performance. Testing effects and interactions are very likely to occur in mental testing, particularly of children who are relatively unfamiliar with test procedures (see

Zigler and Butterfield, 1968). The second problem is common in field studies in which an intervention has been preceded by some preintervention familiarization process or informal intervention. In both cases, the failure to consider the interaction of the intervention with the preceding event or events leads the investigator to interpret the identified relationships in causal terms that are not consonant with the empirical facts of the study.

A third type of threat to construct validity involves subject or investigator biases. Cook and Campbell refer to the interaction of treatments with hypothesis-guessing (a subject bias), evaluation apprehension (a subject bias), and experimenter expectancies (an investigator bias). In the first case, the relationship between independent and dependent variables is due not only to the putative effects of the treatment variable but also to the effect of the subject's knowing the hypothesis of the study and thus what behavior is desired or expected. A second type of interaction occurs when the subject behaves in the way that he does in an assessment as a result not only of the intervention but also of his concerns about how he should behave in the test situation. Evaluation-apprehension effects are common when the test situation involves behaviors that could be viewed as socially inappropriate or questionable, such as aggressive or emotional responses to a situation. The third type of problem occurs when the subject's criterion performance is affected not only by his standing on the independent variable but also by the fact that the investigator was not blind to the subject's standing on the independent variable and knew how he "should" behave. While these problems—hypothesis-guessing, evaluation apprehension, and experimenter expectancies—could each also act as a threat to internal validity (i.e., operate independently of treatment effects), they are more likely to interact with treatments in influencing outcomes.

Finally, and particularly important with respect to research on malnutrition and behavior, are those threats that involve the investigator's failure to fully consider the theoretical implications of the research and, specifically, the processes whereby the independent variable might influence the dependent variables. Such failures may lead the investigator to draw inappropriate conclusions about the relationships between the theoretical constructs. Two threats to construct validity fall into this category. One is what Cook and Campbell have called "confounding construct and level of construct." This occurs when the investigator examines the operation of the independent variable over a specific range and assumes that the relationship "holds" at some other (usually broader) range. This is a common problem when the independent variable involves dosages (as in drug or nutrition studies) or learning trials. It may be that when other ranges of the independent variable are considered, the relationship between independent and dependent variables is different than was originally thought. For example, a relationship that was linear at one level of the independent variable may be curvilinear over a larger range. In this regard, Horton (1982) found that the effects of carbohydrate and protein restriction on the endurance of obese adults were different after 1 and 6 weeks of treatment. While 1 week of restriction resulted in decrements in performance on an endurance task of moderate intensity, 6 weeks of restriction resulted in improved performance, apparently due to the subjects' abilities to use the prolonged period of restriction to adapt to the lowered intake and increase the utilization of lipids (p. 1233). If the data had been analyzed using a correlational approach, duration of dietary restriction and endurance would have been negatively correlated for the period 0–1 week but positively correlated for the period 1–6 weeks, and the two variables would have shown a curvilinear relationship over the duration of the study. The second problem is restricted generalizability across constructs. This occurs when the investigator fails to test for relationships between variables that can help confirm or disconfirm his hypothesis—specifically, when he fails to include in the hypothesized network of relationships between variables constructs that should be included by virtue of their (1) importance to the theory and (2) probable sensitivity to the independent variable. This is a common problem in nutrition research and will be illustrated below.

In this section, I will discuss what are perhaps the most important threats to the construct validity of research on malnutrition and behavior. First is inadequate conceptualization of the treatment variable, leading to interpretations of treatment effects that misconstrue the true nature of the effects. Second is inadequate conceptualization of the behavioral measures, seriously impairing the investigator's ability to make inferences about the theoretical relationships being investigated. Third is confounding of the construct with levels of the construct—the failure to specify, in designing and implementing the research, the levels of the independent variable that are being investigated, whether they represent randomly sampled or fixed levels, and why they, out of all possible levels, were chosen (if they were not randomly chosen). Finally, there is the problem of restricted generalizability across constructs. I will illustrate these problems, providing examples from published research on the behavioral effects of malnutrition.

The problem of inadequate preoperational conceptualization of the independent variable is most evident when the independent variable is a nutritional manipulation, such as nutritional supplementation. In such research, the treatment will necessarily entail more than simply the nutritional intervention. There will be interpersonal contact between researchers and subjects, provision of information and other cognitive stimulation, and possibly social or health supports or both, such as medical checkups or monitoring of nutrient intake. Under such circumstances, it is important that the investigator, prior to the beginning of the study, try to anticipate the role that such factors will play in the intervention and attempt to build them into his conceptualization of "treatment." The investigator may have to reconsider the hypothesized relationships between independent and dependent variables in light of this reformulation of the independent variable. Having carefully defined the independent variable with attention to the aforementioned considerations, the investigator can now attempt to match the operationalization of the variable—that is, the implementation of strategy—with his more fully elaborated conceptualization. Only when such steps are taken will the investigator be in a strong position to hypothesize and test for empirical relationships between variables and draw theoretical inferences consonant with the data.

What frequently occurs in conducting field research is that this type of consideration of the "meaning" of the independent variable is not made, for reasons of either time, convenience, or lack of awareness of its importance. As a result, the investigators may find, in attempting to implement the study, that they are exploring different relationships, testing different hypotheses, than they had anticipated. The nutritional treatment is no longer a nutritional treatment as such, but rather an amalgam of various types of stimulation—social, nutritional, cognitive, and medical. Pollitt and Thomson (1977) show in their review of the literature on malnutrition and behavior how this problem has made difficult the interpretation of the results of the major intervention studies (p. 295).

Another problem related to the requirement for adequate preoperational conceptualization of the independent variable is illustrated by the experiences of the investigators in the INCAP longitudinal study, carried out in rural Guatemala (R. E. Klein, 1979). One of the interpretive issues in analyzing the INCAP data has involved understanding the nature of the dramatic supplemental-calorie effects on growth and development (cf. Barrett *et al.*, 1982; R. E. Klein *et al.*, 1977; R. E. Klein, 1979). The INCAP study was originally designed to compare the effects of two nutritional supplements, a protein–calorie supplement and a calorie-only supplement, in promoting the growth and behavioral development of children from an endemically malnourished population. The protein–calorie supplement contained 11 g protein and 163 cal per cup. The calorie-only supplement contained no protein and approximately one third the calories of the protein–calorie supplement. As the study progressed, it became clear that the most consistent predictor of postnatal physical growth and psychological development, of the relevant diet and supplement nutrition measures, was caloric supplementation, both pre- and postnatal. Supplemental calorie intake was a significant predictor of birth weight (Lechtig *et al.*, 1975), height and weight at preschool age (see Barrett *et al.*, 1982), cognitive test performance at preschool age (R. E. Klein *et al.*, 1977), and social and emotional characteristics at school age (Barrett *et al.*, 1982). However, because of the perfect confounding of supplemental calories with supplemental protein in the protein–calorie-supplemented children, one cannot separate the effects of caloric supplementation from protein supplementation unless one studies only those subjects receiving the calorie-only supplement and examines the implications of individual differences in supplemental caloric intake in that group. However, to do so severely limits the conclusions one can draw about the significance of supplemental calories because the absolute *level* of supplement calories was lower in the calorie-only subjects than in the protein–calorie subjects. A second problem involves the confounding of calorie-only vs. protein–calorie supplement treatment with village membership (i.e., village differences). Thus, if different effects of calorie supplementation are found for children receiving supplemental calories with supplemental protein compared to children receiving supplemental calories without supplemental protein [and such differences have been found (see Barrett *et al.*, 1982)], such differences could be due to (1) presence/absence of supplemental protein, (2) absolute level of caloric supplementation, or (3) village membership. It should be noted with regard to the last point that although the study villages were originally matched at the start of the study on the basis of a number of socioeconomic variables including nutritional status (see R. E. Klein, 1979), by the time the children were born, mothers from the different villages differed significantly in total prenatal caloric intake. A final interpretive problem arises because all children received supplementation beginning prior to birth. In a number of papers, INCAP researchers have attempted to investigate the importance of the timing of supplementation for later development (R. E. Klein *et al.*, 1977; Barrett *et al.*, 1982). The limitation of such analyses is that one can draw conclusions about the timing of supplementation—or, more precisely, the differential importance of relatively high or low supplementation at different periods of development—only in a population supplemented continuously beginning prior to birth. No statement is possible about the effects of supplementation beginning at a particular postnatal period, since no groups of subjects were "staggered" with respect to the initiation of the treatment. In sum, the INCAP study, while it has made a highly significant contribution to our understanding of the importance of chronic energy deficit for later behavior and development, is nonetheless limited in the theoretical contributions that it can make, since the nutritional variable that became the primary (i.e., most consistently used and most consistently significant) predictor of the physical and behavioral outcomes was not identified as such at the outset and therefore could not be completely adequately conceptualized and operationalized.

A second major issue involves the importance of careful, theoretically based conceptualizations of the dependent, behavioral variables. I have addressed this issue at length elsewhere (Barrett, 1982) and will present here only the central argument. In studying the question of how a history of malnutrition influences children's behavior, investigators have frequently chosen their dependent variables on the basis of nontheoretical considerations. For example, it has frequently been noted (Brozek, 1978; Ricciuti, 1981) that the majority of studies on the effects of malnutrition on cognitive development have conceptualized cognitive functioning in terms of "general intelligence" as defined by performance on standardized IQ tests. The problem with such a global and atheoretical conceptualization is not that one will necessarily fail to identify significant nutrition–behavior relationships; indeed, most such studies have shown that malnutrition is related to poor cognitive test performance (for a review, see Pollitt and Thomson, 1977). Rather, the problem is that unless dependent variables are explicit theoretically derived and operationalized, the results of the research are unlikely to clarify the processes whereby malnutrition may influence later behavior. A notable exception in the area of cognitive research is the recent study by Leibel *et al.* (1982), which examines the effects of iron deficiency on specific attention and learning capabilities that have been theoretically linked to iron deficiency and empirically related to one another.

Several studies have attempted to assess the effects of malnutrition on children's social and emotional development. In a number of studies in this area, conceptualization of behavior variables has been weak. The reason may be that in most studies, the primary focus has been on the effects of malnutrition on intellectual functioning, with social and emotional functioning a secondary consideration. Whether or not this is the case, dependent behavior variables in the noncognitive domains have tended to be defined only in very broad terms—for example, social maturity or school behavior problems—and have been operationalized accordingly, that is, in a global fashion (see, for example, P. S. Klein *et al.*, 1975; Lloyd-Still *et al.*, 1974). As was the case with respect to measures of cognitive functioning, it is essential that studies investigating the effects of malnutrition on social and emotional development attempt to develop the criterion measures on the basis of theoretical considerations. One way this can be done is to begin by asking the question: is there prior research that suggests that certain behavioral functions are particularly vulnerable or sensitive to the particular nutritional variable being investigated? In developing the dependent measures for our research on the effects of early nutritional supplementation on children's social and emotional development at school age, we considered both the research on the behavioral effects of protein–calorie malnutrition in animals and the recent research on the behavior characteristics of undernourished human infants. Both sources of information suggested that there was a distinct complex of behavior impairments associated with a history of undernutrition. Specifically, attention, state control, responsiveness to social stimulation, activity level, and tolerance for mild stress or frustration appeared to be affected (for a review, see Barrett *et al.*, 1982). On the basis of these considerations, we formulated a set of hypotheses about the types of problems in social behavior and emotional development that one might expect to see at school age in children with a history of undernutrition. That is, we asked ourselves: if the kinds of behavior impairments identified above—i.e., in the animal and infant research—continued to characterize the child's early development, what would be the likely social and emotional consequences for the child by the time of early or middle childhood? On the basis of this formulation, we began to identify specific classes of behavior function relevant to nutritional history and then operationalized the variables. The research, summarized elsewhere (Barrett, in press; Barrett *et al.*, 1982; Barrett and Yarrow, 1983), has shown significant relationships between early caloric undernutrition and the child's social responsiveness, interest in the environment, capacity for active play with peers, persistence in a frustrating situation, and motor impulse control as measured in structured and nonstructured small-group situations at school age.

One of the most difficult and persistent problems in the research on malnutrition and behavior has been the problem of the confounding of constructs with levels of constructs, specifically, with respect to the conceptualization and measurement of nutritional variables and the inferences drawn on the basis of these operations. For example, none of the major longitudinal intervention studies investigating the effects of nutritional supplementation on child behavior (Chavez and Martinez, 1979; Herrera *et al.*, 1980; R. E. Klein, 1979; McKay *et al.*, 1978) has been able to specify the range of supplementation over which effects of supplementation on specific behavior variables are likely or not likely to be identified. Further, there has been little attention to the question of linearity vs. nonlinearity of supplementation effects with respect to specific behavior outcomes. There are several reasons why this particular problem of not differentiating construct from level of construct may occur. First, there are difficulties that arise solely from practical or ethical considerations or both. In nutritional intervention studies, the investigators often are not in a position to manipulate levels of the independent variable simply because they are interested in identifying the parameters of specific nutrition–behavior relationships; rather, they are constrained by ethical considerations, social–political considerations, and the material resources available. In studies in which nutritional variables are not manipulated, it may be virtually impossible to obtain reliable data on the intake of specific nutrients. Under these circumstances, any plan to make inferences regarding effective levels of nutrient consumption or deprivation with respect to the behavior outcomes of interest may have to be relinquished and more general inferences about nutrition and behavior made. But it is also the case that a major cause of this problem is the failure of researchers to specify for themselves, to know beforehand, whether the levels of the independent variables that are being considered in their studies may be treated as randomly sampled levels from among a specified population of possible levels or whether they represent fixed levels. The difference, of course, is critical with respect to the inferences that one may justifiably draw from the research. It seems reasonable to assume that researchers who are investigating the effects of malnutrition on human behavior by examining the influence of specific nutrient supplements on specific behavior outcomes do not wish to restrict their conclusions to inferences about the effects of one specific level, or perhaps a few discrete levels, of the treatment variable on behavior outcomes. Nonetheless, insofar as there is no attempt to address the issue of how levels of the independent variable were sampled from possible levels, there is no justification for generalizing inferences about nutrition–behavior relationships to levels of the independent variable not included in the study.

A final concern with respect to the construct validity of our interpretations of empirical relationships between nutrition and behavior variables or the lack of such relationships is the problem of restricted generalizability across constructs. This is a potential problem both when a significant relationship between independent and dependent variables has been obtained and the investigator wishes to draw an inference about the theoretical meaning of the relationship and when there is no relationship identified and the investigator wishes to attach a theoretical interpretation to the lack of relationship. In the first instance, it may be the case that the validity of our inference that a particular independent variable has a specified effect on a particular outcome depends not only on a significant empirical relationship between the independent and dependent variables but also on the assumption that other theoretically relevant potential outcome variables would have been similarly related to the independent variable, or perhaps even more strongly related, had they been included in the analyses, and that certain other variables would not have been related. For example, in a study of the effects of nutritional supplementation on children's cognitive functioning, the investigator might examine the relationship between nutritional supplementation and Wechsler Intelligence Scale for Children (WISC) IQ based on the hypothesis that malnutrition has a deleterious effect on attention capacities and that since the WISC includes subtests that appear to tap, to some degree, attentional capacity (e.g., Digit Span and Picture Completion), attention deficits should result in poor performance on the WISC and nutritional supplementation should result in improved WISC performance. If a significant relationship between nutritional supplementation and WISC IQ were obtained, the investigator might wish to interpret this outcome as indicating support for the interpretation that nutritional supplementation had a positive effect on attention and thereby improved WISC performance. However, the assumption that is being made when one makes such an inference is that had a more precise measure of attention—i.e., one specifically developed to measure attention—been included as a dependent measure in the analyses, an even stronger relationship between the nutrition measure and the behavior outcome would have been identified. If it were in fact the case that had such a measure been included *no relationship* would have been identified, and that the significant relationship between the nutritional variable and WISC performance was due not to the effects of nutritional supplementation on attention but rather to the effects of nutrition on some other factor related to WISC performance, the investigator would be making a theoretical inference that was not justified by the data. When we consider the problem in this way, it is clear that restricted generalizability across constructs is a threat to any research study, since it is not possible to include in one's design and analysis all variables that might be expected to bear a statistical relationship to the independent variable under study (nor would it be desirable to try to do so, due to statistical considerations). However, it is also the case that this threat to the construct validity of the interpretation of the study can be minimized if one includes in one's design and analysis one or more variables that on the basis of theoretical considerations could be expected to be strongly related to the independent variable and that if found not to be so related would render questionable the validity of the investigator's interpretation of the relationship under study.

The converse of this situation occurs when an investigator finds no significant relationship between an independent and a dependent variable and concludes that no relationship exists, but fails to recognize that had other, theoretically relevant variables been considered in the data analysis, a different conclusion would have been drawn.

To illustrate this situation, let us again consider the situation in which an investigator is interested in examining the effects of a long-term nutritional intervention on children's growth and development. Let us also assume that the investigator begins with the assumption that before examining the effects of the intervention on behavioral development, it is first necessary to document that the supplementation has affected physical growth. The assumption is that if the supplementation is not significantly affecting growth, then it is not "taking hold" and that therefore it makes no sense to spend time attempting to measure its effects on behavior variables.

This type of reasoning illustrates a common fallacy in nutrition research: that unless a nutritional manipulation—deprivation or supplementation—affects physical growth outcomes, it is not going to have an influence on other (for example, behavioral) functions. For example, in a chapter on malnutrition and behavior, Ricciuti (1974) discussed the findings from the INCAP Guatemala study (p. 184):

> As the authors suggest, these preliminary findings may mean that the amount of additional protein taken in by these children was not sufficiently great to have any appreciable impact on cognitive functioning as reflected in psychological test performance. To shed some further light on this question, it would be very useful to know whether variations in protein intake had any relationship to gains in height from age five to age seven. If the effects of dietary improvement were not sufficiently great to be discernible in improved physical growth during this particular age period, for example, one might wonder whether it is reasonable to expect significant improvement in psychological functioning.

While it may seem reasonable to assume in the case of a long-term intervention study such as the INCAP study that lasting effects on psychological function must necessarily be accompanied by significant effects on physical growth, it is not necessarily the case that a nutritional manipulation must influence physical growth before it can influence other functions and activities of the organism. Under certain conditions, manipulations of nutrient intake will change behavioral functions prior to influencing physical growth. In a condition of nutrient deprivation, body and tissue stores must be used to compensate for the deficiency in nutrient intake. At the stage of depletion of body tissue, the effects on the organism include not only the lowering of nutrient levels in the tissues (which can be measured by biochemistry) and changes in growth (measured by anthropometry) but also changes in enzyme levels and metabolic activities (Robson, 1978). The latter may influence the behavior of the organism including activity level, affect, and attention. Evidence for changes in nutrient levels influencing psychological processes prior to body size comes from studies of specific nutrient deficiencies, for example, iron deficiency, and studies of failure to thrive in infancy. Iron-deficient infants are often characterized, behaviorally, as showing little interest in their physical environments and being irritable. Positive changes in these affective and attentional characteristics occur within days after the institution of iron therapy, and this occurs before there is any significant change in hemoglobin (Oski and McMillan, 1981). Further, Oski and Honig (1978) found that when infants with iron deficiency anemia were treated with iron supplements, there were significant improvements in scores on the Bayley Scales of Mental Development Index, compared to control infants, as well as increased alertness, responsiveness, and sensorimotor functioning. Studies of failure-to-thrive infants, infants who have been nutritionally deprived in the context of inadequate care-giving and present with growth failure and clinical affective disturbances, have documented that improvements in social responsiveness as a result of nutritional intervention often precede changes in growth (cf. Rosenn *et al.*, 1980). In sum, an investigator who fails to examine the behavioral outcomes of a nutritional intervention because the intervention does not appear to have influenced physical growth may thus miss the opportunity to identify significant behavioral effects of the intervention, with the result that his conclusion that the nutritional variable had no effect on the child will be invalid.

6. Conclusions and Recommendations

In this chapter, I have examined potential methodological problems in conducting research on malnutrition and behavior. I have categorized problems using the classification scheme of Cook and Campbell (1979). Cook and Campbell identify four classes of threats to the validity of a research design: *statistical-conclusion validity*, *internal validity*, *external validity*, and *construct validity of putative causes and effects*. In terms of research on the behavioral effects of malnutrition, these four types of validity may be defined as follows: (1) statistical-conclusion validity is the extent to which the investigator can have confidence in his or her inference that an empirical relationship between specified measures of malnutrition and behavior does or does not exist. (2) Internal validity is the extent to which the investigator can have confidence in his or her inference that an identified empirical relationship between the measures of malnutrition and behavior represents, to at least some degree, a causal influence of malnutrition on behavior. (3) External validity is the extent to which the investigator can have confidence in his decision to generalize the nutrition–behavior relationship to the target population, that is, the population of interest. (4) Construct validity of putative causes and effects is the extent to which the investigator can have confidence that (a) his theoretical interpretation of the nutrition–behavior relationship is consistent with his operationalization (measurement) of the variables and (b) his operationalization of the variables is compatible with the underlying theoretical constructs and thus that the results of the study can be interpreted in terms of the theoretical framework in which the research issue was posed. I have discussed and illustrated specific problems relevant to each of these four classes of validity and have given examples from the published research on malnutrition and behavior.

Unfortunately, the researcher investigating the effects of malnutrition on behavior is rarely in a position to control for the many threats to the validity of a study that I have explicated. First, unless the study is a true experiment— i.e., involving random assignments to treatments under the control of the investigator—as are the major nutritional intervention studies, the majority of the threats to the internal validity of the study will be operative. A large majority of the published studies on the effects of malnutrition on behavior have involved retrospective or prospective comparisons of groups with natural histories of malnutrition vs. adequate nutrition, and it is reasonable to assume that such studies will continue to be necessary to address certain issues (i.e., when it is necessary to determine whether a particular potential source of variance in behavior is significant under natural conditions). In such studies, the issue of internal validity must be addressed using the frequently inadequate methods of matching or statistical control (cf. Warren, 1973). Second, in most studies that do involve experimental manipulations, the investigators are not likely to be in a position from which they can completely control all the factors that might influence the utilization of the treatment, due to practical, ethical, and social–political constraints. For example, in the Bogota study (Herrera *et al.*, 1980) discussed earlier, it was necessary for the investigators to provide supplements on a family basis without precise monitoring of individual intake. This resulted in differences among families in utilization of the supplement, which likely resulted in increased heterogeneity of response on dependent variables within treatment groups. In the Guatemala INCAP study (R. E. Klein, 1979), it was felt important to give individuals the choice of whether or not to attend the supplementation centers and with what frequency. This left open the possibility of the confounding of selection and treatment effects, or selection–treatment interactions, or both. Although INCAP investigators have not identified specific selection effects—specifically, maternal background or personality variables predictive of attendance at supplementation centers (Engle *et al.*, 1979)—such effects cannot be ruled out. Third, the investigator studying the behavioral effects of a nutritional intervention may not be in a position, again for ethical reasons, to study the independent variable over a range of levels sufficient to address the important questions of (1) the amount of supplementation necessary to effect behavior change and (2) whether there are nonlinear effects of supplementation on developmental outcomes. The researcher is likely to want to provide supplements beginning at a level of supplementation at which he can be confident that the supplement will have beneficial consequences for the subjects and to try to design the study so that nonfavorable effects of supplementation [as occurred, for example, in the New York supplementation project with the high-protein supplement group (see Rush *et al.*, 1980)] are avoided. Fourth, in many studies of the effects of malnutrition on behavior, researchers are interested in testing complex causal models involving large numbers of independent and dependent variables, not just one independent variable and one dependent variable. Not only are hypotheses frequently couched in terms of multiple predictors or multiple and correlated dependent variables (as in multivariate analysis of variance), but also a large number of significance tests are generally carried out. As a result, there is, first, a seemingly overwhelming number of alternative causal relationships that have to be systematically considered (i.e., ruled out on *a priori*, theoretical grounds or empirically controlled or tested or both) and, second, a substantial increase in the probability of making incorrect probability statements and, specifically, type I errors. Finally, it is not always possible to anticipate, much less recognize, label, and conceptualize, the many components of an intervention as it occurs in the field; i.e., components that were not part of the original theoretical conceptualization of the independent variable. For example, in a nutritional supplementation study that will involve personal interaction between investigators and subjects, it is difficult to know precisely what types of interaction will take place, how these interactions can be expected to influence the subjects, and therefore how one can best conceptualize these components of the treatment so that they can be built into the construct "nutritional intervention." Failure to account for these components will influence the construct validity of the investigator's interpretation of cause–effect relationships. Given these many potential problems of implementation, designing a study that meets even a majority of the threats to the statistical-conclusion, internal, external, and construct validity of a design would seem to represent a major achievement.

In the face of what would appear to be unattainable requirements for validity of research on the effects of malnutrition on behavior, the investigator must attempt to devote his energies to those elements of the design the fulfillment of which is essential if the results of the study are to be theoretically significant. Thus, the position taken here is that although logically the hierarchy of threats to the validity of a research study moves from statistical-conclusion validity (was there a relationship?) to construct validity (how, exactly, may we interpret this relationship with respect to our hypotheses?), the priorities with respect to research design *must be* to identify one's theoretical issues, conceptually derive one's variables, and operationalize those variables in such a way that if and when an empirical relationship is identified, one can, with a high degree of confidence, interpret the relationship in terms of the theory or theoretical constructs that provided the basis for the research. In short, this means that utmost attention should be given to the many issues involving construct validity of putative causes and effects.

Adequate construct validity begins with identifying one's theoretical issues and then operationalizing both independent and dependent variables with a view toward (1) comprehensiveness and accuracy of definitions of independent and dependent variables, with respect to theory and prior research, and (2) careful operationalization so that operations correspond to conceptual definitions. It should be noted that in nutritional intervention studies, adequate conceptualization and operationalization requires giving special attention to the many "hidden" aspects of an intervention (i.e., as it occurs in the "real world") that might be anticipated and, it is to be hoped, built into the investigator's governing conceptual framework. It is at this stage of design also that the investigator must assess for himself the theoretical importance of empirical issues involving levels of treatment, linearity vs. nonlinearity of relationships, and related statistical concerns and decide whether such issues can be addressed in the study. Only when these essential considerations pertaining to the purpose and theoretical implications of the research have been thought through and resolved should the more strictly methodological issues of validity be addressed.

In making specific recommendations about priorities and requirements for research design, I will suggest first several guidelines that pertain solely to experimental nutritional intervention studies. This distinction requires further explanation, because the view is frequently expressed that experimental designs are inherently superior to nonexperimental designs (cf. R. E. Klein *et al.*, unpublished manuscript; Tuddenham, 1974; Warren, 1973). Thus, one might expect that a first "guideline" would be to employ experimental designs whenever possible. As indicated earlier, this argument rests largely on the reduced threats to the internal validity of the study, resulting in greater confidence in causal inferences. I would suggest, in contrast, that the choice of a research design should depend primarily on the purpose of the study and that one should not assume that one type of approach is best suited for all purposes. For example, it is no doubt true that if the sole or primary purpose of the study is to demonstrate, as unequivocally as possible, a causal influence of nutrient intake on a particular developmental outcome, and in particular the possibility for enhancing behavioral development, an experimental design is essential. This is because, first, if the purpose is to unequivocally demonstrate a cause–effect relationship, experimental designs do reduce the number of potential threats to internal validity and, second, *logically*, experimentally manipulating nutrient intake provide the only conceptually valid, or at least potentially conceptually valid, means to address the research issue. On the other hand, for certain purposes, naturalistic, nonexperimental designs are preferable. For example, let us suppose that the primary purpose of the study is to examine two samples of retarded and nonretarded children to determine and compare the different odds ratios for risk of retardation due to such factors as early malnutrition, congenital disorder, parent psychopathology, and socioeconomic status. In such a case, a nonexperimental design would be conceptually more valid than an experimental design; one would not use an experimental manipulation of nutritional intake to address this research question. It is worth noting, in this regard, that under certain conditions, retrospective comparison analyses of data obtained from such a study are more powerful than prospective analyses of the data, a fact often overlooked by researchers in our field. This has been demonstrated empirically by Fleiss (1973). Consider the general case in which (1) associations are sought between a person's probability of being in a particular clinical, c, or nonclinical, $\bar{c}$, category on a presumed antecedent variable A, such as previous history of malnutrition or adequate nutrition, and a person's probability of being in a clinical or nonclinical category on a presumed outcome variable B, such as mentally retarded or of normal intelligence; (2) the prior (unconditional) probability of clinical status on the outcome variable is lower than the prior probability of clinical status on the antecedent variable; that is, $p(B_c) < p(A_c)$; and (3) there is some true association between the variables; that is, $p(B_c/A_c) \neq P(B_c)$. Under these conditions, the probability of identifying, statistically, the association between the two variables for fixed sample size n is greater for a retrospective matched comparison analysis than for a matched prospective comparison analysis (i.e., given the same true relationship between A_c and B_c), and both approaches are more powerful than the naturalistic cross-sectional analysis (see Fleiss, 1973, p. 57).

To conclude this chapter, I suggest guidelines that the researcher should attempt to follow when designing a study to examine the behavioral effects of malnutrition or nutritional supplementation on human subjects. Each of these suggestions represents a conceptual or procedural activity on the part of the investigator to ultimately increase the construct validity, and thus the theoretical importance, of the research.

1. In examining the effects of a nutritional supplement or treatment on behavior, the investigator should attempt to conceptualize the treatment variable with a view toward anticipating and building into the construct those nonnutritional components that will necessarily be included when the study is implemented. Thus, for example, theoretical inferences will be made about the effects of a particular nutritional, medical, social intervention, rather than simply a nutritional intervention.

2. In examining the effects of a nutritional supplement or treatment, the investigator will, to the extent possible, know precisely how levels of the treatment variable were determined (i.e., do they represent fixed or randomly sampled levels?) and, in all cases, seek to include in the study those levels that permit the testing of the hypotheses of interest.

3. In examining the effects of a nutritional supplement or treatment on behavior, the investigator should, to the extent possible, administer treatments uniformly. If it is not possible to ensure that treatments are used in the same way by all subjects, data that would allow the investigator to test for differences in utilization and the effects of such differences on the relationships identified between independent and dependent variables should be collected.

4. It is essential that dependent variables be conceptualized and operationalized in a comprehensive and theoretically meaningful way. It is improper for an investigator to design a study without thinking through the questions of (a) what are the dependent variables of interest: what are the behaviors that on theoretical grounds one should expect to be related to my independent variable; and (b) what are the operations by which I can measure individual differences in these variables so that the measurements correspond to my theoretical constructs?

5. That in testing a hypothesis about causal relationships between malnutrition and particular behavior outcomes, it is desirable to include in one's data analyses variables that, although they are not the primary dependent variables of the study, can on theoretical grounds be presumed to bear significant relationships with the independent variable(s). If such variables are included in the analysis, failure to identify the postulated relationships provides cause for reevaluation of the original theoretical formulation.

6. To the extent possible, threats to the internal validity of the research design should be controlled by employing techniques of randomization and control groups, which significantly reduce these threats. However, when, due to the nature of the study, such rigorous *a priori* control is not possible, and more imperfect statistical procedures must be employed, the investigator should reflect considerably on the likely contribution of such potential confounding influences to the results of the

study and, to the extent that he believes that most or all alternative hypotheses can be handled within the theoretical framework in which he is working (i.e., ruled out *a priori* or given meaningful interpretation within the theoretical framework), should proceed to design and carry out the research with consideration for the many conceptual issues that have here been emphasized.

7. In carrying out the research, attention should be given to the reliability of assessments, both nutritional and behavioral, with recognition that failure to do so may compromise the power of the data analyses and thus the investigator's ability to identify nutrition–behavior relationships. Particular attention should be given to assessing the reliability of behavior measures, since when the meaning of an assessment is different for the investigator and the subjects (as may happen when assessment procedures developed in one culture are used in another), it is unlikely that consistent and meaningful individual differences in responses will be obtained.

These recommendations, if followed, will increase the potential of a study for making useful contributions to theory and research on malnutrition and behavior. It is emphasized that more important than the specific type of design employed is the carefulness and honesty with which the investigator has thought about the meaning of the variables he wishes to investigate, the relationships that are hypothesized, and the broad theoretical network in which these relationships are thought to occur. When this careful conceptual groundwork is laid and the investigator takes the time and effort to translate the theoretically derived constructs into reliably measurable variables, the research that results is most likely to advance the scientific pursuit of knowledge regarding the role of malnutrition in behavior and development.

7. References

Barrett, D. E., 1982, An approach to the conceptualization and assessment of social–emotional functioning in studying nutrition–behavior relationships, *Am. J. Clin. Nutr.* **35**:1222–1227.

Barrett, D. E., 1983, Malnutrition and child behavior: Conceptualization and assessment of social–emotional functioning and a report of an empirical study, in: *Critical Assessment of Key Issues in Research on Malnutrition and Behavior* (B. Schurch and J. Brozek, eds.), Hans Huber, Bern, Switzerland, in press.

Barrett, D. E., and Yarrow, M. R., 1983, Effects of early nutritional supplementation on children's behavior at school age in novel, frustrating and competitive situations, Presented at the meeting of the Society for Research in Child Development, Detroit, April 1983.

Barrett, D. E., Yarrow, M. R., Ziegler, M., Livingston, R. B., and Klein, R. E., 1981, Effects of early malnutrition on children's social emotional functioning at school age, Presented at the meeting of the Association for the Care of Children's Health, Toronto, May 1981.

Barrett, D. E., Yarrow, M. R., and Klein, R. E., 1982, Chronic malnutrition and child behavior: Effects of early caloric supplementation on social and emotional functioning at school age, *Dev. Psychol.* **18**:541–556.

Brozek, J., 1978, Nutrition, malnutrition and behavior, *Ann. Rev. Psychol.* **29**:157–177.

Cairns, R. B., and Green, J. A., 1979, How to assess personality and social patterns: Observations or ratings? *The Analysis of Social Interactions* (R. B. Cairns, ed.), pp. 209–226, Lawrence Erlbaum, Hillsdale, New Jersey.

Campbell, D. T., and Frey, P. W., 1970, The implications of learning theory for the fade-out gains from compensatory education, *Disadvantaged Child*, Vol. 3 (J. Helmuth, ed.), pp. 445–463, Brunner/Mazel, New York.

Campbell, D. T., and Stanley, J. C., 1966, *Experimental and Quasi-experimental Designs for Research*, Rand-McNally, Chicago.

Chase, H. P., and Martin, H. P., 1970, Undernutrition and child development, *N. Engl. J. Med.* **282:**933–939.

Chavez, A., and Martinez, C., 1979, Consequences of insufficient nutrition on child character and behavior, in: *Malnutrition, Environment and Behavior* (D. A. Levitsky, ed.), pp. 238–268, Cornell University Press, Ithaca, New York.

Cook, T. D., and Campbell, D. T., 1979, *Quasi-experimentation: Design and Analysis Issues for Field Settings*, Rand-McNally, Chicago.

Elias, M. F., 1976, Malnutrition in infancy and intellectual development, in: *The Developing Individual in a Changing World*, Vol. 2 (K. F. Reigel and J. A. Meacham, eds.), pp. 454–461, Aldine, Chicago.

Engle, P. L., Irwin, M., Klein, R. E., Yarbrough, C., and Townsend, J. W., 1979, Nutrition and mental development in children, in: *Human Nutrition: A Comprehensive Treatise*, Vol. 1, *Nutrition: Pre- and Postnatal Development* (M. Winick, ed.), pp. 291–306, Plenum Press, New York.

Fleiss, J. L., 1973, *Statistical Methods for Rates and Proportions*, John Wiley, New York.

Frankova, S., 1974, Interaction between early malnutrition and stimulation in animals, in: *Early Malnutrition and Mental Development* (J. Cravioto, L. Hambraeus, and B. Vahlquist, eds.) pp. 202–209, Almqvist and Wikwell, Uppsala, Sweden.

Greenough, W. T., 1975, Experimental modification of the developing brain, *Am. Sci.* **63:**37–46.

Habicht, J. P., Yarbrough, C., and Klein, R. E., 1974, Assessing nutritional status in a field study of malnutrition and mental development, in: *Methodology in Studies of Early Malnutrition and Mental Development* (J. Cravioto, L. Hambraeus, and B. Vahlquist, eds.), pp. 35–42, Almqvist and Wiksell, Uppsala, Sweden.

Herrera, M. G., Mora, J. O., Christiansen, N., Ortiz, N., Clement, J., Vuori, L., Waber, D., De Paredes, B., and Wagner, M., 1980, Effects of nutritional supplementation and early education on physical and cognitive development, in: *Life-span Developmental Psychology* (R. R. Turner and F. Reese, eds.), pp. 149–184, Academic Press, New York.

Horton, E. S., 1982, Effects of low energy diets on work performance, *Am. J. Clin. Nutr.* **35:**1228–1233.

Im, H. S., Barnes, R. H., and Levitsky, D. A., 1976, Effect of early protein–energy malnutrition and environmental changes on cholinesterase activity of brain and adrenal glands of rats, *J. Nutr.* **106:**342–349.

Jensen, A. R., 1974, Cumulative deficit: A testable hypothesis?, *Dev. Psychol.* **10:**996–1019.

Jensen, A. R., 1977, Cumulative deficit of IQ of blacks in the rural South, *Dev. Psychol.* **13:**184–191.

Johnson, S. M., and Bolstad, O. D., 1973, Methodological issues in naturalistic observation: Some problems and solutions for field research, in: *Behavior Change: Methodology, Concepts, and Practice* (L. A. Hamerlynck, L. C. Handy, and E. J. Mash, eds.), pp. 1–67, Research Press, Champaign, Illinois.

Klein, P. S., Forbes, G. B., and Nader, P. R., 1975, Effects of starvation in infancy (pyloric stenosis) on subsequent learning abilities, *J. Pediatr.* **87:**8–15.

Klein, R. E., 1979, Malnutrition and human behavior: A backward glance at an ongoing longitudinal study, in: *Malnutrition, Environment and Behavior* (D. A. Levitsky, ed.), pp. 219–237, Cornell University Press, Ithaca, New York.

Klein, R. E., Irwin, M., Engle, P. L., and Yarbrough, C., 1977, Malnutrition and mental development in rural Guatemala, in: *Studies in Cross-Cultural Psychology* (N. Warren, ed.), pp. 91–119, Academic Press, New York.

Lechtig, A., Yarbrough, C., Delgado, H., Habicht, J. P., Martorell, R., and Klein, R. E., 1975, Effects of moderate maternal malnutrition on the placenta, *Am. J. Clin. Nutr.* **28**:1223–1233.

Lechtig, A., Martorell, R., Yarbrough, C., Delgado, H. and Klein, R. E., 1976, The urea/creatinine ration: Is it useful for field studies?, *J. Trop. Pediatr.* **22**:121–128.

Leibel, R. L., Pollitt, E., Kim, I., and Viteri, F., 1982, Studies regarding the impact of micronutrient status on behavior in man: Iron deficiency as a model, *Am. J. Clin. Nutr.* **35**:1211–1221.

Levitsky, D. A., and Barnes, R. H., 1972, Nutritional and environmental interactions in the behavioral development of the rat: Long-term effects, *Science* **176**:68–71.

Lloyd-Still, J. D., 1976, Clinical studies on the effects of malnutrition during infancy on subsequent physical and intellectual development, in: *Malnutrition and Mental Development* (J. D. Lloyd-Still, ed.), pp. 103–159, Publishing Sciences Group, Littleton, Massachusetts.

Lloyd-Still, J. D., Hurwitz, I., Wolff, P., and Scwachman, H., 1974, Intellectual development after severe malnutrition in infancy, *Pediatrics* **54**:306–311.

Magnussen, D., 1966, *Test Theory*, Addison-Wesley, Reading, Massachusetts.

McKay, H., Sinisterra, L., McKay, A., Gomez, H., and Lloreda, P., 1978, Improving cognitive ability in chronically deprived children, *Science* **200**:270–278.

Mitchell, S. K., 1979, Interobserver agreement, reliability, and generalizability of data collected in observational studies, *Psychol. Bull.* **86**:376–390.

Mora, J. O., Amezquita, A., Castro, L., Christiansen, N., Clement-Murphy, J., Cobos, L., Cremer, H., Dragastin, S., Elias, M., Franklin, D., Herrera, M., Ortiz, N., Pardo, R., de Paredes, B., Ramos, C., Riley, R., Rodriguez, H., Vuori-Christiansen, L., Wagner, M., and Stare, F. J., 1974, Nutrition health and social factors related to intellectual performance, *World Rev. Nutr. Diet.* **19**:205–236.

Oski, F. A., and Honig, A. S. 1978, Effects of therapy on the developmental scores of iron-deficient infants, *J. Pediat.* **92**:21–25.

Oski, F. A., and McMillan, J. A., 1981, Iron in infant nutrition, in: *Textbook of Pediatric Nutrition* (R. M. Suskind, ed.), pp. 153–162, Raven Press, New York.

Payne, F. D., 1980, Children's prosocial conduct in structured situations and as viewed by others: Consistency, convergence, and relationships with person variables, *Child Dev.* **51**:1252–1259.

Piaget, J., 1970, Piaget's theory, in: *Carmichael's Manual of Child Psychology*, Vol. I (P. H. Mussen, ed.), pp. 703–732, John Wiley, New York.

Pollitt, E. and Thomason, C., 1977, Protein–calorie malnutrition and behavior: A view from psychology, in: *Nutrition and the Brain*, Vol. 2 (R. J. Wurtman and J. J. Wurtman, eds.), pp. 261–306, Basic Books, New York.

Ricciuti, R. N., 1974, Assessing the interaction of nutritional and socio-environmental influences on development, in: *Methodology in Studies of Early Malnutrition and Mental Development* (J. Cravioto, L. Hambraeus, and B. Vahlquist, eds.), pp. 182–185, Almquist and Wiksell, Uppsala, Sweden.

Ricciuti, H. N., 1981, Developmental consequences of malnutrition in early childhood, in: *The Uncommon Child* (M. Lewis and L. Rosenblum, eds.), pp. 150–172, Plenum Press, New York.

Robson, J. R. K., 1978, *Malnutrition: Its Causation and Control*, Gordon and Breach, New York.

Rosenn, D., Loeb, L., and Jura, M., 1980, Differentiation of organic from non-organic failure to thrive syndrome in infancy, *Pediatrics* **66**:698–704.

Rosenzweig, M. R., and Bennett, E. L., 1972, Cerebral changes in rats exposed individually to an enriched environment, *J. Comp. Psychol.* **55**:429–437.

Rush, D., Stein, Z., and Susser, M., 1980, A randomized controlled trial of prenatal nutritional supplementation in New York City, *Pediatrics* **65**:683–697.

Stoch, M. B., and Smythe, P. M., 1963, Does undernutrition during infancy inhibit brain growth and subsequent intellectual development?, *Arch. Dis. Child.* **38**:546–552.

Stoch, M. B., and Smythe, P. M., 1967, The effects of undernutrition during infancy on subsequent brain growth and intellectual development, *S. Afr. Med. J.* **41**:1027.

Tuddenham, R. D., 1974, On analyzing longitudinal data relating nutrition to cognitive development, in: *Methodology in Studies of Early Malnutrition and Mental Development* (J. Cravioto, L. Hambraeus, and B. Vahlquist, eds.), pp. 72–82, Almqvist and Wiksell, Uppsala, Sweden.

Valman, H. B., 1974, Intelligence after malnutrition caused by neonatal resection of ilium, *Lancet* **1**:425–427.

Warren, N., 1973, Malnutrition and mental development, *Psychol. Bull.* **80**:324–328.

Zeskind, P. S., and Ramey, C. T., 1978, Fetal malnutrition: An experimental study of its consequences on infant development in two caregiving environments, *Child Dev.* **49**:1155–1162.

Zeskind, P. S., and Ramey, C. T., 1981, Preventing intellectual and interactional sequelae of fetal malnutrition: A longitudinal, transactional, and synergistic approach to development, *Child Dev.* **52**:213–218.

Zigler, E., and Butterfield, E. C., 1968, Motivational aspects of changes in IQ performance of culturally deprived nursery school children, *Child Dev.* **39**:1–14.

2

The Use of Animals for Understanding the Effects of Malnutrition on Human Behavior: Models vs. a Comparative Approach

Susan F. Fleischer and Gerald Turkewitz

1. Introduction

1.1. Malnutrition as a "Social Disease"

Although estimates of incidence vary, there is no doubt that a large proportion of the world's children suffer from some degree of nutritional deficiency, most often protein–caloric in nature. As medical treatment improves, yielding a larger pool of survivors, the question of the possible neurological and behavioral consequences of inadequate dietary intake in early life becomes more pressing. Studies have indicated extensive and diverse consequences of malnutrition for the neural development of the child. Thus, cell number, cell size, lipid content per cell, and dendritic branching may all be affected by early malnutrition (Winick, 1976). In addition to the findings concerning neurogenesis, numerous studies of consequences of malnutrition for behavioral development have been reported (see, for example, the reviews of Brozek, 1978; Schrimshaw and Gordon, 1968). Although these studies generally find markedly retarded intellectual development in malnourished children, interpretation of their findings can be only tentative because nutritional deficiencies are seldom, if ever, found in isolation; rather, nutritional deficiencies occur as one aspect of a complex of privations resulting from poverty (Birch and Gussow, 1970;

Susan F. Fleischer • Department of Psychology, Queens College, City University of New York, Flushing, New York 11367. *Gerald Turkewitz* • Department of Psychology, Hunter College, City University of New York, New York, New York 10021; Departments of Psychiatry and Pediatrics, Albert Einstein College of Medicine, Bronx, New York 10461.

Winick, 1976). Populations suffering from inadequate dietary intake are, for example, also the victims of inadequate housing, insufficient medical care and facilities, a high incidence of disease, and poor-quality education. In fact, it is quite possible that the effects of malnutrition are never independent of those social interactions with which malnutrition is confounded. That is, in a certain sense, malnutrition may be best considered a "social disease" that invariably owes its effects, at least in part, to its influences on the nature of social interaction. Thus, as we shall discuss in greater detail later, malnutrition has effects on the behavioral and physical attributes of the child that can influence both the child's response to others and the response of others to the child. Such altered social relationships may then have major consequences for the child's subsequent emotional, social, and intellectual growth and development.

1.2. Use of Animal Models

Because of the problem of the confounding of social and nutritional variables in human populations, researchers have attempted to develop animal models that would enable the investigation of the influences of malnutrition on behavior without the confounding of social variables present in humans. The use of such animal models has a long and respectable history. Our understanding of human physiology as well as its molecular basis is based in large part on the use of animal models. Indeed, it is difficult to imagine the limitations on our understanding of basic human biological processes had animal models not been available and utilized.

Animal models have not only been useful in elucidating basic physiologic mechanisms, but also have played important roles in contributing to our understanding and treatment of various disease processes. It is therefore not at all surprising that when the enormity of the worldwide problem with malnutrition was recognized and the likelihood that it had consequences for intellectual development was appreciated, investigators began to seek appropriate animal models with which to examine the effects of malnutrition on cognitive and adaptive functions.

2. The Rat as a Model for Human Malnutrition

Although research on a number of animal species [dogs (Platt and Stewart, 1968, 1969), pigs (Barnes *et al.*, 1968), monkeys (Kerr and Waisman, 1968), mice (Howard and Granoff, 1968), guinea pigs (Stern and Bronner, 1970), rabbits (Harrington and Newberne, 1967), and chickens (Collier and Squibb, 1968)] was undertaken, the overwhelming majority of studies relating nutrition and cognitive functioning have been conducted using the rat. The selection of the rat as the animal of choice appears to be largely a consequence of practical considerations such as its easy availability, relatively inexpensive maintenance, the existence of a number of tests of its cognitive capacities, and a great store of knowledge concerning both its behavior and its physiology. Although these are eminently defensible reasons for selecting the rat, we will indicate why we believe that in fact the use of the rat as a model to elucidate the relationship between nutrition and behavioral development in humans is of only limited value.

The effective use of animal models is dependent on the existence of fundamental similarities and a lack of fundamental differences between the human and the model. For example, with regard to basic physiology, the mechanisms for blood transport and renal functioning are highly similar in humans and in the rabbits and guinea pigs in which the mechanisms were explicated. Although there are obvious and marked differences between man and rabbits and guinea pigs, these differences are largely irrelevant with regard to these physiological mechanisms.

However, when dealing with more integrated types of function in which more characteristics of the organism are involved, differences between organisms are more likely to influence the function in question. To the extent that behavior represents an instance of such integrated function, differences between rats and humans may limit the utility of the rat as a model for the effects of malnutrition on human behavior. The essential question therefore becomes whether or not differences between rats and humans result in malnutrition influencing behavioral development via different mechanisms or routes in rats and man. We will approach this question by first examining some of the ways by which malnutrition is known to affect development in the rat and then comparing these effects with those found in human populations. In addition, we will simultaneously examine the related issue of whether rats can be used to simplify the problem of separating social from nutritional effects.

2.1. Different Techniques for Producing Malnutrition in Rats and Their Effects on Mother–Young Interaction

Researchers in different laboratories have used different experimental procedures to induce malnutrition in young rats. The techniques that have most often been employed are (1) raising animals in large litters, (2) maintaining dams on inadequate diets, and (3) periodically separating the litter from the dam.

However, each of these techniques results in extranutritional effects acting on the animals that may themselves either contribute to or obscure subsequent deviations in development. The extranutritional environmental factor that has received the most attention is the young animal's social relations, most particularly, its interactions with the mother, and this is the factor on which we will focus in this chapter. However, the reader should be aware that alterations in other aspects of the environment will also occur as a result of the nutritional treatment. Thus, large-litter rearing, for example, in addition to its effects on the quantity and quality of maternal care, also results in crowding and in increased competition among pups for access to the dam (Frankova, 1972; Leathwood, 1978). Females rearing large litters have been reported to spend less time with their litters (Grota and Ader, 1969) and to be less adequate mothers with regard to retrieving speed, quality of nest construction, and amount of nursing (Seitz, 1958). However, this has not been universally reported. In a more recent study, it was found that relative to controls, females rearing large litters spent more time nursing their pups during the second half of the lactation period and that there were no differences in retrieving or nesting behavior (Crnic, 1980). It was suggested that the discrepancy from earlier results might be due to differences in the protein content of the dam's diet (which was higher than it had been in previous studies) or to differences in housing (in this study, the dam had no escape area in the cage as had been provided in previous studies).

The technique of depriving the dam of an adequate diet has also been found to result in alterations of maternal behavior. Massaro *et al.* (1974) reported that females fed low-protein diets spent more time in the nest than controls did particularly after the first week of lactation. Wiener *et al.* (1977) also reported increased dam–pup contact in the litters of protein-deprived females, a decrease in nest building, and less efficient retrieving by such females. Again, these effects became more marked after the first week of lactation. Hall *et al.* (1979) also found that dams fed low-protein diets spent more time in their nests, actively nursing pups after the first postpartum week. On the other hand, passive nursing, which occurred mostly outside the nest and consisted of the dam's lying on her back while a part of the litter suckled, was actually observed more frequently in controls. Although Hall and colleagues did not find malnourished females to be poor retrievers, it is probable that the measure of retrieving used (frequency rather than latency) is relatively insensitive. Therefore, these findings are quite comparable to those of Wiener and colleagues, as are those of Crnic (1980), who also reported increased nest attendance as well as less rapid retrieving by females fed low-protein diets. Thus, although Frankova (1971) has reported slightly discrepant results, the consensus is fairly strong that although females on low-protein diets may be somewhat slow in retrieving their litters and show decreased nest building, they attend and nurse the young more than do well-nourished controls.

Similar results have been obtained for the female rat deprived of an adequate amount of calories rather than of protein. Smart and Preece (1973) reported that like the protein-deprived female, calorie-deprived females retrieved fewer pups and licked them less than did well-nourished females. In addition, an alteration in the diurnal pattern of nest occupation was found, experimental dams generally spending more time than controls in the nest except around the time of day when they were provided with food. Nest-building was not affected.

The third technique for inducing early nutritional deprivation, periodically separating the litter from the dam, obviously entails maternal deprivation as well as nutritional deprivation. Although pups may be kept warm by being placed in an incubator, the loss of stimulation provided by the mother and littermates cannot be so easily replaced. Furthermore, the stress of being periodically separated from the litter, and the consequences of excess handling received by the pups, may affect the female so that pups receive altered maternal care even during the period when they are with her. Crnic (1980) found that females the pups of which were kept in an incubator for 12 hr daily showed more nest attendance, nesting, and nursing than controls, although they did not differ in retrieving tests.

Each of these by-products of the various experimental techniques listed above (e.g., alterations in the dam–litter relationship, as well as crowding, and competition) may be expected to have marked influences on the behavioral development of the young. In fact, it has even been suggested that in the absence of these extranutritional environmental alterations, early nutritional deficiencies may in themselves have little effect on the subsequent behavior of the rat pup (Slob *et al.*, 1973).

Recognition of these problems has led to attempts to design new techniques for producing malnutrition without such confounding influences. To date, however, such attempts have proven unsuccessful. For example, one frequently cited study improved on the periodic-separation technique by housing pups with maternal but nonlactating females during the separation period (Slob *et al.*, 1973). However, subsequent research using a similar design has shown that both lactating and nonlactating females spent more time in the nest and in a nursing posture and had better-quality nests when caring for malnourished litters than females rearing normal litters. These differences did not become apparent until the second half of the lactation period (Fleischer and Turkewitz, 1981).

Another of the newer techniques consists of the ligation or removal of some of the lactating dam's nipples. Although Codo and Carlini (1979) reported that females with only four or six functional nipples rearing litters of six retrieved and nursed pups no differently than controls, their observations did not extend beyond 12 days after parturition. The literature suggests that differences in maternal behaviors of females rearing well- and poorly nourished litters become more marked toward the end of the litter period as the weight, appearance, and behavior of the pups become more divergent. In fact, a study by Lynch (1976) that examined maternal behavior for 24 days following parturition did indeed report that females in which all nipples were ligated spent more time in the nest and in a nursing posture than did females rearing well-nourished pups. Furthermore, nonligated mothers that cared for the same pups 12 hr daily also showed these alterations in maternal care. Similarly, both Galler and Turkewitz (1977) and Crnic (1980) reported that partial mammectomies resulted in females spending more time nursing than controls. Galler and Turkewitz also reported that such females retrieved fewer pups and built better nests.

It is thus becoming more and more clear that experimentally induced malnutrition in the rat pup results in a variety of changes in nonnutritional elements of the early environment, including most notably dam–pup interactions. The nature of these changes clearly depends on the method used to stunt the pups, on more specific details of the experimental procedure, and on the particular aspect of maternal behavior (e.g., nursing, retrieving, nesting) being measured. However, in general, large-litter rearing does usually (although not invariably) appear to be associated with a decrease in the quantity and quality of maternal care-taking, including all three major components. Paradigms in which the dam's diet is altered (either quantitatively or by protein reduction) are usually associated with decreased efficiency of retrieving, but increased nursing time. Nesting may be decreased or unchanged. Periodic separation of mother and litter (whether pups are housed in an incubator or with a nonlactating female during the separation periods) results in increased mother–pup contact and nursing and nesting during the time they are together, as does the paradigm in which malnutrition is achieved by ligating some or all of the female's nipples. Neither of these techniques appears to be consistently associated with alteration in retrieving.

It is thus possible that in rats, as in humans, many of the long-term behavioral effects previously attributed to early malnutrition are in fact due to alterations in social factors. Since this is the same problem that had plagued research with humans and was one of the reasons for studying rats to begin with, it is obvious that unconfounding of nutritional and social variables is not achieved by studying rats. However, since social and nutritional factors are interrelated in both rats and man, there is still the possibility that the rat may be useful as a model for human malnutrition as it occurs in the "real world." To further investigate this possibility, we need to study in more detail the mechanisms by which malnutrition affects the young rat, particularly how it alters the relationship with the mother.

2.2. Mechanisms through Which Malnutrition Affects Mother–Young Interaction

There is a growing body of literature showing that the maternal rat is exquisitely sensitive to and will alter her behavior as a function of a variety of attributes of the young. Recently, both animal and human developmental research has indicated not only that the mother affects the offspring but also that the offspring affect the mother's behavior (e.g., Bell, 1971; Harper, 1971; Lewis and Rosenblum, 1974). As a result, under typical circumstances, there is usually a synchrony between the needs of the offspring and the maternal care received (Rosenblatt, 1963), although if the early environment is markedly atypical, or if the characteristics of the offspring are markedly deviant, it is likely that disruption of this synchrony will occur.

In view of the sensitivity of the female to the characteristics of the offspring, it is not surprising that any effective experimental method for inducing nutritional deprivation is likely to affect the maternal care received because it will change the stimulus attributes (both physical and behavioral) the treated animal presents to the mother. In addition, those techniques of stunting that involve altering the diet of the dam are likely, by altering her energy stores, to also affect her maternal behavior in a more direct fashion. Thus, as we have already seen, different techniques for inducing nutritional deprivation during the litter period are very likely to be associated with different effects on the mother–offspring relationship, depending both on whether or not the mother is herself deprived and on the specific changes, e.g., amount and timing of stunting, induced in the pups. In this section, we will discuss more specifically how malnutrition affects mother–young interaction in the rat with an eye toward the question of whether similar mechanisms are likely to be operative in the human situation.

2.2.1. Auditory Mechanisms

Many years ago, Beach and Jaynes (1956b) suggested that the maternal behavior of the rat was under multisensory control. In more recent years, investigators have begun to examine in more detail the relationships between specific aspects of the stimulation provided by the offspring and the specific maternal behaviors they may affect. One facet of the pup that has received much attention is the auditory stimulation it provides. Pup calls have been shown to elicit approach behavior on the part of the dam and are probably important for retrieving of the litter and may stimulate other maternal behaviors as well (Noirot, 1964; Smith and Sales, 1980). Furthermore, several studies have shown that relative to controls, the ultrasound production of malnourished rat pups is altered. One study (Hunt *et al.*, 1976) reported that pups stunted by feeding the dams a low-protein diet produced few if any vocalizations, but a second study, using more sensitive recording techniques (Hennessy *et al.*, 1978), found, rather, that such pups showed a delay in the age at which the peak amount of vocalization occurred (from approximately 4 to 11 days old). It is thus possible that changes in the auditory attributes of malnourished pups may play a role both in the decreased retrieving sometimes found in females rearing them and in the increased amount of nursing and nesting during the later parts of the litter period. However, Smotherman *et al.* (1974) have suggested that in fact, pup cries are not the most important cues for stimulating retrieving. As will be discussed later, olfactory stimuli appear to be of greater importance. Moreover, the fact that not all techniques of stunting are associated with diminished retrieving further suggests that changes in pup attributes may not be the most critical factor in the altered retrieving sometimes found in the malnourished litter. That is, it is those techniques in which the dam herself is malnourished by which retrieving is most reliably (and negatively) affected, probably because of diminished maternal energy. Several studies have, in fact, shown that the retrieving deficit of the malnourished female is independent of pup condition, since it will occur even if the female is tested with well-nourished pups (Crnic, 1976; Smart, 1976; Wiener *et al.*, 1977). Retrieving probably requires more energy expenditure than do other maternal activities, and it is not surprising that it should be affected by maternal diet.

Large-litter rearing, as has previously been noted, is also associated with decreased maternal retrieving. In this case, it is likely that maintaining a large number of pups also taxes the female's energy stores. It will be recalled that when females with large litters were given a high-protein diet, they did not evidence decreased maternal care (Crnic, 1980).

2.2.2. Olfactory Mechanisms

2.2.2a. Changes in Odors of Pups. Another modality that is importantly involved in dam–pup interactions is olfaction. It has been suggested that pup odors may be important in stimulating retrieving in the lactating female (Smotherman *et al.*, 1974), and there is some suggestion that at least some stunting treatments alter olfactory-based elements in the mother–pup relationship. Such alterations could contribute to deficits in retrieving and possibly other maternal behaviors. Wiener *et al.* (1976) found that lactating female rats showed diminished responsiveness (as indexed by pituitary–adrenal cortical secretions) when exposed to malnourished pups that had been handled or shocked and then placed in their cages in a basket. It is the olfactory stimuli provided by the pups that was most critical in eliciting these hormonal changes, since females rendered anosmic showed no such response (Zarrow *et al.*, 1972).

In addition to the importance of the litters' nutritional status, there was some evidence that the dietary condition of the female herself was also significant. Protein-deprived females became less responsive to older malnourished pups, while control females showed no such effects. Unfortunately, only the effects of a low-protein dietary regime have been studied using this paradigm, so that it is impossible to determine whether other stunting techniques will also be associated with such alterations in pup stimulation and female responsiveness.

2.2.2b. Changes in the Maternal Pheromone. Not only does the pup's odor affect the dam, but also there is evidence indicating that the dam's odor influences the behavior of the pup. Leon and Moltz (1972) showed that lactating females produce a distinct odor to which pups are attracted. There is a marvelous synchrony in that the dam produces and the liter is responsive to this odor (or pheromone) only between 14 and 27 days after parturition. Moltz *et al.* (1974) have shown that pheromone production by the female depends on the stimulus attributes of her litter, so that giving a female a younger litter will prolong pheromone production. Exactly what pup attributes are responsible for this age effect have not been determined, but it is quite likely that the stunting and infantalizing effects on the pups of an effective nutritional-deprivation treatment will alter pheromone production, possibly shifting the age at which production begins and ends. Pheromone production may also be altered by a more direct route when techniques that involve dietary deficiencies for the female are used. Leon (1975) has shown that lactating females fed diets that are quantitatively reduced produce less pheromone than do well-fed females. Similarly, qualitative changes in the female's diet were found to affect the odorous quality of the pheromone. Furthermore, pups were attracted only to the odor of test females that were fed the same diet as that of the female that nursed them, so that stunting techniques that involve rotation of pups between females fed different diets (e.g., between malnourished and well-nourished females) result in a very complicated stimulus situation for the pups. Unfortunately, the females in Leon's studies, although tested for pheromone production using healthy control pups, were rearing their own malnourished litters, so that it is still possible that it was changes in the pups rather than direct dietary-induced changes in the females that were primarily responsible for alterations in pheromone production.

In any case, the long-range effects on the pups of such changes in the olfactory bond with the dam are still being explored. There is evidence that exposure in early life to specific odors may determine long-range food and even conspecific preferences (Galef and Kaner, 1980; Leon *et al.*, 1977). Also, studies have shown that pups malnourished by a wide variety of techniques (Altman *et al.*, 1971; Fleischer and Turkewitz, 1979a,b; Galler, 1979, 1980; Galler and Seelig, 1981) show both a reduced amount and a delay in the development of homing to the nest from other parts of the home cage. This could in part reflect alterations in maternal pheromone production, since olfactory cues have been found to be critical for such homing. If odor cues are removed from the nest, pups home at much lower frequencies (Fleischer *et al.*, 1981). However, Johanson *et al.* (1980) concluded that alterations in pheromone production are probably not solely responsible for group differences in homing, since pups in different treatment groups responded (or failed to respond) to a standard odor stimulus (shavings from the cage of a 14-day lactator) much as they did to their own home shavings. Furthermore, group differences in homing have been observed to be present before 14 days of age, the supposed beginning of pheromone production (e.g., Fleischer and Turkewitz, 1979b).

It is thus clear that early malnutrition may affect odor-based aspects of the relationship between mother and young via a variety of mechanisms. Some effects result from alterations in the pups, some from alterations in the dam, and some undoubtedly from an interaction of effects on both. Furthermore, different techniques of stunting alter different aspects of this relationship.

2.2.3. Thermal Mechanisms

2.2.3a. Nursing. Another factor that influences mother–pup interaction is thermal stimulation. Leon (1978) demonstrated that the progressive decline in nursing time normally found in female rats as their pups approach weaning age is under thermal control. During a nursing bout, the female and her litter form a huddle that causes the temperature of the ventral portion of the female's body to rise. At a threshold level, the female will leave the huddle and thus terminate nursing. Any variable that increases the rate of heat production in the nursing unit (e.g., increasing the number of pups in the litter, increasing age and thus size of the pups) would therefore decrease the amount of time spent nursing. Contrariwise, any factor that decreases heat production or promotes its dissipation (e.g., small, stunted pups, delayed fur eruption) will result in prolonged nursing. Furthermore, Leon (1978) has shown that heat production by the lactating female is dependent on her elevated metabolic rate, which is in turn due to increased prolactin and adrenal cortical secretions. These hormonal changes are the result of the suckling stimuli provided by the litter. Weak suckling stimulation or reduced ability to thermoregulate on the part of malnourished pups, or both, could thus also contribute to prolonged nursing. Similarly, underfeeding the dam or altering her diet in a qualitative manner could also, by affecting her metabolic rate, hormonal status, or the amount of body fat she has, alter nursing times. Thus, there are a variety of mechanisms by which early malnutrition could disturb the nursing relationship.

As we have seen, techniques that involve alterations of the dam's diet result in longer nursing times, probably because pups are smaller and thus generate less heat and probably also because of changes in the dam's metabolism. Techniques that produce stunting of the pups without any direct dietary effects on the female (e.g., rotating pups between lactating and nonlactating females) also result in increased nursing times because of the decreased ability of smaller, less hairy pups to generate and maintain heat. Finally, large-litter rearing has been found to result in increased or decreased nursing time in different studies. These equivocal findings could be due to an unstable balance of heat-related forces in the large-litter situation. Although the pups are smaller and less hairy, they are also more numerous. Thus, depending on specific details in the situation (e.g., the exact number of pups, the diet of the female), one could predict that females caring for large litters would spend more time than, less time than, or the same amount of time nursing as females with control litters.

2.2.3b. Nest-Building. Nest-building, like nursing, is in part thermally controlled, being related to ambient temperatures and probably the heat generated by the litter (Kinder, 1927; Lehrman, 1961). Therefore, the similar effects of malnutrition on nursing and nest-building may be based on their reliance on common mechanisms. Different techniques of stunting the pups are associated with different effects on nest-building. As discussed with regard to nursing, small, stunted pups are likely to produce less heat, thus stimulating increased nesting by the dam. Furthermore, malnourished pups may leave the nest less frequently, thus permitting better nest maintenance than a more active litter. In those studies in which nonmalnourished females reared malnourished pups, better nests were found, particularly in the later part of the litter period when differences in activity levels and size are most marked (Crnic, 1980; Fleischer and Turkewitz, 1981; Galler and Turkewitz, 1977).

However, the condition of the pup may not be the only factor relevant to nest-building. Like retrieving, nesting may require considerable energy expenditure by the dam, particularly if the nesting material is at a distance or difficult to achieve access to. Thus, nesting may be diminished when the female herself is malnourished. This was found in protein-deprived females (Wiener *et al.*, 1977), but not in females fed low-calorie diets (Smart and Preece, 1973).

Finally, in the case of stunting by large-litter rearing, as discussed previously, although individual pups may produce less heat, there may be more total heat generated. Furthermore, the sheer number of pups may mess the nest even if the female does build an adequate one. Therefore, it is not surprising that reports on nesting in the large-litter situation are inconsistent. Whether or not a decrease is found may depend on specifics of the testing and living situation.

2.2.4. Summary

There is no question that malnutrition in the rat is associated with changes in early social interactions, especially with the mother. We have discussed two general routes by which these changes can be mediated: (1) malnutrition affects the physical and behavioral properties of the pup as a stimulator of the dam and (2) some types of malnutrition produce direct energy changes in the dam. In general, the mechanisms by which these changes are effected are relatively simple: Maternal retrieving is frequently diminished primarily because the malnourished female has less energy, although less effective olfactory and auditory stimulation from the pups may also be involved. Nursing tends to be increased because malnourished pups generate less heat and possibly also because the malnourished female herself produces less heat and has less energy with which to leave the litter. Similarly, nesting also tends to be increased provided the dam is not malnourished (or rearing a large litter) because this behavior, like nursing, is related to the heat generated by the litter and possibly to diminished activity levels. Finally, malnutrition may affect the pheromone production of the dam so that her stimulus characteristics are altered.

There is no question that in the rat, as in the human, malnutrition is a social disease, since these alterations in mother–young interactions can have marked developmental consequences for the young. In this very general sense there is a similarity between rat and man. However, we need to study in more detail the mechanisms by which these social interactions are affected in man to determine whether the rat provides a suitable model.

3. Malnutrition and Mother–Young Interactions in Humans

There are few studies available that are directly concerned with the relationship between malnutrition and early social interaction in humans. Those that are available suggest that in humans, as in rats, well- and poorly nourished infants differ with regard to their physical and behavioral characteristics and mothers are responsive to these differences. Thus, in addition to obvious and defining characteristics such as low body weight, small head circumference, low weight/length index, and reduced amounts of subcutaneous fat, malnourished infants, as compared to better-nourished infants, are also less responsive to environmental stimulation and change in stimulation (Lester, 1975) are more irritable (Als *et al.*, 1976), are less active (Chavez *et al.*, 1975), show less vigorous activity (Graves, 1976), and cry more weakly with more arrhythmia and at higher fundamental frequencies (Lester, 1976). Although unequivocal evidence that mothers of malnourished infants are responsive to these aspects of their infants' appearance and behavior is lacking, there is considerable indirect evidence suggesting that this is likely to be the case. Thus, Graves (1976) found that mothers of malnourished boys showed less interaction with their infants at a distance and were less responsive to their infants, both when they were crying and when they were not, than were the mothers of more adequately nourished boys. These differences either appeared or increased as the children grew older. It therefore seems likely that they reflect the mothers' responding to increasing differences in the children. In a similar but more experimentally controlled study (Chavez *et al.*, 1975), mothers in one group were provided with dietary supplementation beginning on the 45th day of gestation and their infants were provided with dietary supplements beginning at 12 or 16 weeks of age. A matched gorup of infants and mothers was not supplemented. Distinct differences between the groups were found in a number of infant behaviors including activity, time spent sleeping, and time spent playing. Mothers of the infants in the supplemented feeding group cleaned and bathed and changed the diapers of their infants more frequently than did the mothers in the nonsupplemented group. In addition, the investigators reported that children in the supplemented group were given more protection from hazards such as falling from the crib, burns, and accidents in the yard than were nonsupplemented children. Striking differences in the attention paid to supplemented and non-supplemented children are indicated by the investigators, who state:

> At approximately 36 weeks, various attitudes of deference to the child begin to appear, but only towards the supplemented children. They are spoken to more frequently and they receive more praise. The parents feel prouder of their child and they reward him with more presents, clothes and other objects. The supplemented child is given progressively more attention and consideration, he is listened to more, his belongings are better taken care of, and he is given space which, significantly, does not happen to the non-supplemented child. The establishment of greater emotional contact is demonstrated through a more continuous and permanent concern for the needs of the supplemented children.

It is likely that some of the differences in the behavior of the mothers of the two groups stem from direct effects on the mothers of their dietary supplementation during pregnancy. Another obvious basis for the difference in the behavior of the mothers is their different responses to their well- and poorly nourished infants, who look and behave differently. This interpretation is bolstered by the finding that as in the Graves study, differences between mothers in the two groups increased with age, and possibly even more tellingly, the fathers of infants in the supplemented group (who were not themselves given any dietary supplementation) also responded more positively to their infants than did fathers of nonsupplemented children. These men were likely to play with their infants and to interact with them in general more than were the fathers of nonsupplemented infants.

Further evidence in support of the view that adults respond differentially to characteristics of well- and poorly nourished infants is provided in a study by Zeskind and Ramey (1978), who found less maternal involvement with 18-month-old malnourished infants than with a comparison group of better-nourished infants. As was the case in the other studies cited, these differences appeared only at older ages, not being apparent at 6 months.

The effects previously described may stem from changes in both infants' behavioral characteristics and physical appearance. There is evidence that human mothers are responsive to a variety of physical attributes of babies, many of which may vary with the babies' nutritional status. Studies have reported that facial variables such as fatness of cheeks may increase attractiveness ratings of infants (Brooks and Hochberg, 1960; Hildebrandt and Fitzgerald, 1979). Furthermore, adult attentiveness may vary with such perceived attractiveness (Hildebrandt and Fitzgerald, 1978). Newly parturient mothers of premature infants have usually been found to be less responsive to and involved with their babies than mothers of full-term infants (Goldberg, 1979). Kallen (1975) has suggested that bigger infants are likely to be perceived and treated as more mature and more competent. Thus, since malnutrition affects infants' size and fatness, the physical changes in the infant may be as important as the behavioral changes in mediating alterations in social interactions.

As is the case for some of the studies of the effects of malnutrition on mother–young interaction in rats, the mothers in most of the studies involving malnutrition in humans have themselves been malnourished. Although we have been stressing the point that effects on maternal behavior are likely to derive from the mothers' responsiveness to the characteristics of the young, maternal malnutrition with accompanying lassitude, weakness, and increased irritability is also likely to negatively affect mother–infant interactions.

4. Comparison of Effects of Malnutrition in Rats and Humans

4.1. Parallels between Rats and Humans

Comparison of the literature concerning the effects of malnutrition on the development of mother–young relationships in rats and humans indicates some striking similarities in the general picture presented. Malnutrition affects the behavior and appearance of both rat pups and human infants, and rats and human mothers are responsive to these effects. Furthermore, mothers rearing malnourished rats or infants may themselves be malnourished, a condition that is likely by itself to influence mother–young interaction.

Thus, in both rats and people, malnutrition owes its effects to both direct and indirect mechanisms, with one principal mode of indirect effect being via alterations in mother–young interaction. There are also more specific similarities in the way in which malnutrition affects development in rats and humans. Thus, one specific attribute of both human and rat infants that is known to be influenced by malnutrition is vocalization. Furthermore, there is evidence that the behavior of both rat and human mothers can be influenced by changes in the quality of the cry. A second parallel is that in both rat and human young, malnutrition is sometimes associated with decreases in activity levels, which may, in turn, affect both maternal care and decrease the opportunity the young organism has to learn about its environment (Birch and Gussow, 1970; Levitsky *et al.*, 1975). Similarly, in most cases, malnutrition appears to be associated with a decrease in the mother's energy, so that activities that require much energy expenditure (e.g., retrieving in the rat, active interaction or play in the human) will be reduced.

In addition to these relatively clear parallels, studies of rats have suggested several potentially useful areas of exploration that have yet to be carried out in humans. Thus, examination of the literature concerning rat malnutrition indicates that one important way in which malnutrition influences mother–young interaction is by changing the amount of heat produced by the smaller offspring. It is not out of the question that one of the significant stimulative properties of human infants is the amount of heat they produce, which may make babies more or less satisfying to hold or cuddle. To our knowledge, no studies of the relationship between malnutrition and heat production of infants and response of mothers to differences in heat production have been undertaken in man.

Another nutritionally relevant variable that has been found to play a role in mother–young interaction in rats but that has been largely unexplored in humans is the mother's response to the infant's odor. It is not known whether malnutrition affects the odors of human infants or whether mothers would be affected by any such changes in odor. However, the evidence from animal studies suggests that this should be explored.

The rat studies also suggest that it would be fruitful to explore the possibility of a malnutrition-induced change in the obverse relationship, i.e., the odor stimulus provided by the malnourished mother to her infant. A recent study (Macfarlane, 1975) has shown that within the first weeks of life, an infant is responsive to his or her mother's odor, being able to discriminate the odor of the mother's breastpad from that of another nursing mother. Thus, the malnourished human mother, like the rat dam, may provide an altered olfactory environment to her offspring, one to which the child may be responsive.

4.2. Limitations of the Rat Model

Some of the potential uses of the rat for examining the effects of malnutrition on mother–young interaction in humans having been stated, it is necessary that some of the limitations and even drawbacks of the rat as a model be examined. As stated earlier, models are most useful when there are marked similarities in the mechanisms between the model and that which is being modeled and when differences between the two are relatively trivial. In point of fact, we have thus far been ignoring potentially important differences between rats and man that probably make some of the obtained parallels discussed previously superficial. For example, although both rats and humans may respond differently to bigger and smaller infants, the basis for this difference is likely to be quite different in the two. Thus, as discussed previously, rats appear to be responding to differences in the stimulative properties of the bigger and smaller young and spend more time nursing the smaller pups because of the reduced heat provided by them. Although we know of no studies relating nursing behavior and infant size in humans, what studies we do have suggest that mothers of malnourished babies tend to devote less, not more, time to them. In fact, we suspect that this outcome may occur only in women who are themselves the victims of poverty, malnutrition, and other disadvantages. In more advantaged women, we might expect to find a compensation by the mother for the fragility or immaturity of the infant so that increased attentiveness may occur. In short, we are suggesting that in humans, as opposed to rats, a much more complicated set of factors in addition to simply physical characteristics of the baby influences maternal behavior, e.g., mothers' feelings of pride, expectations, and other cognitive factors. In addition, there is a growing body of evidence indicating that in humans, biological variables in general interact with social class, so that factors such as prematurity and complications of pregnancy and delivery are associated with more negative outcomes in children from lower-social-class families than those from upper-social-class families (Richardson, 1968; Sameroff and Chandler, 1975).

Thus, although human mothers may also be affected by direct stimulative characteristics of their babies, to the extent that mother–young interactions in people are affected by social structure, cultural background, and cognitive influences, use of the rat as a model for effects of malnutrition on human mother–young interaction is likely to be more misleading than helpful. That is, it may result in an undue emphasis on direct stimulative influences at the expense of concern for what might be more relevant cognitive and cultural influences. For example, although, as we have previously hypothesized, different heat production by infants of different sizes could influence mother–young interactions in humans as well as in rats, customs involving the use of clothing could markedly reduce any such effect and do so differently in different cultures. Another and probably more important example of an area in which use of the rat as a model could be counterproductive is with regard to the influence of gender on mother–infant relationships. Examination of the literature on the response of the female rat to male and female offspring reveals relatively little difference in maternal behavior as a function of sex of the young, and those differences that do exist in the mother's behavior are likely to reflect a difference in the stimulative properties of the pup, such as sex-related odor difference (Moore and Morelli, 1979). Although it is possible that human adults are responsive to differences in the stimulative characteristics of male and female infants, the overwhelming evidence suggests that differences in the treatment of male and female infants are more culturally determined (Barry *et al.*, 1957; Will *et al.*, 1976). Furthermore, it seems likely that such differences in child-care practices are magnified with regard to the treatment of malnourished boys and girls. For example, Graves (1976) had to restrict her study to boys because the apparent unwillingness of parents to bring their undervalued girls to a nutrition clinic resulted in a sample of girls too small to be meaningful. Since all such societally determined differences in the treatment of males and females are nonexistent in the rat, use of the rat as a model is likely to be misleading.

In view of the important differences between rats and humans with regard to the possible mechanisms mediating nutritionally derived effects on mother–young interaction, it appears likely that the rat represents a poor choice of model for furthering our understanding of the effects of malnutrition on human adaptive functioning. Despite similarities between rat and human of the type Denenberg and Thoman (1976) have identified as essential for a useful model, i.e., similar functional relationships and similar changes with development, the rat provides a useful model for influences of malnutrition on human behavior in only the most broad and general sense. Thus, as Denenberg and Thoman discussed with regard to the rat as a model for at-riskness in general, findings in the area of malnutrition suggest that in both rat and human, there is a reciprocal rather than a unidirectional interaction between mother and young and a role for physical characteristics of the young in determining the mother–young interaction. However, although these parallels are interesting, they are so general as to make them of only limited value in searching for mechanisms by which malnutrition influences human behavior.

5. Alternative Models for Human Malnutrition: Subhuman Primates

In view of the problem with the use of the rat as a model for human malnutrition, it is necessary to consider the question of whether any useful animal model exists. An obvious possibility would be subhuman primates. We will attempt an evaluation of several studies that suggest that such animals may indeed provide a more reasonable model than does the rat.

5.1. Advantages of the Primate Model

As is true for rats and humans, maternal subhuman primates are responsive to the stimulative attributes of their young (Berkson, 1970, 1974; Lindberg, 1969). In addition, however, there is evidence that primate mothers are also influenced by less stimulus-bound aspects of the mother–young situation. Thus, the maternal behavior of primate mothers depends on such characteristics as parity, type of maternal care they received as infants, and opportunity to engage in social contact with other adults (Horvat *et al.*, 1980; Sackett and Ruppenthal, 1974). In addition, they respond to their young as individuals, in a manner that is unlike that of the rat. There is some indication that rats are responsive to the unique attributes of their own young. Beach and Jaynes (1956a) found that female rats tend to retrieve their own pups before retrieving strange pups. However, there appears to be an attachment to a particular individual in some groups of subhuman primates that is more similar to the situation with regard to humans than is the case for rats. For example, in the squirrel monkey, there is evidence of an emotional bond between the mother and her offspring that goes beyond specific stimuli provided by the partners to one another. Studies in which squirrel monkeys are separated indicate that both the mother and infant show an elevation of pituitary–adrenal secretions that can be reduced only be reuniting the pair (Mendoza *et al.*, 1980). Providing another infant to the mother or a caretaking "aunt" for the infant will not eliminate this response For the female rat, on the other hand, one litter is responded to very much like another. Most of the response of the young to separation can similarly be eliminated by providing stimuli such as warmth and food (Hofer, 1975). There is, in other words, no evidence of a specific preference for the individual partner.

These findings suggest that if malnutrition is imposed on the primate, there might be consequences for the mother–infant interaction above and beyond those found in the rat and, further, that these consequences might be similar to those in the human.

5.2. Limitations of the Primate Model

It should be noted that some of the problems previously described with regard to the use of the rat as a model also pertain to the subhuman primate. Many cultural and cognitive factors that are likely to be of major importance for humans are unlikely to play a role in the behavior of any animals, even subhuman primates. For example, we have discussed the fact that social class is likely to moderate the effects of malnutrition in humans. Similarly, clothing of the child may modify the effects of malnutrition. Clearly, there are no animal parallels for such relationships. In addition, although in this chapter we have not dealt with the issue of the types of behavioral functions most likely to be affected by early malnutrition and concomitant alterations of mother–young interaction, we must at least note that serious concern for understanding the possible consequences of malnutrition began when it was recognized that it might have deleterious effects on human cognitive functioning, much of which involves language and reading, precisely those areas for which any animal model is unlikely to be appropriate. Thus, while subhuman primates may provide a somewhat more adequate model than rats, their value, too, is limited. That is, there appear to be fundamental differences as well as similarities in the way in which malnutrition is likely to affect humans and subhuman animals.

6. Comparative Approach for Understanding the Effects of Malnutrition

Despite the rather gloomy prognosis with regard to the use of animals as models, we think it important to indicate that nothing we have said precludes the possibility that the study of malnutrition in animals can be of great value for the understanding of its effects in humans. This apparently paradoxical statement is based on the view that the most appropriate use of animals for understanding complex aspects of functioning may not be as models, but rather as sources of information for comparative study. This comparative approach that we are advocating differs from the use of models in that it seeks to make use not only of similarities across species but also of differences. Rather than viewing any differences uncovered as impediments to the drawing of analogies, the comparative approach relies on the discovery of differences to elucidate underlying mechanisms. Different species and orders of animals are viewed as the equivalent of a series of experiments in nature in which different outcomes of similar treatments can be related to different sensory, neural, or social organizations in these species. Since this is the case, the selection of animals to be studied becomes of great importance. The most important consideration in selecting an organism for study is not its similarity to man, but rather some specific differences from man and the other animals studied that might make it possible to relate any differences in outcome of the common treatment to particular structures, behaviors, or functional organizations.

A comparative study of mechanisms whereby malnutrition might influence mother–young relationships, for example, could derive from the analysis that we have offered here. In that our analysis has sought to focus on differences as well as similarities in the results of studies of rats, primates, and humans, we have used a rudimentary form of the comparative approach. A fuller, more explicit use of the comparative approach in this regard would entail the systematic and preplanned examination of the similarities and differences in the effects of malnutrition on mother–young interaction in animals with different sensory and perceptual characteristics but similar social organizations; other organisms with similar sensory–perceptual and social characteristics but different cognitive capacities; and, finally, organisms with similar sensory–perceptual, social, and cognitive capacities but different cultural organizations. In this way, some of the hypotheses generated in this chapter could be tested.

In our laboratories, we have been using a variation of the comparative approach in which, rather than comparing and contrasting different species, we have focused on both similarities and differences of behavioral outcomes associated with different methods of stunting within a single species. As discussed earlier, each method of stunting entails alterations in nonnutritional aspects of the environment as well as nutritional deprivation, making attribution of behavioral deficits to reduced food intake *per se* difficult. The purpose of the comparative study of different methods of stunting has been to attempt to pinpoint which behavioral effects are due to nutritional deprivation and which to alterations in nonnutritional variables. Those effects that are common to all methods of stunting are likely to be due to reduced food intake, whereas those effects that occur only with some methods of stunting are probably due to extranutritional factors unique to those methods.

For example, we have compared the effects of two different stunting methods on homing to the nest during the litter period and on visual discrimination learning in adulthood (Fleischer and Turkewitz, 1979a,b). Sprague–Dawley rats were stunted either by rearing in large litters of 18–22 pups or by rotation between lactating females (for 8 hr daily) and nonlactating maternal females (for 16 hr). In each case, behavior was compared to appropriate controls: pups reared in small litters of four or pups rotated between two lactating females. Animals were tested on alternate days during the litter period for their ability to return to their nests from other parts of the home cage within 3 min. Both animals stunted by rotation between females and animals from large litters evidenced decreased frequencies of homing to the nest and a delay in the age at which homing was maximal (see Fig. 1). In a subsequent study, it was suggested that stunting might influence homing behavior by altering the development of responsiveness to olfactory and other cues from the nest (Fleischer *et al.*, 1981).

After being weaned, the animals were nutritionally rehabilitated and then tested for visual discrimination performance at 3 months of age using the jump-

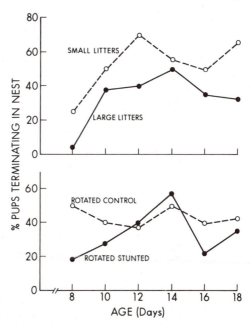

Fig. 1. Comparison of effects of two different methods of stunting on homing behavior.

ing method originated by Lashley. However, because the incentive value of food may be altered in malnourished animals, no food reward was used. Animals were required to jump from an elevated stand to one of two windows in which were hung the stimuli to be discriminated. Choice of the negative stimulus resulted in a fall, while choice of the positive stimulus resulted in a "safe" landing on a platform behind the window. This was the only incentive used. Three descrimination problems of increasing difficulty were presented: black vs. white, horizontal vs. vertical stripes, and circle vs. triangle.

Neither group of malnourished female animals nor the males malnourished by rotation between females showed deficits in visual discrimination learning. However, males from large litters evidenced a marked increase in the number of trials required to reach criterion on both the black vs. white and the horizontal vs. vertical stripes discriminations (see Fig. 2). The basis for the deficit appeared to be primarily emotional rather than perceptual in nature, since they learned the most difficult discrimination as readily as controls. Furthermore, the difficulty appeared to be due to an inability to break position habits, which has been attributed to increased emotionality in the face of a novel situation. In fact, the animals appeared to discriminate the stimuli, since they would jump without prodding when the positive stimulus was on the preferred side, but required prodding when it was on the opposite side.

Evidence of the value of focusing on differences as well as similarities in the outcome of malnutrition produced by a variety of techniques is provided by a comparison of the effects of the treatments just described with the effects of intergenerational malnutrition. When the animals studied by us were compared to those that had a history of 15–17 generations of malnutrition (7.5%

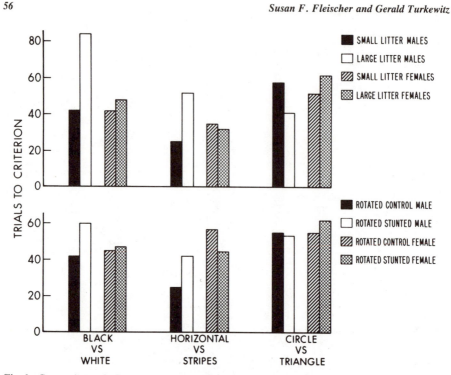

Fig. 2. Comparison of effects of different methods of stunting on visual discrimination learning.

casein diet), the most striking difference was that the intergenerationally deprived animals showed no deficits on the simplest discrimination, but showed
progressive deficits with increasing task difficulty (Galler *et al.*, 1980). These
findings are consistent with those of Turkewitz (1975), who studied a similar
group of animals and suggested that intergenerational malnutrition may result
in true perceptual difficulties. This suggestion needs to be tempered by the fact
that there were strain differences in the animals as well as minor procedural
variations in the various studies. However, we do feel confident in the general
value of the comparative approach as a means of understanding the consequences of malnutrition and the circumstances surrounding it.

Comparison of the results associated with different methods of stunting
has permitted us to suggest which outcomes are probably due to nutritional
factors and which to alterations in nonnutritional variables such as maternal
care. For example, with regard to homing to the nest, the fact that both animals
malnourished by rotation between females and animals reared in large litters
showed similar decreases in homing suggests that this behavioral outcome is
due to nutrition, particularly since extranutritional factors such as maternal
care differ markedly for these kinds of animals. This suggestion is buttressed
by findings of similar homing deficits in still other types of stunted animals
[e.g., hypothyroid (Johanson *et al.*, 1980) and single and intergenerationally
protein-deprived rats (Galler, 1980)].

On the other hand, the fact that visual discrimination deficits were found in males from large litters but not in animals stunted by rotation suggests that this effect may be the result of unique alterations in extranutritional factors in the large-litter situation. Alterations in maternal care, increased competition between littermates, or increased crowding could all contribute to high levels of emotionality in males from large litters. In any case, the effect is unlikely to be due to early nutritional insult *per se*, particularly in view of the finding that intergenerationally malnourished rats show a very different pattern of behavioral deficits in the visual discrimination task.

The comparative analysis of the different methods of stunting, focusing on differences as well as similarities of outcome, has permitted us to draw more definitive conclusions than would studying the consequences of only a single method of stunting or focusing only on similarities of outcome across stunting methods. It should be noted that those measures that yielded different outcomes in animals stunted by different techniques were at least as informative in suggesting underlying mechanisms as were those that yielded similarities. We believe that we have made a beginning at developing a useful intraspecies program of comparison of the effects of different methods of producing malnutrition on behavior, but are convinced that the most productive use of animals as an aid to understanding the effects of malnutrition on human behavior would be to expand to a truly comparative program in which a variety of carefully selected species of animals are studied. Such a program will require examination of both similarities and differences of outcome across species and exploration of the bases for these effects. Although we recognize that this is an extremely difficult and time-consuming approach, we believe that the understanding to be gained by the attempt would, in the long run, be more productive and efficient than would any seemingly more direct approach to the use of animals for the understanding of the effects of malnutrition on human behavior.

ACKNOWLEDGMENT. Preparation of this manuscript was supported in part by City University of New York PSC-BHE Grant No. 12233.

7. References

Als, H., Tronick, E., Adamson, L., and Brazelton, T. B., 1976, The behavior of the full-term yet underweight newborn infant, *Dev. Med. Child Neurol.* **18**:590.

Altman, J., Sudarshan, K., Das, G. D., McCormick, N., and Barnes, D., 1971, The influence of nutrition on neural and behavioral development. III. Development of some motor, particularly locomotor patterns during infancy, *Dev. Psychobiol.* **4**:97.

Barnes, R. H., Moore, A. U., Reid, I. M., and Pond, W. G., 1968, Effects of food deprivation on behavioral patterns, in: *Malnutrition, Learning, and Behavior* (N. S. Scrimshaw and J. E. Gordon, eds.), pp. 203–218, MIT Press, Cambridge, Massachusetts.

Barry, H., Bacon, M. K., and Child, I. L., 1957, A cross-cultural survey of some sex differences in socialization, *J. Abnorm. Soc. Psychol.* **55**:327.

Beach, F. A., and Jaynes, J., 1956a, Studies of maternal retrieving in rats. I. Recognition of young, *J. Mammal.* **37**:177.

Beach, F. A., and Jaynes, J., 1956b, Studies of maternal retrieving in rats. III. Sensory cues involved in lactating female's response to her young, *Behavior* **10**:104.

Bell, R. Q., 1971, Stimulus control of parent or caretaker by offspring, *Dev. Psychol.* **4**:63.

Berkson, G., 1970, Defective infants in a feral monkey group, *Folia Primatol.* **12**:284.

Berkson, G., 1974, Social responses of animals to infants with defects, in: *The Effect of the Infant on its Caregiver* (M. Lewis and L. A. Rosenblum, eds.), pp. 233–249, John Wiley, New York.

Birch, H. G., and Gussow, J. D., 1970, *Disadvantaged Children: Health, Nutrition and School Failure*, Grune and Stratton, New York.

Brooks, V., and Hochberg, J., 1960, A psychophysical study of "cuteness," *Percept. Mot. Skills* **11**:205.

Brozek, J., 1978, Nutrition, malnutrition and behavior, *Annu. Rev. Psychol.* **29**:157.

Chavez, A., Martinez, C., and Yaschine, T., 1975, Nutrition, behavioral development and mother–child interaction in young rural children, *Fed. Proc. Fed. Am. Soc. Exp. Biol.* **34**:1574.

Codo, W., and Carlini, E. A., 1979, Postnatal undernutrition in rats: Attempts to develop alternative methods to food deprive pups without maternal behavioral alteration, *Dev. Psychobiol.* **12**:475.

Collier, G. H., and Squibb, R. L., 1968, Malnutrition and the learning capacity of the chicken, in: *Malnutrition, Learning and Behavior* (N. S. Scrimshaw and J. E. Gordon, eds.), pp. 236–240, MIT Press, Cambridge, Massachusetts.

Crnic, L. S., 1976, Maternal behavior in the undernourished rat (*Rattus norvegicus*), *Physiol. Behav.* **16**:677.

Crnic, L. S., 1980, Models of infantile malnutrition in rats: Effects on maternal behavior, *Dev. Psychobiol.* **13**:615.

Denenberg, V. H., and Thoman, E. B., 1976, From animal to infant research, in: *Intervention Strategies for High Risk Infants and Young Children* (T. D. Tjossen, ed.) pp. 85–106, University Press, Baltimore, Maryland.

Fleischer, S. F., and Turkewitz, G., 1979a, Effect of neonatal stunting on development of rats: Large litter rearing, *Dev. Psychobiol.* **12**:137.

Fleischer, S. F., and Turkewitz, G., 1979b, Behavioral effects of rotation between lactating and nonlactating females, *Dev. Psychobiol.* **12**:245.

Fleischer, S. F., and Turkewitz, G., 1981, Alterations of maternal behaviors of female rats caring for malnourished pups, *Dev. Psychobiol.* **14**:383.

Fleischer, S. F., Turkewitz, G., and Finklestein, H., 1981, Sensory influences on homing of stunted rat pups, *Dev. Psychobiol.* **14**:29.

Frankova, S., 1971, Relationship between nutrition during lactation and maternal behavior of rats, *Act. Nerv. Super.* **13**:1.

Frankova, S., 1972, Influence of nutrition and early experience on behavior of rats, *Bibl. Nutr.* **17**:96.

Galef, B. G., Jr., and Kaner, H. C., 1980, Establishment and maintenance of preference for natural and artificial olfactory stimulation in juvenile rats, *J. Comp. Physiol. Psychol.* **94**:588.

Galler, J. R., 1979, Home orientation in nursling rats: The effects of rehabilitation following intergenerational malnutrition, *Dev. Psychobiol.* **12**:499.

Galler, J. R., 1980, Home orienting behavior in rat pups surviving postnatal or intergenerational malnutrition, *Dev. Psychobiol.* **12**:563.

Galler, J. R., and Seelig, C., 1981, Home orienting behavior in rat pups: The effect of 2 and 3 generations of rehabilitation following intergenerational malnutrition, *Dev. Psychobiol.* **14**:541.

Galler, J. R., and Turkewitz, G., 1977, Use of partial mammectomy to produce malnutrition in the rat, *Biol. Neonate* **31**:260.

Galler, J. R., Fleischer, S. F., Turkewitz, G., and Manes, M., 1980, Varying deficits in visual discrimination performance associated with different forms of malnutrition in rats, *J. Nutr.* **110**:231.

Goldberg, S., 1979, Premature birth: Consequences for the parent–infant relationship, *Am. Sci.* **67**:214.

Graves, P. L., 1976, Nutrition, infant behavior, and maternal characteristics: A pilot study in West Bengal, India, *Am. J. Clin. Nutr.* **29**:305.

Grota, L. J., and Ader, A., 1969, Continuous recording of maternal behavior in the rat, *Anim. Behav.* **17**:722.

Hall, R. D., Leahy, J. P., and Robertson, W. M., 1979, The effects of protein malnutrition on the behavior of rats during the suckling period, *Dev. Psychobiol.* **12**:455.

Harper L. V., 1971, The young as a source of stimuli controlling caretaker behavior, *Dev. Psychol.* **4**:73.

Harrington, D. D., and Newberne, P. M., 1967, Maternal vitamin A deficiency induced hydrocephalus in newborn rabbits: Relationship between maternal blood levels of vitamin A and the incidence of hydrocephalus, *Fed. Proc. Fed. Am. Soc. Exp. Biol.* **26**:519.

Hennessy, M. B., Smotherman, W. P., Kolp, L., Hunt, L., and Levine, S., 1978, Stimuli from pups of adrenalectomized and malnourished female rats, *Physiol. Behav.* **20**:509.

Hildebrandt, K., and Fitzgerald, H., 1978, Adults' responses to infants varying in preceived cuteness, *Behav. Proc.* **3**:159.

Hildebrandt, K., and Fitzgerald, H., 1979, Facial features determinants of perceived infant attractiveness, *Infant Behav. Dev.* **2**:329.

Hofer, M. A., 1975, Infant spearation responses and the maternal role, *Biol. Psychiatry* **10**:149.

Horvat, J. R., Coe, C. L., and Levine, S., 1980, Infant development and maternal behavior in captive chimpanzees, in: *Maternal Influences and Early Behavior* (R. W. Bell and W. P. Smotherman, eds.), pp. 285–311, Spectrum, New York.

Howard E., and Granoff, D. M., 1968, Effect of neonatal food restriction in mice on brain growth, DNA and cholesterol, and on adult delayed response learning, *J. Nutr.* **95**:111.

Hunt, L. E., Smotherman, W. P., Wiener, S. G., and Levine, S., 1976, Nutritional variables and their effect on the development of ultrasonic vocalizations in rat pups, *Physiol. Behav.* **17**:1037.

Johanson, I. B., 1980, Development of olfactory and thermal responsiveness in hypothyroid and hyperthyroid rat pups, *Dev. Psychobiol.* **13**:343.

Johanson, I. B., Turkewitz, G., and Hamburgh, M., 1980, Development of home orientation in hypothyroid and hyperthyroid rat pups, *Dev. Psychobiol.* **13**:331.

Kallen, D. J., 1975, Effects of nutrition on maternal–infant interaction, *Fed. Proc. Fed. Am. Soc. Exp. Biol.* **34**:1571.

Kerr, G. R., and Waisman, H. A., 1968, A primate model for the quantitative study of malnutrition, in: *Malnutrition, Learning and Behavior* (N. S. Scrimshaw and J. E. Gordon, eds.), pp. 240–249, MIT Press, Cambridge, Massachusetts.

Kinder, E. F., 1927, A study of the nest building activity of the albino rat, *J. Exp. Zool.* **47**:117.

Leathwood, P., 1978, Influence of early undernutrition on physical development, behavioral development and learning in rodents, in: *Early Influences*, Vol. 4 (G. Gottlieb, ed.), pp. 187–209, Academic Press, New York.

Lehrman, D. S., 1961, Hormonal regulations of parental behavior in birds and infrahuman mammals, in: *Sex and Internal Secretions*, Vol. II, 3rd ed. (W. C. Young, ed.), pp. 1268–1382, Williams and Wilkins, Baltimore.

Leon, M., 1975, Dietary control of maternal pheromone in the lactating rat, *Physiol. Behav.* **14**:311.

Leon, M., 1978, Thermal control of mother–young contact in rats, *Physiol. Behav.* **21**:793.

Leon, M., and Moltz, H., 1972, The development of the pheromonal bond in the albino rat, *Physiol. Behav.* **8**:683.

Leon, M., Galef, B. G., Jr., and Behse, J. H., 1977, Odor pre-exposure: Effects on ontogeny of pheromonal bonds and diet choice in rats, *Physiol. Behav.* **18**:387.

Lester, B. M., 1975, Cardiac habituation of the orienting response to an auditory signal in infants of varying nutritional status, *Dev. Psychol.* **11**:42.

Lester, B. M., 1976, Spectrum analysis of the cry sounds of well-nourished and malnourished infants, *Child Dev.* **47**:237.

Levitsky, D. A., Massaro, T. F., and Barnes, R. H., 1975, Maternal malnutrition and the neonatal environment, *Fed. Proc. Fed. Am. Soc. Exp. Biol.* **34**:1583.

Lewis, M., and Rosenblum, L. A., 1974, *The Effect of the Infant on its Caregiver*, John Wiley, New York.

Lindberg, D. G., 1969, Behavior of infant rhesus monkeys with thalidomide-induced malformations, *Psychon. Sci.* **15**:55.

Lynch, A., 1976, Postnatal undernutrition: An alternative method, *Dev. Psychobiol.* **9**:39.

Macfarlane, J. A., 1975, Olfaction in the development of social preferences in the human neonate, in: *Parent–Infant Interaction*, CIBA Foundation Symposium No. 33, pp. 103–117, Horth-Holland/Elsevier, New York.

Massaro, T. F., Levitsky, D. A., and Barnes, R. H., 1974, Protein malnutrition in the rat: Its effects on maternal behavior and pup development, *Dev. Psychobiol.* **7**:551.

Mendoza, S. P., Coe, C. L., Smotherman, W. P., Kaplan, J., and Levine, S., 1980, Functional consequences of attachment: A comparison of two species, in: *Maternal Influences and Early Behavior* (R. W. Bell and W. P. Smotherman, eds.), pp. 235–253, Spectrum, New York.

Moltz, H., Leidahl, L., and Rowland, D., 1974, Prolongation of the maternal pheromone in the albino rat, *Physiol. Behav.* **12**:409.

Moore, C. L., and Morelli, G. A., 1979, Mother rats interact differently with male and female offspring, *J. Comp. Physiol. Psychol.* **93**:677.

Noirot, E., 1964, Changes in responsiveness to young in the adult mouse. II. The effect of external stimuli, *J. Comp. Physiol. Psychol.* **57**:97.

Platt, B. S., and Stewart, R. J. C., 1968, Effects of protein–calorie deficiency on dogs. I. Reproduction, growth and behavior, *Dev. Med. Child Neurol.* **10**:3.

Platt, B. S., and Stewart, R. J. C., 1969, Effects of protein–calorie deficiency on dogs. II. Morphological changes in the nervous system, *Dev. Med. Child Neurol.* **11**:174.

Richardson, S. F., 1968, The influence of social-environmental and nutritional factors on mental ability, in: *Malnutrition, Learning and Behavior* (N. S. Scrimshaw and J. E. Gordon, eds.), pp. 346–360, MIT Press, Cambridge, Massachusetts.

Rosenblatt, J. S., 1963, The basis of synchrony in the behavioral interaction between the mother and her offspring in the laboratory rat, in: *Determinants of Infant Behavior*, Vol. 3 (B. M. Foss, ed.), pp. 3–45, John Wiley, New York.

Scakett, G. P., and Ruppenthal, G. C., 1974, Some factors influencing the attraction of adult female macaque monkeys to neonates, in: *The Effect of the Infant on the Caregiver* (M. Lewis and L. A. Rosenblum, eds.), pp. 163–185, John Wiley, New York.

Sameroff, A. J., and Chandler, N. J., 1975, Reproductive risk and the continuum of caretaking casualty, in: *Review of Child Development Research*, Vol. 4 (F. Horowitz, ed.), pp. 187–243, University of Chicago Press.

Scrimshaw, N. S., and Gordon, J. E., 1968, *Malnutrition, Learning and Behavior*, MIT Press, Cambridge, Massachusetts.

Seitz, P. F. D., 1958, The maternal instinct in animal subjects, *Psychosom. Med.* **20**:215.

Slob, A. K., Snow, C. E., and De Natris-Mathot, E., 1973, Absence of behavioral deficits following neonatal undernutrition in the rat, *Dev. Psychobiol.* **6**:177.

Smart, J. L., 1976, Maternal behavior of undernourished mother rats towards well fed and underfed young, *Physiol. Behav.* **16**:147.

Smart, J. L., and Preece, J., 1973, Maternal behavior of undernourished mother rats, *Anim. Behav.* **21**:613.

Smith, J. C., and Sales, G. D., 1980, Ultrasonic behavior and mother–infant interactions in rodents, in: *Maternal Influences and Early Behavior* (R. W. Bell and W. P. Smotherman, eds.), pp. 105–133, Spectrum, New York.

Smotherman, W. P., Bell, R. W., Starzec, J., Elias, J., and Zachman, T. A., 1974, Maternal responses to infant vocalizations and olfactory cues in rats and mice, *Behav. Biol.* **23**:55.

Stern, J. J., and Bronner, G., 1970, Effects of litter size on nursing time and weight of the young in guinea pigs, *Psychon. Sci.* **21**:171.

Turkewitz, G., 1975, Learning in chronically protein-deprived rats, in: *Nutrition and Mental Functions* (G. Serban, ed.), pp. 113–120, Plenum Press, New York.

Wiener, S. G., Smotherman, W. P., and Levine, S., 1976, Influence of maternal malnutrition on pituitary–adrenal responsiveness to offspring, *Physiol Behav.* **17**:897.

Wiener, S. G., Fitzpatrick, K. M., Levin, R., Smotherman, W. P., and Levine, S., 1977, Alterations in the maternal behavior of rats rearing malnourished offspring, *Dev. Psychobiol.* **10**:243.

Will, J. A., Self, P. A., and Datan, N., 1976, Maternal behavior and perceived sex of infant, *Am. J. Orthopsychiatry* **46**:135.

Winick, M., 1976, *Malnutrition and Brain Development*, Oxford University Press, New York.

Zarrow, M. X., Schlein, P. A., Denenberg, V. H., and Cohen, H. A., 1972, A sustained corticosterone release in lactating rats following olfactory stimulation from the pups, *Endocrinology* **9**:197.

Zeskind, P. S., and Ramey, C. T., 1978, Fetal malnutrition: An experimental study of its consequences on infant development in two caregiving environments, *Child Dev.* **49**:1155.

3

Behavioral Consequences of Malnutrition in Early Life

Janina R. Galler

1. Introduction

A large number of studies have focused on the link between malnutrition and behavior. The long-term consequences of early malnutrition are now receiving increasing attention as better public health measures and medical care ensure the survival of many children with childhood malnutrition who would previously have succumbed. The vast majority of these studies have examined the behavioral outcomes of children with severe malnutrition, who are the most easily identified group within the much larger population of children suffering from milder forms of malnutrition. In addition, children with severe forms are more likely to demonstrate aberrant behavioral outcomes than are the milder cases. Those who are less severely malnourished are also likely to be disadvantaged, but the long-term effects in this population remain more difficult to identify by current methodologies.

This chapter has three main purposes. First, it will attempt to review and classify existing studies on malnutrition and behavior according to how long after the episode of severe malnutrition the behavioral consequences have been examined. The effects will therefore be classified as being *concurrent* with the episode of infantile malnutrition; *intermediate*, occurring during recovery from malnutrition, which may last 2–3 years; and finally, *long-term*, persisting many years after the initial episode. In contrast, most reviews, including those of Brozek (1978) and Pollitt and Thompson (1977), have classified these studies only according to the type or degree of malnutrition suffered by the child.

Second, this chapter will analyze the various published studies in terms of the variety of measures of behavioral and intellectual function applied at

Janina R. Galler • Division of Psychiatry, Boston University School of Medicine, Boston, Massachusetts 02118.

each of these time periods. Most studies to date have been restricted to the area of cognitive function, notably intelligence, with the balance of evidence demonstrating long-term impairment of intellectual capacity. However, the majority of these studies do not provide a comprehensive investigation into other aspects of behavioral function such as emotional adjustment, attention span, or behavior at home, all of which may impair the ability of the previously malnourished child to adapt to his environment. Deficits in the latter areas of function are probably more sensitive expressions of the long-term consequences of malnutrition than are deficits in IQ alone.

The final section describes our study of the long-term effects of malnutrition on the behavioral development of Barbadian school children. This study provides a comprehensive overview of many areas of behavioral function in a population of previously malnourished children followed from the first year of life through 11 years of age and demonstrates a clear-cut pattern of behavioral deficits based on identifiable early life experiences.

2. Evidence from Human Studies

2.1. General Considerations

Subsequent behavioral development can be affected both by prenatal or maternal malnutrition and by postnatal malnutrition. The timing of the episode, along with the type of malnutrition and its severity and duration, influence the extent of behavioral and brain development in man and animals (Dobbing, 1973).

Maternal malnutrition can be a major factor in the development of the fetus. In areas where malnutrition is endemic, birth weights of less than 2500 g are prevalent. Low birth weight may result from prematurity or may present as a consequence of fetal growth retardation after pregnancies of normal duration (small-for-dates) as follows: (1) intrinsic growth failure caused by congenital malformations and genetic disease is generally associated with a placenta of normal size; (2) asymmetrical growth failure due to maternal vascular disease, which affects the fetal and placental blood supply and reduces liver size but not brain maturation; and (3) growth failure due to maternal malnutrition, which reduces the growth of all organs proportionately.

Human studies indicate that prenatal malnutrition alone—as may be caused by famine—is unlikely to alter behavioral outcome (Stein *et al.*, 1975); similar findings have been reported in animal studies (Smart and Dobbing, 1971). In human populations chronically exposed to undernutrition, however, weight is often below normal and serious consequences for the behavioral development of the child may result, because maternal malnutrition, which deprives the mother of reserves even before conception, is often coupled with postnatal malnutrition of the infant. For the purposes of this chapter, we shall confine our discussion to the consequences of postnatal malnutrition, though some of these are effects of the combination of prenatal and postnatal malnutrition.

Severe *postnatal malnutrition* during infancy and childhood has been customarily described in terms of two distinct syndromes: kwashiorkor, associated primarily with protein deficiency, and marasmus, an overall reduction in food intake in which total calories are inadequate (Latham, 1974).

Kwashiorkor most often occurs in children between the ages of 1 and 3 years whose diet is grossly deficient in protein, usually as a result of being transferred from breast milk to a starchy diet. The child with kwashiorkor exhibits growth failure, muscle wasting, and edema, including a protuberant abdomen, and there is usually liver enlargement associated with fatty infiltration. Diarrhea and anemia are common, hair is silky, easily plucked, and pale in color, and the skin is depigmented and shows dermatosis. Behavioral characteristics are striking and include marked apathy and irritability. The most striking biochemical abnormality is the reduced level of plasma albumin, but a lowered serum transferrin level is a more sensitive index of severity (McFarlane *et al.*, 1970).

Marasmus, which is due to the insufficient intake of food, occurs most often in children under 1 year of age and accompanies early weaning. Frequently, the food supply is a source of bacterial infection, resulting in gastroenteritis, which then exacerbates the dietary inadequay. The main clinical features include marked growth failure (the child is often less than 60% of normal weight for age, and body length is reduced), muscle wasting, loss of subcutaneous fat, and wizened facial features, contrasting with a protuberant abdomen. As in kwashiorkor, diarrhea and anemia are common, but serum protein levels are nearly normal and there is no edema. Children may be irritable, but apathy is less prominent.

Intermediate forms of severe malnutrition are more common than either kwashiorkor or marasmus. In reality, protein–calorie malnutrition, while primarily due to an inadequate intake of protein and calories, is a complex condition of varied cause and clinical presentation.

The severity of the malnutrition determines the extent and rate of recovery of the child. Biochemical measurements differ in kwashiorkor and marasmus (Olson, 1975), but are not precise enough for classifying severity. Most nutritionists agree that anthropometric measurements (height for age, weight for age, weight for height for age, skinfold thickness, and arm circumference, among others) are the best available indicators of the degree of malnutrition (Gomez *et al.*, 1954a; Jelliffe, 1966; Seoane and Latham, 1971).

In addition to the characteristics of the various types of malnutrition, it is also important to consider the ways in which the episode may modify the child's ability to adapt to the demands of its environment. Such effects may be evident not only in childhood but also throughout the life cycle and may even persist into subsequent generations (Galler, 1983). Certain effects, especially on behavior, may not even appear until later stages of development, when the individual is presented with new challenges, such as child bearing. However, most studies of the long-term effects of early childhood malnutrition in humans extend only to adolescence, and as a result, the total picture is not clear. Nevertheless, by examining the available evidence gathered at different ages, we can identify behavioral patterns resulting from early malnutrition, which modify the child's response to his environment and which are likely to predict later performance.

While the child is acutely malnourished, the behavioral manifestations (*concurrent effects*) are a consequence of the nutritional–metabolic disturbances. These altered behaviors can impair the infant's interaction with his environment and thus impede successful mastery of developmental tasks appropriate to his age group. During the period of recovery from severe malnutrition, which often coincides with the preschool years, rapid metabolic and physiological changes take place as the organism attempts to reestablish normal levels of growth and maturation (Ramos-Galvan *et al.*, 1969). These attempts to "catch up" are only partially successful, as demonstrated by persisting behavioral deficits during this period (*intermediate effects*). Environmental circumstances can play a major role at this time in diminishing or exacerbating these behavioral consequences of malnutrition. Some behavioral effects may be present many years after the episode (*long-term effects*). Although only a few of the studies extend through adolescence, their findings have generally provided the bulk of evidence for permanent behavioral deficits and lags in behavioral development.

Most studies of the long-term effects of malnutrition have been performed in developing countries where this condition is severe and prevalent. Inquiry into the relationship between malnutrition and behavior in these regions of the world has been complicated by the fact that malnourished children most often come from the lowest socioeconomic classes of the population. When reduced function is documented in these children, there has been a tendency to oversimplify this relationship and to attribute behavioral deficits primarily to nutritional factors without taking adequate account of environmental factors, which may have contributed to the etiology of the malnutrition in the first place (Ricciuti, 1981). To disentangle the interrelationships of these two types of factors is virtually impossible in the human condition. One way in which this difficulty presents itself is in the selection of control groups, which are generally more advantaged than the index, malnourished groups and therefore have not had malnutrition. A study that usefully analyzes the determinants of behavioral outcome must look at both these factors in combination, as we have attempted to do in our study of Barbadian school children. On the other hand, studies of severe malnutrition secondary to pyloric stenosis and cystic fibrosis are not confounded by socioeconomic factors but have other limitations, and their results cannot be directly applied to cases of primary malnutrition.

2.2. Concurrent Effects of Malnutrition

There are a limited number of studies of the behavioral and neurological characteristics of children acutely ill with infantile malnutrition. All available studies do point to behavioral changes in children with various types of concurrent malnutrition including intermediate forms, but neurological changes are found primarily in children with marasmus. Published studies of the neurological and behavioral effects in the malnourished child are summarized in Tables IA (neurological), IB (behavior), and IC (mother–infant interaction).

The relationship of neurological changes to intellectual function is unclear in studies of malnutrition in human populations. Nevertheless, some important observations can be made (Table IA) about head circumference, brain weight, and electrophysiological function. Malnourished children, especially those with marasmus, show reduced head circumference in clinical studies, implying a smaller brain size. Autopsies of infants and preschool children who died of several types of malnutrition showed that brain weight is in fact reduced as compared to children who died of other causes (Brown, 1965; Winick and Rosso, 1969; Chase *et al.*, 1974). The Brown study in Uganda examined a large group of malnourished infants ($N = 127$) and control infants ($N = 351$) and is particularly convincing. The two remaining postmortem studies also examined whether the number of brain cells was reduced in the cases of marasmus and came to different conclusions. Thus, Winick and Rosso (1969) reported a reduction in the number of brain cells in their study of Chilean children; Chase *et al.* (1974) found reduced cell size in Guatemalan children. It has been suggested by Dobbing (1973) that the primary brain changes accompanying malnutrition include a reduction in dendritic arborization, thereby reducing connections between brain nerve cells. In the Engsner study in Ethiopia (Engsner *et al.*, 1974), brain size was reduced and the size of the cerebral ventricles was significantly increased in cases of kwashiorkor, but not marasmus, as a result of edema. This deficit subsided after several months of nutritional rehabilitation. Data from electrophysiological studies of the brain may have more direct implications for behavior. Electroencephalograms obtained during the episode of acute malnutrition show evidence of cerebral dysfunction. The earliest studies of EEG abnormalities associated with an episode of malnutrition were those of Gallais *et al.* (1951) in West Africa, Engel (1956), and Nelson (1959). These authors reported that the dominant frequency of the EEG in acutely malnourished children was much lower than that found in healthy children of the same age; this effect persisted for several months after treatment, but eventually vanished. A more recent study in Mexico by Barnet *et al.* (1978) showed abnormal responses to auditory evoked potentials in 26 marasmic infants who were compared with 46 well-nourished infants; clear-cut evidence of cerebral dysfunction in the malnourished cohort was found up to 1 year after discharge.

These observations from human studies have been confirmed and amplified in animal models (Nowak and Munro, 1977). Briefly, postnatal malnutrition in the rat (during the early suckling period) reduces brain weight and protein and DNA content, most notably in the cerebellum, where the neurons are still rapidly multiplying after birth (Chase *et al.*, 1969). The effects are accompanied by deficits in other areas of active brain growth and development, notably the reduction in number of glia and in dendritic arborization. Malnutrition after weaning does not produce such changes, confirming the general hypothesis that there are "critical periods" of brain development (Dobbing, 1973; Dobbing and Smart, 1972) during which the brain is especially vulnerable to a wide variety of insults. For other species, including humans, the same principle

Table IA. Selected Studies of Brain Size and Electrophysiology in Malnourished Children (Concurrent Effects)

Reference	Test	Results	Type of malnutrition[a]	Age studied	Study design
Brown (1965)	Brain weight	*Reduced* in index children. Since body weight increases, brain weight/ body weight ratio increases over time. Children less than 1 year most vulnerable.	Mixed PEM	Birth–15 Yr	Uganda Autopsy data from 124 children compared with data from 351 children who died of other causes.
Winick and Rosso (1969)	Head circumference	*Reduced.*	Marasmus	Birth–27 Mo	Chile Autopsy data from 9 children compared with data from 8 children who died of nonnutritional causes.
and Rosso *et al.* (1970)	Brain weight	*Reduced* relative to controls. Reduced protein/DNA ratio also reported, suggesting lower cell number in these index children. [Later work confirms no effect on cell number for children with kwashiorkor (Winick and Rosso, 1975)].			

Reference	Measure	Results	Type	Age	Population
Chase et al. (1974)	Brain weight	*Reduced* relative to controls. Reduced brain protein content in cerebellum; DNA similar. Results suggest no reduction in cell number, only cell size.	Marasmus and marasmic–kwashiorkor	12–24 Mo	Guatemala. Autopsy data from 6 children compared with data from 5 U.S. children who died of other causes. (Both groups had normal birth weight.)
Vahlquist *et al.* (1971)	Head circumference and Echoencephalography	*Reduced* in children with marasmus only. *Increased* size of the cerebral ventricles in kwashiorkor children only, probably due to edema. [Return to normal size following 6 months of rehabilitation (Engsner et al., 1974).]	Marasmus and kwashiorkor	3.5–22 Mo 18–26 Mo	Ethiopia. 28 Children (18 with marasmus and 10 with kwashiorkor) compared with 38 well-nourished children (from both an orphanage and well-to-do homes).
Barnet *et al.* (1978)	EEG (auditory evoked potential)	*Higher* EPI (composite index) indicating more abnormalities in malnourished children at admission to hospital; these were lower at discharge but still not at control values. *Higher* EPI present even 1 year after discharge.	Marasmus	3–12 Mo	Mexico. 26 Children compared with 46 well-nourished children.

[a] Here and throughout the following tables: (PEM) protein–energy malnutrition.

applies, except that different periods of time are critical. Thus, for man, rapid brain development is most vulnerable to malnutrition from 3 months of gestation through the 2nd year of life.

Striking behavioral changes at the time of malnutrition have been observed in hospitalized children. The most common behavioral characteristic is apathy, which is not always distinguishable in etiology from apathy due to the hospitalization itself. For example, Bowlby (1960) reports apathy in 18- to 20-month-old infants who were abruptly separated from their families by hospitalization. Spitz (1945, 1946) has applied the term marasmus to the emotional syndrome caused by extreme conditions of environmental deprivation. It should be noted, however, that apathy has also been recognized in malnourished children who were not hospitalized (Meneghello, 1949), indicating that malnutrition itself is a causal factor. In addition to apathy, passivity, reduced motor activity, and decreased verbal responses are also present (Trowell *et al.*, 1954; Gomez *et al.*, 1954b; Valenzuela *et al.*, 1959; DeSilva, 1964) such that malnourished children are often described as being "functionally isolated" from their environment. Indeed, recovery from malnutrition is first signaled by alertness and smiling. The observation by Clark (1951) that "once a child can be persuaded to smile he is well on the way to recovery" is often quoted. This clinical picture has been expanded by the application of various tests of behavioral development (development quotient) and by observations of the mother–infant relationship.

Table IB summarizes studies of the behavioral performance of malnourished children measured on a number of developmental scales, including those of Bayley, Griffiths, and Gesell. In virtually all studies, the development of malnourished children is significantly delayed, regardless of the type of malnutrition. Language has generally been recognized to be a highly sensitive indicator of cerebral dysfunction in children. Of particular interest are three studies (Cravioto and Robles, 1965; Guthrie *et al.*, 1976; Cravioto and De-Licardie, 1972) in which language and verbal development were identified as being the functions most affected by malnutrition. Cravioto and Robles (1965) studied 20 children with malnutrition and found reduced performance on all subtests of the Gesell Developmental Schedule but especially in language. In their subsequent prospective study of a Mexican village, Cravioto and De-Licardie (1972) found lower language scores in children even before they developed the clinical picture of severe malnutrition. Guthrie *et al.* (1976), in their study of 17 malnourished Filipino children and 23 comparison children who were 6 months to 3 years of age, observed that verbal scores on the Bayley scale were reduced in the malnourished group.

Another study of particular interest is that of Grantham-McGregor *et al.* (1978), who used hospitalized children as controls. Since children with severe malnutrition are usually studied during hospitalization, it has not always been possible to distinguish the effects of malnutrition from those of hospitalization. Hospitalization *per se*, when accompanied by a lack of stimulation for the child, has been known to impede normal behavioral development and, in extreme cases, to result in death (Spitz, 1945). The study of Grantham-McGregor *et al.*

Table IB. Studies of Developmental Quotients in Malnourished Children (Concurrent Effects)

Reference	Test	Results	Type of malnutrition	Age studied	Study design
Geber and Dean (1956)	Gesell Developmental Schedule	*Lower scores*, though reduction was moderate at time of discharge from hospital.	Kwashiorkor	12–38 Mo	Uganda 25 Children studied during hospitalization.
Cravioto and Robles (1965)	Gesell Developmental Schedule	*Lower scores* in all areas, but lowest in language. All children improved with medical treatment, except for children admitted at 6 mo of age or younger.	Mixed PEM	3–42 Mo	Mexico 20 Children, subdivided by age at admission, were studied every 2 weeks during hospitalization.
Cravioto and DeLicardie (1972)	Gesell Developmental Schedule	*Lower scores* in language development prior to and after PEM had occurred.	Mixed PEM	Birth–60 Mo	Mexico Prospective study of 334 children born during 1 year in a small village. 22 Children developed severe PEM (15 had kwashiorkor and 7 had marasmus) and were compared with 22 well-nourished children (matched by age, birth weight, and Gesell scores).
Graves (1976)	Gesell Developmental Schedule	*No differences.*	Unspecified	7–18 Mo	India 23 Children (60–65% weight/age) compared with 30 well-nourished controls (80–95% weight/age). Neither group hospitalized.

(Continued)

Table IB. (Continued)

Reference	Test	Type of malnutrition	Age studied	Results	Study design
Graves (1978)	Gesell Developmental Schedule	Unspecified	7–18 Mo	*Lower scores* in index children.	Nepal 38 Children (60–75% weight age) compared with 36 well-nourished controls (80–95% weight/age). Neither group hospitalized.
Pollitt and Granoff (1967)	Bayley Scale	Marasmus	11–32 Mo	*Lower scores* in mental development at all stages of illness and recovery, but motor development improved more rapidly.	Peru 19 Children assessed during hospitalization were compared with 8 siblings.
Guthrie *et al.* (1976)	Bayley Scale	Mixed PEM	6–36 Mo	*Lower scores* in index children, especially on verbal items. These infants clung to their mothers throughout examination.	Philippines 17 Children were compared with 23 children who were not severely malnourished. Neither group hospitalized.
Celedon *et al.* (1980) and Celedon and de Andraca (1979)	Bayley Scale and Psychomotor Development Scale (modified for Chile)	Marasmus	5–18 Mo	*Lower scores* especially in younger infants, who were slower to recover. Social, language, and fine motor skills improved more rapidly than gross motor skills.	Chile 24 Children assessed during hospitalization.

Reference	Measure	Type	Age	Sample	Findings
Grantham-McGregor *et al.* (1978), Grantham-McGregor and Stewart (1980)	Griffiths scale	Mixed PEM	6–24 Mo	Jamaica 18 Children were compared with 15 well-nourished controls (who had been hospitalized with conditions other than undernutrition).	*Lower scores* in index and control children, although a larger deficit was present in index children.
Lester (1975)	Cardiac habituation	Mixed PEM	12 Mo	Guatemala 20 Boys compared with 20 well-nourished boys.	*Reduction* of an adequate response by index children, who required many more trials to habituate, suggesting an attention deficit likely to interfere with learning.
and Lester (1976)	Crying patterns			12 boys were compared with 12 well-nourished controls (subsample from 1975 study).	*Abnormal* in index children; similar to patterns seen in children with central nervous system dysfunction.
Juntunen *et al.* (1978)	Crying patterns	Mixed PEM	7–24 Mo	Nigeria 5 Children were compared with 15 well-nourished controls (mostly from Finland).	*Abnormal patterns*, especially in the three marasmic infants, including increased cry frequency and biphonation.

(1978) in Jamaica is among the few that compare malnourished children with children hospitalized for other reasons. They found that both groups of children were developmentally delayed but that the reduction on the Griffith scale was significantly more pronounced in the malnourished children.

Two other areas of behavioral functioning that have been studied in malnourished children are the rate of cardiac habituation to repeated auditory or visual stimuli and crying patterns. Lester (1975, 1976) found that Guatemalan children with active malnutrition took a longer time to habituate to a new stimulus. Ordinarily, heart rate rises and then quickly returns to base-line levels in well-nourished infants presented with a novel auditory stimulus. This decline did not occur as rapidly in the malnourished children, which Lester attributes to lack of attention. These infants may, however, also be more irritable and under more chronic stress, which may also inhibit the habituation response. In a second study, Lester (1976) reported abnormal crying patterns, similar to those seen in cases of central nervous system dysfunction, in 12 malnourished infants compared with 12 well-nourished children. Juntunen *et al.* (1978) studied 5 Nigerian infants with malnutrition and compared them to 15 Finnish infants. He demonstrated abnormal crying patterns resembling brain damage in the children with marasmus and no abnormalities in the child with kwashiorkor. Crying patterns may therefore provide information on the extent of brain damage resulting from malnutrition.

These behavioral abnormalities are likely to alter the interrelationships of the mother and the malnourished child with important implications for the child's further development. Studies of mother–infant interaction are summarized in Table IC (and reviewed extensively in Chapter 8). It has generally been assumed that mothers of malnourished infants are inattentive and not attached to their infants and that this pattern of interaction may lead to the episode of malnutrition. There is no consistent evidence to support this assumption, and, in fact, direct observation of malnourished infants and their mothers may also show increased contact between the mother and child, probably to compensate for the infant's malnutrition. Thus, Graves (1976, 1978), who directly observed mother–infant interaction, noted an altered relationship between malnourished infants and their mothers, including reduced mother–infant reciprocity in his Indian study, but no such reduction in his Nepalese study. In both studies, infant behavior was characterized by more suckling, increased clinging, and greater dependence by the infant on the mother. When malnourished mothers received nutritional supplements during pregnancy and lactation, such behaviors were reversed in a study of Mexican mothers and infants by Chavez and co-workers (Chavez *et al.*, 1974, 1975; Chavez and Martinez, 1975).

Several studies do not include direct observation of the mother–infant interaction, but have looked at the home environment of the malnourished infant, including maternal characteristics. It is generally assumed that these environmental and maternal features were responsible for the infant's malnutrition, though the only prospective study reported is that of Cravioto and DeLicardie (1972). In this study, the characteristics of the homes of children who eventually became malnourished were found to be less favorable than for well-nourished infants. In contrast, Grantham-McGregor and Stewart (1980) and Sheffer *et al.* (1981) in Jamaica found no differences in home environment between their groups of malnourished children and well-nourished children. This study was conducted after the episode of malnutrition had occurred. Such a range of evidence suggests that the determinants of malnutrition are not uniform and may vary from one cultural setting to another. Moreover, there is likely to be considerable variation in the response of mothers to their malnourished infants, which may interfere with or facilitate the infant's ability to overcome the adverse effects of malnutrition.

In summary, the metabolic and structural deficits in brain development during the acute phase of malnutrition have been well documented, especially in marasmus. These are reflected in the behavioral patterns of the sick child and result in marked delays in development. Malnutrition also produces changes in the mother's response to the malnourished infant and vice versa. In the long run, these changes may interfere with behavioral development by reinforcing the infant's state of "functional isolation" from the environment, or they may encourage healthy development. While certain of these concurrent effects disappear during recovery, others persist and form the basis of long-term deficits in behavior.

2.3. Intermediate Effects of Malnutrition

In children who have survived the acute phase of the illness, many of the effects of malnutrition may be overcome once an adequate diet and proper medical care have been provided, but this period of recovery can last for up to several years. Certain metabolic functions may recover rapidly (Ramos-Galvan *et al.*, 1969), whereas other functions, such as physical growth and behavior, may improve gradually over several years and may never reach normal values. We refer to behavioral patterns at this time as *intermediate effects* of severe malnutrition, since they occur after the acute phase has subsided but before recovery is complete or has reached a plateau. Since most cases of severe malnutrition occur when a child is less than 2 years of age, we have limited the discussion of the intermediate effects to studies of preschool children.

Investigators have used several different strategies to examine the intermediate (and long-term) effects of early malnutrition. There are four broad categories of studies. First, *comparative studies* have been made of tall and short children in populations where malnutrition is endemic and is therefore a likely cause of stunted growth [Ashem and Janes (1978) in Nigeria]. Other studies use differences in weight to categorize groups of children. In both instances, it is assumed that differences in height or weight result primarily from previous malnutrition. Since such studies are retrospective, information about the early history of these children, including the age at which the child was malnourished and the degree of malnutrition of these children, is limited.

Table IC. *Studies of Mother–Infant Interaction and Home Environment of Malnourished Children (Concurrent Effects)*

Reference	Test	Results	Type of malnutrition	Age studied	Study design
Kerr et al. (1978)	Psychosocial measures of mothers	*Lower scores* in index mothers, including less stable homes. However, statistics are inadequate to support any conclusions (inappropriate use of χ^2 test with small sample size).	Mixed PEM	9–14 Mo	Jamaica First group included 6 hospitalized children compared with 6 children hospitalized with conditions other than malnutrition. Second group included 5 children (2 hospitalized with malnutrition, 3 had only mild malnutrition) compared with 5 well-nourished children.
Grantham-McGregor and Stewart (1980), Sheffer et al. (1981)	Caldwell Home Inventory Scale	*No differences.* House quality and maternal IQ were lower in index group, but were not found to contribute to child's delayed development.	Mixed PEM	6–24 Mo	See Table IB.
Cravioto and DeLicardie (1972, 1976)	Caldwell Home Inventory	*Lower scores* in homes in which children eventually developed malnutrition or	Mixed PEM	Birth–60 Mo	See Table IB.

| Graves (1976) | Mother–infant interaction | had malnutrition. Index mothers were also more passive, but other than less frequent radio listening, socioeconomic factors did not differ from those of control mothers. *Reduced* attachment behaviors in the child and *reduced* maternal scores associated with a lack of mother–infant reciprocity in index cases. However, time spent in lap was *increased.* | Unspecified | 7–18 Mo | India See Table 1B. |
| Graves (1978) | Mother–infant interaction | Socioeconomic factors similar in both groups. Similar *reduced* attachment behaviors in child as above. However, maternal scores and mother–infant reciprocity were similar in both groups. Time spent in lap and suckling was *increased.* | Unspecified | 7–18 Mo | Nepal See Table 1B. |

Second, *longitudinal studies* of children in developing countries suffering from severe malnutrition are by far the most common. Children were either followed from the time of their hospitalization or identified later in childhood using hospital records. In the earliest longitudinal studies, no control groups were used [Barrera-Moncada (1963) in Venezuela, Cravioto and Robles (1965) in Mexico, and Monckeberg (1968) in Chile]. Later studies compare malnourished children with adequately nourished children, often, though not always, from the same socioeconomic class [Stoch and Smythe (1963) in South Africa, Botha-Antoun *et al.* (1968) in Lebanon, Chase and Martin (1970) in the United States, Brockman and Ricciuti (1971) in Peru, McLaren *et al.* (1973, 1975) in Lebanon, DeLicardie and Cravioto (1973) in Mexico, and Sheffer *et al.* (1981) and Grantham-McGregor *et al.* (1982) in Jamaica]. Some longitudinal studies also make use of sibling controls who were not severely malnourished, but may have had mild malnutrition. Furthermore, differences in parental attention to the index child and his siblings reduce the validity of this approach.

Third, *longitudinal studies* have also been carried out in children with severe malnutrition secondary to organic diseases that impair the utilization of food, notably pyloric stenosis and cystic fibrosis. There are very few studies of this type, and these have all been conducted in developed countries [Lloyd-Still *et al.* (1972, 1974) in the United States].

Fourth, there are *intervention studies* in which the effects of nutritional supplementation with or without environmental enrichment on the behavior of malnourished children were studied. Intervention programs may be especially relevant during the period of recovery. Because such studies are comprehensively reviewed in Chapter 4, we have elected not to discuss these here.

Despite the limitations of the various approaches, studies of the intermediate effects of severe malnutrition are consistent in demonstrating delays in brain and behavioral development. We shall first discuss two published studies on neurological function dealing specifically with the recovery period. Graham (1966) studied head circumference and height in children recovering from severe and mild–moderate malnutrition. As Fig. 1 shows, children who were severely malnourished in the first 12 months of life (Group A) were found to be the least likely to attain normal head circumference in observations extending through 50 months of age. In children with severe malnutrition occurring after the first 12 months of life (Group B), head circumference was not significantly reduced. To clarify these data by correcting differences in the physical growth of children, we calculated head circumference/height ratios when the children were about 38 months of age. A ratio of 0.8 was found for three of the groups (children with *mild malnutrition* hospitalized before or after 12 months of age and children with *severe malnutrition* hospitalized after 12 months of age); in contrast, a ratio of 0.6 was found for the children with *severe* malnutrition hospitalized before 12 months of age. These findings confirm that head size is most affected by malnutrition during the limited "critical period" of development—in fact, brain size is 90% complete by 2 years of age. The second study of neurological function during the recovery period, by Barnet *et al.* (1978), was described in Section 2.2. These authors continued their study

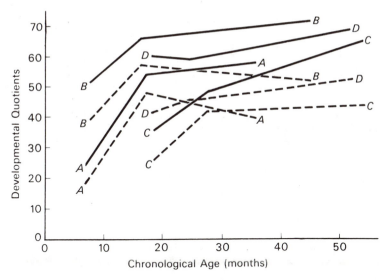

Fig. 1. Evolution of developmental quotients for height and head circumference during recovery from malnutrition. The solid lines correspond to height and the broken lines to head circumference. Lines *A* are the mean of 14 infants under 12 months of age on admission with height quotients of less than 40 percent, and lines *B*, of 14 infants under 12 months of age with height quotients of 40 percent or more. Lines *C* are the mean of 13 children 12 months of age or more on admission with height quotients under 50 percent, and lines *D*, of 12 children 12 or more months of age with height quotients of 50 percent or more. From Graham (1966).

of 26 marasmic children and 46 controls for 1 year after hospitalization and found a persistent increase in the auditory evoked response index, a sign of cerebral dysfunction. Thus, the auditory evoked potential appears to be a more sensitive indicator of the intermediate effects of malnutrition than general EEG measures which return to base-line levels fairly soon after the child begins to recover.

Published studies of the intermediate effects of malnutrition on behavior are summarized in Tables IIA (IQ and developmental quotients) and IIB (home environment). Following nutritional rehabilitation, recovery begins with the reversal of apathy, impaired exploratory behavior, and reduced motor activity that characterize the acute phase and that have been referred to as "functional isolation." Studies during this period have, however, demonstrated continued delays in development as measured on various scales of IQ and developmental quotient (Table IIA). The only *comparative study* of the intermediate effects of malnutrition was undertaken in Nigeria by Ashem and Janes (1978). These authors studied two groups of poor children—urban and rural—who were classified as having been moderately to severely malnourished or adequately nourished on the basis of current height and weight. At 2–6 years of age, the small children performed less well on the McCarthy scale of intelligence than did the taller children within the same group. Urban and rural poor children performed less well than urban children from a higher socioeconomic class. The findings

Table IIA. Studies of Development and Intellectual Performance in Children Recently Recovered from Malnutrition

Reference	Test	Results	Type of malnutrition	Age studied	Study design
Ashem and Janes (1978)	McCarthy Scale of Children's Abilities	*Lower scores* on abstract reasoning, logical and relational thinking in malnourished groups, intermediate scores in poor but well-nourished group, and highest in healthy group.	Unspecified	24–72 Mo	Nigeria 73 Poor children (45 from rural areas and 28 urban) were subdivided by height/weight per age and compared with 45 wealthier children. Malnourished children were <10% of British standards.
Monckeberg (1968)	Gesell Developmental Schedule and Stanford–Binet IQ Test	*Lower scores*, especially in language development. *Lower scores.*	Marasmus (3–11 Mo)	36–72 Mo	Chile 14 Children were assessed.
Stoch and Smythe (1963)	Gesell Developmental Schedule and Merrill–Palmer IQ Test (depending on age of child)	*Lower scores* in index group, which could not be fully attributed to lack of nursery experience. Socioeconomic factors were worse and mothers more negligent in index group.	Marasmus (10–24 Mo)	Studied up to 72 Mo	South Africa 21 Cape colored children (11 girls and 10 boys) who were below the 2.5 percentile for weight were compared with 21 children at or above the 10th percentile (matched by sex and age). Controls were in a nursery crèche. No children were hospitalized.
Brockman and Ricciuti (1971)	Categorization behavior	*Lower scores* in malnourished children (6–8 Mo delay relative to U.S. norms) even after	Marasmus (mean age of 9 Mo or 16 Mo)	12–44 Mo	Peru 20 Children (subdivided by mean age of hospitalization: 9 Mo or

Reference	Test	Diagnosis (age at onset)	Age at test	Results	Sample
McLaren et al. (1973, 1975)	Stanford–Binet IQ Test	Marasmus (2–16 Mo)	36–60 Mo	12 weeks of nutritional rehabilitation. Test scores correlated with the length of rehabilitation. *Lower scores* in all three malnourished groups compared with siblings and controls. Two hospitalized groups had lower scores than outpatient group. No long-lasting effect of stimulation. No sex differences.	Lebanon 29 Hospitalized children (14 not stimulated in hospital and 15 stimulated) and 15 moderately malnourished outpatients were compared with 14, 15, and 15 siblings, respectively, and also with well-nourished controls. 16 Mo) were compared with 19 age- and sex-matched controls and 7 stunted controls.
Botha-Antoun et al. (1968)	Stanford–Binet IQ Test (modified for Lebanon)	Mixed PEM (up to 18 Mo)	54 Mo	*Lower scores* for both verbal and performance items in index group at time of study. Groups, however, were similar before the onset of malnutrition with respect to weight and general functioning. Maternal age and IQ similar in both groups.	Lebanon 22 Children were compared with 22 well-nourished children (matched by age, sex, birth weight, and ethnic group). Sample derived from longitudinal study of 316 children. All children were healthy at birth, but most index children then fell below the 3rd percentile of the Stuart scale, whereas controls did not fall below the 25th percentile.
Chase and Martin (1970)	Yale Revised Developmental Examination	Mixed PEM (2–10 Mo)	23–82 Mo	*Lower scores* in index children, especially in language. Children hospitalized at 4 Mo or older did worse than younger group.	United States 19 Children were compared with 19 well-nourished children.

(Continued)

Table IIA. (Continued)

Reference	Test	Results	Type of malnutrition	Age studied	Study design
DeLicardie and Cravioto (1973)	WPPSI IQ Test and Behavioral styles of children during WPPSI testing.	*Lower scores* in index children. Behavioral patterns of index children and controls matched for lower IQ were similar, showing more passivity than controls not matched for IQ.	Mixed PEM (4–38 Mo)	60 Mo	See also Cravioto and DeLicardie (1972) in Tables IB or IC. 14 Children (of 22) who developed malnutrition over the course of a 5-year prospective study of 334 children were compared with well-nourished controls (matched by IQ at 60 Mo or by birth weight and Gesell score in infancy).
Barrera-Moncada (1963)	Gesell Developmental Schedule	*Lower scores* in children compared to expected values for this age group. Language slower to recover than motor skills.	Mixed PEM (16–72 Mo)	40–96 Mo	Venezuela 60 Children were assessed after hospitalization and 2 years later.
Cravioto and Robles (1965)	Gesell Developmental Schedule	*Lower scores*, especially in language development. Children hospitalized prior to 6 Mo of age were slowest to recover.	Mixed PEM (3–42 Mo)	15–54 Mo	See Table IB. Same children studied 1 year after discharge.
Grantham-McGregor *et al.* (1982)	Griffiths Scale	*Lower scores* in malnourished children continued. Weight/height index reaches that of hospitalized controls.	Mixed PEM (6–24 Mo)	42–60 Mo	See Table IB. Same children studied 3 years after discharge.
Lloyd-Still *et al.* (1972, 1974)	Merrill–Palmer IQ Test	*Lower scores* in index children (40 vs. 70% for siblings).	Undernutrition secondary to cystic fibrosis, ileal atresia, or severe diarrhea (up to 6 Mo)	12–72 Mo	United States 26 Children were compared with 29 siblings.

of this retrospective study are complex, suggesting that both malnutrition and poverty affect performance.

Longitudinal studies of development quotient also report delays in previously malnourished children. The findings of the earlier studies (Barrera-Moncada, 1963; Cravioto and Robles, 1965; Monckeberg, 1968), which did not include control groups, show the greatest deficit to be in the area of language development. As was pointed out earlier, the acquisition of language skills is particularly sensitive to the effects of malnutrition. This result was confirmed by Chase and Martin (1970), in the United States, using the Yale Revised Developmental Examination. In their study of 19 malnourished and 19 well-nourished children, these investigators also reported that children who were hospitalized after 4 months of age had a poorer prognosis than those who were hospitalized prior to 4 months and who were therefore malnourished for a shorter period. The home environment of the previously malnourished group was also found to be disadvantaged at the time of the study (Table IIB). The findings are difficult to interpret, however, because the items in the Home Inventory Scale are not described and also because the authors found no differences on a Family Social Functioning Scale administered at the same time.

In their study of the intermediate effects of malnutrition, Grantham-McGregor *et al.* (1982) in Jamaica found deficits on the Griffiths scale in 18 3- to 5-year-old children with histories of malnutrition compared with 15 children hospitalized for other reasons. This is the only longitudinal study of intermediate effects that permits comparison of test scores at two points in time, namely, at the time of discharge from the hospital and 3 years later. At both points in time, the Caldwell Home Scale did not differentiate between homes of index and control children (Sheffer *et al.*, 1981).

The *longitudinal studies* of IQ [Stoch and Smythe (1963), Botha-Antoun *et al.* (1968), McLaren *et al.* (1973, 1975), DeLicardie and Cravioto (1973), and that of Monckeberg (1968) with no control group] all show reduced performance in small samples of children. Two of the studies of IQ, those by Botha-Antoun *et al.* (1968) and DeLicardie and Cravioto (1973), are prospective and provide information on the development of the children even before the onset of malnutrition. Botha-Antoun *et al.* (1968) in Lebanon studied a sample of 22 children who eventually developed malnutrition before 18 months of age; index children did not differ from 22 control children prior to the onset of malnutrition, but had markedly reduced IQ scores at $4\frac{1}{2}$ years of age. Index and control children studied by Cravioto and DeLicardie (1976) were also similar with respect to developmental quotient and birth weight in infancy. However, in this prospective study of 22 malnourished children and controls in Mexico, deficiencies in the home environments of index children were found even before they developed malnutrition. Thus, any reduction in IQ scores must be attributed both to the episode of malnutrition and to disadvantaged home conditions in this study.

Deficits in IQ were also reported in the Lloyd-Still *et al.* (1972, 1974) *longitudinal study* of children in Boston with infantile *malnutrition secondary to cystic fibrosis* and other severe organic disorders. At 1–6 years, the 26 index

Table IIB. Studies of the Home Environment of Children Recently Recovered from Malnutrition

Reference	Test	Results	Type of malnutrition	Age studied	Study design
Chase and Martin (1970)	Family Social Functioning Scale and Home Inventory	*No differences* between index and control families. *Lower scores* (42 vs. 48) for index homes. Maternal IQ was similar.	Mixed PEM (2–10 mo)	24–82 Mo	See Table IIA.
DeLicardie and Cravioto (1973), Cravioto and DeLicardie (1976)	Caldwell Home Inventory	*Lower scores* in both index and IQ-matched control children compared to normal IQ controls. Similar results reported even prior to the onset of malnutrition (see Section 2.2)	Mixed PEM (4–38 mo)	60 Mo	See Tables IC and IIA.
Sheffer *et al.* (1981)	Caldwell Home Inventory	*No differences* between homes of index, control, and neighborhood children. Similar results reported earlier (see Section 2.2).	Mixed PEM (6–24 mo)	42–60 mo	See Granthem–McGregor *et al.* (1982) in Table IIA.

children had lower scores (40th percentile) than their siblings (whose scores were at the 70th percentile). These authors also examined motor development and found no reduction in this area. The use of siblings as controls does not eliminate differences in parental attention to the affected child, as compared with other members of the family.

In summary, even though the recovery phase has not received much attention in the literature, this period is likely to be important in determining the long-term behavioral consequences of malnutrition in two ways. First, continued delays in behavioral development during this period are likely to impair the child's ability to interact adequately with his environment. Under normal circumstances, the preschool child masters a large number of developmental tasks (Freud, 1963) that lead to increased independence and set the stage for the school years. Delays during this period may have a cumulative effect on the child's development interfacing with the mastery of subsequent developmental tasks. Thus, while average scores on behavioral tests in children of this age group do not predict future performance (McCall *et al.* 1972), marked reductions in test scores are likely to continue (Werner *et al.*, 1971). Second, interventions during this period in particular may have a beneficial effect on long-term behavioral outcome by shortening the period of recovery, making recovery more complete, and by short-circuiting the vicious cycle of delayed development and environmental deprivation (McKay *et al.*, 1973, 1974, 1978).

2.4. Long-Term Effects of Early Malnutrition

The long-term effects of severe malnutrition have been primarily studied in school-aged children, and in some instances, studies have extended through adolescence. There are no systematic data on adults, although presumably early malnutrition may also affect the individual through the life cycle and even into the next generation. For instance, Baird (1947) and his colleagues, in a series of studies on a population of 200,000 births in Aberdeen, Scotland, to define the biological and social factors contributing to a women's competence in childbearing, found that mothers whose heights were in the lowest percentiles of standard were more likely to have delivery complications, including higher fetal mortality rates, and to give birth to smaller infants than taller mothers. Their children were also the most likely to have reduced IQs (Birch *et al.*, 1970). Although other factors may have contributed, undernutrition in the mother's childhood was associated both with her short stature and with abnormal pelvic shapes in this population (Bernard, 1952). These data suggest that a mother's antecedent nutritional history when she was a child can significantly influence the intrauterine growth, as well as the type of delivery and subsequent development of her child.

Research on the long-term effects of malnutrition has focused on various indices of brain growth and function, intelligence, and school performance, including behavior in the classroom. In general, although electrophysiological measures of brain function are essentially normal at this time, deficits in IQ and school performance are prominent. These findings emerge from a wide range of studies, which can be classified according to the four research strategies described earlier. These include *comparative studies* of tall and short children in populations with endemic malnutrition [Cravioto *et al.* (1966) in Guatemala; Winick *et al.* (1975) and Lien *et al.* (1977) in Korean orphans adopted into United States families; and Singh and Anand (1976) and Singh *et al.* (1977) in India]. The second category includes *longitudinal studies* of children in developing countries with an early episode of severe malnutrition. Although the earliest studies compared children to local standards [for example, Cabak and Najdanvic (1965) in Yugoslavia], we will restrict our discussion to studies using control groups. One study in this group examines only neurological evidence [Bartel *et al.* (1979) in South Africa], and three focus on neurological as well as behavioral parameters [Liang *et al.* (1967) in Indonesia, Evans *et al.* (1971) in South Africa, and Stoch and Smythe (1963, 1967, 1976) and Stoch *et al.* (1982) in South Africa]. The remainder are concerned with behavioral and intellectual function [Champakam *et al.* (1968) in India; Hertzig *et al.* (1972) and Richardson *et al.* (1972, 1973, 1978) in Jamaica; Birch *et al.* (1971) and Cravioto and DeLicardie (1971) in Mexico; Fisher *et al.* (1972) in Zambia; Hoorweg and Stanfield (1972, 1976) in Uganda; Pereira *et al.* (1979) in India; and Galler *et al.* (1983a–c) in Barbados]. The third category includes *longitudinal studies* of children with severe malnutrition secondary to an organic disease, primarily cystic fibrosis and pyloric stenosis [Lloyd-Still *et al.* (1972, 1974), Klein *et al.* (1975), and Beardslee *et al.* (1982) in the United States; Berglund and Rabo (1973) in Sweden; Valman (1974) in England; and Ellis and Hill *et al.* (1975) in Canada]. The fourth category includes *intervention studies*, which have been summarized by Rush in Chapter 4 and will not be discussed here.

A number of *longitudinal* studies have studied the physical aspects of brain function in previously malnourished children, and in all of these, cases of marasmus or kwashiorkor were identified from hospital records (see Table IIIA). In the case of marasmus, head circumference was found to be consistently reduced (Hoorweg and Stanfield, 1976; Stoch and Smythe, 1976; Stoch *et al.*, 1982; Branko, 1979), but this was not true for children with kwashiorkor (Branko, 1979; Evans *et al.*, 1971; Hoorweg and Stanfield, 1976). Birch and Richardson (1972) also reported reduced head circumference in children who had been hospitalized in the first 2 years of life with mixed forms of malnutrition. The more severe findings in the case of marasmus may be the result of the fact that this condition generally occurs earlier in life than does kwashiorkor. As discussed in the previous section, rapid brain growth is limited to children less than 2 years of age and may be essentially complete by the time most cases of kwashiorkor have their onset. EEG studies, in contrast to studies of head circumference, do not, for the most part, report any consistent abnormalities in children with various types of malnutrition who were examined at 6–20 years of age. The exception is Bartel *et al.* (1979). Thus, Evans *et al.* (1971), who studied children with mixed types of malnutrition, failed to find any differences between index and control children. Stoch and Smythe (1967) reported lower alpha indices in previously marasmic children, but in a follow-up study, Stoch and Smythe (1976) reported that both index and control children had EEG abnormalities. Bartel *et al.* (1979) analyzed the EEG patterns of South African children with histories of kwashiorkor by computer and found lower alpha activity and more slow-wave activity at 6–12 years of age. This study included three control groups, namely siblings, neighbor children, and upper-class children, and is especially convincing. The findings reported suggest a reduced arousal level in children with previous kwashiorkor. There are no reported studies in this age group of auditory evoked potential, which Barnet *et al.* (1978) found to be abnormal in preschool children with histories of severe malnutrition. Thus, specific tests of electrophysiological function, using computer analysis, should be made, since these are likely to be more sensitive indicators of the long-term effects of severe malnutrition.

In research on the long-term behavioral effects of severe malnutrition, by far the most commonly studied indicator has been IQ, which has been found to be reduced on a variety of scales. As discussed in detail by Pollitt and Thomson (1977), the usefulness of the IQ tests in such studies is limited. First, the test may not be equally appropriate for all cultures and ethnic groups. Second, the functional significance of a reduced IQ is not always clear. Therefore, it is necessary to also examine other aspects of behavioral function that may have more direct relevance to the child's adaptive capabilities, such as school performance and behavior in school.

Studies of the long-term effects of severe malnutrition on IQ are summarized in Table IIIB. The one *comparative study* of intelligence examined IQ scores in 240 Korean orphan girls who were adopted by American families when they were 2 or 2–5 years of age (Winick *et al.*, 1975; Lien *et al.*, 1977). These children were divided into three groups according to the degree of deficit in percentage of expected height for age at the time of adoption; 68 were identified as having been severely malnourished on this basis. At 7–15 years of age, the girls who had been severely malnourished had a significant reduction in performance on four different IQ tests compared with the moderately malnourished and well-nourished girls, although the performance in all groups was well within the average range of United States school-children. These authors also found that in general, children who were adopted at younger ages had better scores than children adopted at later ages. Socioeconomic conditions in the adopting families were similar for all three groups.

Although the conditions of the studies vary widely, the results of *longitudinal studies* of the effects of early malnutrition on IQ are generally consistent in demonstrating reduced performance. The long-term outcome of children with a history of kwashiorkor was examined in four studies, three of which confirm reduced performance (Champakam *et al.*, 1968; Pereira *et al.*, 1979; Birch *et al.*, 1971). In contrast, Evans *et al.* (1971) and Hansen *et al.* (1971) found no reduction in IQ on the New South African Individual Intelligence Scale (NSAIIS) in their study of 40 South African children and their siblings, but they did find lower scores in the malnourished group on the Harris–Goodenough Scale as a result of emotional immaturity. The index children were compared only to the siblings, whose performance was also reduced relative to

Table IIIA. Studies of the Long-Term Effects of Early Malnutrition on Brain Function

Reference	Test	Results	Type of malnutrition	Age studied	Study design
Stoch and Smythe (1967)	EEG	*Maturational lag.* Two markedly abnormal records in index group only. However, index children had poorly formed low-voltage alpha-waves, *lower* alpha indices, and an excess of theta waves.	Marasmus (10–24 mo)	11-Yr follow-up	South Africa 20 Children (11 boys and 9 girls) compared with well-nourished controls from more advantaged homes. At 10–24 mo, index children were below 2.5 percentile for weight of Cape colored children, whereas controls were at or above the 10th percentile.
Stoch and Smythe (1976)	EEG	*EEG abnormalities* reported in 7 index children, but also in 8 controls.		15-Yr follow-up	
Stoch *et al.* (1982)	CT Scan	5 Index children had underdevelopment of the temporoparietal region. These children also had impaired visual–motor perception.		20-Yr follow-up	

Reference	Measure	Age	Nutritional condition	Findings	Sample
Evans *et al.* (1971)	EEG	12 Yr	Kwashiorkor (10–48 mo)	*No differences* in alpha index between malnourished children and their siblings.	South Africa 40 Children compared with 40 like-sexed siblings (within 2 yr of age of index children).
Liang *et al.* (1967)	EEG	6–12 Yr	Mixed PEM (up to 72 mo)	Convulsive susceptibility reported on five tests, and, in 9 children, slowing of brain waves observed. However, there was *no correlation* between these findings and nutritional status.	Indonesia 31 Children (12 of whom also had vitamin A deficiency) compared with 33 "healthy" controls also reported to be small for their age.
Bartel *et al.* (1979)	EEG (computer analysis)	6–12 Yr	Kwashiorkor and marasmic–kwashiorkor up to 27 mo	*Reduced* average frequencies in index children only for both hemispheres and over all EEG leads. *Lower* alpha activity and *more* slow-wave activity in index children.	South Africa 30 Children compared with 30 siblings, 30 neighborhood children, and 30 higher-class white children.

Table IIIB. Studies of the Long-Term Effects of Early Malnutrition on Intellectual Performance

Reference	Test[a]	Results	Type of malnutrition	Age studied	Study design
Winick et al. (1975), Lien et al. (1977)	IQ (one of four group tests)	*Lower scores* in severely malnourished children than in well-nourished children, although they scored in the average range for U.S. school-children.	Unspecified	School age	United States Korean girls adopted by U.S. families at 2 yr or at 2–5 yr (Study I and II, respectively) and subdivided into groups, based on growth deficit: severely undernourished (37 and 31 children), moderately nourished (38 and 62), and well-nourished (37 and 39) children.
Champakam et al. (1968)	IQ (test designed for Indian children in this study)	Markedly *lower scores* in index children, especially in younger ages.	Kwashiorkor (18–26 mo)	8–11 Yr	India 19 Malnourished children compared to 50 well-nourished controls (matched by religion, caste, family size, socioeconomic level).
Evans et al. (1971), Hansen et al. (1971)	NSAIIS	*No difference* between index children and siblings.	Kwashiorkor (10–48 mo)	12 Yr	South Africa 40 Children compared with 40 like-sexed siblings (within 2 years of age of index children).
Birch et al. (1971)	WISC	*Lower scores* (13 pts) in index children. Reduction greater for boys—IQ scores of control boys higher than of control girls, but	Kwashiorkor (6–30 mo)	5–13 Yr	Mexico 37 Children compared with 37 siblings (within 3 years of age of index children).

Pereira *et al.* (1979)	Seguin Form Board and Passalong tests	Kwashiorkor and marasmic–kwashiorkor (12–48 mo)	6–12 Yr	*Lower scores* in index children, who took longer to complete the task—similar at all ages for Seguin test, but most impaired in younger children for Passalong test. Not related to age at hospitalization. same for index boys and girls. Verbal and performance IQs equally depressed.	India 79 Children compared with 142 controls from nearby villages with comparable socioeconomic features.
Stoch and Smythe (1967, 1976)	Individual Scale of National Bureau of Education Research IQ Test or NSAIIS (after 1967)	Marasmus (10–24 mo)	Tested three times at 5-yr intervals through 18 yr	*Lower scores* in previously malnourished children at all ages. Results cannot be attributed to nutritional factors because of different social backgrounds. Reduction greater for boys—control boys had higher IQs than control girls, but similar for index boys and girls. Greatest reduction in verbal scores. Signs of MBD—impaired visual–motor perception.	South Africa 20 Children (11 boys and 9 girls) compared with well-nourished controls from more-advantaged homes (who also attended nursery schools). At 10–24 mo, index children were below 2.5 percentile for weight of Cape colored children and controls were at or above 10th percentile. See also Stoche and Smythe (1963) in Table IIA.

(Continued)

Table IIIB. (Continued)

Reference	Test[a]	Results	Type of malnutrition	Age studied	Study design
Galler et al. (1983a)	WISC (modified for Barbados)	*Lower scores* in previously malnourished children at all ages. Reduction greater for girls—control girls had higher IQs than control boys, but values similar for index boys and girls. Small differences in environmental features between index and control groups did not contribute to these differences.	Marasmus (3–8 mo)	5–11 Yr	Barbados 129 Children (77 boys and 52 girls) compared with 129 well-nourished classmates or neighborhood children (matched by age, sex, and handedness).
Liang et al. (1967)	WISC (modified for Indonesia)	*Lower scores* in index children who also had vitamin A deficiency as compared with controls.	Mixed PEM (up to 72 mo)	6–12 Yr	Indonesia 31 Children (12 of whom also had signs of vitamin A deficiency) compared with 33 "healthy" controls (who were also reported to be small for their age).
Fisher et al. (1972)	Koh's Blocks and Matrix Design Test (both developed in Zambia)	*Lower scores* in malnourished group. Boys performed better than girls.	Mixed PEM (44 Kwashiorkor cases, 12 marasmus, 13 masasmic–kwashiorkor and 3 unclassified at an average of 19 mo)	10–17 Yr	Zambia 72 Children compared to 143 well-nourished controls (neighborhood children matched by family income and size). Both groups more prosperous than those in South African studies.

Reference	Sample	Age	Type of malnutrition	Tests	Results
Hoorweg and Stanfield (1972, 1976)	Uganda 60 Children (subdivided into three groups based on age at hospitalization) compared with 20 well-nourished controls.	11–17 Yr	Mixed PEM (8–27 mo)	WISC (Vocabulary and Math) Porteus Mazes Knox Cubes and Raven Matrices Memory-for-Design WAIS Block Design	*No differences.* *Lower scores* in index children all attributed to "chronic PEM" and not "acute PEM." Results not associated with age at hospitalization.
Hertzig *et al.* (1972), Richardson *et al.* (1978)	Jamaica 74 Boys compared with 38 male siblings and 71 male classmates.	6–10 Yr	Mixed PEM (up to 24 mo)	WISC	*Lower scores* in index than in siblings on verbal and full scale IQs; sibs had lower scores than classmates on performance IQ. Results also associated with amount of intellectual stimulation provided in the home. Results not associated with age at hospitalization.
Berglund and Rabo (1973)	Sweden 180 Men (grouped into four groups on the basis of severity and duration of the undernutrition) selected from medical records between 1923 and 1943.	About 18 yr	Undernutrition secondary to pyloric stenosis (treated medically only)	IQ test (administered to Swedish men at time of induction into military)	*No differences* between groups. No unaffected control group studied.

(Continued)

Table IIIB. (*Continued*)

Reference	Test[a]	Results	Type of malnutrition	Age studied	Study design
Lloyd–Still *et al.* (1972, 1974) and Beardslee *et al.* (1982)	WISC WAIS (depending on age at testing)	*No differences* between groups [although index children had lower scores at 5 yr (see Section 2.3)]	Undernutrition secondary to cystic fibrosis, ileal atresia, or severe diarrhea (up to 6 mo)	5–22 Yr	United States 26 Children were compared with 29 siblings. See also Lloyd–Still *et al.* (1972, 1974) in Table IIA.
Ellis and Hill (1975)	WISC	*Lower scores* on Digit Span subtest in index children, but otherwise *no differences.* At time of study, 9 of the control children were at or below the 3rd percentile of growth and were also undernourished.	Undernutrition secondary to cystic fibrosis (up to 12 mo)	7–10 Yr	Canada 22 Children (13 boys and 9 girls) were compared with 16 "well-nourished" children, also diagnosed as having cystic fibrosis.
Klein *et al.* (1975)	Raven Matrices Peabody Vocabulary Test and WISC Coding WISC Vocabulary	*No differences.* *Lower scores* in severely malnourished group, suggesting deficits in attention and short-term memory.	Undernutrition secondary to pyloric stenosis (up to 2 mo)	5–14 Yr	United States 50 Children (44 boys and 6 girls who were further subdivided by severity of undernutrition—13 had severe PEM) were compared with 44 siblings and 50 well-nourished controls.

[a] (NSAIIS) New South African Individual Intelligence Scale; (WISC) Wechsler Intelligence Scale for Children; (WAIS) Wechsler Adult Intelligence Scale.

South African standards and who may therefore also have been undernourished. This hypothesis is supported in a recent intervention study by the same authors (Evans *et al.*, 1980). When they provided dietary supplementation for the first 2 years of life to younger siblings in families having at least one child diagnosed with kwashiorkor, these authors found higher IQ scores in the supplemented children than in siblings who either *were* or *were not* previously malnourished. The IQ of children with histories of marasmus was examined in two studies, both of which showed reduced performance. Stoch and Smythe (1967, 1976) reported reduced performance, especially in boys, on the NSAIIS in 20 South African children who were compared to controls matched for socioeconomic levels. Galler *et al.* (1983a) found reduced IQ scores in 129 Barbadian boys and girls, aged 5–11, with histories of marasmus who were compared with 129 classmates and neighborhood children who were well-nourished. In this study, girls were more affected than boys. The four remaining longitudinal studies include mixed cases of malnutrition, and these also confirm reductions in IQ among the previously malnourished children. Liang *et al.* (1967), Hoorweg and Stanfield (1972, 1976), and Fisher *et al.* (1972) attribute these findings to the episode of severe malnutrition. However, Richardson *et al.* (1978), in their study of 74 Jamaican schoolboys, their siblings, and well-nourished controls, concluded that deficits in IQ and other behavioral indices resulted from a disadvantaged home environment, rather than from malnutrition *per se*. This study also concluded that children hospitalized at 1 year of age or less did not differ in outcome from children hospitalized between 1 and 2 years of age, in agreement with the results reported by Hoorweg and Stanfield (1972, 1976), but this contrasts with studies of concurrent and intermediate effects that generally report greater deficits in children hospitalized at younger ages.

In contrast to findings from both comparative and longitudinal studies in developing countries, longitudinal studies of malnutrition secondary to organic illness do not show consistent findings. Thus, Ellis and Hill (1975), Lloyd-Still *et al.* (1972, 1974), and the follow-up study by Beardslee *et al.* (1982) of Lloyd-Still's groups all reported no significant reduction in IQ among small samples of previously malnourished children (aged 5–22 years) with various diseases. In an early study, Berglund and Rabo (1973) also found no association between IQ at the time of induction into the military and the severity of malnutrition in 180 Swedish men all of whom had histories of pyloric stenosis. Unfortunately, there was no unaffected control group used. Klein *et al.* (1975), in the best-designed study of this group, compared 50 boys and girls, aged 5–14 years, having histories of pyloric stenosis, with 44 siblings and 50 adequately nourished controls. These authors found deficits suggesting impaired attention and short-term memory in the malnourished group only using several different scales. They suggest that since they found no reductions in overall IQ scores, performance on specific subtests may be more useful measures. Although evidence from these studies has been used to demonstrate that malnutrition *per se* has only limited effects on IQ apart from disadvantaged socioeconomic conditions, this interpretation is not justified. There are too few studies, based on small sample size, and there are major limitations in the design of the research. The most comprehensive study of this series, that of Klein *et al.* (1975), does demonstrate persistent behavioral deficits in conflict with the other studies.

In addition to IQ, intersensory integration, or the ability to transfer information from one sensory modality to another, has also been studied in previously malnourished children (Table IIIC). Reductions in this ability have been associated with reading and writing disabilities (Birch and Lefford, 1964; Birch and Belmont, 1964). The comparative study of tall and short Guatemalan children by Cravioto *et al.* (1966), the longitudinal studies in India by Champakam *et al.* (1968) and Pereira *et al.* (1979), and that in Mexico by Cravioto and DeLicardie (1970, 1971) report delays in the development of intersensory integration in previously malnourished children.

In addition to tests of IQ and intersensory integration, several studies of the long-term consequences of early malnutrition have also examined school performance, which includes academic achievement and classroom behavior. These are summarized in Table IIID. Most of these studies have already been discussed. Since academic achievement is, in part, associated with IQ—Anastasi and Foley (1949) report a correlation of 0.4–0.6 in United States populations—it is not surprising that reduced school performance has been observed in previously malnourished children. In the comparative study of adopted Korean orphan girls (Winick *et al.*, 1975; Lien *et al.*, 1977), scores on various standardized achievement tests were lower in severely malnourished girls as compared with those identified as being moderately nourished or well-nourished. These results were also related to the age of adoption—those children who were adopted at younger ages did better. The other comparative study by Singh and Anand (1976) and Singh *et al.* (1977) with very limited data and the five longitudinal studies by Evans *et al.* (1971), Richardson *et al.* (1972, 1973), Pereira *et al.* (1979), Stoch *et al.* (1982), and Galler *et al.* (1983b,c) also report reduced marks in school or lower achievement test scores in previously malnourished children, regardless of the type of malnutrition. As was the case with IQ, available evidence from longitudinal studies of children malnourished secondary to organic illness fails to establish any consistent impairment in school achievement in these children (Valman, 1974; Ellis and Hill, 1975; Beardslee *et al.*, 1982).

Several studies have also examined classroom behavior, as assessed by the teacher. This appears to be more closely related to school success or failure than IQ. Thus, Richardson *et al.* (1972, 1973) reported reduced attention span, conduct problems, and limited social skills in previously malnourished Jamaican boys compared to both siblings and classmates. Galler *et al.* (1983b) expanded these findings to girls and reported an increased incidence of attention-deficit disorder and reduced socialization in both boys and girls with previous malnutrition. Similarly, Klein *et al.* (1975), in their study of children with histories of pyloric stenosis, report immaturity and hyperactivity in addition to the reduced attention span and memory skills discussed earlier. It is of interest to note the uniformity of these observations, though there are too few studies to permit broad generalizations.

Table IIIC. Studies of the Long-Term Effects of Early Malnutrition on Intersensory Integration

Reference	Test	Type of malnutrition	Age studied	Study design	Results
Cravioto *et al.* (1966)	Intersensory integration	Unspecified	6–11 Yr	Guatemala 143 Rural children (generally of low socioeconomic level) and 120 urban children (higher socioeconomic level) were subdivided into quartiles, based on height/age at the time of the study. The short rural children were assumed to have been malnourished and were compared to tall rural children. Short and tall urban children were also compared.	Short rural children (malnourished) had *more errors* than tall rural children, whereas short and tall urban children did not differ. Environmental features were similar in the two rural groups, although level of maternal education was higher in the tall group.
Champakam *et al.* (1968)	Intersensory integration	Kwashiorkor (18–36 mo)	8–11 Yr	See Table IIIB.	*More errors* among index children, especially in 8- to 9-year-olds.
Cravioto and DeLicardie (1970, 1971)	Intersensory integration	Kwashiorkor (6–30 mo)	5–11 Yr	See Birch *et al.* (1971) in Table IIIB.	*More errors* among index children as compared with their siblings; showed delays in development in index group.
Pereira *et al.* (1979)	Intersensory integration	Kwashiorkor (12–48 mo)	6–12 Yr	See Table IIIB.	*More errors* in index children as compared with well-nourished controls.

Table IIID. Studies of the Long-Term Effects of Early Malnutrition on School Performance

Reference	Test[a]	Results	Type of malnutrition	Age studied	Study design
Winick et al. (1975), Lien et al. (1977)	School achievement tests	*Lower scores* in severely malnourished children compared with moderately and well-nourished children. Children adopted at older ages did worse than those adopted at younger ages.	Unspecified	School age	See Table IIIB.
Singh and Anand (1976), Singh et al. (1977)	Class grades (teacher's grading of day-to-day performance and final exam scores)	*Lower scores* in school associated with degree of malnutrition, even when social class was controlled.	Unspecified	5–9 Yr	India 204 Children were divided into four groups based on weight/age deficits at the time of study (socioeconomic level differed between groups). See Tables IIIA and IIIB.
Evans et al. (1971)	School performance (interview completed by school principal)	*No differences* between index children and their siblings in class grades. *Lower marks* in children with illness at 15–18 mo than at 10–15 mo.	Kwashiorkor (10–48 mo)	11 Yr	See Tables IIIA and IIIB.
Pereira et al. (1979)	Class grades	*Lower marks*, more classroom failure, and a higher drop-out rate in index children compared with siblings.	Kwashiorkor (12–48 mo)	6–12 Yr	See Tables IIIA and IIIB.

Reference	Measure	Nutritional status	Age at test	Notes	Results
Stoch *et al.* (1982)	Educational level	Marasmus (10–24 mo)	Up to 23 yr	See Tables IIA, IIIA, and IIIB.	*Lower*—11/20 index children completed primary school only as compared with 6/20 control children. Educational level associated with IQ and visual-motor perception, which was *reduced* in index cases. Girls had the worst progress in school
Galler *et al.* (1983b,c)	Class grades	Marasmus (3–8 mo)	5–11 Yr	See Table IIIB.	*Lower marks* on 8/9 subjects for index children compared with classmates. Results not accounted for by environmental features.
	School behavior				Index children have attention-deficit disorder, poor social relationships, and emotional lability. These characteristics are largely responsible for the lower marks in school.
Richardson *et al.* (1972, 1973)	WRAT	Mixed PEM (up to 24 mo)	6–10 Yr	See Table IIIB (Hertzig *et al.*, 1972).	*Lower scores* in index children than classmates on all three tests—reading, spelling, and arithmetic. Sibs also had lower scores than their classmates.
	Class grades				*Lower marks* in index children than either classmates or sibs.
	School behavior				Index children had behavioral problems related to classwork—inattention, distractibility, poor social relationships, and more conduct problems than either siblings or classmates.

(Continued)

Table IIID. (Continued)

Reference	Test[a]	Results	Type of malnutrition	Age studied	Study design
Valman (1974)	School progress	*No differences.* Few details provided.	Undernutrition secondary to ileal atresia (8) or cystic fibrosis (13)	3–14 Yr	England 21 Children compared with 26 well-nourished controls; all from high socioeconomic groups.
Ellis and Hill (1975)	WRAT	*No differences* between index children and controls, who also had cystic fibrosis (and many of whom had stunted growth at the time of this study).	Undernutrition secondary to cystic fibrosis	7–10 Yr	See Table IIIB.
Klein *et al.* (1975)	Ottawa School Behavior Checklist	*Poorer performance* in children with onset of malnutrition at 21–30 mo. These children had more immaturity, overactivity, and conduct problems compared with other index children.	Undernutrition secondary to pyloric stenosis	5–15 Yr	See Table IIIB.
Beardslee *et al.* (1982)	School progress	*No differences,* although in some instances index children outperformed their siblings.	Undernutrition secondary to cystic fibrosis, ileal atresia, or severe diarrhea	Up to 22 yr	See Lloyd–Still *et al.* (1972, 1974) in Table IIA and IIIB.

[a] (WRAT) Wide Range Achievement Test.

3. Study of Barbadian Schoolchildren

3.1. General Background

In this section, we shall describe the major findings of our longitudinal study of 129 Barbadian schoolboys and girls aged 5–11 years, who suffered from severe protein–energy malnutrition (marasmus) in the first year of life. The objective of this study was to examine the long-term consequence of malnutrition for a broad range of behaviors including IQ, classroom behaviors, and school performance (Galler *et al.*, 1983a–c). To control for the effects of adverse conditions in the child's environment that may have contributed to the long-term behavioral outcome, we compared the children malnourished in the first year of life with 129 classmates of similar social backgrounds who had no histories of malnutrition.

This study was conducted in collaboration with the Ministry of Health at the National Nutrition Centre in Barbados. Since 1967, the center has provided medical care to all children with malnutrition from the time of diagnosis through 11 years of age. All cases of malnutrition in Barbados must be reported to the National Nutrition Centre; thus, good documentation is available on a population of approximately 2100 malnourished children. Several features of the Barbadian population were favorable to our investigation. First, the delivery and documentation of health are good. Almost all children are born in the Queen Elizabeth Hospital or allied facilities, and records of obstetric and perinatal care are available in almost all cases. Furthermore, children are followed routinely by local clinics, and these records were made available to us. Second, education is almost universal. Nearly all children attended school to 16 years of age, and 95% of the Barbadian population is literate. Comprehensive school records are available for each child. Third, the population is stable—with little immigration or emigration—and is largely homogeneous with respect to ethnic group and socioeconomic level, which is mostly lower middle class. Because of all these advantages, we were able to obtain extensive documentation of index and control children during infancy (including specific details surrounding the episode of malnutrition for index children), early childhood, and at the time of the study. Such data have not been available in most of the other longitudinal and comparative studies of the long-term effects of malnutrition on behavior. Our research also differs from studies published earlier because of the large sample including both boys and girls.

Our index group included *all* children who were hospitalized with malnutrition from 1967 to 1972 who met the following criteria:

1. They were diagnosed on clinical examination by our group as showing the symptoms of nutritional *marasmus*, including severe weight loss (below 75% of expected weight for age) in the absence of edema.
2. Their birth weight was at least 5 lb (to exclude those children exposed to significant fetal growth retardation).
3. They had no evidence of prenatal or perinatal complications, as measured by standard criteria including Apgar scores.

4. They had no history of convulsions, head injury, or loss of consciousness.

Approximately 140 children met these criteria, and 129 (77 boys and 52 girls) were studied. The 129 control children were for the most part classmates of the index children, and were matched to them by age, sex, and handedness. Except that they did not have histories of malnutrition and had normal growth during early childhood, the control children met the same criteria as the index children.

A range of tests of physical growth, cognitive function, and school performance were applied to both groups, and their social and home environments were also evaluated. On each set of tests, children with previous malnutrition were found to be disadvantaged in comparison with adequately nourished control children. Some of the major findings will be discussed below.

3.2. Physical Growth and Development

Heights and weights were measured for all children in the study and their mothers (Galler *et al.*, 1983c). Height for age was lower in the index children than in the control children. We attributed this finding to the early history of malnutrition, especially since mothers of both groups of children did not differ in stature. In contrast, weight for age, when adjusted for height, did not differ for the two groups. This result confirms that none of the children were malnourished at the time of the study. Head circumference was measured but varied widely as a result of the children's hairstyles at the time of the study. A standard neurological examination of both groups of children did not reveal any specific deficits in the index group. We also measured soft neurological signs (SNS) including performance on selected tasks of motor coordination (finger-tapping, successive finger movements, hand pronation and supination, hand-patting, hand flexion-extension, toe-tapping, heel–toe rocking). Children with histories of malnutrition took longer to perform the motor tasks than their matched controls when using the nondominant limb. Moreover, when we calculated a standard performance score (z-score) for the population, approximately 30 children—mostly with previous malnutrition—performed at a level below the standard. Of these children, 5—all with histories of malnutrition— were markedly below standard on the soft neurological examination and also showed serious impairment on IQ, classroom behavior, and academic performance. Thus, the soft neurological signs were valuable in identifying the most severely damaged children. At all levels of impairment, a significant relationship was present between the SNS and cognitive function, especially performance IQ, and to a lesser degree, verbal IQ and the presence of an attention deficit disorder as evaluated by the child's teacher. The SNS were not related to physical growth, height, and weight.

3.3 Intelligence Quotients

We assessed the IQs of both groups of children using a version of the Wechsler Intelligence Scale for Children that was modified to make the test

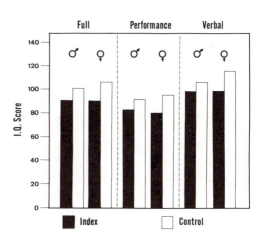

Fig. 2. Mean IQ scores of children with histories of malnutrition ($n = 129$) and their matched controls ($n = 129$).

culturally relevant (Galler *et al.* 1983a). We found that scores on all subtests of the IQ were significantly lower in the previously malnourished children than in the control children. The results of these tests in boys and girls are shown in Fig. 2. As this figure demonstrates, the mean overall IQ of the index children was 12 points lower than the mean IQ for the control children. This was the result of a disproportionate number of index children with low scores. Thus, 17% of the control children had IQ scores of 90 or lower, while 50% of the previously malnourished children scored at or below 90. Similar results were found for scores on the verbal and performance portions of the IQ test. Figure 2 also shows that both boys and girls with histories of malnutrition had similar IQ scores. However, control girls scored higher than control boys—a common finding in this age group. Therefore, the difference in IQ scores between previously malnourished and control children was greater among the girls. The reduction in IQ among previously malnourished children is consistent with findings of earlier studies with generally smaller samples. However, in contrast to the few studies that considered differences between boys and girls, we found a greater reduction in the performance by girls with histories of malnutrition.

It has often been suggested that IQ scores are strongly influenced by socioeconomic factors and by characteristics of the home environment. We obtained extensive information on each of these at the time of the study and, to a certain extent, earlier in the life of the child—for instance, the age at which the mother left school. These data allowed us to examine the relationship between IQ scores and environmental features. Our analysis of the environmental factors showed that on the average, the children with histories of malnutrition were living in homes with fewer modern conveniences and fewer rooms than the control children and that the fathers of the malnourished children held less skilled jobs. When we examined home characteristics, we found several significant differences between the two groups. For example, mothers of previously malnourished children had less education than mothers of control children, although both groups of mothers were literate. Mothers of previously malnourished children also socialized less than mothers of control children, and adults, in general, spent less time with the previously malnourished children. Finally, mothers of previously malnourished children were depressed. We did not directly measure these socioeconomic and home features prior to or at the time of the episode of severe malnutrition, though the presumption remains that these were present earlier and have persisted.

When we controlled for the effects of socioeconomic and home features using multivariate statistical analyses, we found that a history of malnutrition still had a strong association with lower IQ scores. For example, as shown in Fig. 3, children in our population who fell above or below the group mean in terms of socioeconomic status (SES) had similar IQ scores. Within each of these two socioeconomic groups, however, scores of previously malnourished children were consistently lower than those of control children. Thus, our study clearly demonstrates that whereas disadvantaged environmental conditions and maternal characteristics may lead to the episode of malnutrition, it is the episode itself which is associated with reduced IQ performance many years later.

There are several possible explanations for the limited impact of environmental factors and home characteristics in this study once the episode of malnutrition has occurred. First, since the Barbadian population is notably homogeneous, the range of variation was small. Thus, in countries where poverty is more extreme, we would expect to see a wider range of socioeconomic factors and home characteristics, which in turn may be more likely to affect IQ scores. Second, the quality and general availability of health care and education in Barbados may help to compensate for more disadvantaged circumstances of families below the group mean.

3.4. Academic Performance

To assess the academic performance of both children with histories of malnutrition and control children, we obtained grades in all classroom subjects from existing school records. These grades were converted to numbers on a 6-point scale for ease of comparison. The previously malnourished children received marks that were an average of 1 point lower than control children on eight of nine classroom subjects: language arts, mathematics, general science, social science, reading, religion, health, and arts and crafts. They did not differ in grades in writing. As is typical of this age group, girls did better than boys, and this was not related to nutritional history. A number of previously malnourished children (21 boys and 16 girls) were below the expected grade in school and therefore had already been identified as having some degree of school failure. When these were eliminated from the analysis, marks in school were still found to be lower for previously malnourished children, though grades on reading no longer differed. Thus, the children who were most severely impaired in general had the greatest deficits in reading skills.

We next examined the relationship of IQ scores to grades in school and found that these were significantly correlated. The range of the correlation coefficients (0.37–0.57) between IQ and different school subjects was similar

Full Scale IQ:	Household items		Housing		Father's work	
	Malnourished	Control	Malnourished	Control	Malnourished	Control
SES High	91.36	104.45	93.83	103.46	91.30	103.71
SES Low	89.98	100.37	88.95	103.49	89.26	102.92

SES factors:

Results of analysis of variance

	Household items F value	Housing F value	Father's work F value
Nutritional history	40.4[a]	44.2[a]	49.3[a]
SES factor	1.7	1.5	0.5
Interaction	0.5	1.7	0.1

[a] $p \leq 0.001$.

Fig. 3. Relationship between socioeconomic factors and full IQ.

Table IV. Behavioral Groupings Based on Factor Analysis of Teacher Interview

Attention-deficit disorder factor[a]	Social skills factor[a]
Attention	Initiates conversation with teacher
School performance	Peer relationships
Memory	
Distractibility	
Class rank	
Obedience	
Special problems	
Restlessness	
Physical appearance factor[a]	Emotional stability factor[a]
Not getting enough to eat	Poor emotional control
General health	Emotional outbursts
Sleepy in class	
Special problems I factor	Special problems II factor
Soils	Behavioral problems
Wets	Specific problems
School attendance	
Days in school	

[a] Significant ($p < 0.05$) difference between index and control children by Student's t test.

to that reported in the United States (Anastasi and Foley, 1949). Although it is clear that IQ is a major factor in school performance, when we controlled for the effects of IQ, we found that previously malnourished children still received lower scores in language arts, mathematics, and general science than control children.

When we examined the role of socioeconomic factors and home characteristics in academic performance, the results were identical to those found for IQ scores. Our results confirm that children with histories of severe malnutrition perform less effectively in school than do children with no history of malnutrition, as was also reported by other investigators. Such children are clearly at increased risk for school failure.

3.5. Classroom Behavior

Another significant aspect of why a child succeeds or fails at school is classroom behavior. We asked teachers who were not told about the children's past nutritional history to complete questionnaires on classroom behavior for both the previously malnourished children and their controls, who were, for the most part, classmates (Galler *et al.*, 1983b). A public health nurse reviewed the questionnaires in detail with the teachers. These questionnaires were then analyzed, using a factor analysis that groups related questions. This analysis made it possible to compare seven categories of behaviors, shown in Table IV. In four of these categories, previously malnourished children were found to be impaired relative to controls. The first category included items that have been associated with *attention-deficit disorder*, namely, reduced attention span, poor memory, distractibility, lack of cooperativeness, lower class rank, and restlessness, among others. About 60% of boys and girls with histories of malnutrition showed this deficit as compared with about 15% of the control boys and girls. The symptoms of attention-deficit disorder have been associated with abnormal perinatal events other than malnutrition, such as hypoxia and birth trauma. Our results imply that severe malnutrition in the first year of life may be another cause of attention-deficit disorder.

Previously malnourished children, especially boys, were also found to be less emotionally stable than control children, as shown by frequent temper tantrums and crying spells (*emotional stability factor*). When the 37 children who were below the expected grade level in school for their age were eliminated from our analysis, there were no longer any differences in the incidence of emotional instability between previously malnourished and control children. This finding implies that lack of emotional control (and possibly depression) are strongly associated with failure in school.

Two additional categories of behavior were found to be important. Social skills with both adults and children (*social skills factors*) were also reduced in previously malnourished children, including boys and girls. Finally, teachers noted that the previously malnourished children seemed poorly nourished (though this was not confirmed when we examined the children), had poor health, including more colds, and were sleepy in class (*physical appearance factor*). This was true for both boys and girls. Thus, attention-deficit disorder, reduced social skills, and poor physical appearance were all present in previously malnourished children, even when the 37 children below the correct grade for age were eliminated from these analyses; when these were included, emotional instability was also noted in the index group.

We found marks in school to be more highly correlated with classroom behavior (0.56–0.63) than was IQ (0.37–0.57). In fact, when classroom behavior was controlled, differences in nutritional history had no further effect on marks in school. This implies that it is the poor classroom behavior and, to a lesser extent, lower IQ associated with early malnutrition that are responsible for lower marks in school and a greater probability of school failure.

As was the case for both IQ and academic performance, we found no significant contribution to classroom behavior by socioeconomic factors or characteristics of the home environment at the time of the study. Since poor classroom behavior is prominent in previously malnourished children and an important predictor of school failure, it should receive more attention from investigators in this area.

3.6. Overview of the Interrelationships among Malnutrition, School Performance, and Socioeconomic Status in Barbadian Children

As discussed earlier, academic performance was reduced on eight of nine classroom subjects in children with a prior history of malnutrition when compared with age-matched control children (Galler *et al.*, 1984). This index group also had impaired classroom behavior and reduced IQ scores. The interrelationships between these variables are presented in Fig. 4 as correlation coef-

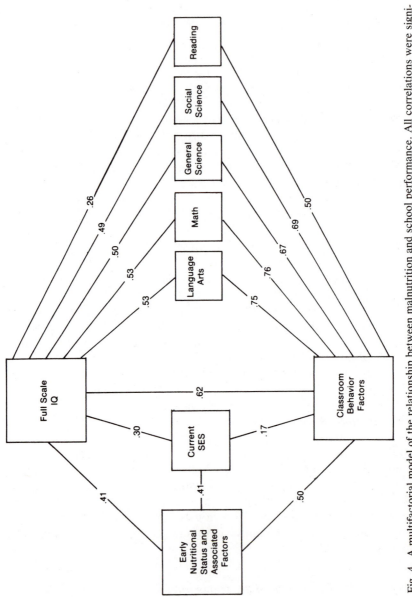

Fig. 4. A multifactorial model of the relationship between malnutrition and school performance. All correlations were significant at the .01 level or better, except the correlation between current SES and classroom behaviors which was not significant. (From Galler *et al.* (1984).

ficients. The figure shows moderate correlations between the Full Scale IQ scores and performance on each of five major classroom subjects (results are similar for the remaining significant subjects, which are omitted to conserve space). Deficits in classroom behavior are also significantly correlated with reduced performance on the school subjects, but as shown in Fig. 4, the *r* values in each case are much higher than those reported for the corresponding correlations between IQ and the school subjects. Thus, it would appear that classroom behavior is a more important contributor to academic performance than IQ. This can be confirmed by a statistical technique in which two factors are compared while the third is kept constant. This approach shows significant correlations between nutritional history and academic performance even when IQ is controlled, whereas controlling for classroom behavior eliminates the correlation of nutritional history with academic performance. Thus, the effects of malnutrition must operate through changes in classroom behavior.

Current socioeconomic status (SES) shows a significant correlation with a prior history of malnutrition, since index families tended to be somewhat disadvantaged. However, as can be seen in Fig. 4, the correlations obtained between socioeconomic conditions and IQ or classroom behavior were small. Thus, even when the socioeconomic factors are controlled statistically, early malnutrition has a major association with IQ and classroom behavior and, as a result, also with academic performance. Conversely, none of these indices was correlated with current SES when nutritional history was kept constant in the analysis. It can be concluded that in our study, early malnutrition results in impaired academic performance mainly through deficits in classroom behavior and, to a lesser degree, through reduced IQ performance, whereas factors in the social environment are primarily associated with the occurrence of the episode of malnutrition in the first place.

Other studies confirm that children with early histories of severe malnutrition perform less effectively in school than do children with no history of malnutrition (Richardson *et al.*, 1973; Richardson, 1980; Singh and Anand, 1976; Evans *et al.*, 1971; Lien *et al.*, 1977; Pereira *et al.*, 1979). These children are at risk of continued school failure and eventually are most likely to become "school dropouts" (Rutter and Madge, 1976). It should be noted, however, that none of the studies cited above has been able to characterize the mediating variables associated with reduced academic performance that we have described in detail. These have important policy implications and suggest that greater attention be given to early malnutrition as a hidden cause of learning failure (Birch and Gussow, 1970).

4. Summary

Some of the deficits in behavior and brain growth observed at the time of the acute episode of severe malnutrition and, later, during the recovery period persist at least as late as adolescence. Among neurological effects, brain size, primarily reported as head circumference, is reduced, especially in children with histories of marasmus or early onset of malnutrition or both. Since brain growth is virtually complete by 2 years of age, this finding is not surprising, though its functional significance is not fully understood. In contrast, EEG findings, which are markedly abnormal at the time of the illness, are marginally abnormal, with some evidence of reduced alpha activity in the intermediate phase and in the long term. Behavioral patterns that characterize the period of illness, namely, apathy, withdrawal, and other behaviors that create the condition of "functional isolation" in the infant, and compensatory alterations in the mother–infant relationship, are no longer observed shortly after recovery begins. Nevertheless, reductions in intellectual performance and delayed behavioral development are prominent in the intermediate period and during the school years. During the intermediate period, retardation in general development and especially in language skills is associated with a history of severe malnutrition. Hospitalization at the time of the episode may compound these delays, though it is clear that malnutrition *per se* also has an independent effect on behavior. These lags in behavioral development may be cumulative, since the failure to master developmental tasks at one age is likely to interfere with later stages of development.

The preceding analysis of the published literature also demonstrates that most of the comprehensive studies addressing the link between malnutrition and behavior have examined long-term outcome many years after the early episode. The majority of comparative and longitudinal studies of children in developing regions of the world report reduced IQ and school performance in previously malnourished groups. Several studies also report aberrant classroom behavior, specifically attention-deficit disorder, which may explain the greater likelihood of school failure in these groups. These outcomes are fairly consistent regardless of the type of malnutrition—kwashiorkor, marasmus, or mixed types. An earlier age at the time of hospitalization has been suggested as a factor in determining a less favorable behavioral outcome, but this hypothesis has not been substantiated. In the few studies that looked at both boys and girls with histories of malnutrition, the findings are inconsistent, and may depend on differences in cultural attitudes locally. The question of whether these effects of malnutrition are irreversible has not yet been answered.

Studies of the socioeconomic and environmental factors require consideration in relation to the long-term effects of malnutrition. In the studies reviewed, the selection of control groups has not provided thorough comparability, probably because complex interactions occur between malnutrition and the environment, independent of behavioral changes. It was hoped that studies of malnutrition secondary to severe gastrointestinal disease in children born in developed countries would help to clarify the contribution of environment. However, studies of long-term outcome in cases of malnutrition resulting from pyloric stenosis or cystic fibrosis have provided conflicting results. These few studies of secondary malnutrition are limited by various methodological problems, and their findings are not conclusive. Furthermore, recent studies in developing countries, using more comprehensive investigation of environmental factors, have substantiated, and allow us to define, the complex relationships among environment, malnutrition, and behavior.

The Barbados study has examined a wide range of measures, physical, cognitive, and behavioral, in a large and especially well-documented population of children with previous marasmus and matched control children. Although physical and neurological indices of growth and development show only small deficits attributable to the early episode of malnutrition, cognitive and behavioral measures consistently show striking reductions in IQ and school performance and aberrant school behavior. These findings were present even when we controlled for differences in socioeconomic factors and characteristics of the home. This research has enabled us to paint a detailed picture of school-aged children with early histories of malnutrition, showing that even in a supportive environment, they do not fully recover and always lag behind children who have not suffered from malnutrition.

Among the notable findings is the increased incidence of attention-deficit disorder in previously malnourished children. Male children also appear to be especially vulnerable, showing symptoms of emotional instability associated with classroom failure. Classroom behavior, in this study, was the major contributor to reduced school performance and therefore would be useful as a screening tool in populations where malnutrition is endemic.

Several implications for clinical practice can be derived from these observations. First, malnutrition should be considered a possible etiological factor in behavioral deficits even in affluent countries. Such deficits include school failure and the presence of an attention-deficit disorder. Accordingly, it is entirely conceivable that even in developed countries, cases of attention-deficit disorder (and emotional instability) may have resulted from nutritional disorders early in life. The clinician should be alert to this possibility and should evaluate the incidence of low birth weight, failure to thrive, and other evidence of nutritional deficit in the patient. Since this is not yet a recognized component of the evaluation of children with behavioral disorders, research programs in developed countries could profitably examine the frequency of this possible cause of behavioral problems in later life. Second, the use of physical growth alone as a measure of recovery from early malnutrition can obscure these more serious consequences. Even though this is a useful indicator in the period immediately after the acute episode, measures of behavioral function are by far more relevant to long-term outcome. Since clinicians make almost exclusive use of indices of physical growth in judging normal development, there is a need to expand these to individual measures of behavioral development. This should include measures of adaptive function, namely, those behavioral characteristics that permit the individual to respond to the demands of his environment.

5. References

Anastasi, A., and Foley, J. P., 1949, *Differential Psychology*, Macmillan, New York.
Ashem, B., and Janes, M. D., 1978, Deleterious effects of chronic undernutrition on cognitive abilities, *J. Child Psychol. Psychiat.* **19:**23–31.

Baird, D., 1947, Social class and foetal mortality, *Lancet* **253**:531–535.

Barnet, A. B., Weiss, I. P., Sotillo, M. V., Ohrlich, E. S., Sakurovich, Z. M., and Cravioto, J., 1978, Abnormal auditory evoked potentials in early infancy malnutrition, *Science* **201**:450–452.

Barrera-Moncada, G., 1963, *Estudios sobre Alteraciones del Crecimiento y del Desarrollo Psicológico del Síndrome Pluricarencial (Kwashiorkor)*, Editora Grafas, Caracas, Venezuela.

Bartel, P. R., Griesel, R. D., Freiman, I., Rosen, E. U., and Geefhuysen, J., 1979, Long-term effects of kwashiorkor on the electroencephalogram, *Am. J. Clin. Nutr.* **32**:753–757.

Beardslee, W. R., Wolff, A. H., Hurwitz, I., Parikh, B., and Shwashman, 1982, The effects of infantile malnutrition on behavioral development: A follow-up study. *Am. J. Clin. Nutr.* **35**:1437–1441.

Berglund, G., and Rabo, E., 1973, A long-term follow-up investigation of patients with hypertrophic pyloric stenosis with special reference to the physical and mental development, *Acta Paediatr. Scand.* **62**:125–129.

Bernard, R. M., 1952, The shape and size of the female pelvis, Transactions of the Edinburgh Obstetrical Society, *Edinburgh Med. J.* **59**(2):1–16 (transactions bound at end).

Birch, H. G., and Belmont, L., 1964, Auditory–visual integration in normal and retarded readers, *Am. J. Orthopsychiatry* **34**:852–861.

Birch, H. G., and Gussow, J. D., 1970, *Disadvantaged Children: Health, Nutrition, and School Failure*, Harcourt, Brace, and World, New York.

Birch, H. G., and Lefford, A., 1964, Two strategies for studying perception in "brain damaged" children, in: *Brain Damage in Children* (H. G. Birch, ed.), p. 46, Williams and Wilkins, Baltimore.

Birch H. G., and Richardson, S. A., 1972, The functioning of Jamaican school children severely malnourished during the first two years of life, in: *Nutrition, the Nervous System and Behavior*, pp. 64–72, PAHO Scientific Publication No. 251.

Birch, H. G., Richardson, S. A., Baird, D., Horobin, G., and Illsley, R., 1970, *Mental Subnormality in the Community: A Clinical and Epidemiologic Study*, p. 17, Williams and Wilkins, Baltimore.

Birch, H. G., Pineiro, C., Alcalde, E., Toca, T., and Cravioto, J., 1971, Relation of kashiorkor in early childhood and intelligence at school age, *Pediatr. Res.* **5**:579–585.

Botha-Antoun, E. Babayan, S., and Harfouche, J. K., 1968, Intellectual development related to nutritional status, *J. Trop. Pediatr.* **14**:112–115.

Bowlby, J., 1960, Separation anxiety, *Int. J. Psychoanal.* **41**:89.

Branko, Z., 1979, Height, weight and head circumference in survivors of marasmus and kwashiorkor, *Am. J. Clin. Nutr.* **32**:1719–1727.

Brockman, L., and Ricciuti, H., 1971, Severe protein–calorie malnutrition and cognitive development in infancy and early childhood, *Dev. Psychol.* **4**:312–319.

Brown, R. E., 1965, Decreased brain weight in malnutrition and its implications. *East Afr. Med. J.* **42**(11):584–595.

Brozek, J., 1978, Nutrition, malnutrition and behavior, *Annu. Rev. Psychol.* **29**:157–177.

Cabak, V., and Najdanvic, R., 1965, Effect of undernutrition in early life on physical and mental development, *Arch. Dis. Child.* **40**:532–534.

Celedon, J. M., and de Andraca, I., 1979, Psychomotor development during treatment of severely marasmic infants, *Early Hum. Dev.* **3**:267–275.

Celedon, J. M., Csaszar, D., Middleton, J., and de Andraca, I., 1980, The effect of treatment on mental and psychomotor development of marasmic infants according to age of admission, *J. Ment. Defic. Res.* **24**:27–35.

Champakam, S., Srikantia, S., and Gopalan, C., 1968, Kwashiorkor and mental development, *Am. J. Clin. Nutr.* **21**:844–852.

Chase, H. P., and Martin, H. P., 1970, Undernutrition and child development, *N. Engl. J. Med* **282**:933–939.

Chase, H. P., Lindsley, W. F. B., and O'Brien, D., 1969, Undernutrition and cerebellar development, *Nature (London)* **221**:554–555.

Chase, H. P., Canosa, C. A., Dabiere, C. S., Welch, N. N., and O'Brien, D., 1974, Postnatal undernutrition and human brain development. *J. Ment. Defic. Res.* **18**:355–366.

Chavez, A., and Martinez, C., 1975, Nutrition and development of children from poor rural areas. V. Nutrition and behavioral development. *Nutr. Rep. Int.* **11**(6):477–489.

Chavez, A., Martinez, C., and Yaschine, T., 1974, The importance of nutrition and stimuli on child mental and social development, in: *Early Malnutrition and Mental Development* (J. Cravioto, L. Hambraeus, and B. Vahlquist, eds.), pp. 211–225, Almquist and Wiksell, Uppsala, Sweden.

Chavez, A., Martinez, C., and Yaschine, T., 1975, Nutrition, behavioral development and mother–child interaction in young rural children, *Fed. Proc. Fed. Am. Soc. Exp. Biol.* **34**(7):1574–1582.

Clark, M., 1951, Kwashiorkor, *East Afr. Med. J.* **28**:299.

Cravioto, J., and DeLicardie, E. R., 1970, Mental performance in school age children, *Am. J. Dis. Child.* **120**:404–416.

Cravioto, J., and DeLicardie, E. R., 1971, Infantile malnutrition and later learning, in: *Progress in Human Nutrition* (S. Margan and N. Wilson, eds.), pp. 80–96, AVI, Westport, Connecticut.

Cravioto, J., and DeLicardie, E., 1972, Environmental correlates of severe clinical malnutrition and language development in survivors from kwashiorkor or marasmus, in: *Nutrition, the Nervous System and Behavior*, pp. 73–94, PAHO Publication No. 251.

Cravioto, J., and DeLicardie, E. R., 1976, Microenvironmental factors in severe protein–energy malnutrition, in: *Nutrition and Agricultural Development: Significance and Potential for the Tropics* (N. S. Scrimshaw and M. Behar, eds.), pp. 25–35, Plenum Press, New York.

Cravioto, J., and Robles, B., 1965, Evolution of adaptive and motor behavior during rehabilitation from kwashiorkor, *Am. J. Orthopsychiatry* **35**:449–464.

Cravioto, J., DeLicardie, E. R., and Birch, H. G., 1966, Nutrition, growth and neurointegrative development: An experimental and ecologic study, *Pediatrics* **38**(2)(Pt. I):319–372.

DeLicardie, E. R., and Cravioto, J., 1973, Behavioral responsiveness of survivors of clinically severe malnutrition to cognitive demands, in: *Early Malnutrition and Mental Development* (J. Cravioto, L. Hambreau, and B. Vahlquist, eds.) pp. 134–154, Almquist and Wiksell, Uppsala, Sweden.

De Silva, C. C., 1964, Common nutritional disorders of childhood in the tropics, *Adv. Pediatr.* **13**:213–264.

Dobbing, J., 1974, The later development of the brain and its vulnerability, in: *Scientific Foundations of Pediatrics* (J. A. Davis and J. Dobbing, eds.), pp. 565–577, W. B. Saunders, Philadelphia.

Dobbing, J., and Smart, J. L., 1973, Early undernutrition, brain development and behaviour, in: *Ethology and Development* (S. A. Barnett, ed.), pp. 16–36, Clinics in Developmental Medicine, No. 47, Heinemann, London.

Ellis, C. E., and Hill, D. E., 1975, Growth, intelligence and school performance in children with cystic fibrosis who have had an episode of malnutrition during infancy, *J. Pediatr.* **87**(4):565–568.

Engel, R., 1956, Abnormal brain-wave patterns in kwashiorkor, *Electroencephalogr. Clin. Neurophysiol.* **8**:489–500.

Engsner, G., Hable, D., Stogren, I., and Vahlquist, B., 1974, Brain growth in children with kwashiorkor, *Acta Paediatr. Scand.* **63**:687–694.

Evans, D. E., Moodie, A. D., and Hansen, J. D. L., 1971, Kwashiorkor and intellectual development, *S. Afr. Med. J.* **45**:1414–1426.

Evans, D., Bowie, M. D., Hansen, J. D. L., Moodie, A. D., and van der Spuy, H. I. J., 1980, Intellectual development and nutrition, *J. Pediatr.* **87**(3):355–363.

Fisher, M. M., Killeross, M. C., Simonsson, M., and Elgie, K. A., 1972, Malnutrition and reasoning ability in Zambian school children, *Trans. R. Soc. Trop. Med. Hyg.* **66**:471–478.

Freud, A., 1963, The concept of developmental lines, *Psychoanal. Study Child* **18**:245–265.

Gallais, P., Bert, J., Corriol, J., and Mileto, G., 1951, Les rhythmes des Noirs d'Afrique (étude des 100 premieres traces sujets normause), *Electroenephalogr. Clin Neurophysiol.* **3**:110.

Galler, J. R., 1983, Developmental and long-term consequences of early malnutrition on behavior, in: *Pediatrics Update* (A. J. Moss, ed.), pp. 115–131, Elsevier Science Publishing Co., Inc., New York.

Galler, J. R., Ramsey, F., Solimano, G., Lowell, W. E., and Mason, E., 1983a, The influence of early malnutrition on subsequent behavioral development. I. Degree of impairment in intellectual performance, *J. Child Psychiatry* **22**:8–15.

Galler, J. R., Ramsey, F., Solimano, G., and Lowell, W. E., 1983b, The influence of early malnutrition on subsequent behavioral development. II. Classroom behavior, *J. Child Psychiatry* **22**:16–22.

Galler, J. R., Ramsey, F., Solimano, G., and Propert, K., 1983c, Sex differences in the growth of Barbadian school children with early malnutrition, *Nutr. Rep. Int.* **27**:503–517.

Galler, J. R., Ramsey, F., and Solimano, G., 1984, The influence of early malnutrition on subsequent behavioral development. III. Learning disabilities as a sequel to malnutrition, *Ped. Research* (in press).

Geber, M., and Dean, R. F. A., 1956, The psychological changes accompanying kwashiorkor, *Courrier* **6**:3.

Gomez, F., Galvan, R. G., Cravioto, J., and Frenk, S., 1954a, Malnutrition in infancy and childhood, with special reference to kwashiorkor, *Adv. Pediatr.* **7**:131–169.

Gomez, F., Velaxo-Alzaga, J., Ramos-Galvan, R., Cravioto, J., and Frenk, S., 1954b, Estudios sobre el nino desnutrido. XVIII. Manifestaciónes psicológicas (comunicación preliminar), *Bol. Med. Hosp. Infant. Mex. (Span. Ed.)* **11**:631–641.

Graham, G., 1966, Growth during recovery from infantile malnutrition, *J. Am. Med. Women's Assoc.* **21**:737–742.

Grantham-McGregor, S. M., and Stewart, M. E., 1980, The relationship between hospitalization, social background, severe protein energy malnutrition and mental development in young Jamaican children, *Ecol. Food Malnutr.* **9**:151–156.

Grantham-McGregor, S. M., Stewart, M. E., and Desaix, P., 1978, A new look at the assessment of mental development in young children recovering from severe malnutrition, *Dev. Med. Child Neurol.* **20**:773–778.

Grantham-McGregor, S. M., Powell, C., Stewart, M. E., and Schofield, W. N., 1982, Longitudinal study of growth and development of young Jamaican children recovering from severe malnutrition, *Dev. Med. Child Neurol.* **24**:321–331.

Graves, P. L., 1976, Nutrition, infant behavior and maternal characteristics, a pilot study in West Bengal, India, *Am. J. Clin. Nutr.* **29**:305–319.

Graves, P. L., 1978, Nutrition and infant behavior: A replication study in the Katmandu Valley, Nepal, *Am. J. Clin. Nutr.* **31**:541–551.

Guthrie, G. M., Masangkay, A., and Guthrie, H. A., 1976, Behavior, malnutrition and mental development, *Cross-Cultural Psychol.* **7**(2):169–180.

Hansen, J. P. L., Freesman, C., Moodie, A. D., and Evans, D. E., 1971, What does nutritional growth retardation mean?, *Pediatrics* **47**:299–313.

Hertzig, M. E., Birch, H. G., Richardson, S. A., and Tizard, J., 1972, Intellectual levels of school children severely malnourished during the first two years of life, *Pediatrics* **49**(6):814–824.

Hoorweg, J., and Stanfield, P., 1972, The influence of malnutrition on psychologic and neurologic development: Preliminary communication, in: *Nutrition, the Nervous System and Behavior*, pp. 55–63, PAHO Publication No. 251.

Hoorweg, J., and Stanfield, J. P., 1976, The effect of protein energy malnutrition in early childhood on intellectual and motor abilities in later childhood and adolescence, *Dev. Med. Child Neurol.* **18**:330–350.

Jelliffe, D. B., 1966, The assessment of the nutritional status of the community, WHO Monograph Series No. 53.

Juntunen, K., Siroio, P., and Michelsson, K., 1978, Cry analysis in infants with severe malnutrition. *Eur. J. Pediatr.* **128**:241–246.

Kerr, M. A., Begues, J. L., and Kerr, D. S., 1978, Psychosocial functioning of mothers of malnourished children, *Pediatrics* **62**(5):778–784.

Klein, P. S., Forbes, G. B., and Nadar, P. R., 1975, Effects of starvation in infancy (pyloric stenosis) on subsequent learning abilities, *J. Pediatrics* **87**(1):8–15.

Latham, M. C., 1974, Protein–calorie malnutrition in children and its relation to psychological development and behavior, *Physiol. Rev.* **54**:541–565.

Lester, B. M., 1975, Cardiac habituation of the orienting response to an auditory signal in infants of varying nutritional status, *Dev. Psychol.* **11**(4):432–442.

Lester, B. M., 1976, Spectrum analysis of the cry sounds of well-nourished and malnourished infants, *Child Dev.* **47**:237–241.

Liang, P. H., Hie, T. T., Jan, O. H., and Giok, L. T., 1967, Evaluation of mental development in relation to early malnutrition, *Am. J. Clin. Nutr.* **20**(12):1290–1294.

Lien, N. M., Meyer, K. K., and Winick, M., 1977, Early malnutrition and "late" adoption: A study of their effects on the development of Korean orphans adopted into American families, *Am. J. Clin. Nutr.* **30**:1734–1739.

Lloyd-Still, J. D., Wolff, P. H., Hurwitz, I., and Shwachman, H., 1974, Studies on intellectual development after severe malnutrition in infancy in cystic fibrosis and other intestinal lesions, in: *Proc. 9th Int. Congr. Nutr., Mexico, 1972*, Vol. 2, pp. 357–364, S. Karger, Basel.

Lloyd-Still, J. D., Hurwitz, I., Wolff, P. H., and Shwachman, H., 1974, Intellectual development after severe malnutrition in infancy, *Pediatrics* **43**(3):306–311.

McCall, R. B., Hogarty, P. S., and Hurlburt, N., 1927, Transitions in infant sensorimotor development and the prediction of childhood IQ, *Am. Psychol* **27**:728–748.

McFarlane, H., Reddy, S., Adcock, K. J., Adeshina, H., Cooke, A. R., and Akene, J., 1970, Immunity, transfering, and survival in kwashiorkor, *Br. Med. J.* **4**:268–270.

McKay, H. E., McKay, A., and Sinisterra, L., 1973, Behavioral intervention studies with malnourished children: A review of experiences, in: *Nutrition, Development and Social Behavior* (D. J. Kallen, ed.), pp. 121–146, DHEW Publication No. (NIH) 73–242.

McKay, H., McKay, A., and Sinisterra, L., 1974, Intellectual development of malnourished preschool children in programs of stimulation and nutritional supplementation, in: *Early Malnutrition and Mental Development* (J. Cravioto, L. Hambraeus, and B. Vahlquist, eds.), pp. 226–233, Almquist and Wiksell, Uppsala, Sweden.

McKay, H., Sinisterra, L., McKay, L., Gomez, H., and Lloreda, P., 1978, Improving cognitive ability in chronically deprived children, *Science* **200**:270–278.

McLaren, D. S., Yaktin, U. S., Kanawati, A. A., Sabbagh, S., and Kadi, Z., 1973, The subsequent mental and physical development of rehabilitated marasmic infants, *J. Ment. Defic. Res.* **17**:273–281.

McLaren, D. S., Yaktin, U. S., Kanawati, A. A., Sabbagh, S., and Kadi, Z., 1975, The relationship of severe marasmic protein–energy malnutrition and rehabilitation in infancy to subsequent mental development, in: *Protein–Calorie Malnutrition* (R. E. Olson, ed.), pp. 107–112, Academic Press, New York.

Meneghello, J., 1979, *Desnutrición en el Lactante Mayor* (*Distrofia Poliarencial*), Central de Publicaciones, Santiago, Chile.

Monckeberg, F., 1968, Effect of early marasmic malnutrition on subsequent physical and psychological development, in: *Malnutrition, Learning and Behavior* (N. Scrimshaw and J. Gordon, eds.), pp. 269–277, MIT Press, Cambridge, Massachusetts.

Nelson, G. K., 1959, The electroencephalogram in kwashiorkor, *Electroencephalogr. Clin. Neurophysiol.* **8**:459.

Nowak, T. S., and Munro, H. N., 1977, Effects of protein–calorie malnutrition on biochemical aspects of brain development, in: *Nutrition and the Brain*, Vol. 2 (R. J. Wurtman and J. J. Wurtman, eds.), pp. 193–260, Raven Press, New York.

Olson, R. E., 1975, Introductory remarks: Nutrient, hormone, enzyme interactions, *Am. J. Clin. Nutr.* **28**:626–637.

Pereira, S. M., Sundararaj, R., and Begum, A., 1979, Physical growth and neurointegrative performance of survivors of protein–energy malnutrition, *Br. J. Nutr.* **42**:165–171.

Pollitt, E., and Granoff, D., 1967, Mental and motor development of Peruvian children treated for severe malnutrition, *Rev. Interam. Psicol.* **7**:93–102.

Pollitt, E., and Thomson, C., 1977, Protein–calorie malnutrition and behavior: A view from psychology, in: *Nutrition and the Brain*, Vol. 2 (R. J. Wurtman and J. J. Wurtman, eds.), pp. 261–306, Raven Press, New York.

Ramos-Galvan, R. R., Mariscal, A. C., Viniegra, C. A., and Perez, D. B., 1969, *Desnutrición en el Niño*, pp. 535–551, Impresiones Modernas, Mexico City, Mexico.

Ricciuti, H. N., 1981, Developmental consequences of malnutrition in early childhood, in: *The Uncommon Child* (M. Lewis and L. A. Rosenblum, eds.), pp. 151–172, Plenum Press, New York.

Richardson, S. A., 1980, The long range consequences of malnutrition in infancy: A study of children in Jamaica, West Indies, in: *Topics in Pediatrics*, Vol. 2, *Nutrition in Childhood* (B. Wharton, ed.), pp. 163–176, Pitman, London.

Richardson, S. A., Birch, H. G., Grabie, E., and Yoder, K., 1972, The behavior of children in school who were severely malnourished in the first two years of life, *J. Health Soc. Behav.* **13:**276–284.

Richardson, S. A., Birch, H. G., and Hertzig, M. E., 1973, School performance of children who were severely malnourished in infancy, *Am. J. Ment. Defic.* **77**(5):623–637.

Richardson, S. A., Koller, H., Katz, M., and Albert, K., 1978, The contributions of differing degrees of acute and chronic malnutrition to the intellectual development of Jamaican boys, *Early Hum. Dev.* **2**(2):163–170.

Rosso, P., Hormazabal, J., and Winick, M., 1970, Changes in brain weight, cholesterol, phospholipid, and DNA content in marasmic children, *Am. J. Clin. Nutr.* **23:**1275–1279.

Rutter, M., and Madge, N., 1976, *Cycles of Disadvantage: A Review of Research*, Heinemann, London.

Seoane, N., and Latham, M. C., 1971, Nutritional anthropometry in the identification of malnutrition in childhood, *J. Trop. Pediatr.* **17:**98–104.

Sheffer, M. L., Grantham-McGregor, S. M., and Ismail, S. J., 1981, The social environment of malnourished children compared with that of other children in Jamaica, *J. Biosoc. Sci.* **13:**19–30.

Singh, M. V., and Anand, N. K., 1976, Scholastic performance in relation to protein calorie malnutrition, *Indian J. Psychometry Education* **7**(1):5–8.

Singh, M. V., Anand, N. K., Dhingra, D. C., and Gupta, S., 1977, Scholastic performance in relation to protein calorie malnutrition, *Indian J. Clin. Psychol.* **4:**15–18.

Smart, J. L., and Dobbing, J., 1971, Vulnerability of developing brain. VI. Relative effects of foetal and early postnatal undernutrition on reflex ontogeny and development of behavior in the rat, *Brain Res.* **33:**303–314.

Spitz, R. A., 1945, Hospitalism: An inquiry into the genesis of psychiatric conditions in early childhood, *Psychoanal. Study Child*, **1:**53–74.

Spitz, R. A., 1946, Hospitalism: A follow up report on investigation described in volume 1, *Psychoanal. Study Child*, **2:**113–117.

Stein, Z., Susser, M., Saenger, G., and Marolla, F., 1975, *Famine and Human Development: The Dutch Hunger Winter of 1944–1945*, Oxford University Press, New York.

Stoch, M. B., and Smythe, P. M., 1963, Does undernutrition during infancy inhibit brain growth and subsequent intellectual development?, *Arch. Dis. Child.* **38:**546–552.

Stoch, M. B., and Smythe, P. M., 1967 The effect of undernutrition during infancy on subsequent growth and intellectual development, *S. Afr. Med. J.* **41:**1027.

Stoch, M. B., and Smythe, P. M., 1976, 15-Year developmental study on effects of severe undernutrition during infancy on subsequent physical growth and intellectual functioning, *Arch. Dis. Child.* **51:**327–336.

Stoch, M. B., Smythe, T. M., Moodie, A. D., and Bradshaw, D., 1982, Psychosocial outcome and CT findings after growth undernourishment during infancy: A twenty-year developmental study, *Dev. Med. Child Neurol.* **24:**419–436.

Trowell, H. C., Davies, J. N. P., and Dean, R. F. A., 1954, *Kwashiorkor*, Arnold Press, London.

Vahlquist, B., Engsner, G., and Sjogren, I., 1971, Malnutrition and size of the cerebral ventricles, *Acta Paediatr. Scand.* **60:**533–539.

Valenzuela, R. H., Hernandez-Peniche, J., and Macias, R., 1959, Aspectos clínicos eletroencefalográficos y psicológicos en la recuperación del niño desnutrido, *Gac. Med. Mex.* **89:**651–656.

Valman, H. B., 1974, Intelligence after malnutrition caused by neonatal resection in ileum, *Lancet* **1:**425–427.

Werner, E. E., Bierman, J. M., and French, F. E., 1971, *The Children of Kauai: A Longitudinal Study from the Prenatal Period to Age Ten.* University of Hawaii Press, Honolulu.

Winick, M., and Rosso, P., 1969, Head circumference and cellular growth of the brain in normal and marasmic children, *J. Pediatr.* **74:**774–778.

Winick, M., and Rosso, P., 1975, Malnutrition and central nervous system development, in: *Brain Function and Malnutrition* (J. W. Prescott, M. S. Read, and D. B. Coursin, eds.), pp. 41–51, John Wiley and Sons, Inc., New York.

Winick, M., Meger, K. K., and Harris, R. C., 1975, Malnutrition and environmental enrichment by early adoption, *Science* **190:**1173–1175.

The Behavioral Consequences of Protein–Energy Deprivation and Supplementation in Early Life: An Epidemiological Perspective

David Rush

1. Introduction

1.1. Aims of this Review

This review focuses on the security of inference that can be drawn from the existing literature on the relationship of protein–energy nutrition during pregnancy and early childhood and subsequent mental function and behavior. [It follows a recent review of the effect of gestational nutrition on fetal growth and survival (Rush, 1982a).] Since only the most limited conclusions can be drawn from most currently available observational studies, because they are unable to dissociate malnutrition from the constellation of other forms of deprivation that almost invariably accompany nutritional insult, the review is limited to those few studies from which strong inference can be drawn. Also, noninterventional studies offer little help in guiding us to the most efficient and humane ways to ameliorate the suffering of all too many of the world's children.

1.2. Theoretical Model for Relating Nutrition to Behavior

Birch (1972) formulated the requirements for a comprehensive understanding of the relationship of nutrition and subsequent behavior. His multivariate model included four terms partitioning variance in behavior into: that due di-

David Rush • Rose F. Kennedy Center, Bronx, New York 10461; Division of Pediatric and Perinatal Epidemiology, Departments of Pediatrics and of Obstetrics and Gynecology, Albert Einstein College of Medicine, Bronx, New York 10461.

rectly to nutrition, that due to nonnutritional factors, a third term that takes into account the correlation between nutrition and nonnutritional factors, and an interaction term, since the interrelationship of nutrition and nonnutritional factors may be different at different levels of these variables. If this interaction between nutritional and nonnutritional factors is quantitatively important, the inferences to be drawn from studies such as that of Stein *et al.* (1975) on the consequences of the Dutch famine of 1944–1945 and the studies reviewed below of malnutrition in infancy secondary to medical conditions would be limited. While these studies can separate the effect of nutrition from that of correlated socioenvironmental insults, the populations are affluent, or at least not deprived, and possibly nutritional insult may matter less among those not exposed to concomitant nonnutritional deprivation.

1.3. Distinctions between Current and Past Malnutrition

Levine and Weiner (1976) criticize much of the research relating nutritional deficiency to behavior in animals, because conclusions are drawn from the performance of acutely malnourished, unrehabilitated subjects. The same caution is appropriate to studies of children. There are discomforts and distractions associated with being starved that hardly allow any reasonable estimate of the child's current or potential performance. We must rather be concerned with the long-term consequences of such starvation, after the child has been rehabilitated.

2. Limitations of Retrospective Observational Studies Relating Malnutrition to Behavior

2.1. Summary of Observational Studies

There is an enormous body of observational literature relating nutritional insult in childhood to later mental development and behavior. Pollit and Thomson (1977) have comprehensively reviewed this literature. They concluded that

> severe but acute protein–calorie deficiency during the second year of life among populations where malnutrition is endemic may, but generally does not, leave measurable retardation in intellectual function as compared to standards from the same population. There are no available current data to determine whether severe but acute episodes of protein–calorie malnutrition with an onset during the second year of life result in any impairment of intellectual function.

Concerning marasmus, they concluded that

> severe protein–calorie deficiency occuring throughout most of the first 12 months of life among populations where malnutrition is endemic results in a severe deficit (one to two standard deviations below the average of 100) in intellectual function as compared to standards from the same population.

These conclusions seem fair and reasonable. However, judgment of the effect of marasmus assumes that the nutritional component of the insult suffered by the marasmic infant can be isolated. If the nutritional component of the child's deprivation cannot be separated from the web of factors that led to and sustained the child's starvation, conclusions about the isolated effect of starvation may be beyond current research techniques. In any case, the data do not allow the assumption that the simple provision of food is likely to prevent, or reverse, the consequences of marasmus.

While all investigators accept the need for clear and unambiguous research design, most rely on the comparison between the developmental status of children previously starved and controls drawn from the same population, either siblings or others, thought or assumed to be similar, particularly in social status.

2.2. Malnutrition as an Index of More General Deprivation

Retrospective studies of the consequences of earlier malnutrition depend on the assumption that control of nonnutritional factors is possible, using what are invariably superficial social and environmental indices. Where the comparability of other relevant circumstances in the child's milieu that distinguish the malnourished child from controls has been directly studied, the conditions of the malnourished child *prior to* the onset of malnutrition are consistently more stressful and unstimulating.

Cravioto and DeLicardie (1975) presented detailed descriptions of the reactions of mothers to their children undergoing psychological testing prior to the diagnosis of clinical malnutrition. Of mothers of infants who later became clinically malnourished, 40% were passive when their children performed tests adequately, and none were appreciative or proud of their children, while all the mothers of control children at least considered that it was normal for their children to perform adequately and easily, and most responded positively to their children's performance. Mothers of children later malnourished were reserved and defensive when responding to their child's questions, while control mothers often answered enthusiastically, and volunteered much information. There was little verbal communication between mothers and children later to become malnourished: only 10% talked with their children in a two-way conversation. A third of the mothers who had malnourished children were not aware of the child's need for attention, support, and comforting, vs. 6% of controls. Only 10% of the mothers of malnourished children were always aware of the child's needs and continually aware of the child's presence, vs. 40% of the control mothers. Thus, the mothers of children later to become malnourished were less interested in the child's test performance and less sensitive to the child's needs, communicated verbally far less, had less sense of their role in the bringing up of the child, were less involved emotionally, expressed less affection, and were less responsive when the child performed well. This malfunction in mothering could have been due to the mother's inability or inclination, or may have been a function of the child's own character and personality, or some combination of characteristics of the mother and child. However, the end result is the same: the child who later came to be diagnosed as clinically malnourished had experienced severe prior emotional and cognitive deprivation as well. We thus cannot argue that mental and behavioral deficit following malnutrition can be ascribed solely to nutritional insult and cannot infer that damage might have been avoided solely by the provision of food. Deprivation of nurturing is not only related to food, but also must be considered as part of a more comprehensive pattern, with clinical malnutrition only one index of a web of insult.

2.3. Sibling-Control Studies as an Attempt to Isolate the Effects of the Deprivation of Food

2.3.1. Limits of Sibling Controls

The use of sibling controls is helpful, but by no means solves all problems. There may be differential deprivation within a family with prejudice specific to some children. It seems unlikely that the starved child is not deprived in other ways as well relative to its siblings, in ways that are likely to make for less affection, concern, and stimulation. Miller (1981) has studied in detail the remarkably higher infant and child death rates among girls in North India. Girls are less valued than boys and more frequently starve and die. If a girl is more likely to starve than her brother, can we assume that any starved child has received care necessary for adequate mental and behavioral development equivalent to that received by a nonstarved sibling?

2.3.2. Strengths of Sibling Controls: Birth Weight and Intelligence

Sibling controls can help us to sort out causal relationships between some supposedly noxious prior event and later performance. There is much literature on the relationship of size in general, and birth weight in particular, to later mental performance. The relationship of birth weight to intelligence is frequently interpreted as demonstrating an effect of "intrauterine malnutrition" on neurological and intellectual function. This is not necessarily so. Record *et al.* (1969) and McKeown (1970) related differences in verbal reasoning scores at age 11 of 2863 sibling pairs to differences in their birth weights. Scores were virtually identical until birth-weight differences were over 1500 g. Even with that difference in birth weight, the reasoning scores were different by only 1.5 points. These authors also studied 857 twin pairs, among whom verbal reasoning was virtually identical between heavier and lighter co-twins, up to a difference in birth weight of 700 g. Over 700 g, there were differences of 3.1 and 3.4 points. They also compared scores for 962 twin pairs with 148 children whose co-twins died in infancy. These survivors were, like all twins, of very low average birth weight. The mean verbal reasoning scores were 95.2 where both twins had survived, but 98.8 with only one survivor, only 0.7 point lower than for single births. Thus, it is far more likely that the low scores of twins follow from the postnatal environment than from intrauterine growth retardation.

2.4. Low Birth Weight and Lowered Intelligence as Results of Prior Environmental Deprivation

A causal relationship between low birth weight and later cognition has been challenged by Douglas's studies of the 1946 British national birth cohort. Douglas (1956) originally concluded that

> premature (<2500 grams birthweight) children scored less than their controls in each of these tests (reading, vocabulary, and intelligence), being proportionately the most handicapped in reading.

Several years later, Douglas (1960) wrote:

> As this survey progressed it became increasingly clear that, in spite of the careful matching of the premature and control pairs at the beginning of the survey for fathers' occupations, there were, in fact, marked social differences between them which have grown larger with time. The fathers of the premature children were more likely to move during the survey to less favorable occupations than the fathers of the controls, and were less likely to improve their occupational status or move into self-employment. Moreover, periods of unemployment were more than twice as frequent among them—6.5% were unemployed on one or more occasions as compared with 3.2% of the fathers of the controls.

He said in summary:

> In a national study of the mental ability and primary school progress of "premature" children a number of striking handicaps were found, which were later shown to be of environmental origin rather than the result of low birthweight per se. . . . These differences are largely explained by the fact that premature birth is not only associated with poor living conditions but also, at each social level, with low standards of maternal care and lack of educational interest. This study casts doubt on the utility of the method of controlled comparisons in socio-medical studies.

Thus, families that produced children of low birth weight were strikingly different, on the average, than those of controls. Whatever the disabilities of these low-birth-weight children, they were as or more likely to have been caused by the less stimulating and supportive environment that their families provided than by the fact of low birth weight.

Similar results were found by Goldman *et al.* (1974) in their follow-up of low-birth-weight infants who had been randomly allocated to high-density or normal-density protein diets in the neonatal period. The authors administered the Wechsler Adult Intelligence Scale (WAIS) to 211 of the mothers of these low-birth-weight infants and found that

> there were more WAIS scores below 90 (35%) and fewer above 110 (18%) than anticipated. The observed distribution probably reflects the socio-economic status known to be associated with premature births.

2.5. Short Stature as a Surrogate for Nutritional Deprivation

Many studies make the dubious assumption that the size of infants and children is a function only of nutrition. Not only is linear growth dependent on many other factors besides nutritional intake (e.g., parental stature, concurrent illness, ethnicity, climate, activity, birth order), but also there are almost surely environmental and familial causes and correlates of chronic, subclinical malnutrition similar to those of clinical malnutrition. For instance, Graves (1976) found that mothers of stunted Bengali children had married and had their first child younger, and expressed a more negative and unwelcoming attitude toward the pregnancy, than mothers of normally grown children.

Thus, it is hazardous to ascribe later differences in intellectual function and behavior of children who are small in infancy or early childhood to nutrition alone. The constellation of factors associated with small stature almost surely also affects later behavior.

2.6. Severity of Organic Illness and Family Function

The aim of this review is to focus on those studies that can best disentangle interrelationships of malnutrition and prior or concurrent insult. Thus, we will review the later behavior of children starved in infancy secondary to severe organic disease, since we assume that the incidence of such organic disease is unrelated to the social, emotional, or intellectual environment of the child's family. Although the incidence of disease can be assumed to be independent of family function, there is evidence that the severity of disease is related to the function of the child's family.

Among children with cystic fibrosis causing infantile malnutrition, more severe growth failure was associated with less parental education, less stable families, worse acceptance of the disease by the child and parents, and higher stress from the disease (Ellis and Hill, 1975). Clearly, some of these differences between families could have *followed* from the greater stress induced by more severe disease. It is more likely, however, that better growth was associated with better circumstances for the care of the child. It is thus nearly impossible to ascribe greater long-term disability to more severe disease, since severity of disease, indicated by growth failure, is almost surely related to characteristics of the family that will also affect later development.

2.7. Summary

We thus cannot presume that low birth weight, or short stature, or more severe organic nutritional disease is caused by isolated, or isolatable, depressed intake of food, or that the effect of malnutrition can, by observational study, be separated from all the conditions that led to and sustained such malnutrition. Efforts to systematically control for differences in social and economic status are generally feeble, since we can only rarely identify and measure those aspects of socioeconomic status or family function that directly influence child development. The indices available to us, such as years of parental education, or income, or occupational status, can be only crude surrogates for factors that influence the behavior of the child, and to match for them still leaves the possibility of vast differences in family function, convincingly signaled by the starvation of the child.

3. Nutritional Differences Unrelated to Family Environment

3.1. Introduction

A few study designs can help distinguish the effects of the malnutrition from the web of precedent and concurrent circumstance.

Since we presume that the incidence of organic disease causing infant malnutrition is independent of family function, the later consequences of such disease may help us clarify the direct effects of nutriton on later behavior. Similarly, when acute famine affects an entire population, or when children or mothers receive different amounts of food in experimental interventions of various sorts, determined by other than the recipient or family being studied (and are thus not self-selected), the comparison and study groups are much less likely to be systematically different in ways that will also affect later behavior and mental development.

Not only can intervention studies help sort out causal relationships, they can also potentially provide models for the amelioration of deficits that might follow from nutritional deprivation. While it may be impossible to fully understand the web of causation relating nutritional insult and mental deficit, it should be possible to determine how much benefit accrues from feeding malnourished or suboptimally nourished pregnant women and children. Such studies thus serve the dual function of strengthening inference and of possibly offering models for efficient and effective intervention programs to prevent either malnutrition or its sequelae.

We will also review studies of emotional and cognitive intervention among malnourished children. Such studies help to clarify both whether prior insult was exclusively, or predominantly, nutritional and whether rehabilitation must include improvement of the child's emotional and cognitive environment, as well as food.

3.2. Undernutrition in Infancy and Childhood Secondary to Organic Disease (Table I)

3.2.1. Pyloric Stenosis

We shall consider five studies, two of hypertrophic pyloric stenosis and three in which the majority of cases were of cystic fibrosis. Berglund and Rabo (1973) traced 180 of a total of 202 males medically treated for pyloric stenosis between the years 1922 and 1942 in one Swedish referral center. They divided cases into four groups (I–IV) based on the severity and duration of weight loss. For the assessment of adult stature, they had 88 control brothers available. As adults, the most severely affected group was 3.5 cm shorter than the Swedish mean, but there was no consistent gradient of height by severity of disease, and the three less severely affected groups were little different from the national mean. On the other hand, the affected sibling was shorter in 46 of the 57 families in which brothers were of different adult stature than cases.

Intellect was assesed at induction into military service for all men born after 1925. There was no regular gradient of intelligence score by severity,

Table I. Studies of Undernutrition in

Authors, disease, age of onset	N	Study group				Control group		
						N	Criteria for choice	Age at evaluation
Berglund and Rabo (1973) Hypertrophic pyloric stenosis, 1922–1942 Onset at 6–20 days Males Nonsurgical therapy only		Arbitrary classification based on visual plot of degree and duration of weight loss				28	Brothers	Assessed at evaluation for military service (only those born after 1925)
		N	Group	Weight deficiency (%)	Mean duration of deficit (wk)			
		56	I	21.0 ± 9.7	—			
		57	II	35.5 ± 8.3	4.9			
		44	III	35.3 ± 4.9	9.7			
		23	IV	47.1 ± 7.5	14.8			
P. S. Klein *et al.* (1975) Hypertrophic pyloric stenosis Onset <3 months of age Neither duration of insult, type of therapy, nor age at therapy stated	44 6	Male Female				44 50	Siblings Community controls, matched for age, sex, father's education	5–15 years (Subjects: mean age = 9 yr 2 mo) (Controls, mean age = 10 yr 1 mo)

although the most severely affected group was lowest. The differences were stated to be nonsignificant (152 cases were tested). Also, the most severely affected group was relatively infertile in later life.

Only limited conclusions can be drawn from this study. No data on duration of hospitalization or early separation from families were presented. Most comparisons are by severity, among cases: if family characteristics that might affect ultimate stature and intelligence are also associated with severity of illness, little can be made of later differences. However, it appears safe to conclude that there were no consistent or severe psychological sequelae to this early insult.

P. S. Klein *et al.* (1975) compared later behavior of 50 cases of hypertrophic pyloric stenosis with onset under 3 months of age with 44 siblings and 50 com-

Infancy Secondary to Medical Conditions

		Results				

Group	Height (cm)[a]	Height vs. brothers (N)		Intelligence[b]	Adaptation[c]	Percentage with children	Comments
		Taller	Shorter				
I	175.9	7	11	5.1	4.5	66.1	"Sons outnumbered the daughters two-
II	177.2	6	14	5.5	4.6	62.5	fold." Fertility not adjusted for age (i.e.,
III	176.1	1	6	5.0	4.1	70.5	if group IV men were younger, they
IV	173.3	1	5	4.8	3.9	45.5	might not have had equal opportunity to
	(Mean	15	46				reproduce). Neither the construct validity
	for Sweden, 176.8)						nor the sensitivity of intelligence test

[a] Regression coefficient of weight deficit and duration, with height, $p \leq 0.01$.
[b] 1 = low, 9 = high. Differences stated to be nonsignificant, $N = 152$.
[c] On subjects with high intelligence only; $N = 67$. Differences stated to be nonsignificant.

known. No data presented on duration of hospitalization or separation.

				Average standardized difference[d]				

				Memory				
Severity	PPVT	Vocab	Coding	Auditory	Visual	Raven	Reading	Arithmetic
High (N = 13)	0.38	0.20	−1.02[f]	−0.75[e]	−0.54	0.43	−0.12	−0.15
Moderate (N = 18)	−0.06	−0.08	0.36	−1.01[g]	−0.65	−0.71	−0.07	−0.14
Low (N = 13)	0.12	0.40	0.25	0.63[e]	0.23	0.13	−0.15	0.23

[d] Calculated from authors' data [Index case − (Sib + Control)/2].
[e] $p < 0.05$.
[f] $p < 0.01$.
[g] $p < 0.001$.

Sex of sibling controls not reported, nor were results controlled for sex. Six community controls omitted without explanation in presentation of case–control differences. Examiners said to be blind to whether case or sib control. However, with unlike-sexed pairs, male almost always the case, female the control. Case–control differences presented by significance level (*t*'s), which, with variable *N*s, do not reflect magnitudes of effect. Most results presented within case group only; comparisons among severity groups problematical (see text). No data presented on duration of hospitalization or separation.

(Continued)

munity controls, matched for age, sex, and years of father's education. Although examiners were said to be blind to whether a member of a sibling pair was a case or a control, since this is predominantly a male disease, in all unlike-sexed pairs, the male would almost always be the case and the female the control. No data were presented on the duration of hospitalization or on the separation of the child from the family. The siblings were not matched by sex, and results were not controlled for sex.

The investigators shifted between case–control comparisons and comparisons within the study group stratified by severity. No mention was made of the data for the six case–community control pairs for whom no sibling control was available. Case–control differences were limited to eight indices (Table I)

Table I.

Authors, disease, age of onset	N	Study group	Control group N	Criteria for choice	Age at evaluation
Lloyd-Still *et al.* (1974) Protracted infant malnutrition (58% with 3° malnutrition)	34 4 3	Cystic fibrosis Protracted diarrhea Ileal atresia	41	Siblings (not matched for sex or ordinal position).	<5 Years of age: 15 Patients 12 Controls 5–21 Years of age: 26 Patients 29 Controls
Valman (1974) Neonatal period	A. 8 B. 13	Resection of ≥45 cm ileum Cystic fibrosis, without severe chest disease	C. 6 D. 26	Siblings One complete suburban classroom	A. 3–14 years B. 7–12 years C. 6–15 years D. 6.5–7 years
Ellis and Hill (1975) Cystic fibrosis	A. 22 B. 19	"Malnourished" (13 male) 3 >40% Deficit in weight 19 20–40% Deficit in weight "Control" (5 male) <20% Deficit in weight		No normal controls	7–10 Years of age

and presented as differences in the t statistic; since the t statistic depends on sample size, as well as effect size, we have transformed these into standardized effects, since cell sizes differed. The only consistent deficit was in auditory memory, which was significantly depressed at all levels of severity. Coding was also significantly depressed, but only among the most severely affected cases. For the majority of functions tested, including child's vocabulary, reading, arithmetic, and ability to do Raven's Matrices, there were no deficits, and it seems unlikely that these children are much impaired, at least on the basis of their test performance.

The consistent deficit in auditory memory is impressive. Firm conclusions would depend on replication, and controls ought to be chosen to have been separated from their families and hospitalized for periods of time comparable to cases.

(Continued)

Test	Cases	Controls	Significance	Comments
<5 Years of age				Although not explicitly stated, since number of controls and cases unequal in stratifying by age, cases probably *not* compared to their own sibs. No information on how sib chosen, if >1 available, or on current nutritional state of cases.
Merill–Palmer IQ (mean percentile)	40th	70th	$p < 0.005$	
>5 Years of age				
WISC, WAIS IQ	102 ± 12	104 ± 10	N.S.	
				Heights and weights between 3rd and 25th percentiles.
				Duration of hospitalization or separation not mentioned.
Draw-A-Man:	A. $\bar{X} = 106.3$, range 78–122			Ileal resection: Fathers all in highest three social classes. Maternal separation from 2 to 38 weeks during first 2 years of life. Patient with 38 weeks' separation had IQ of 71; separation was for "social and not medical reasons."
	B. $\bar{X} = 98.4$, range 66–123			
	C. $\bar{X} = 94.8$, range 97–116			
	D. $\bar{X} = 99.9$, range 75–118			
				Cystic fibrosis: 1–15 months in hospital, first 2 years; 1–32 months total.
				Lack of SES comparability and small numbers make generalization difficult. However, clearly no profound sequelae.
Height and Weight: No differences.				"Parental education was generally less, family stability poorer and acceptance of cystic fibrosis by the index child poorer in the malnourished group (n.s.). . . . In the acceptance of cystic fibrosis by the family, the malnourished group scores were significantly less than the control group, and the control group's family functioning was significantly less stressed by cystic fibrosis than that of the malnourished group."
WISC, WRAT: Group A lower on all 12 scores, significant only for Backward Digit Span.				
				Social differences not accounted for in analysis of outcome.
				No data presented on duration of hospitalization or separation.

3.2.2. Cystic Fibrosis

There are three studies of the long-term consequences to cognition of cystic fibrosis in infancy. Lloyd-Still *et al.* (1974) studied 34 cases of cystic fibrosis, 4 of protracted diarrhea in infancy, and 3 of infantile ileal atresia. They matched these cases with 41 siblings. Sex was not matched (or reported), nor was there matching for ordinal position. No information was given on how a sib was chosen if more than one was available. Also, nothing was reported about the current nutritional state of cases, of some import for those still under age 5 at time of evaluation ($N = 15$). The authors concluded that there was a deficit on the Merrill–Palmer IQ for children assessed who were under 5, but no deficit on the Wechsler Intelligence Scale for Children (WISC) or WAIS for those over 5. The conclusion that there was a deficit under 5 is clouded by several uncertainties. The study design was said to be of matched pairs. However, while there were 15 patients under 5, there were only 12 controls, suggesting that the analysis was not done using the matched pairs (there were, in turn, 3 more controls than patients for the group over age 5). Thus, both the logical strength of the sibling control design (with controls having similar social and genetic circumstances and other similarities) and the statistical power were in all likelihood lost. Given the small numbers of subjects, relatively minor differences in the social circumstances of cases and controls could have had large effects on the observed outcomes. The younger the children, the more likely they were not yet fully nutritionally rehabilitated. Thus, those under 5 might still have been malnourished, and the results could reflect not the residua of malnutrition after rehabilitation, but results for children in less than optimal current protein–energy balance. While the conclusion that there was little observed deficit over the age of 5 is probably correct (the bulk of cases and controls were in this group, and the result is likely to have held if matching was reasonably efficiently done), the result reported for children under 5 can be accepted only with great caution. Also, no account was taken of the duration of hospitalization or separation.

Valman (1974) studied 8 children between 3 and 14 years of age who had more than 45 cm ileum resected in the neonatal period, and 13 cases of infantile cystic fibrosis without severe chest disease, who were 7–12 at the time of study. He compared performance in the Draw-A-Man test with 6 siblings of 4 of the patients with ileal resection and with 26 $6\frac{1}{2}$- to 7-year-olds who formed one complete suburban school classroom. There were no deficits among those with infantile malnutrition; indeed, those with ileal resection did the best of all groups studied (their fathers were all in the highest three social classes, but they did better than their sibs, as well as children with cystic fibrosis and control children). The author reported extensive separation in infancy. The cases of ileal resection were separated from families anywhere from 2 to 38 weeks in the first 2 years, and the cases of cystic fibrosis anywhere from 1 to 32 months total, and from 1 to 15 months during their first 2 years.

The child with ileal resection with the lowest IQ (71) had been separated from her family for 38 weeks during the first year of life for "social and not medical reasons." This profound disruption in normal nurturing again should caution against considering nutritional deprivation as an isolated insult.

The absence of any attempt at judging comparability of socioeconomic status (SES), and small numbers, make generalization from this study difficult. It does illustrate, however, that these children and their families underwent the additional stress of lengthy periods of hospitalization in infancy, besides malnutrition. Nevertheless, there were no profound sequelae.

Ellis and Hill (1975) studied 41 cases of cystic fibrosis, which they characterized as "malnourished" ($N = 22$) or "controls" ($N = 19$). The malnourished group had greater than 20% deficit in expected weight some time during the first year of life. There were no controls without disease. At 7–10 years of age, there were no differences in height and weight between the groups. Twelve psychological tests were performed, and the "malnourished" group was lower in all 12, but only the difference for Backward Digit Span was statistically significant.

Social circumstance differed between the more and the less severely affected children. The severity of the initial illness was strongly related to family circumstances. The authors stated:

> Parental education was generally less, family stability poorer and acceptance of cystic fibrosis by the index child poorer in the malnourished group (n.s.). In the acceptance of cystic fibrosis by the family, the malnourished group scores were significantly less than the control group, and the control group's family functioning was significantly less stressed by cystic fibrosis than that of the malnourished group.

Thus, while more severe disease may have caused more family stress, it is also possible that the better-functioning families were better able to sustain the growth of their children: children may have had less severe disease *because* of better family functioning. This report therefore suggests the strong possibility that the severity of the initial nutritional insult is not independent of factors that almost surely have powerful influence on the child's later psychological function and behavior. Thus, it is not sensible, when comparing the more severely with the less severely affected children, to assume that the determinants of severity are not related to the determinants of later psychological function. The data of Ellis and Hill conform to the general clinical impression that the more competent family can better care for organic problems, as well as prevent retarded or pathological psychological development.

3.2.3. Summary

The later consequences of organic disease causing severe infantile malnutrition are remarkably slight and have not been distinguished from the prolonged hospitalization and separation from the family. In addition, given that initial severity cannot be presumed to be independent of family functioning, studies that omit nonstarved controls, and that only contrast less and more severely affected cases, are suspect.

3.3. Nutritional Interventions Limited to or Beginning during Pregnancy (Table IIA)

The studies in this section are reviewed in the alphabetical order of the senior authors' surnames.

3.3.1. Mexican Village Study

Chavez and colleagues executed a longitudinal intervention in a poor, small Mexican village (Chavez *et al.*, 1974, 1975; Chavez and Martinez, 1975, 1979a,b; Brozek *et al.*, 1977). Numbers were small: there were 17 cases and 17 controls. Cases were recruited 2 years after controls; assessment could not be blinded. While cases were said to be matched to controls, no analyses have been presented using matched-pair analysis. The birth weights of controls have

Authors	Site and study design	Study group	Control group
Chavez *et al.* (1974, 1975), Chavez and Martinez (1975, 1979a,b) (also Brozek *et al.,* 1977)	Poor rural Mexican village Controls born before intervention. Intervention started at 45-days' gestation and continued to 5 years. Cases and controls: all mothers para 1–4, 1.38–1.5 m tall, age 18–34, and healthy. Infants chosen to be healthy and to weigh >2500 g (and, possibly, <3400 g).	*N* = 17 20 Infants chosen from either 39 or 40 pregnant mothers, identified either before or early in pregnancy, and "matched" to controls (2 died and 1 refused to continue). Treatment: For mother: Stated to be both 2–3 glasses of milk/day and 64 g of powdered milk and vitamins and minerals/day. For child: "One bottle" of full fat milk provided at 3 mo, and "gradually increased." Maternal intake of supplements assessed by two 24-hr recalls: At 8 mo gestation: 205 cal/day, 15 g protein; at 6 mo post partum: 305 cal/day, 23 g protein. (No data on child food consumption.)	*N* = 17 20 Recruited from 40 followed pregnancies, 1968. 1 Infant died of measles and 2 developed clinical malnutrition and were removed from study.
Hicks *et al.* (1982)	Rural Louisiana Comparison between siblings, both of whom received WIC benefits. Younger sibling started during third trimester of pregnancy and continued to at least 1 year of age; older sib started after age 1. 19 Black, 2 white.	*N* = 21 Younger sibs, mean age: 6.3 yr	*N* = 21 Older sibs, mean age: 8.8 yr
Osofsky (1975)	Philadelphia, Temple University Hospital Controlled trial of high-protein nutritional supplementation (Meritene) among "low-income" pregnant women. Subjects recruited after completion of control pregnancies.	*N* = 122 Amount prescribed not stated. Intake of 80.3 g protein/day.	*N* = 118 92% black. Intake of 71.8 g protein/day.
Pencharz *et al.* (1983)	Montreal Schoolchildren born at Royal Victoria Hospital. Cases' mothers serviced during pregnancy by Montreal Diet Dispensary, with significant increase in birth weight (Rush, 1981).	*N* = 406 Nutritional assessment, counseling, and supplementation during pregnancy.	*N* = 286 Matched for language group, parity, weight at conception, trimester of registration for care, and year of delivery. *N* = 153 Siblings

Results	Comments

Results	Comments
rom birth to 7 years of age)	Each case said to be paired with a control, but
Somatic growth	closeness of matching not presented, nor are
Physical activity	analyses presented by matched pairs. Unknown
Sleeping time	whether losses from original samples were
Time in crib	matched pairs. No mention of sex matching, or
Time out of house	sex composition of study groups.
Mother–child proximity (in *rebozo*)	Uncertain whether birth weight was matched.
Speed of maternal response	Upper birth weight limit stated for controls
Talking, play	(Chavez and Martinez, 1979b) and implied for
Prenatal protection, attention, speech, praise, emotional content, other stimuli	cases.
Paternal play	Case–control birth-weight differences clouded by
	constrained range of birth weight required for
	selection; reported as both 4% (Chavez and
	Martinez, 1979a) and 8% (Chavez *et al.,* 1974)
	difference. Mean birth weight of controls re-
	ported as both 2.7 and 3.0 kg.
	Cases assessed 2 years after controls: Neither par-
	ents nor assessor blind. Six case–control pairs
	are sibs. Extent of sharing of supplements not
	reported.
arly-supplemented significantly better on Verbal, Performance, and Full Scale Estimated Learning Potential; WISC-R Verbal IQ, Performance IQ, Full Scale IQ, and Digit Span, on Draw-A-Person, and on grade point average in first year of schooling. Also lower score on Behavior Problem Check List, and greater height for age.	Nonblinded cognitive evaluation, except for grade point average.
arly-supplemented had 115 g higher birth weight (N.S.).	11 Firstborn controls, accounting for most of case–control birth-weight difference. Height for age not partialed for birth weight.
	Older sibling had to pass certification screening (i.e., be judged at nutritional risk). Younger sibling was not comparably screened, being *in utero,* with consequent bias to greater severity in older sib.

Measures	Subjects	Controls
irth weight (g)	3005	3119[a]
ength (cm)	49.0	50.0[a]
ead circumference (cm)	33.4	34.1[a]
pgar (5 min)	9.4	9.7[a]
azelton Neonatal Behavioral Assessement		
Response decrement, pinprick	1.4	2.0[a]
Inanimate auditory orientation	5.6	4.9[b]
General tonus	6.1	5.7[b]
Amount activity at peak excitement	6.3	6.7[a]
Rapidly, quiet to agitated state	4.0	5.0[b]
Amount startling	2.3	3.6[b]
Lability of state	2.9	4.1[b]

Subjects had significantly fewer years of school, lower occupational status, and lower initial Hgb, but magnitude of differences not stated. Racial composition of subject group not presented. Results not controlled for these differences.
''The infants in the supplemental group appeared somewhat less active . . . for those dimensions falling within the factor of activity–reactivity: peak of excitement, rapidity of build up, and lability of state.''

^{a} $p < 0.05$. ^{b} $p < 0.01$.

School Performance (in Stanines)^c			
Test	Cases	Controls	Sibs
English-speaking schools			
Reading	5.39 ± 1.87^d (246)	5.88 ± 1.94 (181)	5.72 ± 1.98 (102)
Spelling	4.78 ± 1.77 (239)	5.18 ± 1.98 (176)	5.01 ± 2.07 (102)
Mathematics	4.09 ± 1.52 (244)	4.17 ± 1.62 (181)	4.25 ± 1.74 (102)
French-speaking schools			
English	5.21 ± 2.16 (57)	5.34 ± 2.03 (38)	4.52 ± 2.50 (25)
Mathematics	4.16 ± 1.78 (155)	4.26 ± 1.69 (100)	3.98 ± 1.75 (44)
French	3.88 ± 1.66 (159)	4.08 ± 1.82 (105)	3.82 ± 1.68 (51)

While 83% of those traced were tested, only 39% of the original sample could be traced. Those tested were of somewhat higher birth weight, and their mothers were more likely to have been married, than others.

^{c} ( ) = N; ± = S.D. ^{d} $p < 0.05$.

(Continued)

Authors	Site and study design	Study group	Control group
Rush *et al.* (1980a,b)	New York City Public Prenatal Clinic, 1970–1975. Randomized, partially double-blinded prospective intervention; treatment from registration to term among pregnant black women registered <30 weeks' gestation; <140 lb at conception; plus at least one other risk factor (low weight gain, weight <110 lb at conception, past low birth weight, and/or ingestion of <50 g protein in past 24 hr). No postnatal intervention.	$N = 263$ Supplement (40 g protein, 470 cal/day) $N = 272$ Complement (6 g protein, 322 cal/day). Participants reported taking 74% (326 cal) of Supplement, 20% (65 cal) as replacement to regular diet. For complement, 72% (233 cal) with 11% (26 cal) as replacement.	$N = 272$ Control (routine multivitamin mineral tablets)
Stein *et al.* (1975)	Holland, 1944–1945. Retrospective cohort study of 18-year-old men with intrauterine exposure to acute starvation compared to those born before and after famine, and in areas not exposed.	$N = 36,416$ Conceived or with intrauterine exposure during famine of 10/44–4/45.	$n = 5919$ North control area, same period. $N = 8080$ South control area, same period. All males born in 9 months prior to famine, and 11 months after last famine conception.

Vuori *et al.* (1979), Waber *et al.* (1981)

Bogotá, Columbia. Random allocation of nutritional supplementation (6 mo gestation to 3 years of age) and of intervention to promote mother's cognitive stimulation of child from birth.

		Nutritional supplementation		Cognitive stimulation[h]	
N		6 mo gest. to 6 mo of age	6 mo to 36 mo of age	0–36 mo of age	$N = 54$ Group A No Rx
42 D$_1$		+	+	+	
34 A$_1$		−	−	+	
57 D		+	+	−	
57 C		+	−	−	
60 B		−	+	−	

	Daily Nutritional Supplements for *All* Family Members			
	Age	Cal	Protein (g)	Other
Pregnant and lactating women	—	856	38.4	Vitamin–mineral tablets
Infants weaned	<2 mo	—	—	1 lb/wk dried skim milk
	2–6 mo	—	—	2 lb/wk dried skim milk
All	3–6 mo	—	—	125 g/wk skim milk/veg mixture
	6 mo–1 yr	—	—	1 lb dry milk/wk, 250 g veg mixture/wk, 37.5 mg FeSO$_4$/day.
	>1 yr	523	20	Children: 200,000 units vitamin A every 6 mo

[h] Cognitive stimulation adapted from Infant Education Curriculum, High Score Foundation.

Results	Comments

ignificant overall effect on birth weight, but increase in very early delivery and neonatal death margins of significance in Supplement group. Significant depression of birth weight among pre-tures with Supplement. Increased duration of gestation, and ↓% < 2500 g in Complement •up.

Effects of heavy smoking on maternal weight gain partially reversed, and on birthweight, reversed, by both dietary treatments.

year of age, no residual toxicity, and no treatment differences for somatic growth, Bayley In-t Development scores, Object Permanence (Corman–Escalona), or sophistication of play unt).

Measure	Supplement[e]	Complement[e]	Controls[e]
tion of play isodes (sec)	87.3[f] (144)	59.7 (136)	71.7 (148)
al habituation ope)	−0.84[g] (134)	−0.57[f] (131)	−0.66 (135)
abituation (sec)	4.6[f] (127)	2.2 (123)	1.3 (126)

) = N.
: 0.05, vs. other two groups combined.
< 0.01, vs. other two groups combined.

erinatal
ssociated with any exposure
  ↑ Infant mortality
ssociated with early exposure
  ↑ Perinatal mortality
  ↑ CNS abnormality
  ↑ Later obesity
associated with late-pregnancy exposure
  ↓ Birth weight
  ↓ Later obesity
at age 18
No relationship of famine exposure to any psychological outcome, including Raven's Progressive Matrices and Bennett Mechanical Comprehension, Clerical Aptitude, Language Comprehension, nd Arithmetic Score, or to frequency of mild or severe mental retardation.
irth weight higher, Apgar lower in supplemented boys only; effects N.S. on two-tailed tests, significant on one-tailed tests.
Visual habituation at 15 days of age.

Approximately 400,000 records available, 95% complete (3% exempt from exam; medical reports available on others).
Neither differential fertility nor survival an adequate explanation for results (results for those who conceived before famine, in general, not different from those who conceived during famine).

All publications use one-tailed tests of statistical significance (see text). For prenatal supplementation, all growth differences limited to males, while habituation differences limited to females. Griffiths scores from 4 to 36 mo quite variable; SES not presented, nor is it clear that every subject was tested at every age; 86 cases excluded from analyses; authors state that "comparison of group means at each age indicated that the performance of the 86 excluded subjects did not differ systematically from those included in the longitudinal analysis."

	Supplemented	Unsupplemented	Statistic
tuating (%)	70	56	t = 2.3, p < 0.05
tuation rate			
Girls	0.44 ± 0.42	0.27 ± 0.35	t = 2.3, p < 0.05
Boys	Not given	Not given	N.S.
abituation (%)	38.2	42.0	N.S.
ements during test-	10.2 ± 5.5	11.8 ± 6.9	t = 2.0, p < 0.05
g			
ng and fussing as als continued			
Girls	2.3 ± 3.7	Not given	t = 2.4, p < 0.05
Boys	Not given	Not given	N.S.

riffiths Mental Development Scales, departures from control group mean, at each age group, sexes combined[i].

Group	Age (months)						Mean, all ages
	4	6	12	18	24	36	
D₁	12.3	−2.0	16.0	−1.5	3.3	7.5	5.9
A₁	−2.0	−1.3	11.0	−10.3	−1.3	0.6	0.4
D	01.9	−1.2	11.2	9.5	1.6	9.2	4.7
C	7.9	−0.6	7.9	−2.5	−10.1	4.7	1.1
B	3.7	0.9	12.0	1.0	0.7	3.7	3.6
A (controls)[j]	(100.4)	(105.6)	(92.0)	(87.6)	(98.8)	(88.0)	(95.6)
tal population	3.3	−0.6	9.4	0.1	−1.1	4.4	2.7

culated from authors' data; no S.D.'s given, and no statistical tests. [j] ( ) Mean, control.

been reported to be both 2.7 and 3 kg, and the case–control differences in birth weight to be both 4 and 8%. While infants under 2500 g were excluded, it is unclear whether infants of over 3400 g birth weight were also excluded. Six case–control pairs were sibs, and the extent of sharing of dietary supplementation by the control sibling was not assessed. Although the study was small (17 cases and 17 controls), the only dietary assessments were two 24-hr recalls, one at 8 months' gestation and another for the lactating mother when the child was 6 months of age. No data were gathered on child food consumption, and the amount of supplementation was reported differently in different publications.

With all these caveats, the results reported were striking. Supplemented children were followed from birth to age 7, and were said to have accelerated somatic growth, strikingly increased physical activity, less sleeping time, less time in the crib, more time out of the house, less time in proximity to the mother in the *rebozo*, greater speed of maternal response, more parental talking, play, protection, attention, speech, praise, emotional content, and other stimuli, and more involvement of the father in play. There is little reported cognitive assessment.

3.3.2. Evaluation of WIC Benefits in Louisiana

The study by Hicks *et al.* (1982) must be viewed as an important first step in assessing the effect of the Special Supplemental Food Program for Women, Infants and Children (WIC) on the behavioral and cognitive development of children. In this program, poor pregnant and postpartum women and young children who are also considered at nutritional risk are given vouchers (usually) that can be cashed in to receive about 700–800 cal daily of high-quality food. The program began in 1974 and by the end of the decade was funded nationwide at over $900 million dollars each year. Past evaluation has been meager, and this study stands out for its moderately strong research design. The subjects were children of around 6 years of age from rural Louisiana parishes (counties) whose mothers had received WIC benefits throughout the third trimester of pregnancy and who had themselves received such benefits at least through the first year of life. They were compared to older siblings who had not begun WIC benefits until after the first birthday. Numbers were small: there were only 21 sibling pairs. The early-supplemented children did significantly better than their older sibs on the Verbal, Performance, and Full Scale Estimated Learning Potential Instrument and on the Verbal IQ, Performance IQ, Full Scale IQ, and Digit Span tests of the WISC-R. They also did significantly better on the Draw-a-Person test, and their grade point averages in the first year of schooling were higher. They had significantly fewer behavior problems and greater height for age.

There are several limitations of this study. Not only was it very small, but also the testing psychologist was not blinded to whether the child was a case or a control. (The only cognitive measure that was blinded was the grade point average.) The meaning of the height for age is somewhat clouded because 11 of the older siblings were first births and therefore had lower birth weights (115 g), and the height at the time of assessment was not partialed for birth weight. Also, the older sibling, by definition, had to be considered to be at nutritional risk to be accepted into the program, while the younger sibling was *in utero* at the time of the first benefit. Thus, the older sibling is skimmed from the population, and although only one control had been recruited into the WIC program specifically because of short stature, the control sibs all had to have had characteristics indicative of nutritional risk in order to be certified.

The magnitude of differences that were observed in this study was atypically large, compared to the other reviewed studies, particularly the Montreal study [Pencharz *et al.*, 1983 (see Section 3.3.4)]. Given these caveats, it appears wisest to consider this study one that needs repetition, elaboration, and extension rather than a definitive behavioral and cognitive assessment of the effects of the WIC program.

3.3.3. Trial of High-Protein Supplements during Pregnancy in Philadelphia

Osofsky (1975) executed a controlled trial of high-density protein supplementation (Meritene) during pregnancy in a poor public clinic population. The subjects were recruited after completion of control pregnancies, precluding blinded assessment. The subjects were socially somewhat different from controls, but the magnitudes of these differences were not stated. The racial composition of the subject group was not given. The birth weight, length, head circumference, and Apgar scores of the infants whose mothers received high-protein supplements were all significantly *lower* than those of controls. In addition, the infants in the supplemented group "appeared somewhat less active . . . for those dimensions falling within the factor of activity/reactivity [on the Brazelton Neonatal Behavioral Assessment]." Cases were significantly different from controls on the following items: lower response decrement to pinprick, greater inanimate auditory orientation, greater general tonus, lower amount of activity at peak of excitement, less rapid ascent from quiet to agitated state, less startling, and less lability of state. Given that the author did not control for social differences between cases and controls, and the inability of Brazelton to find any differences associated with supplementation in Guatemala (see section 4), the meaning of these differences is uncertain. There was no later follow-up.

3.3.4. School Performance of Children Whose Mothers Received Dietary Services during Pregnancy in Montreal

Pencharz *et al.* (1983) performed tests of language and mathematical achievement among children whose mothers had received dietary services during pregnancy from the Montreal Diet Dispensary. In addition to siblings, controls were also children whose mothers delivered at the same hospital, and had been matched on several characteristics, for an earlier study of the effect of dietary services on fetal growth (Rush, 1981). While there had been a small but significant effect of nutritional services on birth weight (40 g), no effects on later school performance were detected.

3.3.5. Trial of Nutritional Supplementation during Pregnancy in New York

Rush *et al.* (1980a,b) performed a randomized, partially double-blinded trial of two levels of calorie and protein supplementation during pregnancy in a poor black New York City population. There was evidence of perinatal toxicity from the high-protein Supplement [lowered birth weight (N.S.) and increased very premature delivery and neonatal death, at the margins of statistical significance, and highly significant depression of birth weight among preterm deliveries] and some evidence of benefit from the balanced protein–calorie Complement [increased birth weight (N.S.) and lowered rates of low birth weight and premature delivery ($p < 0.05$)]. Children were assessed at 1 year of age, and there was no evidence of any residual toxicity. Nutritional supplementation was unrelated to Bayley scores, Object Permanence, or sophistication of play. The infants whose mothers had received the high-protein Supplement had significantly longer episodes of duration of play, habituated more rapidly to a visual stimulus, and had greater dishabituation.

3.3.6. Consequences of the Dutch Famine of 1944–1945

Stein *et al.* (1975) studied over 30,000 young men whose mothers were exposed to acute starvation during pregnancy. Controls consisted of other young men born before and after the famine and from the same areas, as well as from areas of the country relatively unaffected by the starvation. While the famine exposure was associated with decreased birth weight, increased infant mortality, and some other somatic effects (including later obesity if exposed in the first trimester of pregnancy and lower rates of obesity if exposed late in pregnancy), there was no relationship of famine exposure to any measured psychological function at time of army induction at age 18, including Raven's Progressive Matrices and Bennett Mechanical Comprehension, Clerical Aptitude, Language Comprehension, and Arithmetic Score. Rates of neither mild nor severe mental retardation were associated with intrauterine famine exposure. Neither differential fertility (i.e., the more well-to-do reproduced more efficiently during the height of the famine) nor differential survival could explain the results observed.

3.3.7. Bogotá Trial of Pre- and Postnatal Nutritional Supplementation and Childhood Cognitive Stimulation

Vuori *et al.* (1979) and Waber *et al.* (1981) reported on the randomized controlled trial of nutritional supplementation and cognitive stimulation among pregnant women and their offspring in Bogotá. Subjects were assigned to combinations of the three treatment conditions: nutritional supplementation from the 6th month of pregnancy to the 6th month of age, from the 6th to the 36th month of age, and intervention to promote increased play and stimulation between the mother and child. Nutritional supplementation was not limited to the propositus, but was given to all family members, to minimize the possibility that food intended for the study individual was diffused among the rest of the family.

The authors have at all times chosen to use one-tailed tests of probability, which is difficult to justify, especially given the results of the Long Island, Philadelphia, and New York trials: it is no longer appropriate to assume that nutritional intervention must be innocuous and might not cause subjects to do less well than controls. There was an approximate 50-g increase in the mean birth weight with intervention, quite typical of trials worldwide (Rush, 1982a). The effect was limited to males. Visual habituation was tested at 15 days of age, and effects were found primarily among girls. Supplemented children habituated significantly more frequently, moved less during testing, and, among girls, habituated faster, and fussed and cried less, as trials continued. The Griffiths Mental Development Scales were applied six times from 4 to 36 months of age. Results reported were unstable over time, with little or no sex differences. We have recalculated these rates from the authors' data, combining the sexes and comparing the results to the mean of the control group. (Since standard deviations were not given, tests of significance cannot be performed.) Any conclusions are impeded by the instability over time. For instance, controls had a mean of 105.6 at 6 months of age, 87.6 at 18 months, 98.8 at 24 months, and 88.0 at 36 months. It is unclear whether this was the same group of children throughout; it probably was, since the authors note that 86 cases were excluded from analysis, presumably because assessment was incomplete, and that these 86 "excluded subjects did not differ systematically from those included in the longitudinal analysis." It appears that intervention to increase maternal cognitive input had little effect, very different from results in Cali and Jamaica (see Sections 5 and 4.2). The groups that received nutritional supplementation throughout, whether with reinforcement to the mother to increase cognitive stimulation (Group D_1) or without (Group D), did best. There was no evidence of such effect from early nutritional intervention (Group C), but possibly some among those supplemented late (Group B). Attentional measures were not reported after the neonatal period.

3.3.8. Summary

The results of these seven studies are hardly consistent, but some generalization appears possible:

1. The increased activity observed by Chavez and colleagues was striking, but probably of greater magnitude than observed in the Cali and Guatamala studies (see Sections 5 and 3.4). Such effects were not observed in New York or Bogotá.

2. Changes in attentional measures were observed in both the Bogotá study (at 15 days of age) and the New York study (at 1 year of age).

3. Prenatal high-protein feeding was associated with malfunction on the Brazelton neonatal exam in Philadelphia [but not in Guatemala (see Section 3.4)].

4. Prenatal feeding was associated with striking improvement in cognitive function and school performance in the Louisiana WIC study, while no effects were observed in Montreal, and hardly any in Bogotá. Whether there were effects in Guatemala is uncertain (see Section 3.4).

5. No behavioral sequelae were observed after intrauterine exposure to the Dutch famine.

3.4. Nutritional Intervention during Pregnancy and Early Childhood with Subjects Partially Self-Selected: Guatemala Study (Table IIB)

The Guatemala study assessed the effects of nutritional supplementation during pregnancy and early childhood on mental function and behavior in over 1000 children (671 born into the study) through the 7th year of life. Twelve publications based on the assessment of these children were reviewed. There are vexing methodological difficulties with this investigation, some of which were noted in reviewing the effects of supplementation on fetal growth (Rush, 1982a). For instance, birth weight was assessed on only 63% of births. In addition, the measure of supplementation during pregnancy is confounded with duration of gestation; i.e., the number of total calories a mother could ingest is systematically related to how long the pregnancy lasted. The relationship of total ingested supplemental calories to birth weight is thus to some extent a function of the systematic correlation between independent and dependent variables, over and above any experimental effect.

The most profound limitation, one that cannot be overcome by analytical elegance, is that in most reports of this study, individuals who chose to supplement themselves, and are therefore self-selected, are compared with others who chose not to be supplemented. Thus, the same limitations to strong inference of comparing the starved and unstarved from the same community also holds when comparing those who have chosen to supplement themselves with those who have chosen not to. The authors have contended with this problem by showing that the social circumstances of the better supplemented were worse than those of the less well supplemented. However, social circumstances probably do not directly affect child development, but rather are surrogates for such behaviors of parents as verbal stimulation, concern, and interchange with the child. Thus, a heightened level of concern, which might have dictated that the mother or child come to the feeding center (in addition to being hungry), might well be correlated with other parental behaviors that would nurture and promote the benefit of the child. Such a possibility can be neither proved nor disproved from available data.

Such a large number of different publications of the results of the same examinations has been daunting to summarize, given the numerous contradictions among them. The investigators concluded early on that the relationship of supplementation to fetal growth was dictated more by the level of calories taken than by the type of supplement. They have therefore, with occasional exceptions, related further outcomes only to the level of ingested calories. However, while in the text Lechtig *et al.* (1975) do not mention differences between those who had received *atole* (a protein-rich gruel) or *fresco* (a protein-free caloric supplement), the *atole–fresco* differences on the 15-month psychological performance, as well as on the measure of physical growth, were considerably stronger than the differences between the highly and less well supplemented (see Table IIB). Irwin *et al.* (1979) did address *atole–fresco* differences, and they were treated as differences across villages by Barrett *et al.* (1982). Given the marked differential effect with the high-protein supplement in the New York trial, there would surely be some value in reassessing these data to determine whether there might have been, on an equicaloric basis, some difference between the high-protein and protein-free forms of supplementation.

"There was no statistically significant relationship between the amount of supplementation of maternal diet and the neonatal performance [on the Brazelton exam]" (Brazelton *et al.*, 1977).

Contradictions between reports on the children as they grew older militate against drawing any firm conclusions. Irwin *et al.* (1979) stated that "boys . . . were affected more strongly and consistently" than girls. On the other hand, while using many of the same analytical procedures, Engle *et al.* (1979) said, "Since the sex differences in test performance were few, data for boys and girls have been combined." R. E. Klein *et al.* (1976) stated that "once gestational supplementation is removed from consideration there is no longer any relevant pattern of association (to cognitive outcome). Moreover, the correlation with prenatal supplementation is not diminished when later supplementation is partialled out." On the other hand, Freeman *et al.* (1980) stated, "It is the amount of supplement consumed by the child that is correlated with cognitive scores." All these statements are contradicted by Barrett *et al.* (1982), who stated, "Performance on tasks which tapped high level cognitive processes was only weakly predicted at best by the nutritional history variables. . . . Only SES was a significant predictor of these variables."

Thus, from various publications, we are told that supplementation did or did not affect cognitive status; was comparable across the sexes or was limited to boys; and was due to prenatal supplementation alone or postnatal supplementation alone. Also, the data, where given, at times suggested a stronger effect of type, rather than amount, of supplementation.

In six of the publications, there was some form of composite cognitive test performance reported, for children aged from 6 months to 8 years. These are summarized in Table IIB. In four of the six, the sexes are presented separately, and in two they are combined. In three of the papers, the amount of variance accounted for was reported; in two, correlation coefficients; and in one, only the statement that the relationships were not significant. For comparability where only correlation coefficients were reported, we squared them to compare with reported variances.

There are several striking inconsistencies in these data. The magnitude of the composite preschool battery is considerably higher as reported by Irwin *et*

Table IIB. Intervention Started during Pregnancy; Subjects Self-Selected

Authors	Site and study design	Study group	Control group
Barrett *et al.* (1982), Brazelton *et al.* (1977), Engle *et al.* (1979), Freeman *et al.* (1977, 1980), Irwin *et al.* (1979), R. E. Klein (1979), R. E. Klein *et al.* (1972, 1974, 1975, 1976), Lechtig *et al.* (1975)	Guatemala Four rural villages *Ad libitum* feeding of high-protein gruel (*Atole*) in two villages and of protein-free liquid caloric supplement (*fresco*) in two other villages.	N = 671 Births; 1639 children tested at one time or another High-supplementation criteria Pregnancy $\geqslant$20,000 kcal Postnatal $\geqslant$10,000 kcal/quarter	Self-selected subset who did no supplement at high levels

Results	Comments

e was no statistically significant relationship between the amount of supplementation of maternal diet and neonate's performance [on the Brazelton exam]" (Brazelton et al., 1977).

logical test performance at 15 months (Lechtig et al., 1975):

	Fresco villages		Atole villages	
	High-supplement	Others	High-supplement	Others
tage with >51% psychological items passed	1.5	7.4	7.1	13.7
	(68)	(162)	(14)	(211)
	5.7%		13.3%	

ated from authors' data
e vs. Fresco, $\chi^2 = 6.97$, $p < 0.01$.
-supplement vs. Others, $\chi^2 = 4.79$, $p < 0.05$ (reported as $p < 0.01$).
ion.
unt of variance explained by caloric intake [(A) total supplementation; (B) prenatal period ($\pm$ lactation) ly; (C) postnatal period only], on test scores, by age and sex.

Composite test performance[a]			Cognitive composite[b]	Cognitive composite[c]		Composite preschool battery[d]	Composite score (sexes combined)[e]			Composite score (sexes combined)[f]	
A	B	C	A	A	B	A	A	B	C	B	C
A. Males											
0.00	0.02[g]	0.00	—	—	—	—	—	—	—	—	—
0.02[g]	0.00[g]	0.01	—	—	—	—	—	—	—	—	—
0.04[h]	0.01[g]	0.01	—	—	—	—	—	—	—	—	—
0.00	0.01[g]	0.01	0.00	0.02[g]	0.02[h]	0.06	0.02[h]	0.01	0.01[i]	—	—
—	—	—	0.01	0.04[h]	0.02[h]	0.05[h]	0.02[h]	0.02[i]	0.01[i]	—	—
—	—	—	—	0.03[h]	0.00	0.06	0.02[h]	—	—	—	—
—	—	—	—	0.03[h]	0.00	0.05	—	—	—	6–8 yr N.S.	N.S.
—	—	—	—	0.01	0.00	0.05	—	—	—	—	—
B. Females											
0.00	0.02[g]	−0.03[h]	—	—	—	—	—	—	—	—	—
0.01	0.05[g]	0.00	—	—	—	—	—	—	—	—	—
0.02[g]	0.01	0.00	—	—	—	—	—	—	—	—	—
0.01	0.00	0.00	0.00	0.02	.02[g]	.05	—	—	—	—	—
—	—	—	0.01	0.02	.02[g]	.01	—	—	—	—	—
—	—	—	—	0.01	.00	.03	—	—	—	—	—
—	—	—	—	0.00	.00	.03	—	—	—	—	—
—	—	—	—	0.02	.01	.07	—	—	—	—	—

ferences: [a] R. E. Klein (1979) (B excludes lactation; squared correlations); [b] Freeman et al. (1977); [c] Freeman al. (1980) (B includes lactation); [d] Irwin et al. (1979); [e] Engle et al. (1979) (B excludes lactation, C to 24 mo; ared correlations); [f] Barrett et al. (1982) (B excludes lactation, C to 4 yr of age).
).05.
).01.
.001.

Comments:

Birth weight assessed only on 63% of sample ($N = 405$).

Total supplementation in pregnancy confounded with duration of gestation (see text).

Data at 15 mo interpreted by Lechtig et al. (1975) to show value of high supplementation, but relationship with atole stronger than with level of supplementation. Although data showed atole–fresco differences to be greater than those associated with level of supplementation, these differences are not mentioned in later reports except by Irwin et al. (1979). "The villages . . . did differ in mean supplement calorie intake" (Barrett et al., 1982); implication of such differences not pursued in analyses. Strong relationship of psychological performance to physical growth, but no analysis of how much and whether effect of nutritional supplementation mediated by effect on physical growth.

To 15 mo, virtually only those born into sample included in presentations. After that, large numbers included with only shorter-term supplementation even though early on R. E. Klein et al. (1974) said that the "feeding program had no important impact on psychological test performance for the period 5–7 years." Data presented as repeated cross-sectional samples, although, at every age, these are mostly the same children.

Results not standardized for social status or compliance measure; high-supplemented said to be of lower social status and to have less adequate home diets (R. E. Klein et al., 1976). "Home nutrition surveys were regularly undertaken. They are not precise enough for use [in analysis of diet and outcome]" (Freeman et al., 1980).

"Once gestational supplementation is removed from consideration there is no longer any relevant pattern of association. Moreover, the correlation with prenatal supplementation is not diminished when later supplementation is

(Continued)

Table IIB. (Continued)

Results	Comments

Social–emotional behavior (Barrett et al., 1982):

p Values for level of supplementation at different time periods associated with social–emotional behavior (only those tests significant $p < 0.05$ across villages included)[a]

	Across villages			*Atole* villages			*Fresco* village		
	A	B	C	A	B	C	A	B	C
Seeks help	<0.01	—	—	<0.01	—	—	—	—	—
Defends self	—	—	<0.05	—	—	—	—	—	<0.05
Involved in group activity	—	<0.05	—	<0.05	—	—	—	—	—
Only watching[b]	<0.05	—	—	<0.05	—	—	<0.05	—	—
Moderate level of activity	—	<0.05	—	—	—	—	—	—	—
Happy, laughs, smiles	—	<0.05	—	—	—	—	—	<0.05	—
Angry, hostile	—	<0.05	—	—	—	—	—	—	—
Anxious[b]	—	<0.05	—	—	—	—	—	—	—

Comments: partialled out" (R. E. Klein *et* 1976). "The data . . . require revising . . . earlier INCAP analysis. . . . It is the amount consumed by the c that is correlated with cognitive scores" (Freeman *et al.*, 1980).

"Boys affected more strongly and consistently" (Irwin *et al.*, 197 "Since sex differences in test performances were few, data for boys and girls have been combined" (Engle *et al.*, 1979)

"Performance on tasks which taps higher level cognitive processes was only weakly predicted at b by the nutritional history variables, . . . only SES was a significant prediction of these variables" (Barrett *et al.*, 1982

[a] 78 Males and 60 females, aged 6 yr 1 mo to 8 yr 3 mo, from three villages, stratified by supplementation (and by height), but unclear if stratification by both variables; 35 social–emotional variables, plus cognitive function, related to prenatal (A), infant (0–2 yr) (B), and young child (2–4 yr) (C) supplementation. Relationship of cognitive function to supplementation said to be weak, but "persistence" significantly related to (C) across and within villages.
[b] Negative correlation with supplementation.

al. (1979) than in any of the other publications, yet only one of ten of these very high proportions of variance was said to be statistically significant. On the other hand, estimates of very low magnitude of Engle *et al.* (1979) were said to be very highly significant. Freeman *et al.* (1980) assert large significant changes, but only among boys, due to postnatal supplementation, but at the last point of testing (7 years of age), differences were no longer significant for either boys or girls, or higher for boys than for girls. Barrett *et al.* (1982) investigated a subgroup of children and found cognitive performance unrelated to any measure of nutritional supplementation. It is confusing, therefore, to attempt to draw conclusions about this experience.

Barrett *et al.* (1982) have taken a new line on the possible consequences of the nutritional supplementation and have assessed 35 indices of social–emotional behavior. They also take into account marked differences across villages, in both geographic and social circumstance. Of the 35 variables, 8 were significantly related (at the $p < 0.05$ level or greater) to pregnancy, postnatal, or combined supplementation. Only 1 of these 8 (a negative relationship with watching only, in a free-play situation) was significantly related to the same period of supplementation in both the *atole* village and the *fresco* village, as well as across villages. All other relationships were significant in either one village or the other, or across villages and, but not significantly, within villages. Inconsistency across villages could be explained by different village characteristics interacting with nutrition, by differences due to the different form of supplementation, by differing power because of variable numbers of subjects or levels of supplementation, or by these being chance findings.

The other social–emotional variables that were associated with one or another of the supplementation variables were, with pregnancy supplementation, the seeking of help by the child; with supplementation between birth and age 2, involvement in group activity, more frequent moderate levels of activity, frequency of being happy, laughing, smiling, angry, and hostile, and lower frequency of anxiety; and with supplementation between age 2 and 4, more frequent defending of self.

The data of Barrett and co-workers demonstrate a problem that has been only minimally addressed: levels of supplementation at various times in the life are highly intercorrelated. The amount the mother took is strongly correlated with amounts ingested by her child after birth. Therefore, it is by no means straightforward to distinguish the time at which supplementation might have had greatest effect even with sophisticated multivariate statistical techniques. An appropriate way to present such data might be to show the joint, as well as the partial, relationships to pre- and postnatal supplementation.

In no paper are data on overall nutritional intake presented, nor are we told what proportion of the child's diet is accounted for by supplementation, nor whether outcome might be more strongly associated with overall nutritional intake than with supplementation. In all reports, data across age are treated as though these were independent cross-sectional samples, when in fact the children are mostly the same for all age groups.

3.5. Nutritional Intervention during the Neonatal Period on Long Island (Table IIC)

Goldman *et al.* (1971, 1974) randomly allocated newborn infants with birth weights under 2000 g to high-protein-density formula (20% of calories as protein) or low-protein-density formula (10% of calories as protein). They found that with the high-protein-density formula, there was increased pyrexia and lethargy and decreased edema and weight gain. Infants receiving high-protein feedings took longer to regain birth weight and to attain discharge weight. The authors reassessed these children at age 3 and again at age 6. There was an interaction between birth weight and protein density on Stanford–Binet scores at age 6. For those infants who weighed under 1300 g, scores were significantly *lower* with high-protein feeding in infancy, and there were significantly more such children with scores under 90. There was no effect of infant formula among children whose birth weights were over 1300 g, and the IQs of infants with birth weights under 1300 g who had received the low-protein-density formula were not different from those of infants with higher birth weights. The authors postulated that much of the relationship of very low birth weight to low intelligence that had been reported in the literature [most prominently by Drillien (1969)] might have been due to neonatal high-protein feeding.

4. Cognitive Intervention among Malnourished Children

4.1. Cognitive Intervention during Hospitalization for Protein–Energy Malnutrition (Table IIIA)
4.1.1. Study in Mexico

Cravioto and Arrieta (1979) reported that children who received cognitive stimulation during hospitalization for protein–energy malnutrition did consid-

Table IIC. Interventions Beginning during Neonatal Period

Authors	Site and study design	Study group	Control group	Results	Comments
Goldman et al. (1971), 1974	Long Island, New York Randomized controlled trial of high-density protein feeding to neonates with birth weight <2000 g.	N = 152 Fed Similac liquid + 2% casein (20% of calories as protein), until weight 2200 g	N = 152 Fed Similac liquid + 2% lactose (10% of calories as protein), until weight 2200 g	Neonatal period: With high-density protein feeding ↑Pyrexia ↑Lethargy ↓Edema ↓Weight gain ↑Time to regain birth weight and to attain 2200 g	Assessors blind to nutritional treatment.

At 6 years of age:

Stanford–Binet Scores

Protein density of neonatal formula

Birth weight (g)	High		Low	
	Mean score	(%)<90	Mean score	(%)<90
<1300	84.5^b	74^b (23)^a	97.6	23 (26)^a
1300–1700	96.0	40 (50)	95.8	32 (41)
1701–2000	99.9	24 (46)	101.1	29 (51)

^{a} ( ) = N. ^{b} p < 0.01, vs. low-protein group.

erably better on the Gesell Infant Development Scale at hospital discharge than those who were not stimulated. The authors did not report numbers tested or much in the way of methodology (neither the method of assignment to stimulation nor the nature of the stimulation was specified). However, differences were marked: 70–90% of stimulated children were functioning at or above age level (the rate depending on the function tested) compared to 9–27% of unstimulated children.

4.1.2. Study in Lebanon

Yatkin and McLaren (1970), Yatkin *et al.* (1971), and McLaren *et al.* (1973) studied 2.5- to 16-month-old children hospitalized with marasmus who were assigned to cognitive stimulation or no stimulation. Controls were said to be matched to cases, but neither differences in the size of the study and control groups nor differences in the sex distribution of groups were explained. The study group, who received "rich environment and warm nurse–child relationship," performed consistently better in each of five subscales of the Griffiths Mental Development Scales at hospital discharge, but there was no residual advantage 1 year later or at age 3–5 years. No posthospital stimulation was attempted.

4.2. Cognitive Intervention during and after Hospitalization for Protein–Energy Malnutrition: Jamaica Study (Table IIIB)

Grantham-MacGregor *et al.* (1979, 1980, 1983) provided 21 children hospitalized with protein–energy malnutrition (PEM) with 1 hr a day of play while hospitalized, and after discharge with 1-hr visits each week to the mother and child to promote continued play at home. They compared these children to 18 controls with PEM. (Controls had been admitted before reintervention group. Assessment could not have been blind.) and to either 14 or 21 children hospitalized with other diagnoses. At six months after discharge the intervention group was significantly ahead of the nonintervention group, and were no longer significantly behind the adequately nourished group. By 1 year after discharge, the cases were essentially normal on the Griffiths Mental Development Scale, at about the same level as the nonstarved controls, while the controls with PEM were markedly retarded: means were 95 for cases, 81 for PEM controls, and 99 for other controls. At 24 months follow-up the results remained essentially unchanged.

4.3. Summary

What is consistent and provocative about these studies is the minimal recovery of malnourished children not stimulated and the near-normal performance of starved children provided cognitive stimulation, in addition to nutritional rehabilitation. The study of Grantham-MacGregor and colleagues is particularly important, since gains were sustained, even under the difficult home conditions to which the children returned.

Table III. Intervention Studies of Cognitive Stimulation in Young Children with Protein–Energy Malnutrition

Authors	Site and study design	Study group	Control group	Results	Comments
				A. Intervention Limited to Period of Hospitalization	
Cravioto and Arrieta (1979)	Mexico Children hospitalized with PEM, <6 mo of age. Some stimulated. Assignment method not given.	*N* = Not given Treatment not specified, but stimulation added to nutritional rehabilitation.	*N* = Not given Nutritional rehabilitation only.	Gesell Infant Development Scale, at hospital discharge (% functioning at or above age level) Group / Motor / Adaptive / Language / Social–personal Stimulated: 70.0 / 90.0 / 70.0 / 70.0 Nonstimulated: 27.2 / 27.2 / 27.2 / 9.0	Neither numbers of subjects nor assignment method reported. Nature of stimulus not specified. No continuation of intervention after discharge, nor posthospital assignment.
Yatkin and McLaren (1970), Yatkin et al. (1971), McLaren et al. (1973)	Lebanon Children hospitalized with marasmus, 2.5–16 mo of age; assignment to stimulation or control group, "matched" for age and sex.	*N* = 17 (11 male) Age: 32.9 weeks IQ: 51 Wt.: 53.0% (of Boston standard) Treatment: "rich environment and warm nurse–child relationship" plus nutritional rehabilitation.	*N* = 13 (8 male) Age: 30.0 weeks IQ: 46 Wt.: 52.9% Treatment: nutritional rehabilitation	Stimulated group performed consistently better on each of five subscales of Griffiths Mental Development Scales during hospitalization (minimum 4 mo). Both groups improved considerably, however.	No residual advantage of stimulation at follow-up 1 year after discharge or at age 3–5 years. (Stimulation limited to period of hospitalization.) Unequal-sized groups not explained, given "matching."
				B. Intervention Continued after Hospital Discharge	
Grantham-McGregor et al. (1979, 1980, 1983)	Jamaica Intervention among hospitalized children (hospitalized with PEM), 6–24 mo of age	*N* = 21 1 hr/day of play in hospital, 1-hr visit each week at home to promote play with mother.	*N* = 18 Hospitalized, with PEM *N* = 14 or 21 Hospitalized, other diagnoses	Griffiths Mental Development Scales (± = S.D.) Time / PEM Cases / Controls / Other controls At admission: 64 ± 20 / 61 ± 20 / 87 ± 17 After discharge 1 month: 93 ± 13 / 83 ± 12 / 109 ± 11 6 months: 96 ± 11 / 82 ± 12 / 106 ± 11 12 months: 95 ± 10 / 81 ± 10 / 99 ± 9	Cases recruited after controls; assessment could not have been blind.

These results are congruent with the studies described above that demonstrated that the starvation of a child is not a random event, but surely occurs in the most deprived, disrupted, and unstimulating families, those least likely to nurture their children in the manifold ways essential to normal growth and development, and not only in the supply of food. These intervention studies are encouraging; they suggest that not overly complex interventions may possibly protect and help the previously starved child to achieve near-normal cognitive function. The same, therefore, might hold for children with chronic, subclinical malnutrition.

5. Cognitive Stimulation and Nutritional Supplementation among Poor Stunted Preschool Children: Cali Study (Table IV)

The key studies of postnatal intervention were those carried out by the McKays and colleagues in Cali, Colombia (McKay *et al.*, 1974, 1978; McKay and McKay, 1981).

In 1973, this group reported on the studies preliminary to their later complex 36-month field trial (McKay *et al.*, 1973). The preliminary studies were intriguing, but the report included only limited data. The investigators intervened for either 4 or 5 months among pre-school-age slum children, both stunted and of normal stature. Neither the numbers of children studied nor their ages were reported. Among stunted children, some were given food and health care alone; some, food and health care, plus physical stimulation; another group, food and health care with cognitive stimulation; and finally, some were left as controls. Also, some children of normal stature were given cognitive stimulation and food and health care, and some, no intervention. Another control group was composed of siblings of program children. Children who were stimulated, whether of short or normal stature, had marked improvement on all cognitive measures, including Knox cubes, sentence completion, and reading, as well as in their sense of security. The only effect among the short children who received only nutrition and health care was increased activity.

It would be helpful to have a more complete report of these studies, including data on the increase in nutritional intake (if known) and other details of study design, execution, and outcome.

The same investigators then carried out a complex intervention study, with random allocation of a group of stunted 4-year-old, slum-dwelling children to anywhere from one to four 9-month periods of cognitive stimulation, nutrition, and health care. Those with longest treatment began at the time of recruitment; those with shortest duration of treatment began later, so that all children ended at the same age. Those receiving 9 months of cognitive intervention (the shortest period) all received at least 9 months of simultaneous nutrition and health care, but were also randomly assigned to prior additional nutrition and health care of anywhere from 0 to 27 months. Children receiving shorter periods of intervention were treated as controls for those with longer intervention. In addition, 72 slum children of normal stature were followed without intervention,

Table IV. Intervention Studies in Pre-School-Age Children

Authors	Site and study design	Study group	Control group	Results	Comments
McKay *et al.* (1973)	Cali, Colombia Experimental study. Random allocation of preschool slum children, both stunted and of normal stature, to: no treatment, nutrition and health care alone, or nutrition and health care plus cognitive or physical stimulation. Treatment lasted 4 or 5 months.	N = Not reported Stunted children Food and health care; physical stimulation with food and health care; or cognitive stimulation with food and health care. Children of normal stature (same neighborhood) Cognitive stimulation with food and health care.	N = Not reported Stunted children Siblings of program children, or unrelated and Children of normal stature (same neighborhood)	Among malnourished children, no effects of nutrition and health care alone on Knox cubes, sentence completion, reading, security. Only effect of nutrition alone was increased activity. With added cognitive or physical stimulation, significant improvement on all measures. Other than for Knox cubes, stunted children with cognitive stimulation passed normal-sized controls on all measures.	Age, criteria for inclusion in stunted group, level of nutritional supplementation and intake, size of study groups not reported. No report on somatic growth with supplementation.

Authors	Site and study design	Study group				Control group	Results	Comments
McKay *et al.* (1974, 1978), Sinisterra *et al.* (1979), McKay and McKay (1981)	Cali, Colombia (2/71–8/74) Experimental study. Random allocation (by area) of stunted 4-year-old slum children to from one to four 9-mo periods of nutrition and health care (NH) and cognitive stimulation (CS). Group with 9 mo of CS randomly assigned to 9, 18, 27, or 36 mo of NH.	**Months**				72 (T_0) not stunted, from same neighborhood 30 (HS) high SES	A. Preschool: "Nutritional supplementation and health care provided [at home, before CS] had not been found to produce any differences [in psychological development]." Profound, and additive, effects of periods of CS on general ability. Gap between HS and T_4 lowered to 58% of HS–T_0 gap. B. 10.5 Years of age: Gap between HS and T_4 increased to 74% of HS–T_0 gap on WISC (R) primarily because of decreased information, vocabulary, and digit span sections by poor children. With omission of those three scales, gap is 64%, and appears stable. No data presented on nutrition and health care alone.	No data on level of nutritional intake collected for those supplemented at home. No data on somatic growth as index of compliance, or on activity, as index of effect of home or school feeding.
		N	CS	NH	Group			
		154	18	to	36 (T_2–T_4)			
		16	9		36 (T_{1a})			
		17	9		27 (T_{1b})			
		16	9		18 (T_{1c})			
		50	9		9 (T_1)			
		During CS, children fed soya–rice gruel (19% calories as protein). "Children fed at home received prepared World Food Program packages with no indigenous food." Program provided 100% protein, 80% caloric needs.						

as were 30 children of high social class. While receiving cognitive stimulation, the children were provided with enough soya–rice gruel to supply 100% of the daily protein and 80% of the daily caloric needs. Children receiving nutrition and health care prior to cognitive stimulation received prepared World Food Program packages with no indigenous foods. Cognitive stimulation had profound cognitive effects, which increased with longer duration of stimulation. When children were reassessed at age 10.5 years, some of the gains, particularly (and not surprisingly) in Information and Vocabulary (but also in Digit Span), were lost, but other gains were maintained, and continued to be a function of the duration of intervention.

The authors reported that nutrition and health care alone had no effect, but presented no quantitative results. Dietary recalls had not been taken, and therefore no estimates of the food intakes of children supplemented at home were available. We cannot conclude, however, that nutrition and health care alone had no effect, without some evidence that the distributed food was actually eaten by the intended child. This could be fairly simply assessed from measures of physical growth. If the growth of the children receiving nutrition without stimulation was accelerated, interpretation of these results would be much clarified: nutritional intervention of the preschool child, even in a stunted Third World population, would be of little cognitive benefit. If there were no acceleration of physical growth among children receiving their food at home, we would have to judge that the issue had not yet been adequately tested.

6. Conclusion

Given the importance, interest, activity, and amount of study on the issue of nutrition and behavior, the conclusions from these studies are sadly sparse. The biases in the study of long-term consequences of acute and chronic malnutrition in childhood are all toward ascribing greater severity to the nutritional insult alone than is probably warranted, since, other than in special circumstances, malnutrition is preceded and accompanied by other dysfunctional characteristics of the family and the environment. In their review, Pollitt and Thomson (1977) concluded, "Except for those cases of severe and chronic undernutrition with an onset during the prenatal period and early postnatal life, protein–calorie deficiency does not arrest [psychological] development." In all likelihood, even the severe deficits associated with chronic protein–calorie malnutrition in earliest life (marasmus) cannot be ascribed solely to the nutritional insult, nor can we know with any confidence that simply supplying food to children who would otherwise have been starved would protect them from long-term deficit.

We must recognize the possibility that where no long-term consequences of early nutritional insult have been observed, those behaviors or mental functions that were assessed may not have been responsive to changes in nutrition, while others, as yet unmeasured, may have been affected. The behavioral study of Barrett *et al.* (1982) suggests that we must be open to the possibility that we have been measuring the wrong variables.

There is one moderately consistent theme in the studies reviewed: that nutritional change seems to be most related to activity and attentional processes. Activity is probably a function of current nutrition; changes in attentional processes, such as those observed in the New York and Bogotá studies, may persist. The investigations in Guatemala and Cali require further data presentation and analysis before judging the effects of nutrition on cognition and behavior. We have almost too much data from Guatemala, and not enough from Cali. Different reports of the Guatemala study present contradictory findings. Did cognitive change take place? If so, for what age children? Did it persist? Did it take place in boys only, or in both boys and girls? Was age of supplementation (whether in prenatal life, postnatal life, or both) associated with change? Some simple further analyses from the Cali group could help us in deciding whether nutritional supplementation was not associated with benefit because of noncompliance with the nutritional regimen; the children may not have received the supplements. If they did have accelerated linear growth or weight, we could feel confident that they did get supplemental food and that the food had little impact on cognition or behavior.

The very small study of the WIC program (Hicks *et al.*, 1982) is tantalizing. Our society is now investing nearly a billion dollars a year in this supplemental feeding program for poor pregnant mothers and infants. Evaluation of this program has been inconsistent in both design and outcome (Rush, 1982b), even on surrogate issues such as birth weight, and we know hardly anything of the effect of this very large program on perinatal and child mortality or on the mental and behavioral development of children. The effects of nutrition on cognition observed in this study are out of line with other intervention studies, particularly the Montreal study (Pencharz *et al.*, 1983) too much seems to have happened to these children to be purely a function of prenatal and early postnatal feeding, especially in comparison to the benefits that have been demonstrated in presumably much poorer populations in the developing world. This work clearly needs repetition, extension, and elaboration. Subsequent research designs could aim for blind assessment and larger numbers of subjects.

Millions, and possibly hundreds of millions, of people have been assumed to be irreversibly crippled by nutritional deprivation. The reality is clearly far more complex, and what damage has been done is surely of multiple etiology, with concurrent social, intellectual, and emotional deprivation probably far more important than nutritional deprivation alone. Also, given nurturing and cognitive stimulation, deficits seem to be reversible. These results are hopeful, since the human mind appears to be resilient and capable of recovering from devastating insult. Recovery and the achievement of maximal potential, however, imply help. If behavior and intellect are plastic, we are morally obligated to try to reverse deprivation for all children, at all ages. To falsely accept that early damage is irreversible is to become complacent and to avoid the obligation to relieve suffering and promote optimal development. The long-term consequences of overt clinical malnutrition can be minimal; despite past neglect, we can in fact help reverse much of the damage that has been done to children.

Whatever the contribution of nutrition to human behavior and intellect, it is almost certainly very small compared to other, more subtle, forms of needed nurture. While no human can live decently or effectively when starved or chronically hungry, we must reject the myth of easy solutions, i.e., food alone. Surely it is unjust that anyone should starve, but we find, yet again, that we do not live by bread alone.

7. Summary of Results

7.1. General Activity Level

Reported to be increased, across a range of ages, neonatal period through early school life, in Mexican village study (Chavez) and in Cali, Colombia (McKay *et al.*, 1973), and possibly in INCAP (Guatemala) study (Barrett *et al.*, 1982). Probably an effect of current, rather than past, supplementation.

7.2. Measures of Attention and Alertness

Increased by gestational supplementation, in neonates [Bogotá (Vuori *et al.*, 1979)] and in 1-year-olds (Rush *et al.*, 1980a,b), but also either no effect in neonates [Guatemala (Brazelton *et al.*, 1977)] or depressed [by high-protein maternal supplementation (Osofsky, 1975)]. Possible effects among 6- to 8-year-olds [Guatemala (Barrett *et al.*, 1982)].

7.3. Cognition

7.3.1. With Cognitive Stimulation

7.3.1a. Among Nutritionally Deprived Children. Dramatic short-term effects during hospitalization [Lebanon (Yatkin and McLaren, 1970), Mexico (Cravioto and Arrieta, 1979)], sustained for at least 1 year postdischarge [Jamaica (Grantham-McGregor *et al.*, 1979, 1980)].

7.3.1b. Among Chronically Malnourished Children. No effect, to 3 years of age [efforts to promote cognitive stimulation by maternal education in Bogotá study (Waber *et al.*, 1981)], but dramatic effects of a comprehensive program among preschoolers in Cali, sustained in large part to age 10 (McKay *et al.*, 1978; McKay and McKay, 1981).

7.3.2. Nutritional Intervention

Dramatic effects of prenatal and infant intervention by WIC program (Hicks *et al.*, 1982); strong effect, some effect, or no effect in INCAP (Guatemala) study, depending on report cited: no effect at age 1 (Rush *et al.*, 1980a,b), at school age (Pencharz *et al.*, 1983), or in young adulthood (Stein *et al.*, 1975) of nutrition during pregnancy. No consistent effect at age 3 of pre-

or postnatal supplementation in Bogotá study (Waber *et al.*, 1981). No effect of preschool supplementation in Cali, but compliance not assessed (McKay *et al.*, 1978). No regular or consistent effects from neonatal starvation secondary to pyloric stenosis or cystic fibrosis.

ACKNOWLEDGMENTS. The author's research reported herein was supported in part by Grants 5 RO1 HD13347 and 5 RO1 HD13370, NICHHD, and Contract 53-3198-987, U.S. Department of Agriculture.

8. References

Barrett, D. E., Radke-Yarrow, M., and Klein, R. E., 1982, Chronic malnutrition and child behavior: Effects of early calorie supplementation on social–emotional functioning at school age, *Dev. Psychol.* **18:**541–556.

Berglund, G., and Rabo, E., 1973, A long-term follow-up investigation of patients with hypertrophic pyloric stenosis—with special reference to the physical and mental development, *Acta Paediatr. Scand.* **62:**125–129.

Birch, H. G., 1972, Summary of Session V: Future research directions, in: *Nutrition, The Nervous System, and Behavior*, Proceedings of the Seminar on Malnutrition in Early Life and Subsequent Mental Development, Mono, Jamaica, January 10–14, 1972, pp. 133–135, Pan American Health Organization Scientific Publication No. 251.

Brazelton, T. B., Tronick, E., Lechtig, A., Lasky, R. E., and Klein, R. E., 1977, The behavior of nutritionally deprived Guatemalan infants, *Dev. Med. Child Neurol.* **19:**364–372.

Brozek, J., Coursin, D. B., and Read, M. S., 1977, Longitudinal studies on the effects of malnutrition, nutritional supplementation, and behavioral stimulation, *Bull. Pan Am. Health Org.* **11:**237–249.

Chavez, A., and Martinez, C., 1975, Nutrition and development of children from poor rural areas. V. Nutrition and behavior development, *Nutr. Rep. Int.* **11:**477–489.

Chavez, A., and Martinez, C., 1979a, Behavioral effects of undernutrition and food supplementation, in: *Behavioral Effects of Energy and Protein Deficits* (J. Brozek, ed.), pp. 216–228, U.S. Department of Health, Education, and Welfare, NIH Publication No. 79-1906.

Chavez, A., and Martinez, C., 1979b, Consequences of insufficient nutrition on child character and behavior, in: *Malnutrition, Environment, and Behavior: New Perspectives* (D. A. Levitsky, ed.), pp. 238–255, Cornell University Press, Ithaca, New York.

Chavez, A., Martinez, C., and Yaschine, T., 1974, The importance of nutrition and stimuli on child mental and social development, in: *Early Malnutrition and Mental Development: Symposia of the Swedish Nutrition Foundation XII* (J. Cravioto, L. Hambraeus, and B. Vahlquist, eds.), pp. 211–225, Almquist and Wiksell, Uppsala, Sweden.

Chavez, A., Martinez, C., and Yaschine, T., 1975, Nutrition, behavioral development, and mother–child interaction in young rural children, *Fed. Proc. Fed. Am. Soc. Exp. Biol.* **34:**1574–1582.

Cravioto, J., and Arrieta, R., 1979, Stimulation and mental development of malnourished infants, *Lancet* **2:**899.

Cravioto, J., and DeLicardie, E. R., 1975, Mother–infant relationship prior to the development of clinically severe malnutrition in the child, in: *Western Hemisphere Nutrition Congress IV* (P. L. White and N. Selvey, eds.), pp. 126–137, Publishing Sciences Group, Acton, Massachusetts.

Douglas, J. W. B., 1956, Mental ability and school achievement of premature children at 8 years of age. *Br. Med. J.* **1:**1210–1214.

Douglas, J. W. B., 1960, "Premature" children at primary schools, *Br. Med. J.* **1:**1008–1012.

Drillien, C. M., 1969, School disposal and performance for children of different birthweight born 1953–1960, *Arch. Dis. Child.* **44:**562–570.

Ellis, C. E., and Hill, D. E., 1975, Growth, intelligence, and school performance in children with cystic fibrosis who have had an episode of malnutrition during infancy, *J. Pediatr.* **87**:565–568.

Engle, P. L., Irwin, M., Klein, R. E., Yarbrough, C., and Townsend, J. W., 1979, Nutrition and mental development in children, in: *Human Nutrition: A Comprehensive Treatise*, Vol. 1, *Nutrition: Pre- and Postnatal Development* (M. Winick, ed.), pp. 291–306, Plenum Press, New York.

Freeman, H. E., Klein, R. E., Kagan, J., and Yarbrough, C., 1977, Relations between nutrition and cognition in rural Guatemala, *Am. J. Public Health* **67**:233–239.

Freeman, H. E., Klein, R. E., Townsend, J. W., and Lechtig, A., 1980, Nutrition and cognitive development among rural Guatemalan children, *Am. J. Public Health* **70**:1277–1285.

Goldman, H. I., Liebman, O. B., and Freudenthal, R., 1971, Effects of early dietary protein intake on low-birth-weight infants: Evaluation at 3 years of age, *J. Pediatr.* **78**:126.

Goldman, H. I., Goldman, J. S., Kaufman, I., and Liebman, O. B., 1974, Late effects of early dietary protein intake on low-birth-weight infants, *J. Pediatr.* **85**:764–769.

Grantham-McGregor, S., Stewart, M., Powell, C., and Schofield, W. N., 1979, Effect of stimulation on mental development of malnourished child, *Lancet* **2**:200–201.

Grantham-McGregor, S., Stewart, M. E., and Schofield, W. N., 1980, Effect of long-term psychological stimulation on mental development of severely malnourished children, *Lancet* **2**:785–789.

Grantham-McGregor, S., Schofield, W. N., and Harris, L., 1983, Effect of psychosocial stimulation on mental development of severely malnourished children: An interim report, *Pediatrics* **72**:239–243.

Graves, P. L., 1976, Nutrition, infant behavior, and maternal characteristics: A pilot study in West Bengal, India, *Am. J. Clin. Nutr.* **29**:305–319.

Hicks, L. E., Langham, R. A., and Takenaka, J., 1982, Cognitive and health measures following early nutritional supplementation: A sibling study, *Am. J. Public Health* **72**:1110–1118.

Irwin, M., Klein, R. E., Townsend, J. W., Owens, W., Engle, P. L., Lechtig, A., Martorell, R., Yarbrough, C., Lasky, R. E., and Delgado, H. L., 1979, The effects of food supplementation on cognitive development and behavior among rural Guatemalan children, in: *Behavioral Effects of Energy and Protein Deficits* (J. Brozek, ed.), pp. 239–254, U.S. Department of Health, Education, and Welfare, NIH Publication No. 79-1906.

Klein, P. S., Forbes, G. B., and Nader, P. R., 1975, Effects of starvation in infancy (pyloric stenosis) on subsequent learning abilities, *J. Pediatr.* **87**:8–15.

Klein, R. E., 1979, Malnutrition and human behavior: A backward glance at an ongoing longitudinal study, in: *Malnutrition, Environment, and Behavior: New Perspectives* (D. A. Levitsky, ed.), pp. 219–237, Cornell University Press, Ithaca, New York.

Klein, R. E., Freeman, H. E., Kagan, J., Yarbrough, C., and Habicht, J. P., 1972, Is big smart? The relation of growth to cognition, *J. Health Soc. Behav.* **13**:219–225.

Klein, R. E., Yarbrough, C., Lasky, R. E., and Habicht, J. P., 1974, Correlations of mild to moderate protein–calorie malnutrition among rural Guatemalan infants and preschool children, in: *Early Malnutrition and Mental Development: Symposia of the Swedish Nutrition Foundation XII* (J. Cravioto, L. Hambraeus, and B. Vahlquist, eds.), pp. 168–181, Almquist and Wiksell, Uppsala, Sweden.

Klein, R. E., Lester, B. M., Yarbrough, C., and Habicht, J. P., 1975, On malnutrition and mental development: Some preliminary findings, in: *Proceedings of the 9th International Congress of Nutrition, Mexico*, Vol. 2, pp. 315–321, S. Karger, Basel.

Klein, R. E., Arenales, P., Delgado, H., Engle, P. L., Guzman, G., Irwin, M., Lasky, R., Lechtig, A., Martorell, R., Mejia Pivaral, V., Russell, P., and Yarbrough, C., 1976, Effects of maternal nutrition on fetal growth and infant development, *Pan Am. Health Org. Bull.* **10**:301–316.

Lechtig, A., Delgado, H., Lasky, R., Yarbrough, C., Martorell, R., Habicht, J. P., and Klein, R. E., 1975, Effect of improved nutrition during pregnancy and lactation on developmental retardation and infant mortality, in: *Western Hemisphere Nutrition Congress IV* (P. L. White and N. Selvey, eds.), pp. 117–125, Publishing Sciences Group, Acton, Massachusetts.

Levine, S., and Weiner, S., 1976, A critical analysis of data on malnutrition and behavioral deficits, Adv. Pediatr. **22:**113–136.

Lloyd-Still, J. D., Hurwitz, I., Wolff, P. H., and Scwachman, H., 1974, Intellectual development after severe malnutrition in infancy, *Pediatrics* **54:**306–311.

McKay, H., and McKay, A., 1981, The long-term effects of preschool nutritional health and educational attention, Prepared for the XVIII Interamerican Congress of Psychology, Santo Domingo, Dominican Republic, June 1981.

McKay, H., McKay, A., and Sinisterra, L., 1973, Behavioral intervention studies with malnourished children: A review of experiences, in: *Nutrition, Development and Social Behavior: Proceedings of the Conference on the Assessment of Tests of Behavior from Studies of Nutrition in the Western Hemisphere* (D. J. Kallen, ed.), pp. 121–154, DHEW Publication No. (NIH) 73-242.

McKay, H., McKay, A., and Sinisterra, L., 1974, Intellectual development of malnourished preschool children in programs of stimulation and nutritional supplementation, in: *Early Malnutrition and Mental Development: Symposia of the Swedish Nutrition Foundation XII* (J. Cravioto, L. Hambraeus, and B. Vahlquist, eds.), pp. 226–233, Almquist and Wiksell, Uppsala, Sweden.

McKay, H., Sinisterra, L., McKay, A., Gomez, H., and Lloreda, P., 1978, Improving cognitive ability of chronically deprived children, *Science* **200:**270–278.

McKeown, T., 1970, Prenatal and early postnatal influences on measured intelligence, *Br. Med. J.* **3:**63–67.

McLaren, D. S., Yaktin, U. S., Kanawati, A. A., Sabbagh, S., and Kadi, Z., 1973, The subsequent mental and physical development of rehabilitated marasmic infants, *J. Ment. Defic. Res.* **17:**273–281.

Miller, B. D., 1981, *The Endangered Sex*, Cornell University Press, Ithaca, New York.

Osofsky, H. J., 1975, Relationships between prenatal medical and nutritional measures, pregnancy outcome, and early infant development in an urban poverty setting. I. The role of nutritional intake, *Am. J. Obstet. Gynecol.* **123:**682–690.

Pencharz, P., Heller, A., Higgins, A., Strawbridge, J., Rush, D., and Pless, I. B., 1983, Effects of nutritional services to pregnant mothers on the school performance of treated and untreated children, *Nutr. Res.* **3(6):**795–804.

Pollitt, E., and Thomson, C., 1977, Protein–calorie malnutrition and behavior: A view from psychology, in: *Nutrition and the Brain*, Vol. 2 (R. J. Wurtman and J. J. Wurtman, eds.), pp. 261–306, Raven Press, New York.

Record, R. G., McKeown, T., and Edwards, J. H., 1969, The relation of measured intelligence to birth weight and duration of gestation, *Ann. Hum. Genet.* **33:**71–79.

Rush, D., 1981, Nutritional services during pregnancy and birthweight: A retrospective matched pair analysis, *Can. Med. Assoc. J.* **125:**567–576.

Rush, D., 1982a, Effects of changes in protein and calorie intake during pregnancy on the growth of the human fetus, in: *Effectiveness and Satisfaction in Antenatal Care: Clinics in Developmental Medicine Series* (M. Enkin and I. Chalmers, Eds.), pp. 92–113, Spastics International Medical Publications, London.

Rush, D., 1982b, Is WIC worthwhile? Editorial, *Am. J. Public Health* **72:**1101–1103.

Rush, D., Stein, Z., and Susser, M., 1980a, *Diet in Pregnancy: A Randomized Controlled Trial of Nutritional Supplements*, March of Dimes Birth Defects Foundation, Vol. XVI, No. 3, Alan R. Liss, New York, 200 + xvii pp.

Rush, D., Stein, Z., and Susser, M., 1980b, A randomized controlled trial of prenatal nutritional supplementation in New York City, *Pediatrics* **65:**683–697.

Sinisterra, L., McKay, H., Gomez, H., and Korgi, J., 1979, Response of malnourished preschool children to multidisciplinary intervention, in: *Behavioral Effects of Energy and Protein Deficits* (J. Brozek, ed.), pp. 229–238, U.S. Department of Health, Education and Welfare, NIH Publication No. 79-1906.

Stein, Z., Susser, M., Saenger, G., and Marolla, F., 1975, *Famine and Human Development: The Dutch Hunger Winter of 1944–1945*, Oxford University Press, New York.

Valman, H. B., 1974, Intelligence after malnutrition caused by neonatal resection of ileum, *Lancet* **1**:425–427.

Vuori, L., Christiansen, N., Clement, J., Mora, J. O., Wagner, M., and Herrera, M. G., 1979, Nutritional supplementation and the outcome of pregnancy. II. Visual habituation at 15 days, *Am. J. Clin. Nutr.* **32**:463–469.

Waber, D. P., Vuori-Christiansen, L., Ortiz, N., Clement, J. R., Christiansen, N. E., Mora, J. O., Reed, R. B., and Herrera, M. G., 1981, Nutritional supplementation, maternal education, and cognitive development of infants at risk of malnutrition, *Am. J. Clin. Nutr.* **34**:807–813.

Yaktin, U. S., and McLaren, D. S., 1970, The behavioural development of infants recovering from severe malnutrition, *J. Ment. Defic. Res.* **14**:25–32.

Yaktin, U. S., McLaren, D. S., Kanawati, A. A., and Sabbagh, S., 1971, Effect of undernutrition in early life on subsequent behavioural development, in: Proc. XIII Internat. Congr. Pediat. Vienna, Vienna Med. Acad., II, **15**:71.

Nutritional Therapy in Children

C. Keith Conners

1. Introduction

Nutrition and feeding behavior have long been of interest to students of child behavior and development. It is now well recognized that early parental transactions with the child during feeding mediate important aspects of bonding and subsequent personality development. The infant's response repertoire and reaction to feeding in turn have profound effects on maternal behavior. The very nature of this continuous transaction has made it difficult to isolate food-related variables from other variables in the developmental process (Ricciuti, 1981; Chase and Martin, 1970; Hepner and Maiden, 1971; Read, 1973; Pollitt *et al.*, 1975; Graves, 1976, 1978).

Extremes of undernutrition and feeding patterns have been known for some time to be antecedents of psychopathology. Inadequate maternal stimulation in the presence of adequate nutrition has been identified as a major cause of the failure-to-thrive syndrome (Spitz, 1945, 1946; Goldfarb, 1955; Bullard *et al.*, 1967). This syndrome of severe weight loss, growth stunting, and food refusal can apparently result from *overconcern* by the parents regarding feeding, producing what Levy (1954) referred to as "the battle of the spoon" (Egan *et al.*, 1980). Nevertheless, these examples of extremes in early feeding patterns may not be relevant to the issue of the *optimal* patterns within a relatively normal context. Moreover, even severe nutritional deprivation and resultant malnutrition have led to considerable dispute regarding effects on the central nervous system. In their monumental review, Dodge *et al.* (1975) emphasize the importance of the interaction of nutrition with rearing practices and other environmental variables, only cautiously acknowledging that undernutrition has permanent effects on brain growth and development during vulnerable periods in development. Some of the known or suspected interactions are depicted in Fig. 1.

C. Keith Conners • Children's Hospital National Medical Center, Washington, D.C. 20010.

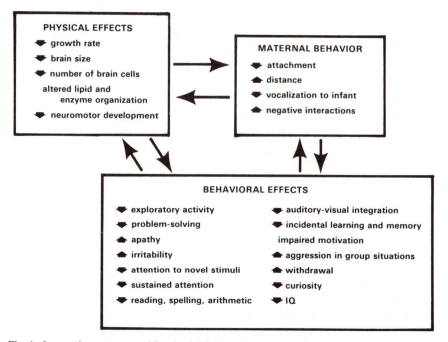

Fig. 1. Interaction among nutritional, physical, and psychological variables. From Read (1973).

2. *Breakfast, Behavior, and Cognitive Function in Children*

Given the uncertainties of causal effects in these extreme situations, it seems likely that milder degrees of undernutrition should pose considerable problems in analysis and interpretation. What of the nutritionally normal child who misses a meal or occasionally eats an inadequate meal? Are there any consequences and, if so, under what conditions? The issue is important because it has been estimated that as many as one third of children in developed countries fail to eat the recommended 25% of daily calories at breakfast (Arvedson *et al.*, 1969).

One might argue on evolutionary grounds that brain function must be well insulated from the vagaries of particular meal contents or timing, yet recent evidence shows that brain neurotransmitters can, under certain circumstances, be affected within minutes by the nutrient composition of a meal (Fernstrom and Wurtman, 1971). Moreover, it seems unlikely that evolution will have prepared children for the sustained demands on cognitive function that modern school systems require. Recent studies of regional cerebral glucose metabolism show wide variations in energy utilization during conscious brain activity, demonstrating a clear and close relationship between the local levels of functional activity and energy metabolism (Sokoloff, 1977). The brain itself has virtually no reserves of glucose, its main energy source, and must constantly replenish it through transport across the blood–brain barrier. Sustained mental work requires large turnover of brain glucose and its metabolic components. Perhaps, then, when placed under the heavy energy demands of mastering a complex technological curriculum, a child of normal intelligence is *not* biologically adapted to wide fluctuations of his internal nutritional milieu. The stresses associated with mental work may be significant, adding to the central load of demand on energy reserves. The range of stresses capable of leading to nutritional pathology is more far-reaching than commonly appreciated (National Dairy Council, 1980).

A number of investigators have studied the effect of breakfast or morning supplements on school behavior and learning (Table I). In one of the earliest studies, Laird *et al.* (1931) concluded that "nervousness" was reduced in children by milk given at 9:30 A.M. in school, and even further by milk supplemented with calcium, phosphorus, maltose, lactose, and vitamin D. Fifty-three "nervous" children in grades 1, 3, and 5 who were physically healthy were selected by teachers. On certain days during the 2 weeks of the study, the children went to the nurse's office, where they received either of the two supplemental feedings or played with toys as a control. The teachers, who were unaware of which children had received the supplement for a given day, rated the children's behavior daily using a checklist of "nervous behaviors." Some of the results are shown in Fig. 2. Although the report has some quaint features (such as using rating items like "Repulsive bearing and physique" or "Has average boy qualities of masculineness"), it is in substantial agreement with later, more sophisticated studies. The background of diet and nutrition is obviously important in the interpretation of any dietary-intervention study. This early study may have been dealing with children who were chronically undernourished. One of the reasons given for the 9:30 A.M. feeding was that "some of the children came to school with only a hasty morsel or two of food as their breakfast . . ." (Laird *et al.*, 1931, p. 495). No data are given regarding the children's physical condition or how much or how often the "morsel or two" was available.

In a study of 4000 schoolchildren, Lininger (1933) found that scholastic progress occurred more frequently in those who received a milk supplement (59%) than in those who did not (24%). However, those who received the supplement also made more gains in *physical* growth, and socioeconomic factors appear to be confounded with the treatment. A much more careful large study, controlled for social class, evaluated the effect of a school lunch and found minimal effects on scholastic progress (Tisdall *et al.*, 1951). The sensitivity of measures in these studies, however, is of necessity quite limited.

In a classic series of studies in the 1950s, Tuttle *et al.* (1954) showed that omission of the morning meal "may result in the lowering of the mental and physical efficiency . . . during the late morning hours" (p. 674). Seven boys were placed on a controlled diet, with all meals taken at a university hospital cafeteria. A basic cereal-and-milk breakfast was served for 4 weeks, and then no-breakfast periods of 2 weeks alternated with breakfast periods of 3 weeks. Thus, the design was a within-subject, own-control A-B-A-B-A design. Although measures of neuromuscular tremor, choice reaction time, and grip-strength endurance were unaffected by missing breakfast, *maximum* work rate

Table I. Summary of Studies on the Effects of Breakfast in Schoolchildren

Author	Sample	Design	Measures	Results
Laird et al. (1931)	48 From grades 1, 3, and 5	Three groups with no special feeding, one with milk, one with milk + calcium over 2 weeks.	Teacher ratings	16% Mean reduction of "nervousness," more careful thinking.
Keister (1950)	133 Nursery-school children	Subjects studied twice when receiving fruit juice, twice with water.	Hyperactivity, withdrawal, hostile behavior, nervous habits measured by 30-sec observation	Less negative behaviors with pineapple juice; no age differences, but males show more reduction of negative behaviors.
Matheson (1970)	100 Fifth-graders from three schools	Each subject observed on day with orange juice supplement, or control, over 10-day period.	Arithmetic and decoding task	Significant improvement on both tasks, not related to normal breakfast habits; most effect at 11:45 and 10:30 feedings.
Dwyer et al. (1973)	139 First-grade males	Half of subjects got liquid supplement in morning, half in afternoon.	Slow tapping test, digit test, block test	No effects on any tasks; no relation to habits of sporadic or regular breakfast.
Tuttle et al. (1954)	12–14-Year-old males	Alternation between breakfast or no breakfast for 17 weeks with total daily nutrient intake kept constant.	Neuromuscular tremor choice reaction time, grip strength, grip endurance, bicycle ergometer	No effects except on work output on ergometer; no difference in type of meals.
Arvedson et al. (1969)	203 Children ages 7–17	40 Subjects divided into two groups, with two high-carbohydrate and two high-protein breakfasts.	Bicycle ergometer	Only one third ate breakfast containing 25% of daily calories; no effects of type of breakfast.

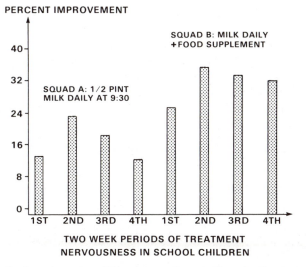

PERCENT IMPROVEMENT

SQUAD B: MILK DAILY
+FOOD SUPPLEMENT

SQUAD A: 1/2 PINT
MILK DAILY AT 9:30

1ST 2ND 3RD 4TH 1ST 2ND 3RD 4TH

TWO WEEK PERIODS OF TREATMENT
NERVOUSNESS IN SCHOOL CHILDREN

Fig. 2. Morning food supplements and "nervousness" as rated by schoolteachers. Children served as own controls in successive 2-week treatment periods. The 2nd week for squad A actually provided the supplement as well as the milk. Data adapted from Laird *et al*. (1931).

and work output were reduced. Teachers made observations and records of attitudes and scholastic progress. The authors comment that "the majority of the boys had a definitely better attitude and a better scholastic record during the period when breakfast was included in the daily dietary regimen than when it was omitted" (p. 677). Apparently, however, no precautions were taken to keep the teachers unaware of the experimental conditions, and the failure to present their observations suggests that the authors' conclusion was based more on impression than on hard data.

Keister (1950) studied the effects of a morning supplement of pineapple juice on the behavior of 76 male and 57 female nursery school children. The study was carried out over four quarters of the school year to control for seasonal effects on behavior. Each quarter, two of the classes (one older, one younger) served as experimental subjects, and two other classes served as controls, receiving only water at the same time as the experimentals received their juice. Groups were crossed over to the other condition in successive quarters of the year. Thus, each class served twice as experimentals and twice as controls. Negative behaviors were observed in the classroom by trained observers using a time-sampling method. Each child was observed at 30-sec intervals for hyperactivity, withdrawing, hostile behavior, and nervous habits. All negative behaviors except withdrawal were significantly less frequent in the experimental condition. Though small, the effects are reliable because of the small error variance of the within-subject design. Unfortunately, this otherwise excellent study also appears not to have kept the observers blind to the treatment condition.

In a doctoral dissertation, Matheson (1970) assessed the effect of a mid-morning juice supplement for 100 fifth graders from three different schools. Measures of arithmetic and letter-symbol decoding tests were collected over a 10-day period. The children served as their own controls. Tests taken after the 10:30 feeding were the most improved for the decoding tasks. Children who generally ate good or poor breakfasts did not differ in their response to the tests, though breakfast data for the day of testing were not obtained. Good and poor breakfast-eaters were determined retrospectively from diet diaries taken *several weeks after* the experiment.

Dwyer *et al.* (1973) found that an instant breakfast (liquid meal) had no effect on tasks of "attention" (slow tapping test, digit test, block test, eye gaze toward work). Effects were the same for "regular" and "sporadic" breakfast-eaters, but again control over actual breakfast intake was not obtained. One also must question the sensitivity of the attentional measures.

Studies by Arvedson *et al.* (1969) on Swedish schoolchildren failed to find effects of high-protein vs. high-carbohydrate breakfasts on either physical or mental work, the latter consisting of continuous mental arithmetic in consecutive 3-min periods. Curiously, postprandial blood glucose was higher on the protein-rich than on the carbohydrate-rich diet; glucose levels were not related to subjective ratings of tiredness and hunger. Part of the explanation for lack of effects may lie in the subject sample: the subjects were 11- to 18-year-old boys attending the same boarding school who had a rigorous athletic program during the 2 weeks of study. Improvements from the program may have masked any effects of the experimental variables.

A critical review of the literature on school feeding programs in the United States (Pollitt *et al.*, 1978a) finds that studies of the long-term effects of feeding programs on scholastic and behavioral functioning are methodologically weak and inconclusive. On the other hand, the few short-term studies appear to suggest "that the provision of breakfast may both benefit the student emotionally and enhance his/her capacity to work on school-type tasks" (Pollitt *et al.*, 1978a, p. 481).

The studies reviewed here suggest that future research must control what children eat prior to testing. They also suggest that the effects of eating or missing breakfast may change over the course of the morning, being more crucial for behavior and learning at some times than at others. All the studies have been restricted to purely behavioral measures of outcome and have contributed relatively little to knowledge of the specific effects of breakfast on brain information processing. In our view, it would be profitable to consider foods and blood glucose as though they were drugs, the dose–time–action characteristics of which need to be investigated with the same sophistication applied to the study of purely pharmacological agents. In one such study, it was observed that eliminating breakfast markedly alters cardiac deceleration during a warned reaction-time task (Conners *et al.*, 1982), indicating that phasic orienting response is affected. In this same study, there were marked effects of missing breakfast on latencies and amplitudes of visual cortical evoked re-

Table II. Classification of Intentional Additives[a]

Additive	Number
Preservatives	33
Antioxidants	28
Sequestrants	45
Surface active agents	111
Stabilizers, thickeners	39
Bleaching and maturing agents	24
Buffers, acids, alkalies	60
Food colors	34
Nonnutritive and special dietary sweetners	4
Nutritive supplements	117
Flavorings	
Synthetic	1610
Natural	502
Miscellaneous: yeast foods, texturizers, firming agents, anticaking agents, enzymes	157
Total number of additives	2764

[a] Compiled from data gathered by the National Science Foundation in 1965.

sponses and concurrent effects on a computerized arithmetic task (Conners and Blouin, unpublished).

It has been traditional in our society to stress the value of "three square meals a day." Further studies will determine whether this folk wisdom is soundly based in human psychophysiology or is just another example of an arbitrary cultural pattern sanctified by tradition. If studies confirm the impact of missing breakfast on the microprocesses of attention and learning, not only will generations of motherly advice be vindicated, but also a simple and powerful dietary therapy will be further mandated for vulnerable children. Perhaps those children whose nervous systems are delicately balanced in meeting school demands or who are already performing at full capacity are those most subject to disruption by dietary factors. The absence of breakfast may not be critical for a child with good compensatory reserves, but critical for a child with borderline central nervous system function. Similarly, the effects of *excessive* amounts of certain nutrients (such as carbohydrates) may matter only when the margin of safety is narrowed by any of the multiple factors known to affect neurodevelopmental sequences and capacity.

3. Food Toxicity

3.1. Intentional Food Additives

One of the most important and controversial theories of behavior disorders in children was proposed by Feingold (1973). He called attention to the fact that over 2000 "intentional additives" are present in modern food supplies (Table II) and that these additives have a causal role in producing behavior

Table III. Substances Eliminated in the Feingold Diet

Part I.	Artificial colors and flavors contained in food
	Butylated hydroxytoluene
	(BHT)
	Butylated hydroxyanisole
	(BHA)
Part II.	Selected foods with natural salicylates

Almonds	Peaches
Apples (cider and cider vinegar)	Plums or prunes
	Tangerines
Apricots	Cucumbers and pickles
All berries	Green peppers
Cherries	Tomatoes
Currants	Cloves
Grapes or raisins (wine and wine vinegar)	Coffee
	All teas
Nectarines	Oil of wintergreen
Oranges	

disorders in children. On the basis of a presumed cross-reactivity of yellow food dye (tartrazine) with acetylsalicylic acid (aspirin), he also implicated so-called natural salicylates. He proposed an elimination diet excluding artificial colors, flavors, and salicylates as a treatment for hyperkinetic and learning-disabled children (Table III). Clinical observations appeared to confirm that the elimination diet was "dramatically effective" within 24–48 hr in reducing the many behavioral and learning disturbances. The appeal of this theory to the lay public was of great magnitude, resulting in tens of thousands of members in "Feingold societies."

Several controlled studies comparing the elimination diet with the control diet, and comparing challenges with artificial colors to placebo, were carried out (Table IV). Conners et al. (1976) reported the first controlled diet trial. They compared the Feingold diet with a control diet matched for nutrient content, compliance, and difficulty in shopping. Carefully diagnosed hyperactive children were observed by teachers and parents for 1 week of baseline and then randomly assigned to one of two diet orders, followed by the other diet after a 4-week observation period. Clinical examinations were made without knowledge of the diet condition. The results of this trial showed that teachers rated the children as significantly improved while on the experimental diet compared to the control phase. Parents also reported improvement; however, the effects were not significant. Clinical global judgments showed a significant improvement in favor of the Feingold diet period. Most of the improvement was associated with the changes occurring in the group that received the control diet first followed by the experimental diet. This somewhat limits the results of the study and suggests explanations other than response to the diet as the source of the changes. Moreover, one could not be certain that the parents were not biased in favor of one diet and communicated this bias to teachers.

Table IV. *Summary of Clinical Trials on Feingold's Hypothesis*

Author	Subjects	Description	Results
Spring *et al.* (1976)	6 Hyperactive boys	Subjects on Feingold diet were challenged twice for 3-day periods with 13 mg color-containing cookies or placebo cookies. Ratings with APQ and global.	Of 6 subjects, 1 became worse on the active challenge, but results could not be replicated.
Stine (1976)	2 Hyperactive preschool boys	KP diet instituted as a last-resort clinical treatment.	Significant improvement in alleviation of hyperkinetic symptoms for both children, especially motoric overactivity and extreme impulsiveness.
Conners *et al.* (1976)	9 Males, 4 females, variable criteria	Double-blind challenge–placebo crossover with 1-week periods and close parent monitoring after cookie ingestion.	Significant effect on parent rating.
Levy *et al.* (1977)	19 Male and 3 female hyperactives	Subjects tested before and after 4 weeks on KP diet, after yellow dye (tartarzine), placebo, and 4-week base line in double-blind crossover.	Diet and base-line effects significant on ratings, but not on objective tests. Challenge effect nonsignificant, but 13 children with 25% reduction of symptoms on the diet phase showed a significant challenge effect on parent ratings.
Harley *et al.* (1978)	36 Hyperactive school-aged boys, 10 preschoolers	Counterbalanced crossover comparing KP and control diet.	KP diet superior only in parent ratings; order effect. Significant parent rating for preschoolers. No objective laboratory effects.
J. I. Williams *et al.* (1978)	24 males, 2 females; 7 probably not hyperactive	Random assignment to 24 possible orders of drug–placebo, active–inert cookie challenge combinations.	Significant drug and diet effects for teacher ratings, borderline effects for parent ratings.

(Continued)

Table IV. (Continued)

Author	Subjects	Description	Results
Goyette *et al.* (1978)	15 Males, 1 female; hyperactives with some not meeting strict criteria	Double-blind challenge–placebo crossover study over four 2-week periods.	No teacher or parent rating effects. Hint of effects on visual–motor tracking task; later confirmed with multiple challenge–placebo replication.
Harley, Matthews, and Eichman (1978b)	9 Males who were best responders from earlier study	Double-blind crossover challenge with active or placebo cookies.	No effects on any measures.
Weiss *et al.* (1980)	22 1- to 7-Year-olds reporting improvement on Feingold diet	Eight challenges with 35.6 mg color, double-blind, during 77-day period; target symptoms, phone ratings, direct observation, behavior counts obtained from mothers.	Consistent challenge results in one child only.
Conners *et al.* (1980)	9 of best responders from previous trials	Paired-associate learning, activity level, and behavior ratings during sessions hourly after eating cookies with active or inert ingredients.	Some suggestion of dose–time effect on first day of testing, but nonsignificant differences.
Levy and Hobbes (1978)	7 Males, 1 female; mean age of 5 years 2 months	Challenge or placebo with cookies containing 1 mg tartrazine over 14 days (4 mg/day).	13% Increase in symptoms on the active challenge, but this was not statistically significant for this sample size.
Rose (1978)	2 8-Year-old hyperactive females who had been on Feingold diet at least 11 months	Double-blind BAB challenge with behavioral ratings in school over 6-week period.	Artificial food colors increased the duration and frequency of hyperactive behaviors. Absence of both a placebo effect and a differential sensitivity of the dependent variables to the challenge effects.

Rapp (1978)	24 Hyperactive children	Challenge of sublingual foods and dyes followed by 7-day diet omitting milk, wheat, eggs, cocoa, corn, sugar, and food coloring, followed by individual challenges of the same previously excluded food items.	12 Children improved to a marked degree during 7-day diet. Improvement persisted in these children at 12-week follow-up.
Conners *et al.* (1980)	22 Males, 8 females; 7.6 years mean age	Double-blind, double crossover in 1-week periods with challenge by active or inert cookies.	No parent or teacher rating effects. No relationship of diet or challenge response to food or color sensitivity as measured by cytotoxic test.
Swanson and Kinsbourne (1980)	20 Hyperactive children, 20 nonhyperactive children	All children on diet free of artificial food dyes and other additives for 5 days. Oral challenges of 100 mg FDC-approved food dyes or placebo administered on days 4 and 5.	Performance of hyperkinetic children on paired-associate learning test on day they received dye blend impaired relative to placebo. Nonhyperactive group showed no effect from food dye challenge.
Mattes and Gittelman (1981)	11 Children already maintained on the Feingold diet.	Double-blind crossover challenge in 1-week periods with active or placebo cookies.	Evaluations of parents, teachers, psychiatrists, and psychological testing yielded no evidence of food-coloring effect.
Adams (1981)	18 School-age responders from Feingold diet	Double-blind crossover design using 14 objective measured in related areas of activity level, auditory, visual and on-task attention, fine motor and gross motor skills to assess differences between diet infraction and noninfraction conditions.	No significant differences between diet infraction and noninfraction condition.
Rogers and Hughes (1981)	9 Male, 1 female hyperactives with some not meeting strict criteria	Double-blind, random assignment using actometers to measure activity rates pre- and posttreatment for 9 weeks of KP vs. reduced-sugar diet.	Mean activity rate on the KP diet significantly lower than on the reduced-sugar regimen.

A study with the same design and many of the same measures was carried out by Harley *et al.* (1978a) at the University of Wisconsin. They took extra precautions to ensure that the parents could not distinguish the experimental and control foods. They provided all foodstuffs to the families in unlabeled containers and made sure that the children followed the diet at parties and at school by donating foods to the classrooms. They even had some foods that appeared to be on the additives list but were not, and in a variety of other ways they ensured that the families were unaware of when the children were eating the additive-containing foods and when they were not. They also added direct classroom observations and a variety of laboratory measures of behavior and cognitive performances. Their results, like ours, indicated that the most improvement occurred in the group that received the control diet first. The improvement was most apparent in ratings by parents and less apparent in the direct and laboratory measures. In general, they felt that the Feingold hypothesis was not supported. However, one important and unexplained finding was that among the younger children of preschool age, 10 of the 10 mothers and 4 of the 7 fathers of children of preschool age rated the children as improved while on the experimental diet. Since such careful precautions were taken to ensure the blindness of the treatment conditions, this finding clearly suggests an effect of the additives on behavior among the younger children. If the findings were due simply to the breaking of the double-blind, it is hard to understand why effects were not found in the older sample, at least for parent report.

Aware of the shortcomings of comparison with a control diet, various investigators carried out a series of studies using a challenge approach to testing. With this approach, it is possible to preselect children who show improvement in behavior in an open trial on the Feingold diet and keep them on the diet throughout the study, while challenging them intermittently with the suspected offending agents. We completed three consecutive studies (Conners, 1980). In each, a similar design was employed. First, diagnosed children were placed nonblind for 1 month on the Feingold diet, with careful monitoring of compliance and with continued surveillance by a nutritionist. Parent and teacher reports were collected during the base-line period and subsequent challenge periods. Those children who improved by at least 25% over a 1-week period of base-line observation prior to the open diet trial were selected for further study. The children were randomly assigned to placebo or active challenge conditions and then crossed over to the opposite condition. These studies thus assume that if the children improve on the nonblind diet phase, they should continue to remain improved during the placebo challenge and worsen during challenge with artificial colors. The colors were consumed in the form of a specially prepared chocolate cookie containing 13 mg of artificial colors, blended in the proportions thought to be consumed in the population. Neither parents, teachers, children, nor investigators were aware of the days on which the placebo or challenge cookies were given.

In the first study, there were 16 children between 4.7 and 11.8 years old with a mean of 8.3 years. These 16 children were selected from a pool of 58 children who had completed 1 week of base-line observation followed by 3 weeks of observation on the diet. We selected for further study those who showed a 25% reduction in symptoms during the diet phase. Of 27 children who started the challenge phase, 11 did not finish. Interestingly, many of these were children who were allergic to the chocolate in the cookies, but the adverse reactions occurred equally often in the placebo and active phases of the trial. Nine children were assigned to a sequence of active-placebo-active-placebo, and seven started with a placebo-active-placebo-active sequence. Each phase was 1 week in duration. During the base-line week, the mean parent rating was 19 (15 being two standard deviations above normal), and at the end of the nonblind diet period of 3 weeks, the average score dropped to 7.9. But there was no difference for either parent or teacher ratings between the active and placebo phases of treatment. There was, however, a decline in a visual–motor tracking task that appeared to follow an acute time course, with performance worse during the 1st hour after cookie ingestion and gradually returning to base line over a 4-hr period, suggesting an acutely acting, rapidly disappearing effect that may have been missed by parents or teachers who made observations only for an entire day and only on Monday, Wednesday, and Friday of each week.

A further study was conducted to take account of these findings. This study lasted 7 weeks. The first 2 weeks were a base-line period during which observations were collected on Mondays, Wednesdays, and Fridays. The children were then placed on the Feingold diet for 3 weeks, after which those with a 25% reduction in symptoms were randomly assigned to receive either the placebo followed by the active cookies or vice versa. Each challenge condition was 1 week long, and data were collected 5 days a week. Parents were instructed to make their ratings during a 3-hr period immediately following ingestion of the cookies at the evening meal. Of the 37 children who began the study, 19 terminated for various reasons, leaving 18 who completed the diet phase and began the challenge phase. Of these, 5 failed to complete the trial, leaving 13 subjects who completed all phases of the study. The results showed a significant effect: parents rated their children's behavior as significantly worse during the week they were receiving the active cookies compared to the week of placebo cookies. The order of treatments had no effect in this study. Thus, for this study, there was an unequivocal worsening of the children's behavior during the period when they received the additive-loaded cookies.

A third study sought to replicate the effect, again with a double-blind crossover, this time with a double crossover of 1-week trials as in the first challenge study. The children were a somewhat more heterogeneous group, since less strict entrance criteria were employed; the mean age was also higher. Of the 92 children volunteering for the study, 54 began the diet, and 30 completed both the diet and challenge phases of the study. This study included a 3-week base-line period, a 3-week period on the diet, and a 4-week period broken into alternating 1-week challenge and placebo phases. This trial failed to replicate the previous study and showed no consistent effects due to the food additives. A virtually identical study was carried out by the Harley group again, and they also failed to find any effects, even though their sample consisted of the 9 best

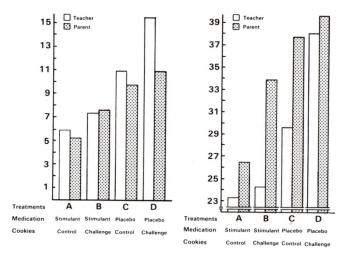

Fig. 3. Effects of combining drug treatment with food-additive challenge. From J. I. Williams *et al.* (1978).

responders from previous trials. Only 1 of their 9 patients seemed to show any effect from the food-additives effect (Harley *et al.*, 1978b).

In summary, the results of the pseudo-diet controlled studies as well as double-blind challenge studies have given quite inconsistent results. Nevertheless, it is difficult to dismiss the Feingold hypothesis entirely, though neither the numbers of children nor the size of the effects is as dramatic as originally claimed, even in those studies with positive results.

One of the claims made by Feingold was that improvement in behavior on the elimination diet was so striking that children no longer needed medication. A study to test the relative merits of methylphenidate and the elimination diet was carried out by J. I. Williams *et al.* (1978). In this study, 26 children were randomly assigned to treatment with active or placebo medications in combination with challenge cookies or control cookies without additives. Each child received all four of the possible combinations, with one child randomly assigned to each of the 24 possible orders of treatment. Double-blind assessments by teachers and parents were made with teacher and parent checklists. Results are shown in Fig. 3. Of most interest is that both parents and teachers reported the most symptoms during the active challenge while the children were on the medication placebo and the fewest when they were on the active medication and control cookie. This is the order one would predict if both the medication and the diet were affecting behavior. Clearly, the stimulant drug has a much more potent symptom-reducing effect, but its effect is optimized when the child is additive-free. The results clearly do not support the contention that the children can forego their medication, but they also argue for a combination of the two treatments.

Several reviews of these studies have now appeared (Conners, 1980; Lipton and Wheless, 1981; National Advisory Committee, 1980; Mattes and Gittelman, 1981). Findings have been somewhat inconclusive, although clearly unsupportive of the claims for *dramatic* changes in behavior as a result of an additive-free diet. Interpretations have been clouded by the strong ideological fervor of opposing camps representing either food industry or "naturalist" biases.

Methodological problems in studying this issue have been substantial. Initial claims of a global nature suggested that many, or even most, hyperactive children were affected by artificial colors and flavors. As negative results of controlled studies accumulated, the issue of detecting specifically vulnerable subjects became more prominent. However, careful studies of children thought to be clearly responsive on clinical grounds have produced both negative (Mattes and Gittelman, 1981; Adams, 1981) and positive (Rogers and Hughes, 1981) findings. In one study, 11 children whose parents were members of the local Feingold society and who were firmly convinced of the diet's efficacy participated in a controlled challenge trial. Children in the experimental group consumed a total of 78 mg of additives a day (Mattes and Gittelman, 1981), yet absolutely no effects of additive-containing cookies were observed. Many of the earlier studies have been criticized for accepting food industry standards for the acceptable daily intake of food additives (Feingold, 1981), but results of studies with larger dosages have been both negative (Mattes and Gittelman, 1981; Weiss *et al.*, 1980) and positive (Swanson and Kinsbourne, 1980).

Age may be a factor in sensitivity to the artificial food dyes. One of the differences we have noted between a wholly negative and a clearly positive double-blind challenge study was the younger age of the sample in our positive study (Conners, 1980). Harley *et al.* (1978a) performed a careful experimental vs. control diet study that eliminated peeking through the double-blind by extraordinary measures to disguise the foods used. Whereas the results for the older sample were entirely negative, all the mothers and most of the fathers of the preschool part of the sample rated the children as more improved while on Feingold's diet than on the control diet. The authors conclude that "while we feel confident that the cause–effect relationship asserted by Feingold is seriously overstated with respect to school-age children, we are not in a position to refute his claims regarding possible causative effect played by artificial food colors on pre-school children" (p. 827). Weiss *et al.* (1980) used a within-subject repeated measures approach to study 22 young children in a challenge study carried out over 77 days, with challenges on 8 of the days. Only the youngest child in the study, a female 34 months of age, showed consistent reactivity to the challenge at a statistically acceptable level.

3.2. Food Allergies

Food allergy is relatively common in the population at large, having an incidence of 2–3% of the adult population and of 10–15% of all children, with another 30–40% of children showing minor manifestations (Humphrey and White, 1970; Prausnitz and Kustner, 1921; Speer, 1975). The most common allergens include animal and plant proteins, inhalants, and drugs. Foods most commonly found to be allergenic are cow's milk, chicken eggs, and cereals. A large but uncontrolled literature suggests that hypersensitive individuals will develop symptoms to these foods, these symptoms including nasal stuffiness, hives, eczema, rashes, migraine headaches, nausea, diarrhea, asthma, and anaphylactic shock in response to certain foods. Psychological disturbances and symptoms of minimal brain dysfunction have also been described (Clarke, 1950; Rowe, 1931; Baldwin *et al.*, 1968), and others have claimed that learning disabilities are often caused by allergies (H. G. Rapaport and Flint, 1976).

Trites *et al.* (1980) carried out a study of three groups of children: 90 hyperactives, 22 children with specific learning disabilities who were not hyperactive, and 8 children who were "restless on an emotional basis." All the children were examined for food allergies using the radioallergosorbent test (RAST) to determine whether specific reaginic (IgE) antibodies were present. Sera from the children were tested for specific IgE antibodies against 43 food extracts. RAST scores for each food extract ranging from 0 (no response) to 4 (very strong response) were based on a serially diluted reference serum. A total allergy score was calculated for each child by adding the scores for the individual allergens. The incidence of allergy in the learning-disabled group was 77%, which was significantly higher than in the hyperactives or emotionally disturbed group (47 and 38%, respectively). In addition, there was a significant positive correlation between the hyperactivity score on the teacher questionnaire and the number of allergies. When those hyperactive children whose neuropsychological profiles indicated brain dysfunction and learning disabilities were examined separately, Trites and co-workers found quite strong correlations with the allergy score. On tests such as the Tactual Performance Test, the Grooved Pegboard Test, the Trail-Making Test, and the Right-Left Orientation Test, all of which are generally presumed to be sensitive to brain dysfunction, performance was impaired. Performance was generally poorer in children who had a high number of allergies.

This same study examined the effect of diets eliminating the allergic foods. Although no significant differences were found among placebo, elimination, and treatment groups, definite changes were found in individual children. The authors indicated that the group results mask the improvements found in a small number of the hyperactive children. Overall, this study lends support to a long-held clinical belief that food allergies play a role in the etiology of learning and behavior disorders for a certain number of children.

The cytotoxic test is a controversial *in vitro* diagnostic test used to determine allergies to foods and chemicals. The test is based on the observation that living leukocytes undergo destruction as a result of the food antigen–antibody reaction. Black (1956) developed this test during the 1950s. It was subsequently refined and popularized by Bryan and Bryan (1960, 1972). Although the reliability of this test has been questioned, revisions of the laboratory method have increased its reliability and sensitivity. We recently reported an overall test–retest reliability of 0.85 (Conners, 1980). Ninety-five adults and children were tested at least twice, and scoring of the test was done double-blind. We compared the cytotoxic test findings with the reaction of the children in our third study ($N = 30$) to the challenge with artificial colors. There was no greater tendency for those children with abnormal cytotoxic tests to colors to show behavioral deterioration when given the food-color challenge. Those few children who did appear to show clear deterioration in behavior when challenged were not the ones with abnormal cytotoxic findings. Thus, it is unlikely that their worsening in response to food colorings is based on an allergic mechanism.

O'Shea and Porter (1981) performed double-blind intradermal and sublingual challenges with food allergens to provoke symptoms in hyperkinetic children. The children improved in subsequent dietary exclusion trials.

Similar work with sublingual provocative testing using artificial food colors in children has been reported by Rapp (1978). While of considerable interest, these clinical trials have numerous methodological flaws that render them inconclusive. Nevertheless, there continues to be strong presumptive evidence that dietary management of food allergies may have some therapeutic value in children, especially in those with a long history of allergic predisposition. Adequate controlled trials remain to be carried out.

3.3. Unintentional Food-Additive Toxicity

3.3.1. Lead and Heavy Metals

It is now well known that small elevations of lead in the blood pose a significant hazard for the cognitive and behavioral capacities of children. Generally, estimates of 20–40 $\mu g/100$ ml blood have been used as the critical threshold for identifying toxic effects. Less well known, however, is the fact that much lower levels of blood lead have been implicated in cognitive and learning impairments of children (Thatcher *et al.*, 1980).

In other important studies, Pihl (1979) and Pihl and Parkes (1977) have shown that hair lead levels are elevated among learning-disabled (LD) children. Their finding that 98% of the LD children could be differentiated from matched controls on the basis of hair analysis of trace minerals and heavy metals is of obvious importance in screening programs for behavioral and learning disturbances. The Centers for Disease Control have previously established a level of 30 $\mu g/100$ ml serum lead as the lowest level at which significant impairment could be expected. In light of the findings that behavioral effects appear at doses as low as 4–5 μg, this position needs to be reevaluated.

Other heavy metals such as cadmium have also been implicated in LD. Indeed, in one study, a combination of five metals other than lead accounted for most of the impairment effects (Pihl, 1979).

Hyperactive children are also alleged to have elevated levels of lead. David *et al.* (1972) found that hyperactive children without other clear bases for their symptoms (such as perinatal insult) showed increased body lead burden. In a treatment study of lead chelation in hyperactive children, David *et al.* (1976) found a behavioral improvement following chelation. Although these studies have been criticized for lack of double-blind and placebo controls, there appears to be strong presumptive evidence that lead and hyperactivity are linked for a subgroup of children. As Rutter (1980) has pointed out, unequivocal establishment of causal direction has not been possible from the correlational studies available. Most studies attempt to control the important social class effects, but not always successfully. We have encountered families in clinical practice in which a steady diet of wall plaster is a long-standing subcultural tradition.

3.3.2. Diet and Heavy Metals

Much of the literature on the role of lead in behavior and cognition of children has ignored dietary variables, despite clear evidence from animal studies that lead toxicity is affected by them (Mahaffey, 1980). In particular, dietary deficiencies of calcium and iron enhance the absorption and toxicity of lead. Dietary supplementation with iron and ascorbic acid prevents the usual lead-induced growth retardation, reduction of food intake, and anemia associated with increased tissue lead burden in rats. Iron supplements appear to reduce accumulation of lead in rat pups, and a combination of iron and calcium reduces lead toxicity. Since children most at risk for lead exposure are also those most at risk for dietary deficiencies, lead exposure and dietary deficiencies may be especially problematic for those children from low-income and poor urban housing environments. Little has been done in the way of systematic dietary prophylaxis for heavy-metal toxicity, but the animal evidence, together with the fact that iron deficiency anemia is extremely common among children in this country and worldwide (Gill and Schwartz, 1972; Haughton, 1963), strongly suggests that the effects of programs of preventive therapy would be significant.

3.4. Sugar and Behavior

Per capita consumption of refined sugar (sucrose) has shown a dramatic increase in the American diet in this century: since 1910, consumption has increased almost 150%, while total carbohydrate consumption has declined by 25%. This means that decline in consumption of starches has been more than offset by increases in sucrose consumption. The peak of sugar consumption occurs for males at about 10 years of age (Cantor, 1975). Thus, dietary sucrose represents a significant new factor in the diets of American children and adults.

While the effects of dietary sugars on dental caries, cardiovascular disease, and other health problems are well known (Cantor, 1975), their role in behavioral disturbances is more controversial and less understood. A wide array of behavioral, autonomic, and neurological symptoms have been attributed to *reactive hypoglycemia*. Symptoms are alleged to include nervousness, irritability, fatigue, weakness, cold sweats, depression, vertigo, headaches, gastrointestinal complaints, insomnia, trembling, tachycardia, allergies, blurred vision, suicidal intent, and others (Phillips, 1959). The diversity of symptoms has prompted some to regard the condition as an "epidemic non-disease" (Yager and Young, 1974; Cahill and Soeldner, 1974). It is certainly true that many studies have failed to find a relationship between plasma insulin or glucose in patients with so-called reactive hypoglycemia. Nevertheless, clinical experience shows that large numbers of such patients continue to be referred for symptoms attributed to ingestion of sugar. Pediatric allergists have commented on the suspected relationship between sugar and behavior for many years (Randolph, 1947; Rinkel *et al.*, 1951; Hoffman, 1971; Rapp, 1976).

Children with hyperactivity have also been suspected of being overly sensitive to refined sugar. Crook (1975) conducted a survey of 45 of his patients with hyperactivity and found that parents reported sugar as the most frequent stimulus for uncontrolled behavior. He later reported that 28 of 48 showed increased hyperactivity to a sugar challenge after a sugar-elimination diet (Crook, 1976). In a controlled study, O'Banion *et al.* (1978) showed that sucrose provoked screaming and destructiveness in a violent autistic child. Rapp (1978) reported increases in irritability, hyperactivity, and gastrointestinal symptoms in blind challenges with 7 of 21 patients reactive to sublingual tests. One recent study claimed that 88% of hyperactive children had abnormal glucose tolerance tests (Langseth and Dowd, 1978), but the study had no normal controls and loose diagnostic criteria. Furthermore, children's norms show that Langseth and Dowd's 1-hr normal range values of glucose are actually above the 90th percentile for children of the same age (Josefsberg *et al.*, 1976). In an important study, Prinz *et al.* (1980) recently showed strong correlations between dietary intake of carbohydrates and behavior of hyperactive children. Seven-day diet diaries of 28 hyperactive and normal 4- to 7-year-olds were correlated with reliable observations of hyperactive behaviors. Amount of sugar products, ratio of sugar to nutritional foods, and carbohydrate/protein ratios were all significantly associated with amounts of destructive–aggressive and restless behaviors during free play. Highly significant effects were also found for normals, but the pattern of correlations was different, showing more gross motor activity and less aggression than hyperactives. Although this was a well-conducted study, the results are entirely correlational, and thus one cannot rule out the possibility that increased aggression or motility leads to increased carbohydrate intake, rather than the reverse.

As noted earlier, we compared the Feingold diet to a control diet, finding improved behavior in hyperactives. Since the Feingold diet also inadvertently reduced carbohydrate intake (Conners *et al.*, 1976), this has led some to speculate that decreased sugar accounted for the improvement. Another recent study also found that the Feingold diet inadvertently restricted more refined sugars than the comparison diet group (Rogers and Hughes, 1981).

Several possible mechanisms exist whereby simple sugars could affect the central nervous system. Temporary states of brain hypoglycemia might occur as a result of lowered transport of glucose to the brain in individuals with chronic *hyper*glycemia who have temporary falls in blood sugar levels. It has recently been shown that rats made chronically hyperglycemic have as much as a 20% reduction in blood-to-brain glucose transport when peripheral glucose is returned to normal levels, possibly due to adaptive changes in the cerebral capillary endothelium (Gjedde and Crone, 1981). A positive feedback or "addiction" mechanism could cause even occasional ingestion of high sugar loads to increase intake sufficiently to stimulate these hyperglycemic conditions

(Rezek *et al.*, 1978). High sugar intake need not lead to obesity, since increased activity level has been shown to offset the increased caloric availability of a high-sucrose diet (Kennedy, 1950). Indeed, rats raised on a high-sucrose diet have *lower* fasting blood glucose levels, illustrating how a functional hypoglycemia could result from chronic high intake of sugars rather than a disease process (Kanarek and Marks-Kaufman, 1979).

Another possible mechanism for direct effects of carbohydrate on brain function has been provided by Fernstrom and Wurtman (1971, 1972, 1974). They showed that animals fed a diet rich in carbohydrates have increases in plasma tryptophan concentration, with associated increases in brain serotonin. The ratio of tryptophan to other dietary amino acids determines the rate at which brain serotonin is modified by dietary intake. Serotonin is known to inhibit locomotor activity and aggressive behavior in animals and also has widespread brain inhibitory functions (Fuxe and Ungerstedt, 1970).

Reduction of brain serotonin by drugs or lesions aggravates shock-induced fighting and predatory killing in mice (Breese and Cooper, 1975; Sheard and Davis, 1976). Four to six days of a tryptophan-free diet induces killing in rats that are traditionally nonkillers (Gibbons *et al.*, 1979). Dietary supplementation with tryptophan reverses this effect, which is apparently not modified by increasing other dietary amino acids.

Coleman (1971) has reported that serotonin concentrations in whole blood of hyperactive children is below normal. Greenberg and Coleman (1976) also reported low blood 5-hydroxyindole levels in hyperactive, institutionalized mentally retarded patients. Contrary results were reported by others (J. L. Rapoport *et al.*, 1975; Wender *et al.*, 1974; Shetty and Chase, 1976). In the best study to date, Ferguson *et al.* (1981) found no differences in total and free plasma tryptophan and plasma cortisol levels between children diagnosed as hyperactive or LD and normal siblings. However, hyperserotonemia has been implicated in autism and retardation in a number of studies (Hanley *et al.*, 1977), and Irwin *et al.* (1981) recently identified a subgroup of hyperserotonemic children with attention-deficit disorder (ADD). They had significantly lower levels of plasma total and protein-bound tryptophan and a higher percentage of free tryptophan than those ADD children with normal serotonin levels (Fig. 4). Concentrations of serotonin did not change with amphetamine or methylphenidate treatment. Approximately 25% of their sample of 53 ADD children were hyperserotonemic. Plasma kynurenine was normal, eliminating the possibility of both a blockade in the kynurenine pathways and an imbalance in the metabolic pathways for tryptophan leading to excessive hydroxylation. Thus, there is a possibility that the findings relate to increased tissue uptake and use of tryptophan. Unfortunately, dietary intake of carbohydrates was not monitored in the study.

Arieffe *et al.* (1974) have developed quite a different model to explain the mechanism by which dietary sugar intake could influence brain function. This hyperosmolality model proposes that insulin causes a change in electrolyte balance in the brain, leading to influx of fluids and temporary brain edema— an effect that is reversed by increased levels of glucose relative to insulin. This

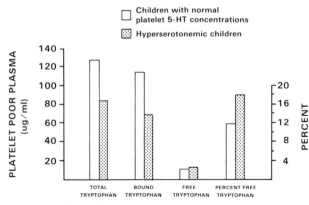

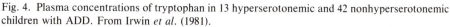

Fig. 4. Plasma concentrations of tryptophan in 13 hyperserotonemic and 42 nonhyperserotonemic children with ADD. From Irwin *et al.* (1981).

model is consistent with the finding that insulin/glucose *ratios* are high during periods when EEG abnormalities and symptoms are present (Hudspeth *et al.*, 1981). Thus, at both high and low absolute glucose levels, insulin/glucose ratios could be elevated, leading to behavioral symptoms. Such symptoms could be quickly reversible as normal insulin/glucose ratios are restored.

The pronounced effects of D-amphetamine in the reduction of activity level and in the improvement of attention in hyperactive children are well known. Amphetamine has highly specific effects on local cerebral glucose utilization, primarily in dopamine-producing or dopaminoceptive brain structures (Wechsler *et al.*, 1979). Changes in availability of cerebral glucose due to dietary factors therefore might be expected to interact at specific brain loci with amphetamine effects. Animal studies indicate that stimulants increase free tryptophan and thus make it more available for brain uptake (Brase and Loh, 1975). No studies have yet examined interactions between dietary sugar and amphetamine, but such an interaction would have important implications for the therapeutic application of amphetamine with ADD and hyperactive children.

4. Deficiency States

4.1. Iron

Feeding practices for infants are highly influenced by maternal–child bonding. Psychological availability of the mother influences how long the child stays on the bottle or breast and can thus contribute indirectly to the development of iron deficiency. Children who rely on bottle feeding exclusively and drink excessive amounts of milk because maternal care is lacking have many illnesses and behavioral problems (Werkman *et al.*, 1964). Figure 5 shows hypothesized interactions among economic, psychosocial, and physical factors in iron deficiency.

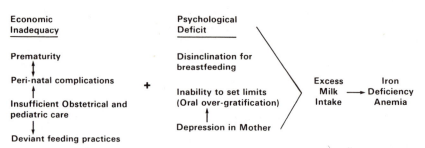

Fig. 5. Hypothesized interactions among economic, psychosocial, and physical factors in iron deficiency. From Werkman *et al.* (1964).

As one of the world's most prevalent health problems, iron deficiency has received considerable attention in the literature (Table V). Leibel *et al.* (1981) point out that although it was formerly believed that iron-dependent enzymes were not affected until later in the development of iron deficiency anemia, more sophisticated measurements have now shown this not to be the case. There is a real possibility that enzyme deficiencies may precede frank clinical appearance of anemia (Siimes *et al.*, 1980). Many anecdotal observations as well as controlled studies have supported the notion that iron deficiency states lead to shortened attention span, irritability, pica, delayed cognitive development, lowered scholastic performance, delayed sensorimotor competence, and impaired learning and memory (Webb and Oski, 1973a,b, 1974; Pollitt and Leibel, 1976; Oski and Honig, 1978; Oski, 1979; Liebel, 1977; Lozoff *et al.*, 1979). However, much of this work uses *ex post facto* designs and fails to control for equality of experimental and control groups with respect to associated environmental problems. When such controls are introduced, results are less conclusive (Deinard *et al.*, 1981). Nevertheless, the reversibility of cognitive deficits with iron therapy provides a powerful argument that such deficits are due to iron deficiency.

Leibel *et al.* (1981) matched 15 iron-deficient children for IQ with iron-replete controls and administered learning tests before and after several months of iron therapy. The treated children showed significant improvement in tasks measuring attentiveness and information reception. One problem with the study is that the controls were iron-replete at the beginning of the study and had significantly *better* pretreatment learning scores than controls. This means that regression to the mean of the cognitive measures in the treatment group could account for the results. What is necessary is a study in which iron-deficient children are randomly assigned to treatment or control status and then remeasured. Given the presumptive role of iron in cognitive deficits and the known somatic effects of deficiency states, such a study would be unethical.

As noted earlier, iron displaces heavy metals that, even in low levels, produce behavioral toxicity. Since socioeconomic factors are strongly associated with prevalence of iron deficiency as well as risk of lead exposure, it is precisely among the poorest segments of the population that nutritional adequacy is most important.

Table V. Summary of Studies of Iron Deficiency

Author	Subjects	Description	Results
Werkman *et al.* (1964)	28 Iron-deficient and 28 non-iron-deficient children	Administration of semistructured questionnaire to mother of each subject.	Iron-deficient children seemed to have more feeding and behavior problems; also, they seemed to have mothers who were inattentive, causing the children to be orally fixated, drinking excessive amounts of milk.
Sulzer *et al.* (1973)	230 Male and female children (ages 4–5) (anemic and nonanemic)	Administration of tests to two subgroups. Tests included global IQ test, vocabulary test, measures of moral development, grouping behavior, reaction time, and attentive recall.	Anemics performed more poorly on vocabulary tests, latency, and associated reaction measures. Younger children showed signs of permanent damage.
Webb and Oski (1973a,b; 1974)	92 Anemic and 101 nonanemic adolescents (ages 12–14)	Comparison of test results between two groups (Iowa Test of basic skills, teacher ratings of personality disturbance, visual afterimage task).	Anemics had lower composite scores on the Iowa test and significantly longer latency in the visual afterimage task. Anemic males exhibited more conduct problems than nonanemic males. Anemic male test scores showed progressive decline from ages 12 to 14. Anemic female test scores remained consistently poor.
R. J. Cantwell (1974)	61 Children studied from birth to 7 years; 32 developed iron deficiency between 6 months and 18 months of age	Neurological exam was given to each subject by age 7 (examiners did not know which subjects were formerly anemic).	Formerly anemic subjects had more "soft signs" of permanent minimal brain dysfunction and were inattentive and hyperactive with lower average IQ scores than the nonanemics.
Webb and Oski (1974)	74 Anemic and 36 nonanemic junior high students	Administration of behavior problem checklist.	Anemics tended to have more conduct disturbances.
Voorhees *et al.* (1975)	9 Male and 2 female iron-deficient infants and children (ages 10 months to 13 years)	Measurement of subjects' urinary excretion of dopamine, norepinephrine, metanephrine-normetanephrine, 3-methoxy-4-hydroxymandelic acid before and after treatment with intramuscular iron.	Subjects excreted increased quantities of norepinephrine before treatment, perhaps as a result of MAO deficiency. Urinary excretion of norepinephrine returned to normal levels within 5 days of treatment with intramuscular iron.

(Continued)

Table V. (Continued)

Author	Subjects	Description	Results
Pollitt *et al.* (1978b)	46 Children (mean age = $3\frac{1}{2}$ years) iron-deficient or iron-sufficient	Administration of behavioral tests to two subgroups; between-group (double-blind) comparison before and after treatment with iron or placebo.	Iron deficiency had adverse effects on attention and memory control processes. Deficits were eliminated after treatment with iron.
Oski and Honig (1978)	24 Infants (ages 9–26 months) with iron-deficiency anemia	Random assignment to treatment with intramuscular iron or placebo groups; administration of Bayley Scale of Infant Development before and after treatment or placebo.	Test scores improved somewhat for placebo group, but performance improved more with iron treatment, especially on the index of mental development.
Lozoff *et al.* (1979)	40 Anemic and 43 nonanemic infants (ages 6–24 months)	Double-blind randomized block design; treatment with oral iron or placebo; Bayley Scale of Infant Development before and after treatment with iron or placebo.	Anemics had significantly lower pretest scores than nonanemics; remained low on physical and mental scales. Short-term oral iron therapy did not correct developmental deficits.
Oski *et al.* (1981)	33 Nonanemic intants (ages 9–13 months)	Divided into four groups: normals, iron-depleted, iron-deficient (with MCV > 70fl), and iron-deficient (with MCV ⩽ 70fl). Bayley Scale of Infant Development administered before and after treatment with intramuscular iron.	Iron-deficient groups exhibited alterations in behavior that were readily reversible with treatment with iron, as shown by greater test score improvement than normals.
Deinard *et al.* (1981)	101 Male and 111 female nonanemic infants (ages 11–13 months)	Subjects were divided into three groups: mildly iron-deficient, severely iron-deficient, and iron-replete. Administration of Bayley Scale of Infant Development and index of attending behavior.	There were no statistically significant differences in overall performance levels between the iron-depleted and the iron-repleted groups.
Lozoff (1983)	28 Anemic and 40 nonanemic infants (ages 6–24 months)	Subjects were assigned to treatment or placebo group; Bayley Scale of Infant Development was administered pre- and posttreatment with intramuscular iron or placebo.	Nonanemics had higher mean scores in pretest. Mental score deficits in anemics became more marked with age. Scores of both treatment and placebo groups improved by 6 points in the posttest.
Pollitt and Liebel (1982)	19 Anemic and 20 nonanemic children (ages 4–5 years)	Subjects assigned to treatment or placebo group. Psychological tests of discriminate learning and oddity learning administered before and after treatment with intramuscular iron or placebo.	No differences were evident between groups on the discriminate learning test. Anemics performed less well on the oddity learning test pretreatment, but had identical scores as the control group posttreatment.

4.2. Possible Role of Essential Fatty Acids

Polyunsaturates are fatty acids that contain more than one double bond. Those that are biologically active and must be supplied in the diet are known as essential fatty acids (EFAs) (Mead and Fulco, 1976; WHO/FAO, 1977). Like vitamins, EFAs cannot be made by the body and must be taken in through food. Linoleic acid, the principal active EFA constituent, is found mainly in vegetable oils such as corn, safflower, and sunflower. Because linoleic acid is relatively inert biologically, it must be converted to γ-linoleic acid (GLA) by the enzyme δ-6-desaturase (Frankel and Rivers, 1978). A number of factors, particularly the effects of food processing, may block formation of GLA from linoleic acid. Linoleic acid starts off in the *cis* form, but hydrogenation and other processing procedures convert much of this biologically useful *cis*-linoleic acid to the more stable *trans* form, which actually competitively blocks formation of GLA, thereby leading to deficiency states. In the process, serum cholesterol is elevated (Anderson *et al.*, 1961; Holman and Aaes-Jorgensen, 1956).

A major role of EFAs is their conversion to prostaglandins (PGs). Pgs have an enormous range of biological effects the relevance of which to neurophysiology and psychiatry is just beginning to be explored (Rotrosen *et al.*, 1980; Gross *et al.*, 1977). It has been demonstrated that PGs function as mediators or modulators of neuronal transmission in the central nervous system (Gross *et al.*, 1977).

Horrobin (1980) has developed an extremely interesting hypothesis regarding the role of EFAs and PGs in alcoholism. He postulates that alcohol dependence and its associated physical complications could be due to the combined effect of alcohol in enhancing conversion of dihomo-γ-linolenic acid (DGLA) to PGE_1, and to the blocking of the δ-6-desaturase enzyme necessary for replenishment of DGLA stores from dietary precursors, concluding that (p. 929)

> the acute effect of ethanol is therefore an increased production of PGE1 but chronic consumption will lead to depletion of DGLA and PGE1. . . . Alcoholics may drink to maintain a normal PGE1 level, something which will require more and more ethanol as DGLA is depleted.

The genetic link between hyperactivity in children and alcoholism in first- and second-degree relatives has been supported by several studies (Morrison and Stewart, 1971, 1973; Goodwin *et al.*, 1975; D. Cantwell, 1972, 1975). Hyperactive children and adolescents are at considerable risk for alcoholism in later life, and while it has generally been assumed that the alcoholism–hyperactivity relationship is a genetic rather than a familial one, no explanation of the genetic mechanism has ever been put forth. Horrobin's hypothesis provides just such a mechanism. It also provides a possible mechanism to link certain food allergies to hyperkinesis and behavior disorders.

PGE_1 is known to be important in the functioning of T lymphocytes, which are necessary in the maintenance of a normal immune system. It is interesting that virtually all the ''dramatic'' cases of improvement on the Feingold diet were patients who had initially presented with allergies (Feingold, 1968, 1974). T lymphocytes are defective in patients with asthma and eczema (Lovell *et al.*, 1981).

It has also been known for over 50 years that impaired fat absorption in children can lead to erratic mood, irritability, and dysphoria (Parsons, 1932). Gluten of wheat and α-casein of milk produce opioidlike peptides (exorphins) in the intestine (Ziourdrou *et al.*, 1979). While these are normally rapidly digested, they may enter the bloodstream, where they inhibit PG synthesis. Wheat and milk sensitivities are quite common in children and thus might be responsible for behavior disorders through the inhibition of PG synthesis.

The fact that males require several times the amount of EFAs that females do (Pudelkewicz *et al.*, 1968) is also consistent with the hitherto unexplained high male/female ratio of hyperkinesis. Inborn errors of the rate-limiting enzyme δ-6-desaturase required for conversion of GLA to DGLA would be more likely to affect males than females. This enzyme is apparently quite vulnerable to blockade (Brenner, 1977, 1980). A recent study lends some credence to this novel hypothesis (Colquhoun and Bunday, unpublished). A survey of over 200 hyperactive children in England found, consistent with Feingold's suggestions, that tartrazine (yellow food dye) was the factor most often implicated by parents in exacerbation of their children's problem behavior. Horrobin (personal communication) has found that tartrazine inhibits PGE_1 formation when EFA precursors are absent. Three quarters of the 200 children were male. The most prominent health problems in the histories were infantile colic, eczema, asthma, rhinitis, and repeated chest and ear infections. Hair samples in 46 of the children showed that 24 of 31 boys and 7 of 15 girls had zinc values below the normal range. Several of the children who were already on Feingold's diet showed behavioral improvement after milk and wheat products were eliminated from the diet. One unexplained finding was "a striking preponderance" of fair- and ginger-haired children in this group. This observation is of interest because of the role of melatonin in mobilizing free DGLA (Cunnane and Horrobin, 1983). Epidemiological studies have shown increased prevalence of behavioral problems in children born in winter months, especially in years when the average temperature in the first trimester was unusually high (Pasamanick and Knobloch, 1966; Gruenberg and Turns, 1975). Seasonal variation in melatonin could thus be connected with changes in the ability to metabolize EFA. An open trial was carried out with 25 of the children by treating them with daily doses of evening primrose oil, one of the few natural sources of EFA that also has substantial quantities of GLA. Half the children responded with improved behavior, even though they had already shown improvement from Feingold's diet. Thus, one might assume that any placebo-related improvement had already occurred.

Without appropriate double-blind controls, such observations are only of hypothesis-generating value. But given the powerful integrative nature of the hypothesis for a number of the unexplained aspects of the hyperkinetic syndrome, it deserves careful testing. The findings of Pihl and Parkes (1977) that LD children have excessive lithium levels and deficient zinc levels are also consistent with the hypothesis. Insulin interactions with the conversion pathway from *cis*-linoleic acid to GLA provide a possible link with the putative effects of sugar on behavior disorders. Much work along these lines deserves to be carried out.

Diets that rely heavily on corn products, as well as the effects of food processing, have also been implicated as a source of B vitamin and ω3-EFA deficiencies. The latter have been hypothesized as a major source of mental illness, analogous to the niacin–tryptophan syndrome of pellagra (Rudin, 1981). Rudin proposed that the ω3 EFA required for the synthesis of the 3-series of PGs is effectively removed by food processing or pure corn diets. He described 12 cases of severe, chronic mental disorders whose physical signs of EFA deficiency remitted after treatment with linseed oil. Eight of the cases showed moderate to marked improvement in mental status along with the physical change. Again, while lacking the necessary evidence of controlled studies, such hypotheses appear to foreshadow a new phase of research into mental illness.

4.3. Vitamins

Orthomolecular psychiatry is a school based on the tenets of biochemical individuality described by R. J. Williams (1956) and further elaborated by Pauling (1968). Pauling's hypothesis that large doses of vitamins may be necessary for optimal function were first applied to patients with schizophrenia. This and subsequent orthomolecular work has been thoroughly reviewed (American Psychiatric Association Task Force, 1973; Lipton *et al.*, 1979).

Among the few modest positive findings from controlled studies, the work with children is perhaps the most important. Rimland *et al.* (1978) carried out a double-blind crossover study with 16 autistic children from a pool of 200 selected from an open trial with vitamin B_6. Subjects who showed replicable improvement and relapse were further tested on individualized doses (75–800 mg/day) of B_6. Of 15 who completed the trial, 11 showed a statistically significant improvement compared with the placebo period. Multivitamin supplements (B_3, B_5, B_6, and C) were administered in a high-protein, low-carbohydrate, sugar-free diet to learning-disabled children without effect (Kershner *et al.*, 1977). Greenbaum (1970) found no effect of megadoses of niacinamide in children diagnosed as childhood schizophrenics.

5. Summary and Conclusions

The most severe impact of diet on the brain occurs as a result of early and chronic malnutrition. Behavioral development is severely compromised by the loss of essential nutrients required for brain growth, especially protein and fats. Beyond this, little is known with certainty regarding nutrition–behavior interactions in children. Chronic undernutrition of a milder degree has been so confounded by psychosocial variables that epidemiological studies are almost wholly inconclusive. Nevertheless, studies of occasional lapses of feeding at breakfast, as well as studies of short-term breakfast and morning food supplements, support the conclusion that higher cognitive functions, and perhaps emotionality, are subtly affected by the amount and timing of meals.

Adverse reactions to food additives have received mixed support as an explanation of behavior and learning disorders in children. There appears to be justification for concern about artificial colors in the diet of preschool children and a small number of predisposed school-age children. Adverse reactions, possibly allergic in nature, have also been weakly implicated in studies with children, but definitive studies have not been done. No firm basis has been established for identification of children sensitive to particular foods.

Behavioral toxicity to unintentional food additives, particularly heavy metals such as lead and cadmium, is well established. Important suggestive evidence exists from animal studies that dietary sources of micronutrients such as iron and zinc may exert strong control over the absorption, distribution, and eventual toxicity of heavy-metal exposure from the environment. Dietary adequacy of micronutrients is probably important for optimal attentional and higher cognitive development. Nutritional adequacy is closely tied to psychosocial aspects of maternal–child interactions.

Neurotransmitter precursors (tryptophan, tyrosine, choline, lecithin) have been studied in adult psychiatric conditions with some promise, but little evidence of their role in children is available. Evidence suggests that subgroups of hyperactive children who are hyperserotonemic may exist. If confirmed, this would open the way for dietary trials involving manipulation of tryptophan.

Deficiency states involving essential fatty acids appear to be worth further investigation as antecedents of behavior disturbances in children. Modest evidence suggests that a subgroup of autistic children responds to treatment with large doses of vitamin B_6. Correlational studies indicate that sugar may have a causative role in behavior disorders, but direct evidence is largely anecdotal and uncontrolled. Hypoglycemia has not received adequate study in children, though interesting animal models suggest that rapid alterations in brain glucose and fluids can be influenced by dietary and hormonal interactions.

Optimizing cognitive and behavioral development in children by nutritional and dietary manipulation, and consideration of nutrition in the prophylaxis and treatment of mental disorders, are exciting new frontiers that deserve careful, systematic research.

6. References

Adams, W., 1981, Lack of behavioral effects from Feingold diet violations, *Percept. Mot. Skills* **52**:307.

American Psychiatric Association Task Force on Vitamin Therapy in Psychiatry, 1973, Megavitamin and orthomolecular therapy in psychiatry, American Psychiatric Association, Washington, D.C.

Anderson, J. T., Grande, F., and Keys, A., 1961, Effects on serum cholesterol in man of fatty acids produced by hydrogenation of corn oil, *Fed. Proc. Fed. Am. Soc. Exp. Biol.* **20**:96.

Arieffe, A. I., Doerner, T., Zelig, H., and Massry, S. G., 1974, Mechanisms of seizures and coma in hypoglycemia: Evidence for a direct effect of insulin on electrolyte transport in the brain, *J. Clin. Invest.* **54**:654.

Arvedson, I., Sterky, G., and Tjernstrom, K., 1969, Breakfast habits of Swedish school children, *J. Am. Diet. Assoc.* **55**:257.

Baldwin, D. G., Kittler, F. J., and Ramsay, R. G., 1968, The relationship of allergy to cerebral dysfunction, *South. Med. J.* **61**:1039.

Black, A. P., 1956, A new diagnostic method in allergic disease, *Pediatrics* **17**:716.

Brase, D. A., and Loh, H. H., 1975, Possible role of 5-hydroxytryptamine in minimal brain dysfunction, *Life Sci.* **16**:1005.

Breese, G. R., and Cooper, B. R., 1975, Behavioral and biochemical interactions of 5,7-dihydroxytryptamine with various drugs when administered intracisternally to adult and developing rats, *Brain Res.* **98**:517.

Brenner, R. R., 1977, Metabolism of endogenous substrates by microsomes, *Drug Metab. Rev.* **6**:155.

Brenner, R. R., 1980, Nutritional and hormonal factors influencing desaturation of essential fatty acids, Presented at the Golden Jubilee Congress on Essential Fatty Acids and Prostaglandins, Minneapolis.

Bryan, W. T. K., and Bryan, M. P., 1960, The application of *in vitro* cytotoxic reaction to clinical diagnosis of food allergy, *Larnygoscope* **70**:810.

Bryan, W. T. K., and Bryan, M. P., 1972, Clinical examples of resolution of some idiopathic and other chronic disease by careful allergic management, *Laryngoscope* **82**:1231.

Bullard, D. M., Glaser, H. M., Heagerty, M. C., 1967, Failure to thrive in the neglected child, *Am. J. Orthopsychiatry* **37**:680.

Cahill, G. F., and Soeldner, J., 1974, A non-editorial on non-hypoglycemia, *N. Engl. J. Med.* **291**:905.

Cantor, S. M., 1975, Patterns of use, in: *Sweeteners—Issues and Uncertainties*, National Academy of Sciences Academy Forum—Fourth of a Series, pp. 19–35.

Cantwell, D., 1972, Psychiatric illness in the families of hyperactive children, *Arch. Gen. Psychiatry* **70**:414.

Cantwell, D., 1975, Genetic studies of hyperactive children: Psychiatric illness in biological and adopting parents, in: *Genetic Research in Psychiatry* (R. R. Fieve, D. Rosenthal, and H. Brills, eds.), pp. 273–280, Johns Hopkins University Press, Baltimore.

Cantwell, R. J., 1974, The long term neurological sequelae of anemia in infancy, *Pediatr. Res.* **8**:342.

Chase, H. P., and Martin, H. P., 1970, Undernutrition and child development, *N. Engl. J. Med.* **282**:933.

Clarke, J., 1950, The relation of allergy to character problems in children: A survey, *Psychiatr. Q.* **24**:21.

Coleman, M., 1971, Serotonin concentration in whole blood of hyperactive children, *J. Pediatr.* **78**:985.

Conners, C. K., 1980, *Food Additives and Hyperactive Children*, Plenum Press, New York.

Conners, C. K., Goyette, C. H., Southwick, D. A., Lees, J. M., and Andrulonis, P. A., 1976, Food additives and hyperkinesis: A controlled double-blind experiment, *Pediatrics* **58**:154.

Conners, C. K., Goyette, C. H., and Newman, E. B., 1980, Dose-time effect of artificial colors in hyperactive children, *J. Learn. Disabil.* **13**:9, 512–516.

Conners, C. K., Blouin, A. G., and Seidel, W. T., 1982, The effects of breakfast on the cardiac response and behavior of children, Presented at the meeting of the Society for Psychophysiological Research, Minneapolis, October 1982.

Crook, W., 1975, Food allergy—the great masquerader, *Pediatr. Clin. North Am.* **22**:227.

Crook, W., 1976, Learning disabilities and hyperactivity in children due to foods, Presented at the International Food Allergy Symposium, Toronto.

Cunnane, S. C., and Horrobin, D. F., 1980, The vascular response to zinc varies seasonally: Effect of pinealectomy and melatonin, *Chronobiologia* **7**(4):493–503.

David, O. J., Clark, J., and Voeller, K., 1972, Lead and hyperactivity, *Lancet* **2**:900.

David, O. J., Hoffman, S. P., Sverd, J., Clark, J., and Voeller, K., 1976, Lead and hyperactivity: Behavioral response to chelation—A pilot study, *Am. J. Psychiatry* **133**:1155.

Deinard, A., Gilbert, A., Dodds, M., and Egeland, B., 1981, Iron deficiency and behavioral deficits, *Pediatrics* **68**(6):828.

Dodge, P. R., Prensky, A. L., and Feigin, R. D., 1975, *Nutrition and the Developing Nervous System*, C. V. Mosby, St. Louis.

Dwyer, J. T., Elias, M. F., and Warren, J. H., 1973, Effects of an experimental breakfast program on behavior in the late morning, Department of Nutrition, Harvard School of Public Health.

Egan, J., Chatoor, I., and Rosen, G., 1980, Non-organic failure to thrive: Pathogenesis and classification, *Clin. Proc.* **36**(4):173.

Feingold, B. F., 1968, Recognition of food additives as a cause of symptoms of allergy *Ann. Allergy* **26**:309.

Feingold, B. F., 1973, *Introduction to Clinical Allergy*, Charles C. Thomas, Springfield, Illinois.

Feingold, B. F., 1974, Hyperkinesis and learning difficulties (H-LD) linked to the ingestion of artificial colors and flavors, Presented at the American Medical Association annual meeting, Section on Allergy, Chicago.

Feingold, B. F., 1981, Dietary management of behavior and learning disabilities, in: *Nutrition and Behavior* (S. A. Miller, ed), pp. 235–246, Franklin Institute Press, Philadelphia.

Ferguson, H. B., Pappas, B. A., Trites, R. L., Peters, D. A. V., and Taub, H., 1981, Plasma-free and total tryptophan, blood serotonin and the hyperactivity syndrome: No evidence for the serotonin deficiency hypothesis, *Biol. Psychiatry* **16**(3):231.

Fernstrom, J. D., and Wurtman, R. J., 1971, Brain serotonin content: Increase following ingestion of carbohydrate diet, *Science* **174**:1023.

Fernstrom, J. D., and Wurtman, R. J., 1972, Brain serotonin content: Physiological regulation by plasma neutral amino acids, *Science* **178**:414.

Fernstrom, J. D., and Wurtman, R. J., 1974, Nutrition and the brain, *Sci. Am.* **230**:84.

Frankel, T. L., and Rivers, J. P. W., 1978, The nutritional and metabolic impact of gamma-linoleic acid on cats deprived of animal lipid, *Br. J. Nutr.* **39**:227.

Fuxe, K., and Ungerstedt, U., 1970, Histochemical and functional studies on central monamine neurons after acute and chronic amphetamine administration, in: *International Symposium on Amphetamines and Related Compounds* (E. Costa and S. Garattini, eds.), pp. 257–288, Raven Press, New York.

Gibbons, J. L., Barr, G. A., Bridger, W. H., and Leibowitz, S. F., 1979, Manipulations of dietary tryptophan: Effects on mouse killing and brain serotonin in the rat, *Brain Res.* **169**:139.

Gill, F. M., and Schwartz, E., 1972, Anemia in early infancy, *Pediatr. Clin. North Am.* **19**:841.

Gjedde, A., and Crone, C., 1981, Blood–brain glucose transfer: Repression in chronic hyperglycemia, *Science* **214**:456.

Goldfarb, W., 1955, Psychological deprivation in infancy and subsequent adjustment, *Am. J. Orthopsychiatry* **15**:247.

Goodwin, D., Schulsinger, F., Hermensen, L., Guze, S., and Winokur, G., 1975, Alcoholism and the hyperactive child syndrome, *J. Nerv. Ment. Dis.* **150**:349.

Goyette, C. H., Conners, C. K., Petti, T. A., 1978, Effects of artificial colors on hyperkinetic children: A double-blind challenge study, *Psychopharmacol. Bull.* **14**:39–40.

Graves, P. L., 1976, Nutrition, infant behavior and maternal characteristics: A Pilot study in East Bengal, India, *Am. J. Clin. Nutr.* **29**:305.

Graves, P. L., 1978, Nutrition and infant behavior: Replication study in the Katmandu Valley, Nepal, *Am. J. Clin. Nutr.* **31**:541.

Greenbaum, G. H. C., 1970, An evaluation of niacinamide in the treatment of childhood schizophrenia, *Am. J. Psychiatry* **127**:129.

Greenberg, A. S., and Coleman, M., 1976, Depressed 5-hydroxyindole levels associated with hyperactive and aggressive behavior: Relationship to drug response, *Arch. Gen. Psychiatry* **33**:331.

Gross, H. A., Dunner, D. L., Lafleur, D., Meltzer, H. L., Muhlbauer, H. L., and Fieve, R. R., 1977, Prostaglandins, *Arch. Gen. Psychiatry* **34**:1189.

Gruenberg, E. M., and Turns, D. M., 1975, Epidemiology, in: *Comprehensive Textbook of Psychiatry*, 2nd ed. (A. M. Freedman, H. I. Kaplan, and B. J. Sadock, eds.), pp. 398–413, Williams & Wilkins, Baltimore.

Hanley, H. G., Stahl, S. M., and Freedman, D. X., 1977, Hyperserotonemia and amine metabolites in autistic and retarded children, *Arch. Gen. Psychiatry* **34:**521.

Harley, J. P., Ray, R. S., Tomasi, L., Eichman, P. L., Matthews, C. G., Chun, R., Cleeland, C. S., and Traisman, E., 1978a, Hyperkinesis and food additives: Testing the Feingold hypothesis, *Pediatrics* **61:**818.

Harley, J. P., Matthews, C. G., and Eichman, P. L., 1978a, Synthetic food colors and hyperactivity in children: A double-blind challenge experiment, *Pediatrics* **62:**975.

Haughton, J. G., 1963, Nutritional anemia in infancy and childhood, *Am. J. Public Health* **53:**1121.

Hepner, R., and Maiden, N. C., 1971, Growth rate, nutrient intake and "mothering" as determinants of malnutrition in disadvantaged children, *Nutr. Rev.* **29:**219.

Hoffman, M. S., 1971, The nutritional aspects of learning disabilities, Presented at the 7th Annual Convention of the Texas Association for Children with Learning Disabilities, Dallas.

Holman, R. T., and Aaes-Jorgensen, E., 1956, Effect of *trans* fatty acid isomers upon essential fatty acid deficiency in rats, *Proc. Soc. Exp. Biol. Med.* **93:**175.

Horrobin, D. F., 1980, A biochemical basis for alcoholism and alcohol-induced damage including the fetal alcohol syndrome and cirrhosis: Interference with essential fatty acid and prostaglandin metabolism, *Med. Hypotheses* **6:**929.

Horrobin, D. F., and Manku, M. S., 1980, Possible role of prostaglandin E_1 in the affective disorders and in alcoholism, *Br. Med. J.* **280:**1363.

Hudspeth, W. J., Peterson, L. W., Soli, D. E., and Trimble, B. A., 1981, Neurobiology of the hypoglycemia syndrome, *J. Holistic Med.* **3**(1):60.

Humphrey, J. H., and White, R. G., 1970, *Immunology for Students of Medicine*, 3rd ed., Blackwell, London.

Irwin, M., Belendiuk, K., McCloskey, K., and Freedman, D. X., 1981, Tryptophan metabolism in children with attentional deficit disorder, *Am. J. Psychiatry* **138**(8):1082.

Josefsberg, Z., Vilunski, E., Hanukuglu, A., Bialik, O., Brown, M., Karps, M., and Laron, Z., 1976, Glucose and insulin responses to an oral glucose load in normal children and adolescents in Israel, *Isr. J. Med. Sci.* **12:**189.

Kanarek, R. B., and Marks-Kaufman, R., 1979, Developmental aspects of sucrose-induced obesity in rats, *Physiol. Behav.* **23:**881.

Keister, M., 1950, Relation of mid-morning feeding to behavior of nursery school children, *J. Am. Diet. Assoc.* **26:**25.

Kennedy, G. C., 1950, The hypothalamic control of food intake in rats, *Proc. R. Soc. Lond.* (*Biol.*) **137:**535.

Kershner, J., Grekin, R., Hawke, W. A., Darwish, H., and Cutter, P., 1977, Pilot studies of high protein, high vitamin, low carbohydrate, sugar-free diet in learning disabled children, *Can. Med. Assoc. J.* **117:**212.

Laird, D. A., Levitan, M., and Wilson, V. A., 1931, Nervousness in school children as related to hunger and diet, *Med. J. Rec.* **134:**494.

Langseth, L., and Dowd, J., 1978, Glucose tolerance and hyperkinesis, *Food Cosmet. Toxicol.* **16:**129.

Leibel, R. L., 1977, Behavioral and biochemical correlates of iron deficiency, *J. Am. Diet. Assoc.* **71:**398.

Leibel, R. L., Pollitt, E., and Greenfield, D. B., 1981, Methodologic problems in the assessment of nutrition–behavior interactions: A study of effects of iron deficiency on cognitive function in children, in: *Nutrition and Behavior* (S. A. Miller, ed.), pp. 299–301, Franklin Institute Press, Philadelphia.

Levy, D., 1954, Oppositional syndrome and oppositional behavior, in: *Psychopathology of Childhood* (P. H. Hoch and J. Zubin, eds.), pp. 204–226, Grune & Stratton, New York.

Levy, F., and Hobbes, G., 1978, Hyperkinesis and diet: A replication study, *Am. J. Psychiatry* **135**(12):1559–60.

Levy, F., Dumbrell, S., Hobbes, G., 1977, Hyperkinesis and diet: A doubleblind crossover trial with tartrazine challenge. *Med. J.* **1**:61–64.

Lininger, F., 1933, Relation of the use of milk to the physical and scholastic progress of undernourished school children, *Am. J. Public Health* **23**:555.

Lipton, M. A., and Wheless, J. C., 1981, Diet as therapy, in: *Nutrition and Behavior* (S. A. Miller, ed.), pp. 213–233, Franklin Institute Press, Philadelphia.

Lipton, M. A., Mailman, R. B., and Nemeroff, C. B., 1979, Vitamins, megavitamin therapy and the nervous system, in: *Nutrition and the Brain*, Vol. 3 (R. J. Wurtman and J. J. Wurtman, eds.), pp. 183–264, Raven Press, New York.

Lovell, C. R., Burton, J. L., and Horrobin, D. F., 1981, Treatment of atopic eczema with evening primrose oil, *Lancet* **1**:278.

Lozoff, B., Brittenham, G., Viteri, F. E., Wolf, A. W., Urritia, J. J., 1979, Developmental test deficit in infants with iron deficiency anemia, *Pediatr. Res.* **13**:334.

Lozoff, B., Brittenham, G. M., Viteri, F. E., Wolf, A. W., Urritia, J. J., 1983, Developmental deficits in iron-deficient infants: Effects of age and severity of iron lack. *Jrl. Ped.* **101**(6):948–952.

Mahaffey, K. R., 1980, Nutrient–lead interactions, in: *Lead Toxicity* (R. L. Singhal and J. A. Thomas, eds.), pp. 425–460, Urban and Schwarzenberg, Baltimore.

Matheson, N. E., 1970, Mid-morning nutrition and its effects on school type tasks, Ph.D. dissertation, University of Southern California (unpublished).

Mattes, J. A., and Gittelman, R., 1981, Effects of artificial food colorings in children with hyperactive symptoms, *Arch. Gen. Psychiatry* **38**:714.

Mead, J. F., and Fulco, A. J., 1976, *The Unsaturated and Polyunsaturated Fatty Acids in Health and Disease*, Charles C. Thomas, Springfield, Illinois.

Morrison, J., and Stewart, M., 1971, A family study of the hyperactive child syndrome, *Biol. Psychiatry* **3**:189.

Morrison, J., and Stewart, M., 1973, The psychiatric status of the legal families of adoptive hyperactive children, *Arch. Gen. Psychiatry* **28**:888.

National Advisory Committee on Hyperkinesis and Food Additives, 1980, Final report to the Nutrition Foundation.

National Dairy Council, 1980, Nutritional demands imposed by stress, *Dairy Council Dig.* **51**(6), ISSN 0011-5568.

O'Banion, D., Armstrong, B., Cummings, R. A., and Stange, J., 1978, Disruptive behavior: A dietary approach, *J. Autism Child. Schizophr.* **8**:325.

O'Shea, J. A., and Porter, S. F., 1981, Double-blind study of children with hyperkinetic syndrome treated with multi-allergen extract sublingually, *J. Learn. Disab.* **14**(4):189.

Oski, F. A., 1979, The nonhematologic manifestation of iron deficiency, *Am. J. Dis. Child.* **133**:315.

Oski, F. A., and Honig, A. S., 1978, The effects of therapy on the developmental scores of iron-deficient infants, *J. Pediatr.* **92**:21.

Oski, F. A., Honig, A. S., Helu, B. M., and Howanitz, P. H., 1981, Effect of iron deficiency without anemia on infant behavior, *Pediatr. Res.* **15**:583.

Parsons, L. G., 1932, Celiac disease, *Am. J. Dis. Child.* **43**:1293.

Pasamanick, B., and Knoblock, H., 1966, Retrospective studies on the epidemiology of reproductive casualty: Old and new, *Merrill–Palmer Q. Behav. Dev.* **12**:7.

Pauling, L., 1968, Orthomolecular psychiatry, *Science* **160**:265.

Phillips, K., 1959, Clinical studies on the hypoglycemic syndrome: A correlation between clinical and laboratory findings, *Am. Pract. Dig. Treat.* **10**:971.

Pihl, R. O., 1979, Recent evidence of physical abnormalities in children with learning disabilities, *McGill Med. J.* **14**:353.

Pihl, R. O., and Parkes, M., 1977, Hair element content in learning disabled children, *Science* **198**:204.

Pollitt, E., and Liebel, R. L., 1976, Iron deficiency and behavior, *J. Pediatr.* **88**(3):372.

Pollitt, E., and Liebel, R. L., eds., 1982, *Iron Deficiency, Brain Biochemistry, and Behavior*, Raven Press, New York.

Pollitt, E., Eichler, A. W., and Chan, C. K., 1975, Psychosocial development and behavior of mothers of failure to thrive children, *Am. J. Orthopsychiatry* **45**:525.

Pollitt, E., Gersovitz, M., and Gargiulo, M., 1978a, Educational benefits of the United States school feeding program: A critical review of the literature, *Am. J. Public Health* **68**(5):477.

Pollitt, E., Greenfield, D., and Leibel, R., 1978b, Behavioral effects of iron deficiency among preschool children in Cambridge, Mass., *Fed. Proc. Fed. Am. Soc. Exp. Biol.* **37**:487.

Pollitt, E., Mueller, W., Leibel, R. L., 1982, The relation of growth to cognition in a well-nourished preschool population, *Child. Devel.* **53**(5):1157–1163.

Prausnitz, C., and Kuster, H., 1921, Studies on supersensitivity, *Zentralbl. Bakteriol.* **86**:160.

Prinz, R. J., Roberts, W. A., and Hantman, E., 1980, Dietary correlates of hyperactive behavior in children, *J. Consult. Clin. Psychol.* **48**(6):760.

Pudelkewicz, C., Seufert, J., and Holman, R. T., 1968, Requirements of the female rat for linoleic and linolenic acids, *J. Nutr.* **64**:138.

Randolph, T. G., 1947, Allergy, a cause of fatigue, irritability and behavior problems in children, *J. Pediatr.* **31**:560.

Rapoport, H. G., and Flint, S. H., 1976, Is there a relationship between allergy and learning disabilities?, *J. Sch. Health* **46**(3):139.

Rapoport, J. L., Quinn, P. O., Scribanu, N., and Murphy, D. L., 1975, Platelet serotonin of hyperactive school age boys, *Br. J. Psychiatry* **125**:138.

Rapp, D., 1976, Double-blind study in relation to the role of foods and dyes to hyperactivity, Presented at the International Food Allergy Symposium, Toronto.

Rapp, D., 1978, Does diet affect hyperactivity?, *J. Learn. Disab.* **11**(6):56.

Read, M. S., 1973, Malnutrition, hunger and behavior, *J. Am. Diet. Assoc.* **63**:379.

Rezek, M., Havlicek, V., and Hughes, K. R., 1978, Paradoxical stimulation of food intake by larger loads of glucose, fructose and mannose: Evidence for a positive feedback effect, *Physiol. Behav.* **21**:243.

Ricciuti, H. N., 1981, Adverse environmental and nutritional influences on mental development: A perspective, *J. Am. Diet. Assoc.* **79**:115.

Rimland, B., Callaway, E., and Dreyfus, P., 1978, The effect of high doses of vitamin B_6 on autistic children: A double-blind crossover study, *Am. J. Psychiatry* **135**:472.

Rinkel, H., Randolph, T. G., and Zeller, M., 1951, *Food Allergy*, Charles C. Thomas, Springfield, Illinois.

Rogers, G. S., and Hughes, H. H., 1981, Dietary treatment of children with problematic activity level, *Psychol. Rep.* **48**:487.

Rose, T. L., 1978, The functional relationship between artificial food colors and hyperactivity. *Jrl. Appl. Behav. Analysis* **11**(4):439–446.

Rotrosen, J., Miller, A. D., Mandio, D., Traficante, L. J., and Gershon, S., 1980, Prostaglandins, platelets and schizophrenia, *Arch. Gen. Psychiatry* **37**:1047.

Rowe, A. H., 1931, *Food Allergy: Its Manifestations, Diagnosis and Treatment*, Lea and Febiger, Philadelphia.

Rudin, D. O., 1981, The major psychoses and neuroses as omega-3 essential fatty acid deficiency syndrome: Substrate pellagra, *Biol. Psychiatry* **16**(9):837.

Rutter, M., 1980, Raised lead level and impaired cognitive/behavioral functioning: A review of the evidence, *Dev. Med. Child Neurol. (Suppl.)* **42**(22):1.

Sheard, M. H., and Davis, M., 1976, Shock-elicited fighting in rats: Importance of intershock interval upon the effect of *p*-chlorophenylalanine (PCPA), *Brain Res.* **111**:433.

Shetty, T., and Chase, T. N., 1976, Central monoamines and hyperkinesis of childhood, *Neurology* **26**:1000.

Siimes, M. A., Refino, C., and Dallman, P. R., 1980, Manifestations of iron deficiency at various levels of dietary iron intake, *Am. J. Clin. Nutr.* **33**:570.

Spring, C., Sandoval, J., 1976, Food additives and hyperkinesis: A critical evaluation of the evidence, *Jrl. Learn. Disab.* **9**(9):28–37.

Sokoloff, L., 1977, Regional metabolic rate in central nervous system as related to function, in: *Alcohol and Aldehyde Metabolizing Systems*, Vol. 3 (R. T. Thurman, J. R. Williamson, H. R. Drott, and B. Chance, eds.), pp. 453–470, Academic Press, New York.

Speer, F., 1975, Multiple food allergy, *Ann. Allergy* **34**:71.

Spitz, R., 1945, Hospitalism, an inquiry into the psychiatric conditions of early childhood, *Psychoanal. Study Child.* **1**:53.

Spitz, R., 1946, Hospitalism, a follow-up report, *Psychoanal. Study Child.* **2**:113.

Stine, J. J., 1976, Symptom alleviation in the hyperactive child by dietary modification: A report of two cases, *Am. Jrl. Orthopsychiatry* **46**(4):637–645.

Sulzer, J. L., Wesley, H. H., Leonig, F., 1973, Nutrition and behavior in Head Start children: Results from the Tulane study, in: *Nutrition, Development and Social Behavior* (D. J. Kallen, ed.), publication 73–242, Department of Health, Education, and Welfare.

Swanson, J. M., and Kinsbourne, M., 1980, Food dyes impair performance of hyperactive children on a laboratory learning test, *Science* **207**:1485.

Thatcher, R. W., Lester, M. L., Ignasias, S. W., and McAlaster, R., 1980, Intelligence and lead toxins in rural children, Presented at the U.S. Department of Agriculture Conference in Atlanta, 11/13/80.

Tisdall, F. F., Robertson, E. C., Drake, G. H., Jackson, S. H., Fowler, H. M., Long, J. A., Brouha, L., Ellis, R. G., Phillips, A. J., and Rogers, R. S., 1951, Canadian Red Cross school meal study, *Can. Med. Assoc. J.* **64**:477.

Trites, R. L., Tryphonas, H., and Ferguson, H. B., 1980, Diet treatment for hyperactive children with food allergies, in: *Treatment of Hyperactive and Learning Disordered Children* (R. M. Knights and D. J. Bakker, eds.), pp. 151–163, University Park Press, Baltimore.

Tuttle, W. W., Daum, K., Larsen, R., Salzano, J., and Roloff, L., 1954, Effects on school boys of omitting breakfast: Physiologic responses, attitudes and scholastic attainment, *J. Am. Diet. Assoc.* **30**:647.

Voorhees, M. L., Stuart, M. J., Stockman, J. A., and Oski, J. A., 1975, Iron deficiency anemia and increased urinary norepinephrine excretion, *J. Pediatr.* **86**:542.

Webb, T. E., and Oski, F. A., 1973a, The effect of iron deficiency anemia on scholastic achievement, behavioral stability and perceptual sensitivity of adolescents, *Pediatr. Res.* **7**:294.

Webb, T. E., and Oski, F. A., 1973b, Iron deficiency anemia and scholastic achievement in young adolescents, *J. Pediatr.* **82**:827.

Webb, T. E., and Oski, F. A., 1974, Behavioral status of young adolescents with iron deficiency anemia, *J. Spec. Ed.* **8**:153.

Wechsler, L. R., Savaki, H. E., and Sokoloff, L., 1979, Effects of *d*- and *l*-amphetamine on local cerebral glucose utilization in the conscious rat, *J. Neurochem.* **32**:15.

Weiss, B., Williams, J. H., Margen, S., Abrams, B., Caan, B., Citron, L. J., Cox, C., McKibben, J., Ogar, D., and Schultz, S., 1980, Behavioral response to artificial food colors, *Science* **207**:1487.

Wender, P. H., Epstein, R. S., Kopin, I. J., and Gordon, E. K., 1974, Urinary monoamine metabolites in children with minimal brain dysfunction, *Am. J. Psychiatry* **127**:1411.

Werkman, S. L., Shifman, L., and Skelly, T., 1964, Psychosocial correlates of iron deficiency anemia in early childhood, *Psychosom. Med.* **26**(2):125.

WHO/FAO, 1977, Dietary fats and oils in human nutrition, a report of an expert consultation, UN Food and Agriculture Organization, Rome.

Williams, J. I., Cram, D. M., Tausig, F. T., and Webster, E., 1978, Relative effects of drugs and diet on hyperactive behaviors: An experimental study, *Pediatrics* **61**(6):811.

Williams, R. J., 1956, *Biochemical Individuality: The Basis for the Genetotropic Concept*, John Wiley, New York.

Yager, J., and Young, R. T., 1974, Non-hypoglycemia is an epidemic condition, *N. Engl. J. Med.* **291**:907.

Zioudrou, C., Streaty, R. A., and Klee, W. A., 1979, Opioid peptides derived from food proteins: The exorphins, *J. Biol. Chem.* **254**:2446.

6

Vitamin Deficiencies and Excesses: Behavioral Consequences in Adults

M. W. P. Carney

1. Introduction

1.1. Vitamin Deficiency

Vitamins are organic chemical substances that the body requires for its metabolism and cannot synthesize for itself. They are therefore needed in the diet. Deficiency may cause physical, mental, and behavioral effects. Vitamins are traditionally divided into fat-soluble and water-soluble, and it is probably true that psychological abnormalities encountered in clinical practice are more often associated with deficiences of water-soluble vitamins than with deficiencies of fat-soluble vitamins. Some substances are grouped as vitamins, but deficiency of these apparently does not occur spontaneously. These substances will be considered separately.

It is arguable that in recent years, most progress has been made in this area in understanding the contributions of folic acid and, to a lesser extent, vitamin B_{12} to the genesis of psychiatric and neurological disease. Special emphasis will therefore be given to these vitamins.

Vitamin-deficiency disorders are due to impaired supply of the appropriate vitamin to the site of action as a consequence of deficiency in the diet, reduced intake due to anorexia and defective absorption, impaired metabolism, e.g., liver disease, or the action of drugs or alcohol. Fortunately, however, the grosser clinical manifestations of vitamin deficiency are now rare in the Western societies. Thus, the rickets that was prevalent among European urban schoolchildren at the end of the last century is rarely seen among the indigenous population, but can still occur sporadically. Indeed, it would be incorrect to

M. W. P. Carney • Northwick Park Hospital and Clinical Research Centre, Harrow, Middlesex HA1 3UJ, United Kingdom.

assume that gross disease is no longer clinically relevant. Apart from the under-developed nations, where severe vitamin-deficiency diseases, like other forms of malnutrition, are still endemic, severe disease still occasionally occurs among deprived, underprivileged, and vulnerable sections of the population. For example, classic wet beri-beri has been recorded in the United Kingdom among alcoholics (Carney, 1971; Riding, 1975) and in a depressed patient (Carney, 1970). However, in this section, evidence will be presented that more subtle clinical and biochemical vitamin deficiency is still widespread, particularly among predisposed groups. There should be increased awareness of this fact among physicians, and it emphasizes the need for screening of susceptible populations.

The elderly, especially when physically or mentally handicapped or when living alone in poor conditions, are very prone to dietary deficiency, particularly of folic acid, vitamin B_{12}, ascorbic acid, and the other B vitamins. Poor cooking facilities, loss of interest in food, and lack of heating exacerbate the situation. Batata *et al.* (1967) showed that patients with chronic medical diseases had a high incidence of abnormal folate levels. Recent investigations have underlined the prevalence of vitamin deficiencies among the mentally ill, especially of folate and B_{12}, and latterly of thiamine, riboflavin, and pyridoxine, particularly when the mental illnesses are complicated by anorexia, poor food intake, and drugs and alcohol. The complex relationships between folate deficiency and psychiatric symptoms in these patients have been systematically explored, so that it may soon be possible to separate which conditions have caused and which have been caused by specific vitamin deficiencies. It has long been recognized that Wernicke's encephalopathy and Korsakoff's psychosis may be due to thiamine deficiency, but the possibility that certain forms of endogenous depression and organic psychoses may be specific manifestations of deficiencies of individual B vitamins suggests rational therapy based on vitamin supplements for the disorders. The associations between cause and effect in the vitamin-deficient mentally ill are complex. Avitaminosis is often secondary to anorexia and impaired food intake; it may also be primary, or the two may be bound in a vicious circle—anorexia leading to malnutrition, avitaminosis, and mental symptoms that in turn give further anorexia and undernutrition.

Other groups with poor dietary habits may also develop the behavioral, emotional, and cognitive symptoms of vitamin deficiency. These include food faddists, religious, social, and health groups on restricted diets, the physically ill, alcoholics, deprived immigrant groups, the poor, and the drug-addicted. The poverty, inappropriate clothing, religious and dietary taboos, and cooking habits of many Afro-Asian immigrants living in the slum sections of European and North American cities place them particularly at risk. Thus, lack of sunlight, voluminous clothing, drain of pregnancy, and poor dietary intake are crucial in precipitating vitamin D deficiency in dark-skinned Muslim women in northern cities. Of course, similar factors may produce vitamin D deficiency in the malnourished alcoholic denizens of darkened bars.

Vitamin B deficiency may also arise among urban adolescents who derive most of their calories from "junk" carbohydrate foods relatively poor in vitamins.

Many drugs interfere with vitamin metabolism and may also reduce intake by their sedative effects or interfere with vitamin assays (see Sections 1.2 and 1.3).

Alas, too few systematic surveys of vitamin deficiency among the various sections of the healthy and deprived population have been done. Until this lack of research has been remedied, it will be difficult to form an accurate assessment of the prevalence of these disorders or plan an effective strategy to deal with them. There also needs to be further elucidation of the etiology of these deficiencies. This is not only of considerable academic importance in assisting the unraveling of the various mechanisms underlying abnormal behavior and mental processes, but also raises therapeutic issues that may transcend the purely academic ones. More rationally based therapies in psychiatry are urgently required. In the treatment of depressive illnesses, the disadvantages of antidepressant drugs in terms of their unwanted and toxic effects are widely acknowledged, and electroconvulsive therapy, though recognized to be life-saving in severe endogenous depression, is deeply unpopular. Correcting vitamin deficiency underlying abnormal mental mechanisms offers a new approach to an old problem, and properly controlled trials of vitamin supplements in these disorders is one of the more urgent needs of contemporary psychiatry.

The results of most recent epidemiological work suggest that most of the vitamin lack in psychiatric populations is the result of poor nutrition consequent on reduced appetite. Substances like drugs and alcohol may also interfere with intake and metabolism as described elsewhere. It is by no means certain that all vitamin deficiencies have a primary role in the genesis of psychiatric or behavioral abnormality, or that where such a primary role is likely, a particular deficiency causes a characteristic syndrome specific to that deficiency. Though it is obviously highly desirable to identify and correct causal vitamin deficiencies, it should not be forgotten that remedying secondary deficiency may indirectly improve the patient's mental condition by restoring his general condition and morale.

Vitamins may also have implications for public health authorities. Already, in several countries, milk and bread are fortified by the addition of vitamin supplements. In Australia, where there is widespread concern regarding brain damage resulting from excessive beer drinking, there is considerable agitation to have thiamine added to beverages as a prophylactic measure.

1.2. Drugs and Vitamin Deficiency

Drugs may depress serum levels of vitamins *in vivo*—in the case of barbiturates, within a matter of hours—or interfere *in vitro* with assay organisms. In these ways, they may give falsely low results in the absence of deficiency. Drugs may also cause deficiency by their metabolic effects. Perhaps the oldest example is alcohol, which is one of the most common associations of vitamin deficiency. Oral contraceptives are also frequently linked with pyridoxine deficiency and, possibly, folate deficiency (Shulman, 1978). Pyridoxine deficiency may also arise during drug treatment with antitubercular drugs like isoniazid (Cameron, 1978).

Many of the drugs used to treat psychiatric and behavioral disorders may be associated with deficiency, including barbiturates, phenothiazines, tricyclic antidepressants, monoamine oxidase inhibitors, and chlordiazepoxide (Carney, 1967; Carney and Sheffield, 1970). Anticonvulsants may also produce folic acid deficiency (Reynolds, 1967; Ziegler, 1978), probably by their antifolate action. Cytotoxic agents and antibiotics also have antifolate effects, while methotrexate, trimethoprim, and aminopterin seem to depend on their specific antifolate action for their therapeutic effect. Rivkin (1979) listed ACTH, phenothiazines, boric acid, and oral contraceptives as being associated with low riboflavin status. However, no differences in the incidence of thiamine, riboflavin, and pyridoxine deficiencies between inpatients taking psychotropic drugs and those not taking these drugs were found in a survey of 172 recently admitted psychiatric patients (Carney *et al.*, 1982).

Other vitamin–drug interactions, for instance, with vitamins C, D, and B_6, are listed under the appropriate headings.

Most of the reports of drug–vitamin effects have so far been anecdotal. This is clearly an unsatisfactory situation that should provoke systematic inquiries into the effects of drugs, for example, by comparing vitamin-deficient and nondeficient groups for drug intake.

1.3. Problems of Assessment

In assessing vitamin status in the clinical context, a number of adventitious influences have to be excluded. The effects of drugs *in vivo* and *in vitro* (as described in Section 1.2) have to be taken into account. Some indices of vitamin deficiency are relatively nonspecific. Thus, blood pyruvate levels can be raised in the presence of multiple sclerosis, hepatic cirrhosis, subacute combined degeneration, congestive heart failure, and thyroid disorders, as well as in thiamine deficiency.

A number of questions will then arise: Is the biochemical evidence (possibly an abnormal serum level or red cell concentration) connected with the patient's condition, or is it a chance finding? If it is related, is it associated with all the clinical features, or are some consistent with a vitamin deficiency and others due to other complications of the primary condition, e.g., signs of thiamine deficiency and hepatic cirrhosis occurring simultaneously in an alcoholic? If it is possible to establish which clinical features are related to the vitamin finding and which are not, the nature of the association between it and the symptoms has still to be unraveled. Is the vitamin deficiency causal or merely a result of the associated condition? Or are they linked in a sort of vicious circle: deficiency producing symptoms that in turn impair appetite and intake and aggravate the deficiency? Such a mechanism has been adduced to explain the relationships between depression and folate deficiency. It is apparent that in investigating the link between vitamin lack and clinical symptoms, information must be gathered about every aspect of the patient's life—e.g., diet, drugs, alcohol intake, and social functioning. Another complicating factor is that multiple vitamin deficiencies are common, possibly more common than single deficiencies. Thus, in addition to resolving the foregoing questions, it will be necessary to disentangle the clinical associations of each vitamin from those of the others.

The problems of ascertainment must also be considered further. The most suitable tests for screening a population for a given vitamin deficiency may not be the best for specifically identifying patients with *that* particular vitamin deficiency and nothing else. The reason is that a good screening test is a net that is cast widely to include all those who *might* prove to have the deficiency, e.g., serum concentrations or the hemogram in folic acid and B_{12} deficiency. Consequently, a large number of false–positives will be included, to be excluded later by further investigation. On the other hand, a test or battery of tests designed to identify a case with absolute specificity, i.e., to exclude all the false-positive cases, will cause a large number of atypical cases to be discarded along with the negatives.

Another problem is that different investigations have differing thresholds of sensitivity for a vitamin deficiency. Thus, one will give a high yield of positives, while the other will give a much smaller total. Further, different investigators may not be measuring the same thing, e.g., direct measurements of vitamin concentrations and indices of indirect concomitants of a vitamin deficiency, e.g., blood thiamine and red cell transketolase. This is not to say that the indirect measure is less valid for the task in hand than the direct measure. In the example just adduced, the blood thiamine assay is cumbersome and erroneous and therefore likely to give *less* valid results than the indirect measure of thiamine status.

A further problem concerns ethical considerations. For example, is it ethically and economically justifiable to perform serum or red cell folate and serum B_{12} estimations on all psychiatric inpatients because 10–20% will yield abnormal values? It could be argued that in view of the possibility of severe sequelae, i.e., dementia, subacute combined degeneration, suicidal depression, and serious illness requiring hospitalization, all such patients should be ascertained and given appropriate replacement therapy. Or could we compromise and say that only the more seriously ill patients should be so screened? If so, where is one to draw the line? Is there unequivocal evidence that the incidence of low values rises with increasing severity of illness? Or a case could be made for performing estimations on those more chronically ill. Again, where is the line to be drawn? Is there sufficient foundation for the supposition that the longer a deficiency goes untreated, the higher is the risk of an irreversible, severe effect like dementia? Since the cerebrospinal fluid (CSF) folate is presumably a closer measure of what is going on in the brain, are lumbar punctures to be recommended or justifiable for all these patients?

A lot then depends on final proof that such and such a vitamin deficiency causes such and such a disease. It may not be possible to secure this proof without double-blind-controlled trials. Here, difficulty may arise in allocating vitamin-deficient seriously ill patients to placebo. Though in a trial of this nature one does not directly intend giving inert pills to a suicidal depressive–he has a 50% chance of getting the active agent—the possibility of giving such a patient an inert preparation is disturbing to many staff members involved in these trials. It may well be that patients participating in this kind of investigation should be admitted to hospital, where close observation and full nursing care are possible. Informed consent, of course, will be required, and it is highly doubtful that psychiatric patients, legally detained against their will, would be considered capable of giving valid consent. Again, if a trial of B_{12} is contemplated, is one justified in giving a placebo to a patient in whom subacute combined degeneration may be precipitated or worsened? And is one ever entitled to give folate without vitamin B_{12} to folate-deficient patients without full knowledge of their B_{12} status in view of the acknowledged risk of precipitating neurological sequelae in them?

2. Fat-Soluble Vitamins

2.1. Vitamin A (Retinol)

Vitamin A is required for the integrity of mucosal surfaces and in the form of aldehyde for vision, especially in dim light. Deficiency results in skin and eye lesions and in reduced resistance to infection. Hypervitaminosis also has severe consequences. Vitamin A is not normally excreted, but is stored in the liver.

Systemic clinical signs of A deficiency include xerosis of the skin, follicular hyperkeratosis, xerosis conjunctivae, keratomalacia, Bitot's spots, and night blindness. Classic examples of Vitamin A intoxication have been reported in Arctic explorers who have eaten polar bear liver (1 lb of liver may contain over 20 times the normal human liver reserve of A). Megavitamin A therapy is considered in Section 4.

2.2. Vitamin D

Vitamin D is essential at all ages for maintenance of calcium balance and the integrity of the skeletal system. It is obtained by ingestion of Vitamin D_2 (ergocalciferol) or D_3 (cholecalciferol) and by exposure to ultraviolet light, which converts the 7-dehydrocholesterol of the skin to vitamin D_3. Most foods of animal origin contain vitamin D and D-fortified milk is an important source for urban dwellers. Excessive amounts of vitamin D are potentially dangerous and may lead to hypercalcemia and its complications. Deficiency of vitamin D classically gives rise to active rickets (and subsequently bony deformities) and osteomalacia with localized or generalized skeletal deformity. Other clinical features of D deficiency include bone pain, waddling gait, muscle weakness, and anorexia.

Vitamin D deficiency is nowadays more commonly seen in patients taking anticonvulsants and with abnormal hepatic excretory function. Low levels are also seen in nephrotic syndrome, those on repeated peritoneal dialysis, and in some chronic alcoholics (Velentzas *et al.*, 1977). Rickets was reported in 24 children of black Muslim women—some of them vegans—living in polluted urban surroundings (Bachrach *et al.*, 1979). Dark voluminous clothing, lack of sunlight, and diet were thought to have been contributory. Bleeding into the costochondral junctions of the ribs constitutes the characteristic "rachitic rosary."

In the psychiatric context, though vitamin D deficiency has been found in chronic alcoholics (King, 1977), presumably the foregoing factors together with the habit of spending the daylight hours in darkened bars were significant. An unusual form of behavior disturbance colloquially known as "Piblotko," thought to be due to hypocalcemia, and to be associated with D deficiency and conditioned by lack of sunlight and all-enveloping clothing, was described in Eskimos (Foulks and Katz, 1977). This took the form of screaming, shouting, running around, and tearing off the clothes. Carpopedal spasm and hypocalcemia were also seen. Serum calcium among Eskimos has been found to be low, and its effects are presumably exacerbated by absence of sunlight and the habit of hyperventilating, said to be customary among Eskimo women as a form of emotional expression.

Overdosage of any form of vitamin D, occasionally seen in food cranks and the victims of megavitamin therapy, is characterized by hypercalcemia and can lead to generalized calcinosis. Calcium phosphate is deposited in the joints, synovial membranes, kidneys, myocardium, pulmonary alveoli, parathyroids, pancreas, skin, and other sites (see also Section 4).

2.3. Vitamin K

Deficiency of vitamin K producing mental and behavioral effects does not occur spontaneously. However, ingestion of excessive quantities occasionally occurs as a form of self-induced illness (see Section 4).

2.4. Comment on Fat-Soluble Vitamins

Though little survey information is available, deficiency of these substances is not of epidemiological importance in Western psychiatry. However, many more data are required before definitive conclusions can be drawn.

3. Water-Soluble Vitamins

3.1. Ascorbic Acid (Vitamin C)

Vitamin C is essential for normal wound-healing and resistance to infection. It has various biochemical functions including collagen synthesis, catabolism of certain amino acids, participation in the formation of norepinephrine from dopamine and 5-hydroxytryptophan from tryptophan, and conversion of folic to folinic acid, so that a macrocytic anemia may result in deficiency states, a rôle in hemopoiesis, and reducing propensities. A principal end product of catabolism is oxalic acid, which may accumulate in megavitamin therapy with undesirable effects, precipitating renal stone formation in predisposed subjects and causing acidosis in patients in chronic renal failure.

Since primates cannot synthesize ascorbic acid, deficiency will result in those with inadequate dietary intake and in alcoholics with cirrhosis of the liver and in other forms of cirrhosis. In animals, depletion of vitamin C also delayed drug metabolism, but so far there has been little evidence of this in man (editorial, *British Medical Journal*, 1977). However, antipyrine half-lives were significantly longer in those patients with liver disease who had the lowest leukocyte ascorbic acid concentrations (Beattie and Sherlock, 1976). Lack of sufficient vitamin C in species unable to synthesize it leads to scurvy. The minimum daily dose to prevent scurvy is 10 mg daily (Beeson and McDermott, 1975, p. 1356), but this does not permit tissue saturation. Ascorbic acid is supplied by fruits, especially citrus fruits, and by most fresh vegetables.

General clinical manifestations are malaise, weakness, lassitude, perifollicular hemorrhages, particularly on the posterior thighs, anterior forearms, and abdomen, hemorrhages in subcutaneous tissue, muscles, or joints, and swollen, inflamed, bleeding gums. Loss of appetite, listlessness, and irritability are early signs of infantile scurvy. Subperiosteal hemorrhages lead to tenderness, swelling, and pain on movement. With respect to mental symptoms, anergia and apathy have traditionally been early signs of scurvy. Specific mental changes reported by Walker (1968) included depression. In experimentally induced scurvy, Kinsman and Hood (1971) described personality changes. Elevated scores on the "neurotic triad" of the Minnesota Multiphasic Personality Inventory—hypochondriasis, depression, and hysteria scales—were noted, as well as impairment of psychomotor and physical performance, motivation, and alertness.

Spring (1979) evaluated the vitamin C status of 39 mentally retarded children with a variety of neurotic, emotional, and physical problems and 9 upper-middle-class children of similar ages. In the two groups, 26% of the retarded children and none of the control children were found to be low excretors of vitamin C. Patients with cirrhosis of the liver, especially chronic alcoholics, may be deficient in ascorbic acid. Furthermore, in the latter, the intake of vitamin C may be reduced by anorexia and vomiting, as well as by malabsorption, pancreatic disease, and abnormalities of vitamin metabolism and storage (editorial, *British Medical Journal*, 1977).

3.2. Comment on Ascorbic Acid Deficiency

No specific psychological or behavioral conditions have been linked with C deficiency, but depression may be an association. However, much more epidemiological and etiological investigation is required to throw light on these questions.

3.3. Thiamine (Vitamin B₁)

Historically, thiamine deficiency, or beri-beri, was described in South East Asia, where polished rice provided a diet low in thiamine and high in carbohydrate. Deficiency became clinically manifest in Japanese prisoners of war when the daily intake fell below 0.3 mg 1000 cal, an acute episode often being precipitated by dysentery or another infectious disease. Beri-beri is still ocasionally seen in these rice-eating people or where there is interference with the ingestion or absorption of thiamine, e.g., in alcoholics. The more florid clinical syndromes are beri-beri, Wernicke's encephalopathy, and Korsakoff's psychosis. Milder cases of beri-beri are characterized by paresthesia, altered reflexes, pains resembling muscle ischemia, and nocturnal muscle cramps. In well-developed cases, there are tachycardia, cardiac enlargement, and dependent edema. The most common symptoms in Japanese prisoners of war in World War II were tachycardia, symmetrical foot- and wrist-drop, muscle tenderness, and mild disturbances of sensation. Most of these symptoms have been reproduced experimentally in 30–120 days in adult humans given less than 0.2 mg thiamine 1000 cal daily. The acute fulminating form of beri-beri, called Shoshin, is dominated by acute circulatory insufficiency. The chronic dry atrophic form, with foot- and wrist-drop, seen in older adults, responds less well to treatment. Infantile beri-beri due to thiamine lack in the mother's milk is also described.

Thiamine is widely distributed in foods and readily available, but is easily destroyed by roasting, stewing, and boiling.

In thiamine deficiency, pyruvate acid tends to accumulate in the tissues and the blood pyruvate is raised. Thiamine pyrophosphate is a coenzyme for transketolase, and reduced transketolase activity can be detected in red cells of patients with low dietary thiamine and also in normal subjects given thiamine-deficient diets (Word *et al.*, 1980).

The characteristic pathological lesions of thiamine deficiency are degeneration of peripheral and cerebrospinal nerves and the heart. In Wernicke's encephalopathy and Korsakoff's psychosis (cerebral beri-beri), histological changes are chiefly found in the mammillary bodies, in the wall of the third ventricle and aqueduct, and in the tegmentum of the medulla. Wernicke's syndrome is characterized by disturbed ocular motility, ataxia, mental changes, and, occasionally, polyneuropathy. Nystagmus is a constant feature, stance and gait are affected, and there is usually an intention tremor. Though sometimes the patient in the acute phase is floridly delirious, with fluctuating awareness, shifting orientation, delusions, colorful hallucinosis (Blackstock *et al.*, 1972), and tremor, more often he is apathetic, listless, indifferent, and disorientated in time and place, and misidentifies those about him. It is generally thought that most of these patients recover fully, but a fair proportion—according to Victor *et al.* (1971), as many as 155 of 186 survivors from Wernicke's encephalopathy—progress to a more chronic amnestic confabulatory syndrome known as Korsakoff's psychosis, with chronic cognitive impairment, deficiencies of abstract and conceptual thinking, and a tendency in the more established cases to fill in memory gaps with false ones (confabulation). Features of Wernicke's and Korsakoff's states may coexist in the same patient. Not all patients with thiamine deficiency develop Wernicke's encephalopathy, and an inborn error of transketolase metabolism has been suggested in those who develop these syndromes (Blass and Gibson, 1977).

In the psychiatric setting, thiamine deficiency has been most amply documented in the context of chronic alcoholism (Reilly, 1979). Thus, Fennelly *et al.* (1964) reported that 86.2% of the alcoholics with peripheral neuropathy and 44% of those without were thiamine-deficient, while Delaney *et al.* (1966) reported that 37% of a group of chronic alcoholics showed evidence of thiamine deficiency.

A major difficulty has been that until recently, efficient methods of estimating thiamine status were not available. Blood thiamine was cumbersome and erroneous to perform. Estimating blood pyruvate (Joiner *et al.*, 1950) had the disadvantage of being relatively nonspecific, other medical conditions also giving abnormally high readings. However, Phillips *et al.* (1952) confirmed by this method that thiamine lack caused Wernicke's encephalopathy.

More recently, in a study of B and C deficiency in 35 patients with alcohol-related disease, Barnes (1978) found 31% to be thiamine-deficient by means of red cell transketolase and 55% by means of pyruvate tolerance. McLaren *et al.* (1980), using various indices, assessed the thiamine status of 73 patients with an alcohol problem, 9 of them with Wernicke–Korsakoff syndrome and 10 with peripheral neuropathy. They found a 70% deficiency rate judged by a 7-day dietary recall measure and concluded that red cell transketolase was an efficient test of thiamine status.

Irritability, aggressive behavior, and personality changes have also been described in thiamine-deficient adolescents whose diet consisted largely of high-calorie carbohydrate "junk" foods (Lonsdale and Shamburger, 1980). Schwartz *et al.* (1979) described 42 physically healthy psychiatric inpatients in whom red cell transketolase estimations were done. In 16 with thiamine deficiency, this seemed to be linked with anorexia and dietary factors rather than a single psychiatric diagnosis.

Two larger studies of recently admitted psychiatric inpatients have been carried out by Carney and colleagues. In the first (Carney *et al.*, 1979), serum pyruvate estimations were done soon after admission on 154 patients with a history of poor diet, 74 of whom also had red cell transketolase measurements. By these criteria, 58 were considered to be thiamine-deficient. Significantly more of the low-thiamine as compared with the normal-thiamine patients showed clinical signs of malnutrition, i.e., weight loss exceeding 7 lb, dependent edema, and characteristic signs of B deficiency, or were diagnosed as chronic alcoholics, drug addicts, or functional psychotics. It was concluded that most of the thiamine deficiency was the result rather than the cause of the mental symptoms, but in some patients, it undoubtedly caused the mental symptoms, e.g., in 2 patients with Wernicke's encephalopathy. In the other study (Carney *et al.*, 1982), 172 unselected patients successively admitted to a district general hospital psychiatric unit were investigated. Psychiatric, physical, drug, and dietary evaluations, together with specific signs of B deficiency, weight loss, and other nutritional changes, were noted on a checklist. Red cell transketolase as well as riboflavin and pyridoxine status were assessed. Of these patients, 53% were deficient in one or more vitamin (41% in one, 12% in more than one), 30% of them in thiamine. There were significant excesses of schizophrenics and alcoholics among the low-thiamine and multiple-vitamin-deficiency patients compared with the remaining patients. It was again concluded that in this psychiatric setting, thiamine deficiency was usually secondary to anorexia in mental illness, rather than a primary factor. However, no systematic study of vitamin-replacement therapy has been carried out, and this clearly needs to be done.

3.4. Riboflavin (Vitamin B_2)

Riboflavin intake must be low for months before symptoms of deficiency appear. Ariboflavinosis tends to be precipitated by periods of stress, e.g., pregnancy, lactation, growth in children, infections. It commonly occurs with niacin and protein deficiency. Characteristic signs appear after 3–8 months' deprivation. In developing countries, mild signs appear in low-income groups and severe manifestations with pellagra conditioned by alcoholism, long-standing debilitating diseases, or food faddism.

Symptoms appear with burning of the lips, mouth, and tongue, photophobia, lacrimation, and burning of the eyes. Fissures appear at the corners of the mouth, and the lips become dry and rough. However, angular stomatitis and cheilosis may also be present in niacin or pyridoxine deficiency. There may be seborrheic dermatitis and erythema, especially on the face, scrotum, and vulva. The tongue is characteristically purplish red and deeply fissured. In chronic states, the papillae may become atrophied. However, the glossitis cannot be distinguished from that found in niacin, folic acid, or B_{12} deficiencies. The eyes may be swollen and red with conjunctivitis and injection and proliferation of the limbic capillaries.

Riboflavin is widely distributed in leafy vegetables, in meat and fish, and in milk. It is water-soluble and may be destroyed by exposure to sunlight or cooking. Riboflavin is the coenzyme for the active prosthetic groups of flavoproteins concerned with oxidative processes. Concentrations of free riboflavin, flavin mononucleotide, and flavin dinucleotide in the blood are easily measured, but are not reliable indicators of deficiency. For this purpose, the 24-hr urinary excretion or the activity of red cell glutathione reductase is more satisfactory, the latter having the advantage of being readily measured.

Little is known of the role of riboflavin lack in the genesis of mental illness and behavioral disturbance. Deficiency can be caused by inhibition of its contribution to its biologically active coenzyme derivatives by hypothyroidism, ACTH and aldosterone deficiency, phenothiazines, boric acid, oral contraceptives, and alcohol (Rivkin, 1979). However, it is difficult to disentangle any unwanted psychiatric effects of these substances from the signs of riboflavin deficiency precipitated by them. In a study of 210 New York adolescents (ages 13–19 years) of low economic status (but not so poor as to require public assistance) and mixed Hispanic-American, black, and white origins, Lopez *et al.* (1980) found an incidence of riboflavin deficiency, as measured by estimating the urinary excretion, of 17%. None had relevant acute or chronic medical conditions or was pregnant, and only 4 were on oral contraceptives. However, when deficiency was defined in terms of elevated red cell glutathione reductase activity coefficient values, which the authors believed to be more sensitive than other methods, then 23.3% were deficient, or 26.6% after those taking multivitamin preparations were excluded. Abnormal values were significantly related to milk consumption of less than one cup per day. The authors considered that in this group, the level of daily milk consumption was a good indicator of adolescent nutrition.

Riboflavin deficiency has also been recorded among alcoholics. Thus, in a series of 35, abnormal red cell glutathione reductase values were found in 23% (Barnes, 1978). In a study of 13 alcoholics, 2 were found to be riboflavin-deficient (Bayoumi and Rosalki, 1976). It is thought that vitamin B_2 deficiency in alcoholism is due to dietary deficiency rather than a direct toxic effect of ethanol (Bonjour, 1980b).

This author also felt that riboflavin deficiency might also be seen in women during pregnancy and lactation, in those on oral contraceptives, and during periods of physiological and pathological stress. As part of an investigation to ascertain the incidence of common B deficiencies in general psychiatric in-patients, Carney *et al.* (1982) carried out red cell glutathione reductase estimations on 172 newly admitted patients. Patients were regarded as deficient if they showed activity levels below, or vitamin effects or activity coefficient values or both above, the normal reference ranges for the test. By these criteria, 29% were riboflavin-deficient. Deficient patients did not show more clinical signs of deficiency or weight loss exceeding 7 lb than nondeficient patients; neither did they differ on age, sex, and length of illness distributions. Riboflavin- and pyridoxine-deficient patients, however, were alike in being associated with affective illness, endogenous depression, and neurotic depression, while thiamine-deficient patients had a variety of conditions in which undernutrition seemed to be a common factor. There was no evidence that consumption of any particular drug or alcohol was responsible for the B_2 deficiency. It was concluded that in these patients, affective change was likely to be an expression of riboflavin deficiency rather than a primary factor in its causation. Riboflavin deficiency has also been linked with depression by Nobbs (1974).

3.5. Niacin (Nicotinic Acid, Nicotinamide, Vitamin B_3)

Good dietary sources of niacin are liver, other meat, fish, whole grain cereal, enriched bread, peas, nuts, and peanut butter. The adequacy of the niacin intake is dependent on the availability of its metabolic precursor, tryptophan, 60 mg of which will be converted to 1 mg niacin. The daily recommended allowance in niacin equivalents (1 niacin equivalent = 1 mg niacin or 10 mg tryptophan) per 1000 calories is 6.6.

Pellagra is the disease resulting from dietary deficiency of niacin. Pellagra historically has been associated with maize diets because this cereal is low in niacin and tryptophan. It does not occur, however, where maize diets are supplemented by sufficient legume production. Though common in the United States in the 1930s, it now occurs only sporadically in Western industrialized societies, except under certain conditions, e.g., alcoholism, malabsorption, metabolic deviation of tryptophan metabolism in Hartnup disease and carcinoid, and in the course of chemotherapy with isoniazid, 6-mercaptopurine, 5-fluorouracil, and chloramphenicol. Pellagra is still common in India in populations subsisting on a diet of jowar. Here, its appearance is conditioned by an excess of leucine as well as by low niacin and tryptophan in this cereal. Sufficient intake of good-quality vegetable or animal protein will supply enough tryptophan to prevent pellagra even when dietary niacin content is low.

Experimental niacin and tryptophan deficiency in human subjects duplicates most but not all the features naturally occurring in pellagra, because in the latter there are nearly always complicating diseases and deficiencies. Cheilosis, angular stomatitis, proctitis, scrotal dermatitis, and vaginitis may be more related to concomitant deficiencies of B_2 and B_6 than to deficiencies of niacin. Deficiency of folic acid probably accounts for the megaloblastic anemia. The neurological and mental symptoms may be related to poor absorption of B_{12}. A Korsakofflike mental picture often seen in pellagra may be due to associated thiamine deficiency. The effects of concomitant protein, methionine, and choline deficiencies due to fatty liver must also be considered.

Pellagra is occasionally encountered in patients with alcoholism (Ishii and Nishilsara, 1981), cirrhosis of the liver, chronic infections, and metabolic and diarrheal disease. Pellagrinous dermatitis was also reported in malignant carcinoid, a rare tumor that may convert 60% of the body's tryptophan into serotonin instead of the usual 1%.

The early signs of pellagra are nonspecific—lassititude, anorexia, digestive symptoms, weakness, anxiety, irritability, and depression. Soreness of the tongue progresses to severe inflammation of all mucus membranes, which become bright red. It is painful to take nutriment by mouth, and severe weight loss ensues. There is dermatitis difficult to distinguish from extensive exposure to sunlight. In chronic pellagra, the skin becomes thickened, scaly, hyperkeratinized, and pigmented. Depression may progress to clouding of consciousness, delusions, and hallucinations. The patient may be hyperactive and manic or apathetic, lethargic, even stuporous.

Nicotinamide in the body functions as a component of two coenzymes, diphosphopyridine nucleotide and triphosphopyridine nucleotide, that contribute to the oxidation–reduction enzyme systems necessary for tissue respiration, glycolysis, and fat synthesis. The combined excretion of the niacin metabolites in N1-methylnicotinamide and pyridoxine rarely exceeds 2 g daily. The skin lesions are hyperkeratotic, with vesicles containing erthyrocytes, fibrin, and melanin. Pathological changes are also found in the brain, spinal cord, and peripheral nerves.

Spivak and Jackson (1977) analyzed the case histories of 18 patients with pellagra seen over a period of 4 years at the Johns Hopkins Hospital. Most were alcoholics. Some had had gastrointestinal surgical procedures or malabsorption. Central nervous system (CNS) abnormalities included Wernicke's syndrome, confusion, stupor, dementia, nystagmus, ataxia, and diffuse slow-wave activity in the EEG. Skin changes and diarrhea were also constantly found.

3.6. Pyridoxine (Vitamin B₆)

Vitamin B_6 in the form of pyridoxal phosphate has been shown to play a part in carbohydrate, fat, and protein metabolism. It is believed to be an essential cofactor for many enzyme reactions in the metabolism of the biogenic amines thought to determine normal mood. Deficiency of pyridoxine may thus theoretically cause mood distrubance. Vitamin B_6 comprises several naturally occurring pyridoxines: pyridoxine, pyridoxal, and pyridoxamine. Good natural sources are liver, other meat, whole grain cereal, soya beans, peanuts, corn, and many vegetables.

Pyridoxine deficiency has rarely been reported in nature. It occurred in children fed a proprietary formula in which B_6 had been destroyed by sterilization, and this outbreak was characterized by restlessness, irritability, and convulsions (Beeson and McDermott, 1975, p. 1357). Pyridoxine lack has also been blamed for some cases of peripheral neuropathy and for the effects of isoniazid overdosage occasionally seen in tuberculous Canadian Indians under treatment with this drug. Clouded consciousness and epileptic fits occurred (Cameron, 1978). It may also be responsible for anemia in adults (Beeson and McDermott, 1975) and was reported in uremia (Stone *et al.*, 1975) and in a geriatric group (Hoorn *et al.*, 1975).

The results of measuring pyridoxine in psychiatric patients have been conflicting and confusing. This is partly due to the multiplicity of methods available and the difficulty in making valid comparisons between the results obtained thereby. There seems to be some agreement, however, that deficiency occurs in chronic alcoholism, though little explanation as to why this should be so. Using a microbiological method, Davis and Smith (1974) reported that 20 of 50 alcoholics were deficient, and Bonjour (1980a), also stating that pyridoxine lack occurred in this condition, felt that this was due to liver damage. However, Lumeng and Li (1974) found 35 of 66 alcoholics with normal liver function to be pyridoxine-deficient. Moreover, none of the patients found to be low in pyridoxine showed abnormal liver function in the group of patients studied by Carney *et al.* (1979). On the other hand, Barnes (1978), measuring red cell aspartate transaminase levels, found no evidence of pyridoxine deficiency in his group of alcoholic patients. One explanation that may reconcile these apparently conflicting results is that aspartate transaminase may be a relatively insensitive test except in profound deficiency (Shane and Contractor, 1975).

Reinker *et al.* (1972) also used red cell aspartate transaminase to demonstrate pyridoxine deficiency in epileptics treated with anticonvulsants.

Depressed women receiving oral contraceptives have been shown to be lacking in pyridoxine (Adams *et al.*, 1973). These investigators, in a double-blind placebo-controlled trial of pyridoxine in deficient women, showed that it was superior to placebo in treating depressive symptoms. However, in another study (Livingston *et al.*, 1978), vitamin B_6 concentrations were measured by the erythrocyte oxaloacetic transaminase activation test in 40 nonpregnant women of reproductive age, 20 postpartum nondepressed women, and 24 postpartum depressed women. Symptoms were assessed on the Beck depression inventory and the depression adjective checklist. No differences in mean vitamin concentrations were found among the groups. The depression scores on both rating scales were higher among the depressed than among the nondepressed women, but no evidence of B_6 deficiency was found among these depressed females.

In a study of 64 psychiatric inpatients selected for a history of inadequate diet before admission, on whom aspartate transaminase estimations were performed, Carney *et al.* (1979) found that 28 (44%) were deficient, and 19 of these had an affective illness. Moreover, abnormal aspartate transaminase was found in 15 (79%) of 19 endogenous depressives and only 13 (29%) of the remaining 45 patients, a significant difference. This link between pyridoxine deficiency and affective illness was subsequently confirmed (Carney *et al.*, 1982).

3.7. Comment on Thiamine, Riboflavin, Niacin, and Pyridoxine Deficiency

The classic B-deficiency diseases, beri-beri, Wernicke's encephalopathy, and pellagra, are rarely seen in most Western industralized countries. Only three cases of Wernicke's syndrome were encountered among 326 newly admitted patients whose vitamin status was ascertained. Two of these had thiamine deficiency and one riboflavin deficiency (Carney *et al.*, 1979, 1982). However, lesser degrees of vitamin deficiency—biochemical and subclinical or with very mild symptomatology—are common.

Thiamine deficiency is particularly frequent among alcoholics and others with poor nutrition, e.g., schizophrenics and drug addicts. Though further work needs to be done to establish the precise implications of biochemical B deficiency, there is little doubt that neglecting to correct thiamine deficiency in alcoholism carries a risk of Wernicke–Korsakoff states with permanent or semipermanent brain damage. Indeed, in Australia, where thiamine deficiency due to excessive drinking is a serious problem, there is considerable agitation to have thiamine added to beer as a prophylactic measure.

Riboflavin and pyridoxine deficiencies are not so obviously linked with anorexia, but may arise as a result of the use of various drugs, especially oral contraceptives. These risks should be borne in mind by physicians prescribing these preparations, and it would seem prudent to keep such patients under some degree of medical observation. Riboflavin and pyridoxine deficiencies seem to be more clearly linked with a single mode of psychiatric presentation, affective illnesses, notably depression, than is the case with thiamine, and this finding suggests that deficiency of these vitamins may play a primary part in the genesis of depressive illness. However, this would not exclude a role for dietary deficiency, in terms of either low intake or impaired absorption and utilization, possibly conditioned by drugs or another physical condition. Vitamin deficiency and affective illness may be joined in a vicious circle—deficiency leading to depression, which in turn produces anorexia, impaired intake, and further deficiency.

The classic nicotonic acid deficiency triad of dementia, dermatitis, and diarrhea is now no longer epidemic in Western countries, but occurs sporadically in alcoholism, malabsorption following gastrointestinal operations, and with certain drugs. Jolliffe's encephelopathy (Jolliffe, 1960) is still seen in the elderly. Pellagra may also complicate rare metabolic conditions such as carcinoid and Hartnup's disease. Mental changes include, in milder cases, affective changes and, in more severe cases, clouding of consciousness progressing to delirium. These mental symptoms are also seen in megatherapy in the treatment of schizophrenia. The prevalence of mild forms of niacin deficiency in hospital populations and the community has yet to be explored.

3.8. Cyanocobalamin (Vitamin B_{12})

Vitamin B_{12} occurs predominantly in foodstuffs of animal origin. There is very little present in vegetables. Thus, vegetarians may be at risk from deficiency. Most deficiency, however, arises in those suddenly predisposed by virtue of infections, malabsorption, absence of intrinsic factor (pernicious anemia), or impaired intake due to anorexia and apathy due to other disease.

B_{12} is essential for the function of all cells, but particularly for those of the bone marrow, CNS, and gastro-intestinal tract. It facilitates reduction reactions and the transfer of methyl groups, as well as participates in nucleic acid and folate metabolism. Deficiency of B_{12} results in the development of megaloblastic anemia, neurological signs, psychiatric disorder, sore tongue, paresthesias, and amenorrhea. As little as 1 μg daily will cure most cases of megaloblastic anemia, but higher doses are needed to replenish body stores.

That pernicious anemia could have nervous complications was recognized in the last quarter of the last century, and neurological descriptions of these complications were refined and extended until 1926, when Minot and Murphy (1926) instituted liver therapy for this disease. In 1948, cyanocobalamin was synthesized (Rickes *et al.*, 1948; Smith and Parker, 1948), and there was little further development of knowledge of its clinical and metabolic effects on the CNS until recent years.

It was known that some cases of megaloblastic anemia were due to folate deficiency and that giving folic acid to pernicious anemia patients could precipitate or cause rapid deterioration of symptoms of subacute combined degeneration. Furthermore, there is a poor correlation between the hematological and neurological complications of B_{12} deficiency (Reynolds, 1976). Administration of folate to pernicious anemia patients will result in hematological re-

Table I. Surveys of Serum B_{12} in Psychiatric Inpatients

Authors	Number	With low B_{12}	Comments
Edwin *et al.* (1965)	396	5.8% below 100 pg/ml	1 Case of pernicious anemia (PA); no specific psychosyndrome.
Carney (1967), Carney and Sheffield (1970)	374	14.2% below 150 pg/ml	Consecutively admitted; no case of PA; organic psychosyndrome.
Carney and Sheffield (1978)	272	26.1% below 150 pg/ml	Selected for poor nutrition, alcoholism, severe psychosis; no case of PA; organic psychosyndromes.
Shulman (1967b)	117	8.5% below 150 pg/ml	5 Had PA; all thought to be due to poor diet.
Hallstrom (1969)	90	6.7% below 170 pg/ml	About half had organic cerebral signs.

sponse, but relapse is very frequent. Thus, there are good grounds for saying that the CNS mechanisms of folic acid and B_{12} deficiency are very different and, on the other hand, that there is little obvious association between the hematological and neurological effects of either vitamin (assuming that folate deficiency has effects on the CNS). In fact, very little is known of the brain metabolism of either vitamin. It is believed, however, that in megaloblast formation, B_{12} deficiency interferes with the metabolism and transport of 5-methyltetrahydrofolate, which is the major circulating form of folic acid (Reynolds, 1976). One action of vitamin B_{12} deficiency may be to cause a block in folate metabolism and thus give relative folate deficiency (Chanarin, 1973; Perry *et al.*, 1976). It is also suggested (Reynolds, 1976) that the differences between the development of hematological and neurological complications of B_{12} deficiency are attributable to the different rates of progress of these complications and the very different characters of neurons and blood-forming cells. In addition, the operation of an efficient blood–brain barrier (Spaans, 1970; Shaw *et al.*, 1971; Mattson *et al.*, 1973; Chanarin *et al.*, 1974) would tend to make neuropsychiatric differences between B_{12} and folate deficiencies likely.

    Carney (1967), in a survey of 377 patients successively admitted to a mental hospital and a general hospital psychiatric unit, found an incidence of 14% of B_{12} values less than 150 pg/ml (Table I). Organic psychosis was the most common diagnosis among them, followed by depression and schizophrenia. This diagnostic distribution was significantly at variance with that of the patients with serum B_{12} exceeding 150 pg/ml. Medically prescribed drugs of various kinds—phenothiazines, benzodiazepines, trycyclic antidepressants, and monoamine oxidase inhibitors as well as barbiturates—were found to be linked with low B_{12} (Carney and Sheffield, 1970).

    In a further survey of 272 newly admitted psychiatric patients (Carney and Sheffield, 1978), partly selected because they were considered to be at risk from folic acid and B_{12} deficiency, B_{12} of less than 150 pg/ml was found in

21.3%. Again, organic psychotics were significantly overrepresented among them, and these patients had a lower mean B_{12} than the remaining patients. The low-B_{12} patients had more hematological abnormalities than the normal-B_{12} patients.

Other surveys of serum B_{12} found low concentrations to be frequent in psychiatric patients (Edwin *et al.*, 1965; Shulman, 1967a,b; Hallstrom, 1969). Shulman (1967a) reported that low B_{12} was causally linked with cerebral abnormality, notably memory defect, a finding that would appear to be in harmony with the aforecited observations of a linkage between B_{12} deficiency and organic psychosis.

More recently, another large survey of vitamin B_{12} concentrations was done on consequtive psychiatric admissions (Elsborg *et al.*, 1979). Among 835 patients, 10% were found to have levels less than 200 pg/ml. By far the most common causal factor was thought to be dietary insufficiency. Good feeding produced improved B_{12} levels in 75%. The only patients to show persistently low B_{12} concentrations were those with organic psychosis, in whom it was thought that this was due to apathy and impaired food intake consequent on loss of appetite. It was also felt that while most patients recover without B_{12} supplements, in some, nervous system damage may be exacerbated unless such supplements are given. However, no specific mention of the result of B_{12} supplementation was made.

In an investigation of the neurological and psychiatric sequelae of 84 patients selected on the basis of megaloblastic bone marrow changes, Shorvon *et al.* (1980) found that the 50 patients with B_{12} deficiency showed organic mental change (26%), subacute combined degeneration of the cord (16%), evidence of peripheral neuropathy (40%), and affective disorders (20%). This diagnostic pattern was very much at variance with that of the low-folate patients, in whom affective disorders predominated (see Section 3.10), peripheral neuropathy was much less common, and subacute combined degeneration of the cord was not seen (though 25% of each group showed organic mental signs).

All these folate- and B_{12}-deficient patients were given appropriate replacement therapy in normal clinical doses, and in the follow-up study extending over a year of those who showed psychiatric disorder and were willing to attend the psychiatric department, only 50% of 16 B_{12}-deficient patients showed complete psychiatric remission compared with 8 of 9 folate-deficient patients (Carney, 1979).

3.9. Folic Acid (Pteroylglutamic Acid)

Folic acid is found in nearly all natural foods, especially yeast, meat, and green leafy vegetables, but it is heat-labile and rapidly destroyed by cooking. Thus, a pregnant woman whose daily diet contains neither uncooked fruit nor a lightly cooked vegetable and all of whose food is heated at 100°C for periods exceeding 15 min can be assumed to be folate-deficient (Beeson and McDermott, 1975, p. 1405). Dietary folate deficiency is also present in many alcoholics because they spend little on food.

Folic acid in the nervous system is involved in nucleoprotein synthesis, as a methyl donor in transmethylation, in monoamine (notably serotonin) metabolism, and in synaptic events, possibly as a neurotransmitter. Folate has excitatory and inhibitory properties and counteracts the inhibitory effects of γ-aminobutyric acid. The neuropsychiatric effects of B_{12} deficiency may well be due to a block in folic acid metabolism (Reynolds, 1976). Folic acid also has a role in purine metabolism.

Deficiency may arise from inadequate dietary intake, impaired absorption, excessive demands by the tissues of the body, metabolic derangements, and interference in folate metabolism by drugs. The clinical effects of folic acid deficiency comprise megaloblastic anemia, a variety of neurological syndromes, and psychiatric disability. These may coexist or appear quite independently over a protracted period of time. However, it must also be remembered that folate body stores are rapidly depleted by reduced dietary intake—much more rapidly than B_{12} is depleted—and deficiency states can arise acutely.

Unlike B_{12} deficiency, the concept that folate deficiency could be associated with nervous system involvement, in particular subacute combined degeneration, has only slowly been accepted. However, the possibility that folate deficiency may well have effects on the nervous system is suggested by several lines of evidence. The concentration of folate in the CSF and in the brain is much higher than in serum, and there is an efficient blood–brain-barrier mechanism to maintain this situation. An obstacle to accepting that folate is incriminated in nervous disease has been the observation that folate deficiency may be present without any evidence of CNS involvement, e.g., in folate-deficient megaloblastic anemia. However, evidence of the existence of nervous system involvement in folate-deficient systemic states is not lacking.

Reynolds *et al.* (1973) reported neurological disease in folate-deficient medical patients. They compared the neurological status of 24 patients with severe folate deficiency and 21 patients with normal folate levels. A significant increase in organic brain syndromes and spinal-tract damage was found in the vitamin-deficient group, this increase being independent of the degree of anemia or the presence of alcoholism. Though both groups contained a number of patients with medical conditions known to produce neurological disease by nonfolate mechanisms, when these were excluded, 5 of 6 remaining folate-deficient patients had neurological signs compared with 1 patient (depressed) of 6 in the normal-folate group. Grant *et al.* (1965) reported 7 cases of unexplained neurological disease associated with folate-deficient, megaloblastic anemia, probably nutritional in origin. Four of these patients had spastic paraplegia and did not improve with folic acid; the other three with peripheral neuropathy did show improvement with replacement therapy. Other isolated examples of apparently wholly or partially reversible neurological disease linked with megaloblastic anemia due to folate deficiency include two elderly women with dementia (Strachan and Henderson, 1967), a female with motor neuropathy and cerebellar signs (Dérot *et al.* 1967), a patient with dementia and a condition indistinguishable from subacute combined degeneration (Pincus *et al.*, 1972) and another possible case of sub-acute combined degeneration (Ahmed, 1972).

Evidence that drug-induced folate deficiency in epileptics receiving long-term anticonvulsant therapy might lead to neuropsychiatric complications was put forward by Reynolds (1967). Ziegler (1978) listed clouding of consciousness, stupor, hallucinations and a variety of neurological signs due to diphenyl hydantoin toxicity, nearly all mediated by its effects on folate metabolism. The majority of reports on this topic supported the Reynolds findings, though Frenkel *et al.* (1973) contended that these effects were due to B_{12} deficiency and Rose and Johnson (1978) concluded, after surveying 102 epileptics with mental symptoms, 96 of whom were receiving long-term anticonvulsants, that the mechanism of macrocytosis in these patients was mainly by nutritional lack other than folate deficiency.

Folate deficiency in geriatric patients with mental symptoms was also reported to be common (Sneath *et al.*, 1973; Read *et al.*, 1965; Shulman, 1967b). The common mental disturbance in these patients is dementia, and it was assumed that the folate deficiency was always secondary to poor nutrition due to their mental disturbance. Subsequent evidence (see below) has thrown doubt on this simple hypothesis. Batata *et al.* (1967) also reported an elevated incidence of folate deficiency and mental symptoms in a series of patients with chronic medical ailments.

There have been a number of reports of improvement in folate-deficient neurological and psychiatric patients with folate-replacement therapy. Thus, Reynolds (1967) reported mental improvement in 22 of 26 patients with treatment lasting 1–3 years. Neubauer (1970) reported improvement in 22 of 50 epileptic children given folate and B_{12}. Carney and Sheffield (1970) reported a significant improvement in terms of condition at discharge and length of stay in 39 folate-treated folate-deficient inpatients compared with 63 untreated folate-deficient inpatients. The folic acid was given fortuitously on a virtually randomized basis, and improvement was confined to patients suffering from schizophrenia and endogenous depression. Patients with neuroses, personality disorders, and dementia were not found to have improved. In another investigation (Carney, 1979), 8 of 9 depressed patients with folate-deficient megaloblastic anemia treated with folate therapy made a virtually complete mental recovery compared with a much smaller proportion of similarly treated depressed patients with B_{12} megaloblastic anemia. Other favorable results with folate-replacement treatment have been reported by Botez *et al.* (1976) and Manzoor and Runcie (1976).

Perhaps the most clear-cut illustrations of the etiological role of folic acid in mental disorder are seen in the inborn errors of folate metabolism associated with mental retardation and neuropsychiatric features (Reynolds, 1976), e.g., congenital malabsorption of folate (Luhby *et al*, 1961; Lanzkowsky, 1970; Santiago-Borrero *et al.*, 1973), formino transferase deficiency syndrome (Arakawa, 1970), methyltetrahydrofolate transferase deficiency (Arakawa *et al.*, 1967), cyclohydrolase deficiency (Arakawa *et al.*, 1967), dihydrofolate reductase de-

Table II. Surveys of Serum Folate in Psychiatric Inpatients

Authors	Number	Subnormal serum folate (%)	Associations
Carney (1967, 1970), Carney and Sheffield (1970)	423	22	Endogenous depression, organic psychoses, chronic illness, physical illness; drug therapy and malnutrition in 75%
Hunter *et al.* (1967)	75	50	35% Taking alcohol, barbiturates, or anticonvulsants
Hallstrom (1969)	90	10	Alcohol, barbiturates, and anorexia
Kallstrom and Nylof (1969)	115	20.9	—
Reynolds *et al.* (1970)	100	24	High depression scores; poor response to antidepressant drugs
Carney and Sheffield (1978)	272	21	Depression, malnutrition, physical illness, and anorexia

ficiency (Taurot *et al.*, 1976), and homocystinuria associated with deficiency of 5,10-methylene-tetrahydrofolate reductase (Freeman *et al.*, 1975).

Another striking example is the occasional accounts of neurological damage and encephalopathy in children given intrathecal administration of the folate antagonist methotrexate in the treatment of childhood leukemia. Other examples followed treatment by the antifolate drugs trimethoprim and aminopterin.

There are many reports of folate deficiency in the psychiatric context (Table II). Herbert (1962) noted insomnia, forgetfulness, and irritability in the course of his self-imposed dietary deficiency of folic acid. Carney (1967) surveyed serum folate levels in psychiatric patients admitted under his care to two hospitals, one a mental hospital and the other a general hospital, over a period of a year. Of 423 patients, 105 (25%) had low folate levels (less than 2 mg/ml), significantly more than a control group of 62 normal patient subjects. Low serum folate was associated with endogenous depression and organic psychoses, a pattern at variance with that found in the normal-folate patients. In an examination of contributory factors, Carney and Sheffield (1970) found that low folate levels were associated with drugs—barbiturates, phenothiazines, and antidepressants—in the 3 weeks preceding admission in 75%, with physical illness in 18.6%, with malnutrition in 22%, and with chronic illness (duration more than 1 year) in 62%. No such extrinsic factors were found in 25% of low-folate patients. Of low-folate patients, 31.4% had hematological abnormalities.

In a further study of 272 psychiatric patients admitted to the general hospital unit over 2 years, Carney and Sheffield (1978) substantially confirmed these findings. Of these patients, 21.3% had serum folate less than 2 mg/ml. Low-folate status was significantly associated with depression, malnutrition, physical illness, and hematological abnormality. There was an excess of alcoholics among those with macrocytosis.

Other investigators have reported a higher than expected incidence of low folate values among psychiatric patients (Edwin *et al.*, 1965; Hunter and Mathews, 1965; Shulman, 1967a,b; Hallstrom, 1969). However, serum folate does not always reflect true folate deficiency because of the *in vitro* and *in vivo* effects of drugs and other factors (Herbert *et al.*, 1965; Carney and Sheffield, 1970; Shulman, 1972). Nevertheless, the association of low folate with hematological abnormality and distinctive patterns of diagnosis seen in several of these studies indicated that these results were not merely artifactual.

An investigation that started with inpatients shown to have megaloblastic anemia due to folate and B_{12} deficiency and surveyed the incidence of various psychiatric and neurological disorders in them was reported (Shorvon *et al.*, 1980). The neuropsychiatric status of 34 patients with folate deficiency and 50 patients with B_{12} deficiency presenting with megaloblastosis in a general hospital was examined and compared. Abnormalities of the nervous system were found in two thirds of both groups. The most common conditions among the folate-deficient patients were depression (56%) and organic mental change (25%). Only 18% showed evidence of peripheral neuropathy and none of subacute combined degeneration of the cord. By contrast, peripheral neuropathy was the most common disorder in the low-B_{12} group, 25% of whom had organic mental change and 16% evidence of subacute combined degeneration. No relationship between the neuropsychiatric findings and hematological abnormalities was found. Of these patients, 25 cooperated in psychiatric follow-up; 9 showed low folate and 16 low B_{12}. All had been given the appropriate vitamin and completed psychometric investigations as well as periodic psychiatric assessment. As noted in Section 3.9, 8 of 9 low-folate but only 50% of the low-B_{12} patients were found to have made a virtually complete recovery with respect to their psychiatric symptoms at the end of the 12 months.

An interesting comment on the role of diet in the folic acid deficiency of psychiatric patients comes from Thornton and Thornton (1978). They compared the folate status and diet of groups of psychiatric inpatients and normal control subjects. Folate deficiency was found only in the psychiatric patients (30%). There being no dietary difference between the groups, the authors were unable to account for the finding on this basis. Another attempt to control contributory factors was the work of Ghadirian *et al.* (1980), who studied three groups of patients, depressed, psychiatric nondepressed, and medical, who were kept on a standard diet without drugs or added vitamins for a period of 1 week. At the end of this time, the mean serum folate of the depressed patients was lower than the means of the other groups, and subjects from this group received significantly higher ratings on certain items of the Hamilton rating scale for depression than the others. These results indicated that depression was associated with folate deficiency not obviously consequent on dietary insufficiency.

The questions raised in Section 1 are very relevant to consideration of the relationship of mental state and folate results. For reasons already given, it is likely that the low-folate results were not merely artifacts of the effects of drugs either *in vivo* or *in vitro*. With regard to the question of whether there is a significant association between measures of folate status and mental state, the answer must be affirmative. Can the psychiatric associations of low folate be clearly distinguished from the psychiatric effects of other variables, for example, drugs, alcohol, diet, physical illness, and chronicity of illness? Again, the evidence listed above is positive. If, then, a clear association between folate and neuropsychiatric features has been established, the next step is to determine whether the folate findings are the cause or the result of the psychiatric condition or whether they are linked in a vicious circle as previously described. Much of the evidence suggests that a good deal of folate deficiency is the result of impaired nutrition consequent on the anorexia found in mental illness. However, this by no means accounts for all the psychiatric findings described. Again, the vicious-circle mechanism undoubtedly plays a part, but it is very hard to demonstrate with certainity that it does. A lot of circumstantial evidence points to folic acid being a primary factor in neuropsychiatric disease. Certainly it is accepted as being such in the inborn errors of folate metabolism and the nervous sequelae of these drugs that work through antifolate mechanisms in leukemia, malignant disease, and epilepsy.

However, final proof of a causal link must depend on long-term prospective studies such as that of Shorvon *et al.* (1980) and, if ethical objections can be overcome, placebo-controlled trials of folic acid in folate-deficient psychiatric and neurological patients.

3.10. Comment on Vitamin B_{12} and Folic Acid Deficiency

Though the role of vitamin B_{12} deficiency in precipitating neuropsychiatric damage has long been recognized, its more precise effects have only recently been worked out. On the other hand, the primary role of folate deficiency in these conditions has only lately been recognized, but knowledge of it has advanced so rapidly that we probably have a more comprehensive grasp of its effects on the CNS than is the case with B_{12}. The prevalence of deficiency of each of these vitamins in psychiatric in-patients probably varies from 10 to 20%. The most common associated psychiatric disorders are depressive illness and dementia, the former being characteristic of folic acid deficiency and the latter, of B_{12} deficiency. However, organic psychosis does occur in folic acid deficiency, and may represent a later stage in its characteristic neuropsychiatric syndrome, and affective symptoms in B_{12} deficiency. The psychiatric effects of both deficiencies may develop independently of hematological or neurological signs, but equally, all these manifestations may develop simultaneously, though in a minority of patients. B_{12} deficiency is more commonly linked with neurological abnormality than folic acid. The B_{12} deficiency seen in psychiatric patients is rarely associated with pernicious anemia. Other associations of folate deficiency are poor nutrition, drugs (mainly phenothiazines, tricyclic antidepressants, monoamine oxidase inhibitors, and barbiturates, but also anticonvulsants, antibiotics, and cyctoxic agents) given in the 3 weeks before admission, concomitant physical disease, and a history of psychiatric illness of more than 3 years. There is good evidence that B_{12} deficiency causes brain and neurological damage in about half the patients with established deficiency, e.g., with megaloblastosis. Neurological abnormalities linked with folate deficiency, however, are relatively uncommon.

Treatment with folic acid supplements in deficient patients appears to give good results in the presence of depression, but less good where there are organic mental signs. Replacement with B_{12}, however, gives less complete improvement in deficient patients. The outcome over 1–2 years after treatment in the two deficiencies therefore seems to differ markedly, that with the folate-deficient patients being better than that with the B_{12}-deficient patients.

While the etiological role of vitamin B_{12} deficiency in neurological and organic brain symptoms is beyond reasonable doubt, that of folate deficiency in its associated psychosyndromes is not yet finally clear. Long-term prospective studies and, if ethical problems can be overcome, controlled trials are needed. However, there is no doubt about the primary role of folate deficiency in the organic brain syndromes seen in the rare inborn errors of metabolism. Nevertheless, most folate and B_{12} deficiency seen in psychiatric patients is probably secondary to anorexia and impaired intake symptomatic of the underlying psychiatric conditions.

3.11. Other Water-Soluble B Vitamins

3.11.1. Pantothenic Acid

Pantothenic acid is of biological importance because it is part of a coenzyme on which acylation, acetylation, and other processes depend. Deficiency has been induced in human volunteers by feeding them the pantothenic acid antagonist ω-methylpantothenic acid together with a diet deficient in the vitamin. This resulted in a few months in reduced antibody formation and serious clinical signs and symptoms.

Neuropathy associated with low serum pantothenic acid has been observed in alcoholics consuming very deficient diets. Deficiency of this substance was also thought to have been largely responsible for the burning feet syndrome. However, the vitamin is so widespread in meat and vegetable foods that deficiency seems not to arise spontaneously.

3.11.2. Biotin

Biotin is essential for the activity of many enzymes in animals and probably man. The vitamin is widely distributed in food, so deficiency does not naturally occur in man. Experimental deficiency in man induced by feeding the metabolic antagonist avidin, found in egg white, results in scaly dermatitis, extreme lassitude, anorexia, muscle and precordial pain, insomnia, and slight anemia.

3.11.3. Choline

Choline serves as a source for labile methyl groups in the body and is considered necessary for nerve function and lipid metabolism. Deficiency in humans, however, has not been demonstrated.

4. Megavitamin Dosage

Megadoses of vitamin may be prescribed in the course of nonpsychiatric medical treatment, self-prescribed by food faddists and those who wish to induce artifactual illness in themselves presumably to attract medical attention, and also to treat a variety of psychological problems and learning difficulties in children.

Recently, vitamin A intoxication has appeared in children given megavitamin A therapy for learning difficulty, schizophrenia, organic brain syndrome, and a variety of other psychological and behavioral problems (Shaywitz *et al.*, 1977). Though in uncontrolled trials of megatherapy (Cott, 1971; Krippner and Fisher, 1973) improvement has been reported, a grave variety of toxic manifestations were encountered (Oliver, 1958; Persson *et al.*, 1965; Rubin *et al.*, 1970; Feldman and Schlezinger, 1970)—skin changes, pruritis, sore tongue and mouth, clubbing of fingers, brittle nails, pigmentation, fatigue, insomnia, low-grade fever, hypoplastic anemia, leukopenia, and mental changes. Also reported were conditions suggestive of brain tumor with symptoms of increased intracranial pressure, e.g., drowsiness, hydrocephalus with protrusion of the fontanelles, and vomiting (Beeson and McDermott, 1975, p. 1375). The Committee on Nutrition of the American Academy of Pediatrics (1976) concluded that the use of megavitamin A for psychiatric problems is not justified. A further problem is the diagnosis of previous megavitamin therapy in the presence of these baffling symptoms on account of the frequently poor documentation of such therapy in the case records. A diagnostic clue in a puzzling case was given by an abnormal bone scan (Shaywitz *et al.*, 1977).

Excessive amounts of vitamin D taken by health faddists, or given in the treatment of medical disorders or as megavitamin therapy to psychologically normal patients, are potentially dangerous and may lead to hypercalcemia and its complications.

Megadoses of vitamin C taken in the course of megavitamin therapy for childhood psychiatric conditions or for the overenthusiastic prevention of respiratory ailments may precipitate renal acidosis and, after cessation, scurvy (Moran and Greene, 1979).

Megadoses of niacin given to schizophrenics may produce niacin toxicity, characterized by clouding of consciousness, fever, flushing, dryness of the skin, increased pigmentation, nausea, diarrhea, and abnormal liver function.

The anticoagulant vitamin K has occasionally been taken to induce bleeding and bruising as a way of attracting medical attention. Stafne and Moe (1951) described a young nurse who secretly took dicoumarol and whose hemorrhagic tendencies defied all forms of investigation and therapy until she was told that if it continued she would be dismissed. A similar patient, also a nursing sister, took dicoumarol secretly for several years, producing numerous hemorrhages, after being abandoned by her doctor-fiancé (Carney and Brozowic, 1978).

Though the place of megavitamin therapy (if any) in psychiatric treatment is yet to be finally evaluated, clearly it may be dangerous.

5. Summary

In Western countries, the classic syndromes of gross vitamin deficiency are now not seen except sporadically in special circumstances of deprivation or in conjunction with drug use. However, recent surveys of psychiatric and other institutional populations show that biochemical and subclinical vitamin deficiencies of the B group are widespread. Little systematic work on the prevalence of avitaminoses A and D has been done, but it does not appear that deficiency of these vitamins is epidemiologically significant. Neither has much systematic study of the prevalence of the water-soluble vitamin C been carried out. Recent systematic work on the prevalance of folic acid and B_{12} deficiencies suggests that these occur in 10–20% of hospitalized psychiatric patients and are then associated with depressive and organic mental syndromes, poor nutrition, drug therapy, concomitant physical illness, and long duration of mental illness. Biochemical thiamine, riboflavin, and pyridoxine deficiencies are also common. Thiamine lack is associated with poor nutrition, while riboflavin and pyridoxine deficiencies seem to be linked with depressive illness. The prevalence of subtle clinical forms of niacin deficiency is unknown, but its milder forms seem to be characterized by depression.

Epidemiological studies can throw little light on the question of a possible causal role for these avitaminoses in associated psychiatric and behavioral disorders. Deficiency in many cases is probably an effect of the poor food intake secondary to the psychiatric disorder. However, in some cases it may be primary. Much longer prospective studies of vitamin-deficient patients and controlled trials of replacement therapy in such patients are required to confirm the causal role of individual avitaminoses in specific syndromes. Drugs associated with one or more avitaminoses include nearly all the psychotropic drugs, anticonvulsants, oral contraceptives, cytotoxic agents, antileukemic drugs, antibiotics, and isoniazid.

The claims for megavitamin therapy are unconfirmed by controlled clinical trials, and the high incidence of toxicity associated with its use suggests that the dangers linked with it outweigh its possible benefits.

6. References

Adams, P. W., Wynn, V., Rose, D. P., Seed, M. Folkard, J., and Strong, R., 1973, Effects of pyridoxine hydrochloride (vitamin B_6) upon depression associated with oral contraception, *Lancet* **2**:867–904.

Ahmed, M., 1972, Neurological disease and folate deficiency, *Br. Med. J.* **1**:181.

Arakawa T., 1970, Congenital defects in folate utilization, *Am. J. Med.* **48**:594–598.

Arakawa, T., Narisawa, K., Tanno, K., Ohara, K., Higashi, O., Honda, Y., Tamura, T., Wada, Y., Mizuno, T., Hayashi, T., Hirooka, Y., Ohno, T., and Ikeda, M., 1967, Megaloblastic anemia and mental retardation associated with hyper-folic-acidaemia probably due to $N5$ methyl-tetrahydrofolate transferase deficiency, *Tohoku J. Exp. Med.* **93**:1–22.

Bachrach, S., Fisher, J., and Parks, J. S., 1979, An outbreak of vitamin D deficiency rickets in a susceptible population, *Pediatrics* **64**:871–877.

Barnes, M., 1978, Detection and incidence of B and C vitamin deficiency in alcohol-related illness, *Ann. Clin. Biochem.* **15**:307–312.

Batata, M., Spray, G. H., Bolton, F. G., Higgins, G., and Wollner, L., 1967, Blood and bone marrow changes in elderly patients with special reference to folic acid, vitamin B_{12}, iron and ascorbic acid, *Br. Med. J.* **2**:667–669.

Bayoumi, R. A., and Rosalki, S. B., 1976, Evaluation of methods of co-enzyme activation of erythrocyte enzymes for detection of deficiency of vitamins B_1, B_2 and B_6, *Clin. Chem.* **22**:327–335.

Beattie, A. D., and Sherlock, S., 1976, Ascorbic acid deficiency in liver disease, *Gut* **17**:571–575.

Beeson, P. B., and McDermott, W., 1975, *Text Book of Medicine*, Sanders and Co., London, 1368 pp.

Blackstock, E. E., Gath, D. H., Gray, R. C., and Higgins, G., 1972, The role of thiamine deficiency in the aetiology of hallucinatory states complicating alcoholism, *Br. J. Psychiatry* **121**:356–364.

Blass, J. P., and Gibson, G. E., 1977, Abnormality of a thiamine requiring enzyme in patients with Wernicke–Korsakoff syndrome, *N. Engl. J. Med.* **297**:1367–1370.

Bonjour, J. P., 1980a, Vitamin and alcoholism. III. Vitamin B_6, *Int. J. Vitam. Nutr. Res.* **30**:215–230.

Bonjour, J. P., 1980b, Vitamins and alcoholism. V. Riboflavin, *Int. J. Vitam. Nutr. Res.* **50**:425–430.

Botez, M. T., Cadotte, M., Beaulieu, R., Pichette, L. P., and Pison, C., 1976, Neurological disorders responsive to folic acid therapy, *Can. Med. Assoc. J.* **115**:217–223.

Cameron, W. M., 1978, Isoniazid overdose, *Can. Med. Assoc. J.* **118**:1413–1415.

Carney, M. W. P., 1967, Serum folate values in 423 psychiatric patients, *Br. Med. J.* **4**:512–516.

Carney, M. W. P., 1970, Serum vitamin B_{12} values in psychiatric inpatients, *Dis. Nerv. Syst.* **31**:566–569.

Carney, M. W. P., 1971, Beri-beri in Blackpool *Br. Med. J.* **2**:109–110.

Carney, M. W. P., 1979, Psychiatric aspects of folate deficiency, in: *Folic Acid in Neurolgy, Psychiatry and Internal Medicine* (M. T. Botez and E. H. Reynolds, eds.), pp. 475–582, Raven Press, New York.

Carney, M. W. P., and Brozowic, M., 1978, Self-inflicted bleeding and bruising, *Lancet* **1**:924–925.

Carney, M. W. P., and Sheffield, B. F., 1970, Associations of subnormal serum folate and vitamin B_{12} and effects of replacement therapy, *J. Nerv. Ment. Dis.* **150**:404–412.

Carney, M. W. P., and Sheffield, B. F., 1978, Serum folic acid and B_{12} in 272 psychiatric in-patients, *Psychol. Med.* **8**:139–144.

Carney, M. W. P., Williams, D. G., and Sheffield, B. F., 1979, Thiamine and pyridoxine lack in newly-admitted psychiatric patients, *Br. J. Psychiatry* **135**:249–254.

Carney, M. W. P., Ravindran, A., Rinsler, M., and Williams, D. G., 1982, Thiamine, riboflavin and pyridoxine deficiency in psychiatric in-patients, *Br. J. Psychiatry* **141**:271–272.

Chanarin, I., 1973, New light on pernicious anaemia, *Lancet* **2**:538–539.

Chanarin, I., Perry, J., and Reynolds, E. H., 1974, Transport of folate into cerebrospinal fluid in man, *Clin. Sci. Mol. Med.* **46**:369–373.

Committee on Nutrition, American Academy of Pediatrics, 1976, Megavitamin therapy for childhood psychoses and learning disabilities, Pediatrics, **58**:910–912.

Cott, A., 1971, Orthomolecular approach to the treatment of learning disabilities, *Schizophr. Bull.* **395**:105.

Davis, R. E., and Smith, B. K., 1974, Pyridoxal and folate deficiency in alcoholics, *Med. J. Austr.* **2**:357–560.

Delaney, R. L., Lankford, M. G., and Sullivan, J. F., 1966, Thiamine, and plasma magnesium lactate abnormalities in alcoholic patients, *Proc. Soc. Exp. Biol. Med.* **123**:675–679.

Dérot, M., Castaigne, P., Morel-Maroger, A., and Leclerq, A., 1967, Syndrome neurologique nutritionel complexe réversible par administration d'acide folique, *Bull. Mem. Soc. Med. Paris* **118**:867–874.

Editorial, 1977, Liver disease and vitamin C, *Br. Med. J.* **1**:735–736.

Edwin, E., Holten, K., Norman, K. R., Schrumpf, A., and Skaag, O. E., 1965, Vitamin B_{12} hypovitaminoses in mental diseases, *Acta. Psychiatr. Scand.* **45:**19–36.

Elsborg, L., Hansen, T., and Rafaelson, O. J., 1979, Vitamin B_{12} concentrations in psychiatric patients, *Acta Psychiatr. Scand.* **59:**145–152.

Feldman, M. H., and Schlezinger, N. S., 1970, Benign intracranial hypertension associated with hypervitaminosis A, *Arch. Neurol.* **22:**1–7.

Fennelly, J., Frank, O., Baker, H., and Leery, C. M., 1964, Peripheral neuropathy of the alcoholic: I. Aetiological role of aneurin and other B-complex vitamins, *Br. Med. J.* **2:**1290–1292.

Foulks, E., and Katz, S. H. 1977, in: *Malnutrition, Behaviour and Social Organisation,* p. 224, Academic Press, London.

Frenkel, E. P., McCall, M. S., and Sheehon, R. G., 1973, Cerebro-spinal fluid folate and vitamin B_{12} in anti-convulsant induced megaloblastosis, *J. Lab. Clin. Med.* **81:**105–115.

Ghadirian, A. M., Anouth, J., and Engelsmann, F., 1980, Folate deficiency and depression, *Psychosomatics* **21:**926–929.

Grant, H. C., Hoffbrand, A. V., and Wills, D. G., 1965, Folate deficiency and neurological disease, *Lancet* **2:**763–767.

Hallstrom, T., 1969, Serum B_{12} and folate concentrations in mental patients, *Acta Psychiatr. Scand.* **45:**19–36.

Herbert, V., 1962, Experimental nutritional folate deficiency in man, *Trans. Assoc. Am. Physicians* **75:**307–320.

Herbert, V., Gottlieb, C. W., and Altschule, M. D., 1965, Apparent low B_{12} levels associated with chlorpromazine, *Lancet* **2:**738.

Hoorn, R. K., Flikweert, J. P., and Westernink, 1975, Vitamin B_1, B_2 and B_6 deficiencies in geriatric patients measured by co-enzyme stimulation of enzyme activities, *Clin. Chim. Acta* **61:**151–162.

Hunter, R., and Matthews, D. M., 1965, Mental symptoms in vitamin B_{12} deficiency, *Lancet* **2:**738.

Hunter, R., Jones, M., Jones, T. G., and Matthews, D. M., 1967, Serum B_{12} and folate concentration in mental patients, *Br. J. Psychiatry* **113:**1291–1295.

Ishii, N., and Nishikara, Y., 1981, Pellagra among chronic alcoholics: Clinical and pathological study of 20 necropsy cases, *J. Neurol. Neurosurg. Psychiatry* **44:**209–215.

Joiner, C. L., McArdle, B., and Thompson, R. H. J., 1950, Blood pyruvate estimations in the diagnosis and treatment of polyneuritis, *Brain* **73**(29):431–452.

Joliffe, N., Bowman, K. N., Rosenblum, L. A., and Fein, H. D., 1960, Nictonic acid deficiency encephalopathy, *JAMA* **114:**307–320.

Kallstrom, B., and Nylof, R., 1969, Vitamin B_{12} and folic acid in psychiatric patients, *Acta Psychiatr. Scand.* **45:**137–152.

Kershaw, P. W., 1967, Blood thiamine levels in alcoholism and confusional states, *Br. J. Psychiatry* **113:**387–393.

King, C. E., 1977, Vitamin D, alcoholism and dialysis, *Ann. Intern. Med.* **86:**830.

Kinsman, R. A., and Hood, J., 1971, Some behavioral effects of ascorbic acid deficiency, *Am. J. Clin. Nutr.* **24:**455–464.

Krippner, S., and Fisher, S. A., 1973, A study of neurological procedures and megavitamin treatment for children with brain dysfunction, *Orthomolecular Psychiatry* **1:**121–132.

Lanzkowsky, P., 1970, Congenital malabsorption of folate, *Am. J. Med.* **48:**580–583.

Livingston, J. E., Moreland, P. M., and Applegarth, D. A., 1978, Vitamin B_6 status in women with post partum depression, *Am. J. Clin. Nutr.* **31:**886–891.

Lonsdale, D., and Shamburger, R. J., 1980, Red cell transketolase as an indicator of nutritional deficiency, *Am. J. Clin. Nutr.* **33:**205–211.

Lopez, C., Schwartz, J. V., and Cooperman, J. M., 1980, Riboflavin deficiency in an adolescent population in New York City, *Am. J. Clin. Nutr.* **33:**1283–1286.

Luhby, A. L., Eagle, F. J., Roth, E., and Copperman, J. M., 1961, Relapsing megaloblastic anemia in an infant due to a specific defect in gastrointestinal absorption of folic acid, *Am. J. Dis. Child.* **102:**482–483.

Lumeng, L., and Li, T. R., 1974, Vitamin B_6 metabolism in chronic alcohol abuse: Pyridoxal phosphate levels in plasma and the effects of acetaldehyde in pyridoxal phosphate synthesis and degradation in human erythrocytes, *J. Clin. Invest.* **53:**693–704.

Manzoor, M., and Runcie, J., 1976, Folate responsive neuropathy—report of 10 cases, *Br. Med. J.* **1**:1176–1178.

Mattson, R. M., Gallagher, B. B., Reynolds E. H., and Glass, D., 1973, Folate therapy in epilepsy, *Arch. Neurol.* **29**:78–81.

McLaren, D. S., Docherty, M. A., and Boyd, D. H. A., 1980, Assessment of thiamine status of patients with an alcohol problem, *Nutr. Soc., Abstracts of Communications* **39**:241.

Minot, G. R., and Murphy, W. P., 1926, Treatment of pernicious anemia by a special diet, *J. Am. Med. Assoc.* **87**:470–476.

Moran, J. R., and Greene, H. L., 1979, The B vitamins and vitamin C in human nutrition, *Am. J. Dis. Child* **133**:308–314.

Neubauer, H., 1970, Mental deterioration in epilepsy due to folate deficiency, *Br. Med. J.* **2**:759–761.

Nobbs, B. T., 1974, Pyridoxal phosphate status in clinical depression, *Lancet* **1**:405–406.

Oliver, T. K., 1958, Chronic vitamin A intoxication, *Am. J. Dis. Child.* **95**:57–67.

Perry, J., Lumb, M., Laundy, M., Reynolds, E. H., and Chanarin, I., 1976, Role of vitamin B$_{12}$ in folate co-enzyme synthesis, *Br. J. Haematol.* **32**:263–268.

Persson, B., Tunell, R., and Ekengren, K., 1965, Chronic vitamin A intoxication during the first half year of life, *Acta Paediatr. Scand.* **54**:49–60.

Phillips, R. B., Victor, M., Adams, R. D., and Davidson, C. S., 1952, A study of the nutritional defect in Wernicke's syndrome, *J. Clin. Investig.* **31**:859.

Pincus, J. H., Reynolds, E. H., and Glazer, G. H., 1972, Subacute combined system degeneration with folate deficiency, *J. Am. Med. Assoc.* **221**:496–497.

Read, A. E., Gough, K. R., Pardoe, J. L., and Nicholas, A., 1965, Nutritional studies on the entrants to an old peoples home with particular reference to folic acid deficiency, *Br. Med. J.* **2**:843–848.

Reilly, T. M., 1979, The value of thiamine replacement in chronic alcoholism: A reminder, *Br. J. Addict.* **74**:205–207.

Reinker, L., Hohenauer, L., and Ziegler, E. E., 1972, Activity of red blood cell glutamic oxalacetic transaminase in epileptic children under antiepileptic treatment, *Clin. Chim. Acta* **36**:270–271.

Reynolds, E. H., 1967, Schizophrenia like psychoses of epilepsy and disturbances of folate and vitamin B$_{12}$ metabolism induced by anti-convulsant drugs, *Br. J. Psychiatry* **113**:911–919.

Reynolds, E. H., 1976, The neurology of vitamin B$_{12}$ deficiency: Metabolic mechanisms, *Lancet* **2**:832–833.

Reynolds, E. H., Pierce, J. M., Bailey, J., Coppen, A., 1970, Folate deficiency in depressive illness, *Br. J. Psychiatry* **117**:287–292.

Reynolds, E. H., Rothfeld, P., and Pincus, J., 1973, Neurological disease associated with folate deficiency, *Br. Med. J.* **2**:398–400.

Rickes, E. L., Brink, N. G., Koniuszy, F. R., Wood, T. R., and Folkes, K., 1948, Crystalline vitamin B$_{12}$, *Science* **107**:396–397.

Riding, J., 1975, Wet beri-beri in an alcoholic, *Br. Med. J.* **3**:79.

Rivkin, R. S., 1979, Hormones, drugs and riboflavin, *Nutr. Rev.* **37**:241–245.

Rose, M., and Johnson, I., 1978, Reinterpretation of the haematological effects of anti-convulsant treatment, *Lancet* **1**:1349–1350.

Rubin, E., Florman, A. L., and Degman, T., 1970, Hepatic injury in chronic hypervitaminosis A, *Am. J. Dis. Child* **119**:132–136.

Santiago-Borrero, P. J., Santini, R., Perez-Santiago, E., Maldano, N., Millán, S., and Coll-Camález, G., 1973, Congenital isolated defect of folic acid absorption, *J. Pediatr.* **82**:450–455.

Schwartz, R. A., Gross, M., Lonsdale, D., and Shamburger, R., 1979, Transketolase activity in psychiatric patients, *J. Clin. Psychiatry* **40**:427–429.

Shane, B., and Contractor, S. F., 1975, Assessment of vitamin B$_6$ status: Studies on pregnant women and oral contraceptive users, *Am. J. Clin. Nutr.* **28**:739–747.

Shaw, D. M., McSweeney, D. A., Johnson, A. L., O'Keefe, R., Naidoo, D., Macleod, D. M., Jog, S., Preece, J. M., and Crowley, J. M., 1971, Folate and amine metabolites in senile dementia: A combined trial and biochemical study, *Psychol. Med.* **1**:166–171.

Shaywitz, B. A., Siegel, N. J., and Benson, H. A., 1977, Megavitamins for minimal brain dys-function, *J. Am. Med. Assoc.* **238**:1749–1750; *Medical Letter*, May 22, 1980.

Shorvon, S. D., Carney, M. W. P., Chanarin, I., and Reynolds, E. H., 1980, The neuropsychiatry of megoloblastic anaemia, *Br. Med. J.* **281**:1036–1043.

Shulman, R., 1967a, Psychiatric aspects of pernicious anaemia: A prospective controlled inves-tigation, *Br. Med. J.* **3**:266–270.

Shulman R., 1967b, A survey of vitamin B_{12} deficiency and psychiatric illness in an elderly psy-chiatric population, *Br. J. Psychiatry* **113**:241–251.

Shulman, R., 1972, The present status of vitamin B_{12} and folic acid deficiency in psychiatric illness, *Can. Psychiatr. Assoc. J.* **19**:205–216.

Shulman, R., 1979, An overview of folic acid deficiency and psychiatric illness, in: *Folic Acid in Neurology, Psychiatry, and Internal Medicine* (M. I. Botez and E. H. Reynolds, eds.), pp. 463–474, Raven Press, New York.

Smith, E. L., and Parker, L. F. J., 1948, Purification of anti-pernicious anemia factor, *Biochem. J.* **43**:viii (abstract).

Sneath, P., Chanarin, I., Hodkinson, M. H., McPherson, C. K., and Reynolds, E. H., 1973, Folate status in a geriatric population and its relation to dementia, *Age Ageing* **2**:177–182.

Spaans, F., 1970, No effect of folic acid supplement on folate and serum vitamin B_{12} in patients on anticonvulsants, *Epilepsia* **11**:403–411.

Spivak, J., and Jackson, D. L., 1977, Pellagra: An analysis of 18 patients and a review of the literature, *Johns Hopkins Med. J.* **140**:295–309.

Springer, N. S., 1979, Ascorbic acid status of children with developmental distribution, *J. Am. Diet. Assoc.* **75**:425–428.

Stafne, W. A., and Moe, A. E., 1951, Hypoprothrombinemia due to dicoumarol in a malingerer: A case report, *Ann. Intern. Med.* **35**:910–911.

Stone, W. J., Warnock, L. G., and Wagner, C., 1975, Vitamin B_6 deficiency in uremia, *Am. J. Clin. Nutr.* **128**:950–957.

Strachan, R. W., and Henderson, J. G., 1967, Dementia and folate deficiency, *Q. J. Med.* **36**:189–204.

Taurot, G. P., Danks, D. M., Rowe, P. B., Van Der Weyden, M. G., Schwartz, M. A., Collins, V. L., and Neal, B. W., 1976, Dihydrofolate reductase deficiency causing megaloblastic anemia in two families, *N. Engl. J. Med.* **294**:466–470.

Thornton, W. E., and Thornton, B. P., 1978, Folic acid, mental function, and dietary habits, *J. Clin. Psychiatry* **April**:315–322.

Velentzas, C., Oreopoulos, D. G., Brandes, L., Wilson, D. R., and Marquez-Julio, A., 1977, Abnormal vitamin D levels, *Ann. Intern. Med.* **86**:198.

Victor, M., Adams, R. D., and Collins, G. H., 1971, *The Wernicke–Korsakoff Syndrome*, Black-well, Oxford.

Walker, A., 1968, Chronic scurvy, *Br. J. Dermatol.* **80**:625–630.

Word, B., Gysbert, A., Grode, A., Davis, S., Mulholland, J., and Breen, K., 1980, A study of partial thiamine restriction in human volunteers, *Am. J. Clin. Nutr.* **33**:848–861.

Ziegler, D. K., 1978, Toxicity to the nervous system of diphenylhydantoin: A review, *Int. J. Neurol.* **11**:383–400.

7

Mechanisms of Nutrient Action on Brain Function

Loy D. Lytle, Carla S. Whitacre, and Mark F. Nelson

1. Introduction

It has long been known that inadequate nutrition during pregnancy and early development can have profound effects on the subsequent maturation of the brain and other tissues. For example, chronic consumption of foodstuffs that lack sufficient nutrient quality (i.e., diets that are inadequate in proteins, carbohydrates, fats, vitamins, or minerals) or energy density (i.e., calories) may produce permanent structural changes in the fetal and early postnatal growth and development of the brain (Winick and Noble, 1966; Winick and Rosso, 1969; Shoemaker and Bloom, 1977; Nowak and Munro, 1977). The structural changes seen in the brains of animals and humans exposed to inadequate nutrition early in life include reductions in the size and weight of the brain, decreased numbers of neurons and glial cells, retardations in the development of dendritic and axonal neuronal processes, alterations in the morphology of synapses, and reductions in nerve-cell packing density in various regions of the brain (Shoemaker and Bloom, 1977).

Interestingly, the evidence is now quite convincing that some of the changes in brain structure produced by malnutrition are more severe if the faulty nutrition is experienced during the period of time [called the vulnerable critical period (Dobbing and Smart, 1973)] in mammalian development when the cells in the brain are proliferating at their most rapid rate (i.e., during the "brain growth spurt" period). Moreover, the extent to which brain structure is affected is greatly influenced by the duration and severity of the inadequate nutrition. Hence, severe malnutrition incurred during the critical period of rapid

Loy D. Lytle, Carla S. Whitacre, and Mark F. Nelson • Laboratory of Psychopharmacology, Department of Psychology, University of California, Santa Barbara, California 93106; Kellogg Co., Battle Creek, Michigan 49106.

brain growth produces permanent structural alterations that apparently cannot be reversed with subsequent improvements in nutritional status (Shoemaker and Bloom, 1977; Nowak and Munro, 1977).

The extreme vulnerability of brain structure to the untoward effects of malnutrition incurred during the stages of late gestation and early postnatal development in altricial mammals is striking and provides indirect support for the notion that except for this period, the brain may be relatively impervious to either major or relatively minor fluctuations in nutritional status. Partial support for this notion is also derived from a variety of studies that show that brain cells are quite efficient at automatically maintaining a constant metabolic and nutritional internal milieu as long as sufficient supplies of oxygen, glucose, and other nutrients are made available to them from peripheral organs via the circulatory system. Finally, several studies conducted quite some time ago provide convincing evidence that the structural composition of the brain is maintained and defended, even at the cost of tremendous wasting in the other organs of adult animals, when they are subjected to chronic or extreme protein or calorie starvation (Jackson and Stewart, 1920). Taken together, then, these data lend support for the hypothesis that brain cells may be relatively impervious to the normal capricious fluctuations occurring in the daily nutritional status of most mammals. Hence, it is generally agreed that brain *growth* and *structure* are most sensitive to nutritional insult only during the vulnerable growth period of the brain, which occurs early in development. However, recent evidence now indicates quite convincingly that certain aspects of brain *function*, which depend on the maintenance of a proper chemical milieu, may always be sensitive to, and are capable of being influenced by, alterations in the quality or quantity of foods eaten by organisms throughout their entire life-spans, even when these changes result from differences in the types of foods eaten on a meal-by-meal basis. These more pervasive dietary influences on brain function may result, at least in part, from direct nutrient influences on the relative abilities of brain and peripheral nerve cells to communicate with one another and with other cells and organs vis-à-vis the release of neuro-transmitter compounds. These latter findings, although relatively new and still requiring further study and elucidation, nevertheless provide a theoretical and empirical guide to future research aimed at understanding how various bio-chemical, physiological, or behavioral changes might result from some types of nutritional-deficiency disease states. Moreover, studies of the possible ways in which nutrition might influence brain and peripheral neurotransmission offer the possibility for the implementation of new therapeutic approaches in which nutritional manipulations might play a primary or secondary role in treatment strategies for various diseases afflicting humans.

2. Brain Neurons, Neurotransmitters, and Nutritional Status

A specialized cell, called the neuron, is the basic functional unit and work-horse of the central nervous system (CNS) and peripheral nervous system

(PNS). We still do not know precisely how many nerve cells exist in the nervous systems of mammals; rough estimates for the number of neurons in the human brain alone range from approximately 10^{10} such cells to several times this number. Moreover, there appear to be an almost limitless variety of nerve cells when one considers their different types of shapes, sizes, and functions. Even though they differ morphologically, most nerve cells share certain homologous features, which allow them to carry out their specialized functions of receiving, conducting, and transmitting information to and from other cells in the body. Most neurons have dendrites (filamentous outgrowth that usually radiate out from the cell body), which function much like receiving stations. The dendrites are the specialized portion of the neuron containing receptors highly sensitive to changes in the amounts of specific chemicals, called chemical neurotransmitter compounds, released by neurons or other cells. The dendrites convey this coded neurochemical information that they receive from other cells to a specialized extension of the neuron called the axon. The axon is the principal effector unit of the cell and is highly specialized to rapidly transmit a bioelectrical impulse, or action potential, from the cell body and dendrites down its length to other portions of the cell called the terminals or synaptic boutons. The synaptic boutons are the places from which chemical neurotransmitter compounds are released onto the receptive portions of other neurons, muscle cells, gland cells, or organs. Similar to other cells in the body, neurons also contain a cell body, or soma, that carries out all those metabolic processes utilizing oxygen and nutrients in the production of energy, as well as in the manufacture of structural components of the neuron essential for its functional integrity.

Communication between nerves and other cells appears for the most part to be a neurochemically mediated event across small gaps, or synapses, that spatially separate the terminals of a nerve cell from the receptors of the target cell—that is, the cell the subsequent activities of which are controlled by the nerve cell. Hence, intercellular communications between nerve cells and target cells involves the release of various amounts of chemical neurotransmitter compounds that ultimately cause a biological response in the target cell if sufficient quantities of neurotransmitter compound interact with enough target-cell receptors. The precise ways in which a neurotransmitter compound effects changes in the receptors of the postsynaptic target cell are still poorly understood. However, if enough neurotransmitter molecules are released by transmitting nerve cells, then the postsynaptic target cell will become depolarized and will be more likely to conduct an action potential, or will become hyperpolarized (or less likely to conduct an action potential), if the postsynaptic cell is another neuron.

Functionally, there are thought to be two basic types of neurotransmitter compounds. One type is called excitatory, because release of sufficient amounts of neurotransmitter from presynaptic terminals will cause increased electrical activity in postsynaptic nerve target cells, or increased synthesis and release of a hormone if the target cell is glandular, or contraction of a muscle fiber if the target cell is a muscle, or increased target-organ activity. The other type of neurotransmitter compound is inhibitory; here, release of presynaptic neurotransmitter molecules makes it less likely that an action potential will be conducted in the postsynaptic neuron, or less hormone will be released, or the functions of the target organs will be diminished. It should be noted that a particular neurotransmitter compound might have excitatory effects at one location in the body, but might be found to be inhibitory at another synapse located elsewhere. The precise changes in the target cell following release of a neurotransmitter compound depend on several different factors, including the nature of the chemical neurotransmitter compound released by the presynaptic nerve, the location of the synapse within the nervous system, the particular types of target cells innervated by the presynaptic neuron, the number of presynaptic terminals innervating the target cell that release excitatory or inhibitory neurotransmitter compounds, the frequency with which presynaptic nerves innervating a target cell release sufficient amounts of neurotransmitter compounds per unit time, and so forth.

Even though each neurotransmitter compound known at present appears to play quite different functional roles in maintaining the organismic biochemistry, physiology, and behavior, these biologically active chemicals appear to share a common general metabolic fate. Each is synthesized from a precursor compound, or is cleaved from or activated by additional biochemical reactions on a prohormone, or is taken up directly by nerve cells from the extracellular fluid. Most neurotransmitter molecules are stored within distinct intraneuronal functional pools in the terminal (which may or may not be associated with subcellular morphologically distinct organelles), released after nerve-cell depolarization into synapses, complexed with pre- or postsynaptic receptors (which, in turn, cause changes in ionic fluxes or in other metabolic processes in the receptive target cell), and inactivated (usually by enzymes or by reuptake into the neuron of origin or into other cells). Effective neural communication with other cells is accomplished when enough molecules of the released neurotransmitter compound interact with receptors located postsynaptically on target cells that share the same synaptic junction with the presynaptic terminal processes of the nerve cell. Although it is undoubtedly true that alterations in nutritional status might influence neurotransmission at any of the major steps involved in this process (i.e., by altering neurotransmitter uptake or synthesis, storage, release, receptor interaction, or inactivation), the bulk of the evidence gathered to date has focused on the possible influences of nutrients on neurotransmitter biosynthesis and release.

Good evidence exists that several compounds localized within nerve endings function as neurotransmitters in the CNS or PNS. Each of these compounds is released into synapses following nerve depolarization and produces on local application a biological response in the postsynaptic effector cell identical to that seen when the nerves thought to contain them endogenously are electrically stimulated. Some of the criteria used to judge whether a given compound found in nerve cells actually does function as a neurotransmitter have been met only indirectly in studies of the CNS because of several technical difficulties. Nevertheless, good evidence has accumulated that dopamine, norepinephrine, epinephrine, serotonin, acetycholine, γ-aminobutyric acid (GABA), glutamate, glycine, histamine, and several small peptide compounds (substance P, somatostatin, the enkephalins) most likely function as neurotransmitter substances in a variety of brain or peripheral nerve cells or both. Most of the neurotransmitter compounds identified thus far are themselves nutrient components found in various foodstuffs (e.g., the nonessential amino acid neurotransmitters glycine, aspartate, and glutamate, and GABA) or are relatively simple, direct metabolic by-products of dietary constituents (the indoleamine neurotransmitter serotonin is derived from the essential amino acid L-tryptophan; the catecholamine transmitters dopamine, norepinephrine, and epinephrine are synthesized from the amino acid phenylalanine or tyrosine; acetylcholine is synthesized from acetyl-CoA and choline; and histamine is derived from the amino acid histidine). In a similar vein, although the synthetic pathways for the newly discovered peptide neurotransmitters are at present poorly understood, they probably involve synthesis from amino acids on intracellular polyribosomal subunits that may generate the peptide neurotransmitter directly or that may produce a prohormone molecule from which the peptide transmitter is evolved by enzymatic cleavage of, or substitution on, the parent prohormone. Hence, on the basis of these considerations alone, it would appear likely that neurotransmitter biosynthesis in select populations of brain or peripheral nerve cells might be dramatically influenced by the relative availability of nutrients involved either directly or indirectly in transmitter metabolism.

The factors that make neurotransmitter biosynthesis most susceptible to nutrient influences have not yet been well described under all conditions, but an impressive array of data has been gathered to show that the relative availability of nutrients may directly alter the rates at which CNS or PNS transmitter compounds are synthesized in axon terminals. Furthermore, other data indicate that nutritional state might also influence transmitter metabolism indirectly by altering the relative availabilities of cofactor substances necessary for optimal enzymatic neurotransmitter biosynthetic or catabolic activity. Finally, some evidence has also been gathered to show that nutritional status might effect changes in other aspects of neurotransmitter metabolism, or in the integrity of the neurons that contain these compounds, in developing organisms in such a manner as to influence neural function in a more permanent fashion, which may cause subsequent deleterious effects on physiological or behavioral maturation or both (for a review, see Shoemaker and Bloom, 1977).

Space does not allow us to detail all reports of nutrient-induced changes in the several different putative neurotransmitter compounds thought to mediate the actions of mammalian central and peripheral neurons, nor does sufficient evidence yet exist to fully assess the effects of nutritional status on neural function vis-à-vis an action on general neurotransmitter metabolism and fate. However, the location, metabolism, and functions of one type of neurotransmitter-containing nerve cell, called the serotoninergic neuron, has been well characterized using a variety of neuroanatomical, electrophysiological, neurochemical, and neuropharmacological approaches. Moreover, a rather substantial amount of research has been carried out that allows for a detailed assessment of the possible ways in which changes in nutritional status might influence the activities of nerve cells that release serotonin as their neurotransmitter. Although a great deal of this research has been conducted in infraprimate mammals such as the rat and cat, sufficient evidence regarding the interactions between nutrition and neurotransmitter function has been gathered to provide tantalizing parallels that may have relevance for the human condition.

3. Serotoninergic Neurotransmission: An Overview

A set of formal criteria has evolved for judging whether a given endogenous chemical might play a role as a neurotransmitter compound in discrete portions of the nervous system: (1) the compound of interest must be localized within nerve terminals; (2) it must be released into synapses following electrical or chemical depolarization of the neuron; and (3) it should produce, following its local application, a biological response in a target cell that is identical to that observed when the nerves thought to contain the neurotransmitter compound are electrically stimulated. Although many suspected neurotransmitter compounds discovered in the PNS have been studied adequately to satisfy these criteria, several difficulties have been encountered in studies of the CNS that prevent direct satisfaction of these criteria. For example, the dense neuropil encountered in the CNS oftentimes does not allow one to isolate single neurons in synapses in such a manner as to fulfill the requirements of neuronal localization, release, and functional homology. Nevertheless, strong, albeit sometimes indirect, evidence has accumulated over the years supporting the neurotransmitter function of serotonin molecules in discrete portions of the CNS.

Histochemical fluorescent visualization techniques have been developed to map the location of serotoninergic neuronal cell bodies in mammals. These nerve-cell bodies appear confined to a discrete set of midline nuclei, collectively called the raphe nuclei, which are found in the mesencephalon, pons, and medulla oblongata (Fuxe *et al.*, 1968; Ungerstedt, 1971; Björklund *et al.*, 1971). Two of the more anteriorly localized nuclei, called the dorsal and median raphe nuclei, send rostrally long axons that innervate cells in virtually all aspects of the diencephalon and telencephalon, including the hypothalamus, hippocampus, amygdala, and cerebral cortex. In addition, more posteriorly localized cell bodies also send long axons that descend in the brainstem and spinal cord to innervate other neurons in the dorsal and ventral horns of the lumbar and sacral segments of the spinal cord. Hence, brain serotoninergic neurons occupy strategic locations throughout the CNS, which may allow them a potentially important role in modulating a wide range of sensory and motor functions, as well as in mediating a substantial number of physiological and behavioral responses.

Serotonin molecules have also been identified in a variety of other cells located outside the CNS (including the enterochromaffin cells of the gastrointestinal tract, the parenchymal cells of the pineal gland as well as in the terminals of the postganglionic, sympathetic neurons that innervate the gland, and

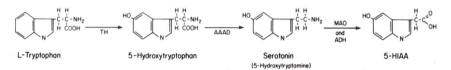

L-Tryptophan   5-Hydroxytryptophan   Serotonin   5-HIAA
(5-Hydroxytryptamine)

Fig. 1. The sequential metabolic steps in the synthesis and degradation of the brain neurotransmitter serotonin (5-hydroxytryptamine). TH = tryptophan hydroxylase; AAAD = aromatic *l*-
amino acid decarboxylase; MAO = monoamine oxidase; ADH = aldehyde dehydrogenase; 5-
HIAA = 5-hydroxyindoleacetic acid.

in blood platelets and mast cells), but these molecules do not appear to function
as neurotransmitters in these tissues (Gershon, 1968; Wurtman *et al.*, 1968a;
Mustard and Packham, 1970).

Brain serotoninergic neurons appear to contain all the substrate and cofactor compounds, as well as the enzymes, involved in the synthesis and degradation of the neurotransmitter (Fig. 1). The rate-limiting step in the biosynthesis of this neurotransmitter is the 5-hydroxylation of the indispensable amino
acid L-tryptophan (α-amino-β-3-indolepropionic acid), following its uptake into
the nerve. The enzyme that catalyzes this reaction is tryptophan hydroxylase,
and the best evidence indicates that this enzyme is not normally saturated with
respect to its amino acid substrate (Kaufman, 1974). (This latter observation
is important, especially when one considers the possible effects of nutritional
status on serotoninergic neurotransmission. The various implications of these
facts are discussed at greater length in Section 4.) Once tryptophan is converted
into 5-hydroxytryptophan, molecules of the latter amino acid are rapidly decarboxylated to form 5-hydroxytryptamine (serotonin), in a reaction catalyzed
by a rather nonspecific, ubiquitously localized enzyme, aromatic L-amino acid
decarboxylase (Lovenberg *et al.*, 1962).

Once synthesized, the majority of intracellular serotonin molecules appear
to be bound to particulate matter, although some molecules are also found in
a soluble form within the cytoplasm of the nerve terminal (Marchbanks, 1966;
Shields and Eccleston, 1973). Electrical stimulation of the raphe nuclei containing serotoninergic cell bodies causes a small reduction in the concentration
of the neurotransmitter in brain regions that contain the terminals of these cells
(Aghajanian *et al.*, 1966). Since levels of the inactive, deaminated metabolite
of serotonin, 5-hydroxyindoleacetic acid (5-HIAA), increase after stimulation,
these data are taken to mean that changes in the neurotransmitter and its metabolite result from increased release of serotonin into synapses following neuronal depolarization. More recently, increased extracellular molecules of serotonin have also been detected (indicating enhanced release of the
neurotransmitter) following a variety of electrophysiological, pharmacological,
and environmental manipulations thought to increase the electrical activities
of brain serotoninergic nerve cells. Unfortunately, the precise mechanisms
involved in the release of the neurotransmitter are at present unknown.

Synaptic molecules of serotonin interact with receptors on pre- or postsynaptic cell membranes, and at least two distinct classes of serotonin-receptor

binding sites have been described in the brain (Haigler, 1981; Peroutka *et al.*, 1981). Once released, small amounts of the neurotransmitter appear to be deaminated by an extraneuronally localized enzyme, monoamine oxidase. However, the major mechanism for the inactivation of synaptic serotonin involves its reuptake into the presynaptic nerve terminal, where it may be re-stored in protected compartments, or where it is deaminated intracellularly by the mitochondrial enzyme monoamine oxidase, and then is subsequently oxidized to form the major serotonin metabolite, 5-HIAA (Fuller, 1972).

The rather crucial anatomical localization of serotonin-containing neurons in the brain and spinal cord suggests the possibility that these nerve cells may be important for basic sensory and motor function, as well as play a role in more global aspects of physiology and behavior. These suggestions have been borne out in a host of different empirical studies designed to assess the relative functional importance of these nerve cells. Although precise conclusions about the exact role played by serotoninergic neurons in physiology and behavior are oftentimes controversial (see, for example, Messing *et al.*, 1978), strong evidence has implicated their essential importance in maintaining normal patterns of sleep and waking (Jouvet, 1968), pain sensitivity (Mayer and Price, 1976; Messing and Lytle, 1977; Harvey and Simansky, 1982), sexual behavior (Meyerson, 1964; Sicuteri, 1974), mood (Murphy *et al.*, 1974), and the responses to novel stimuli (Davis and Sheard, 1974), as well as their participation in the synthesis and release of different hormones, including pituitary growth hormone (Smythe *et al.*, 1975; Martin *et al.*, 1978) and prolactin (Meltzer *et al.*, 1976; Lawson and Gala, 1978; Rowland *et al.*, 1978). Because of the potential importance of serotoninergic nerve cells for physiology and behavior, it appears possible that diet-induced changes in the functioning of these brain nerve cells might well cause direct alterations in the response functions of animals, including humans.

4. Nutrition-Induced Changes in Brain Serotoninergic Neurotransmission

In any well-studied population of brain nerve cells, the rate at which a neurotransmitter is synthesized intracellularly appears to be normally coupled at least to some extent to the rates at which the transmitter is released into synapses, thus ensuring that relatively constant supplies of neurotransmitter molecules are maintained despite rather wide-ranging moment-to-moment changes in the depolarization rates of individual neurons. When neuronal depolarization and neurotransmitter release increase in frequency, for example, the rates of synthesis of many different neurotransmitter compounds are accelerated; conversely, lower rates of neurotransmitter synthesis are observed when neuronal firing rates and neurotransmitter release decline. The precise mechanisms important for neurotransmitter synthesis—release coupling appear to be different in individual populations of neurons and depend to some extent on the biochemical pathways involved in the synthesis of a particular neurotransmitter compound, the nature of its synaptic connections with other nerve cells, and so forth. For example, in central or peripheral noradrenergic neurons (those nerves that release noradrenaline as their neurotransmitter), the rates of neurotransmitter synthesis and release appear to be coupled by compensatory regulation of the activity of tyrosine hydroxylase, the enzyme thought to rate-limit the overall synthesis of norepinephrine from its amino acid precursor, L-tyrosine (Spector *et al.*, 1967). The rate of hydroxylation of L-tyrosine to form the intermediate catecholamino acid, L-3,4-dihydroxyphenylalanine, by tyrosine hydroxylase is influenced by alterations in the net electrical activity, or impulse flow, of noradrenergic neurons. Chronic elevations in the electrical activities of these cells accomplished by direct electrical stimulation (Weiner, 1970; Roth *et al.*, 1975), by exposing animals to stressful stimuli (Thoenen, 1974), or by the administration of drugs that increase neurotransmitter release (Weiner, 1970; Black, 1975; Reis *et al.*, 1975) are all associated with increases in tyrosine hydroxylase activity. As one might anticipate, tyrosine hydroxylase activity declines when noradrenergic neuronal electrical activity diminishes (for a review, see Moskowitz *et al.*, 1978). The increases or decreases in tyrosine hydroxylase activity produce corresponding alterations in the synthesis of norepinephrine and presumably also alter the relative amount of neurotransmitter that is available for release.

A variety of different mechanisms have been proposed to account for neurotransmitter synthesis–release coupling, including possible changes in the absolute amount or activity of synthetic enzyme protein available intracellularly (Reis *et al.*, 1975); allosteric changes in the configuration of the enzyme that alter its affinity for its substrate, tyrosine, as well as for its cofactors, oxygen and pteridine (Roth *et al.*, 1975); or alterations in the relative availability of norepinephrine neurotransmitter molecules, which might influence the accessibility of cofactors necessary to guarantee optimal tyrosine hydroxylase enzyme activity (Spector *et al.*, 1967; Weiner, 1970). Other mechanisms proposed to play a role in norepinephrine synthesis–release coupling include the influence of presynaptically (Langer, 1979) or postsynaptically (Thoenen, 1974; Aghajanian and Bunney, 1974) localized receptors that provide direct or indirect, multineuronal feedback control over the tyrosine hydroxylase reaction following their interaction with norepinephrine molecules released into synapses.

The precise mechanisms that make serotonin and other neurotransmitters potentially sensitive to alterations in nutrient status have not yet been described completely, but an impressive array of evidence has been accumulated to suggest that the relative intracellular availability of the precursor amino acid, L-tryptophan, to brain neurons might exert powerful influences over the rate at which the indoleaminergic neurotransmitter is synthesized in these nerve cells. Wurtman and colleagues (Wurtman and Fernstrom, 1976; Growdon, 1979; Wurtman *et al.*, 1980) recently outlined some of the basic conditions thought to be necessary for precursor control of neurotransmitter synthesis, as well as some of the possible ways in which nutritional status might influence neurotransmitter synthesis and function vis-à-vis its effects on precursor availability. We will briefly review the major lines of evidence showing that brain serotonin

L-TRYPTOPHAN

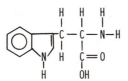

(∝-AMINO-β-3-INDOLEPROPIONIC ACID)

Fig. 2. The chemical structure of the indispensable indoleamino acid, L-tryptophan (α-amino-β-3-indolepropionic acid), the ultimate precursor compound involved in the synthesis of the brain neurotransmitter serotonin.

synthesis is controlled, at least in part, by the relative availability of precursor amino acid, and then consider recent observations suggesting that nutritional state might influence transmitter biosynthesis by altering precursor availability, as well as by causing other possible changes in neurotransmitter metabolism and function.

In brief, precursor control of neurotransmitter biosynthetic rate appears possible when: (1) brain tissue is not normally capable of synthesizing the precursor compound directly from nutrients or other substances that are in plentiful supply; (2) the intraneuronal brain enzymes that synthesize neurotransmitters from precursor compounds are not saturated with their substrates under normal physiological conditions; and (3) the brain enzymes that participate in the synthesis of the neurotransmitter are not subject to the influence of other control mechanisms that couple the rate of transmitter synthesis to its rate of release; i.e., they should not be subject to end-product feedback inhibition, pre- or postsynaptic-receptor-mediated alterations in synthetic enzyme activity, availability of enzyme protein, or the like (for a review, see Wurtman *et al.*, 1980). The synthesis of serotonin from its precursor amino acid, L-tryptophan, appears to satisfy each of these criteria.

Tryptophan (Fig. 2) belongs to the class of amino acids called "essential" or "indispensable" because no mammal studied thus far apparently possesses the metabolic machinery to synthesize this compound *de novo* from other dietary or biochemical constituents. Hence, all tryptophan molecules, including those found in brain tissue, ultimately derive from the diet. In addition, tryptophan is one of the most limited of all amino acids found in dietary protein (Table I), amounting to approximately 1% of the total amino acid content of protein-containing foods (Wurtman, 1970). Concentrations of tryptophan are relatively low in many protein-rich foods, and it is available in only trace amounts in a variety of other foodstuffs (Table I). Hence, only those individuals consuming protein-rich, well-balanced diets are guaranteed adequate amounts of this particular amino acid. As a consequance of its limited supply in foodstuffs, blood levels of this particular amino acid rise to a much more limited extent than most other amino acids following the ingestion of protein-containing foods. Other bodily sources for blood tryptophan do exist in normal, healthy adults, and free pools of the amino acid can be found in muscle and other tissues, either as free or bound molecules of the amino acid that are incorporated into the basic structure of protein or peptides. Hence, bodily reserves of the amino acid can be made available for a limited amount of time during

Table I. Tryptophan Concentrations in Various Foods $(g/100\ g)^a$

Milk	23–114
Cheese	80–491
Eggs	164–235
Beef	160–292
Lamb	168–233
Pork	6–219
Chicken	250–259
Fish	148–209
Legumes	170–526
Nuts	138–471
Grains	61–196
Fruits	1– 61
Vegetables	5– 97

[a] Adapted from Williams (1973).

periods of dietary tryptophan insufficiency caused by malnutrition or starvation. These residual pools of the amino acid can be rapidly depleted under conditions of malnutrition (see, for example, Gal *et al.*, 1961; Thomas and Wysor, 1967; Fernstrom and Wurtman, 1971c; Lytle *et al.*, 1975). Moreover, these other bodily reserves of the amino acid appear to be rather small and may not normally play a significant role in determining the relative availability of tryptophan to brain tissue on a meal-by-meal basis. As far as can be determined, then, brain neurons do not have the capacity to synthesize tryptophan from other constituents and must obtain supplies of the amino acid for serotonin biosynthesis as well as other biochemical processes directly from the periphery. Ultimately, all tryptophan molecules available to the blood derive from dietary sources, free amino acid pools, or the lysis of protein.

The second factor important for precursor control of brain serotonin synthesis is that tryptophan hydroxylase does not appear to be saturated with respect to its substrate amino acid, L-tryptophan. Controversy still exists about the precise K_m value of the enzyme for its substrate; the estimated K_m values of tryptophan hydroxylase for tryptophan vary, depending on the method of preparation (Jequier *et al.*, 1969; Friedman *et al.*, 1972a; Kaufman, 1974). However, there appears to be little doubt that the rate at which tryptophan is hydroxylated to form 5-hydroxytryptophan serves as the rate-limiting step in the overall synthesis of the serotonin end-product neurotransmitter. It should also be understood here that the rate at which 5-hydroxytryptophan is decarboxylated to form serotonin does not seem to be the rate-limiting step in the synthesis of the neurotransmitter, because this reaction occurs at a rate approximately 70 times faster than the rate at which tryptophan is hydroxylated to form 5-hydroxytryptophan (see, for example, A. R. Green and Grahame-Smith, 1975). Since tryptophan hydroxylase is not saturated normally by tryptophan molecules, best estimates of the K_m of the enzyme and the brain concentration of the amino acid are approximately 25 μM (Lovenberg *et al.*, 1968; Carlsson and Lindqvist, 1972, 1978; Hamon and Glowinski, 1974); the rate at which tryptophan is hydroxylated to form 5-hydroxytrytophan is generally believed to limit the overall formation of serotonin neurotransmitter molecules.

Since tryptophan hydroxylase is not saturated by tryptophan molecules, a rather wide variety of evidence indicates that serotonin biosynthesis can be characterized by an open-loop type of system controlled in large part by the relative availability of its precursor amino acid, tryptophan (Wurtman *et al.*, 1980). The potential susceptibility of serotonin biosynthesis to nutrient influences has not yet been described under all possible conditions; however, an impressive array of evidence has now accumulated to suggest that the synthesis of this neurotransmitter is directly influenced by the relative intracellular availability of tryptophan molecules. Unlike the synthesis of norepinephrine, for example, that of brain serotonin molecules does not appear to be controlled by direct feedback mechanisms such as end-product inhibition, because a variety of manipulations that cause large increases in the concentrations of serotonin do not influence the rate at which tryptophan is hydroxylated to 5-hydroxytryptophan, nor do they alter the rate of conversion of 5-hydroxytryptophan to serotonin (Lovenberg *et al.*, 1968; Friedman *et al.*, 1972a; Jacoby *et al.*, 1975). Because the availability of tryptophan to brain cells is thought to be the primary mechanism that influences the rate of brain serotonin synthesis under normal conditions, most (Lovenberg *et al.*, 1968; Fernstrom and Wurtman, 1971a–c; Gessa and Tagliamonte, 1974; Perez-Cruet *et al.*, 1974; Wurtman and Fernstrom, 1976; Growdon, 1979; Fernstrom, 1979b; Wurtman *et al.*, 1980), but not all (see, for example, A. R. Green, 1978), manipulations that change its concentrations in the brain produce corresponding alterations in the synthesis of the neurotransmitter end product. It should also be pointed out in this context that the bulk of the experimental data indicate that once tryptophan molecules gain entrance into the extracellular fluid bathing nerve cells in the brain, they are rapidly taken up by those cells.

Details regarding the precise mechanisms for transporting tryptophan from the extracellular fluid into the intracellular compartment of the neuron are at present not well understood, although estimates of this transport system gathered from *in vitro* preparations of brain slices (Kiely and Sourkes, 1972) and synaptosomes (Logan and Snyder, 1972; Parfitt and Grahame-Smith, 1974) indicate that the process is an active one that may cause 4-fold greater concentrations of the amino acid intracellularly compared to extracellularly. Under *in vitro* conditions, tryptophan uptake into nerve cells may occur via both a high- and a low-affinity active transport system. Interestingly, synaptosomal uptake of the amino acid appears inhibited by many of the same amino acids (leucine, tyrosine, phenylalanine, and methionine) that also compete with it for transport into the brain from the blood (Grahame-Smith and Parfitt, 1970; Parfitt and Grahame-Smith, 1974).

Since the intraneuronal accessibility of tryptophan molecules to brain cells appears to be a prime determinant of the rate of serotonin synthesis, a great deal of research has been aimed at identifying and delineating the various factors that might influence uptake of the amino acid from the blood into the brain.

It is important to now consider these various factors in order to understand how nutritional status might influence serotoninergic nerve function.

4.1. Tryptophan Metabolism and Fate

Once tryptophan-containing foods are ingested, the amino acid is rapidly absorbed from the gastrointestinal tract. Plasma concentrations of the amino acid exhibit characteristic rhythmic fluctuations related to the time of day when food is consumed, as well as to the nutritional quality of the food that is eaten (Wurtman *et al.*, 1968b; Wurtman, 1970). For example, humans who consume a high-protein diet (50 g protein consumed each meal during the daytime) show peak increases in plasma concentrations of tryptophan during the early evening hours following the dinnertime meal; plasma tryptophan concentrations reach a low during the early morning hours prior to the ingestion of the first daytime meal (breakfast) (Fernstrom, 1979b). Humans fed a protein-free, carbohydrate diet at similar times have plasma tryptophan concentrations that fall during the daytime relative to nighttime values; however, the nighttime and early morning peaks in tryptophan concentrations are relatively small and only in the approximate ranges of the nadir values for the amino acid seen in those humans fed the high-protein diets (Fernstrom, 1979b). When humans ingest a moderate amount of protein (75 g per day, consumed in three meals of 25 g each), slight elevations in the blood concentrations of the amino acid are observed without the corresponding daytime reductions typically seen after consumption of protein-free, carbohydrate-rich diets (Fernstrom, 1979a,b). Similar changes are seen in rodents fed a casein-based protein diet, and these daily alterations are related to the time of day when the animals eat as well as to the quality of food that they ingest. Rats are nocturnal animals and consume approximately 70–80% of their daily ration of food during the nighttime (LeMagnen and Tallon, 1966). Peak increases in plasma levels of tryptophan in rats consuming a protein diet are observed during the nighttime, with nadir values observed when the animals eat very little food (during the daytime) (Ross *et al.*, 1973). In contrast, the food-consumption-associated rise in plasma tryptophan concentrations is abolished in animals fed a protein-free, carbohydrate-based diet (Ross *et al.*, 1973) or in humans who are forced to consume small amounts of food at equally spaced intervals during a given 24-hr period (Young *et al.*, 1969). Hence, alterations in plasma tryptophan concentrations in animals or humans appear to depend on the type of food consumed during individual meals, as well as on the time of day or night when food is eaten.

The major route of tryptophan catabolism in the periphery is via the hepatic enzyme tryptophan pyrrolase (L-tryptophan 2,3-dioxygenase) (Rapoport *et al.*, 1966; Fuller, 1970; Ross *et al.*, 1973), which catalyzes the metabolism of tryptophan to kynurenine and, ultimately, to nicotinic acid (Knox and Mehler, 1950; Knox, 1951; Curzon and Green, 1968, Sourkes *et al.*, 1970). Several investigators (Mandell, 1963; Curzon, 1965; Curzon and Green, 1968) have hypothesized that alterations in the activity of this enzyme might influence tryptophan concentrations in plasma as well as the availability of the amino acid to brain cells. The activity of this enzyme is increased by corticosteroid hormones as well as by tryptophan (Knox, 1951), and several different types of manipulations that increase or decrease the activity of tryptophan pyrrolase oftentimes cause corresponding decreases or increases, respectively, in the concentrations of plasma tryptophan (Sourkes *et al.*, 1970; A. R. Green *et al.*, 1975). However, the inverse relationship between tryptophan pyrrolase activity and plasma tryptophan concentrations is not perfect, and the physiological importance of the interaction between activity of the hepatic enzyme and the concentrations of plasma and brain tryptophan has been questioned (Fernstrom, 1979a,b). For example, tryptophan pyrrolase activity is greatly enhanced in livers of animals subjected to prolonged starvation, apparently as a result of starvation-induced release of glucocorticoid hormones from the adrenal glands (A. R. Green and Curzon, 1968; Yuwiler *et al.*, 1971). Somewhat surprisingly, however, plasma tryptophan concentrations tend to increase, rather than decrease, during these periods of starvation (Tagliamonte *et al.*, 1973; Bloxam *et al.*, 1974). Although the precise interrelationships between hepatic tryptophan pyrrolase activity and changes in plasma concentrations of the amino acid remain to be determined under other nutritional conditions, starvation-induced elevations in plasma concentrations of tryptophan might be caused by the breakdown of proteins into their amino acid constituents during the period of starvation (Munro, 1964).

Once taken up from the gastrointestinal tract into the bloodstream, tryptophan molecules circulate in plasma in two different metabolic compartments. This amino acid is the only one identified thus far that in part is normally bound to plasma albumin molecules (McMenamy and Oncley, 1958; Moir, 1971); the remaining fraction of the amino acid, which amounts to approximately 20% of the total, exists in the free, unbound form. Some investigators believe that pharmacological or dietary alterations in the ratio of free tryptophan relative to bound tryptophan may be physiologically important, inasmuch as it has been proposed that changes in the circulating pool of unbound tryptophan may produce corresponding changes in the amount of tryptophan that is available for uptake into brain tissue and that is ultimately metabolized to form serotonin (Knott and Curzon, 1972; Tagliamonte *et al.*, 1973; Curzon *et al.*, 1973, 1975; Gessa and Tagliamonte, 1974; Curzon and Knott, 1974; Hutson *et al.*, 1976; Knott *et al.*, 1977; Bloxam and Curzon, 1978; Bloxam *et al.*, 1980). In contrast, other investigators (Fernstrom and Wurtman, 1972, 1974; Fernstrom *et al.*, 1973; Madras *et al.*, 1973, 1974; Wurtman and Fernstrom, 1975, 1976; Etienne *et al.*, 1976; Pardridge, 1977; Yuwiler *et al.*, 1977; Woodger *et al.*, 1979; Bloxam *et al.*, 1980) believe that the amount of tryptophan taken up by brain tissue depends on the concentrations of total (free + albumin-bound) plasma tryptophan relative to the blood levels of other large neutral amino acids (tyrosine, phenylalanine, leucine, isoleucine, valine) that compete with tryptophan for a saturable transport mechanism (Guroff and Udenfriend, 1962; Blasberg and Lajtha, 1965; Oldendorf, 1971; Pardridge, 1977). According to this latter view, tryptophan molecules circulating in plasma while bound to albumin are still available for transport into the brain because a hypothesized carrier mechanism for transporting tryptophan from plasma into the brain has a greater affinity for the amino acid than does albumin (Madras *et al.*, 1974).

There is great debate about whether total blood tryptophan concentration, disruptions in the bulk equilibrium conditions between free and bound molecules of tryptophan circulating in the blood, or alterations in the ratio of plasma tryptophan relative to other amino acids that share the same brain transport mechanisms might better predict the rate of brain tryptophan uptake (see, for example, A. R. Green, 1978; Fernstrom, 1979a; Wurtman *et al.*, 1980). There is generally good agreement that a rather wide variety of manipulations that alter the brain availability of tryptophan produce positively correlated, corresponding alterations in the rate of brain serotonin biosynthesis (Gessa and Tagliamonte, 1974; A. R. Green, 1978; Fernstrom, 1979a; Wurtman *et al.*, 1980). For example, injections of low doses of the amino acid, in amounts as low as 1/20 of the estimated normal daily total intake of tryptophan consumed by rats fed a protein diet, produce time-related increases in brain tryptophan, as well as in the concentrations of the neurotransmitter serotonin in fasted animals (Fernstrom and Wurtman, 1971a). Even larger increases in plasma and brain tryptophan, and in brain serotonin and its deaminated metabolite, 5-HIAA, occur when larger doses of the amino acid are administered systemically (Ashcroft *et al.*, 1965; Moir and Eccleston, 1968; Fernstrom and Wurtman, 1971b; Fernstrom and Hirsch, 1975; Lytle *et al.*, 1975; Howd *et al.*, 1975; Carlsson and Lindquist, 1978). In addition, a variety of other drugs or manipulations that increase or decrease brain concentrations of tryptophan also produce corresponding alterations in the rate of serotonin synthesis (Gessa and Tagliamonte, 1974; Costa and Meek, 1974), although the relationship is not always perfect (see, for example, A. R. Green, 1978).

4.2. Nutritional Changes in Brain Serotoninergic Neurotransmission: Meal Effects

Wurtman and colleagues (Fernstrom and Wurtman, 1974; Wurtman and Fernstrom, 1975; Growdon and Wurtman, 1979; Growdon, 1979; Wurtman *et al.*, 1980) have also outlined some of the conditions that may make the synthesis of brain neurotransmitter compounds such as serotonin susceptible to nutritional influences affecting the relative availability of their precursor compounds: (1) it should be possible to show that the concentration of the precursor compound in blood normally varies as a function of changes in the quantity or quality of nutrients ingested, or as a result of hormonal or other metabolic changes associated with the consumption, absorption, distribution, or excretion of nutrients in foods that might influence the uptake of the precursor compounds from the blood into the brain; and (2) the amount of the precursor compound actually available to brain neurons should vary as a direct function of its concentration in the blood or as some function of its blood concentration relative to those of other plasma compounds that influence its rate of uptake into the brain from the blood.

Recent experiments confirm the notion that alterations in the quality or quantity of food consumed by animals or humans cause changes in circulating concentrations of tryptophan that then influence serotonin neurotransmitter biosynthesis by altering the relative availability of the amino acid precursor to brain cells. The relationship between blood and brain tissue changes in tryptophan concentrations are complex, and alterations in plasma tryptophan do not always cause corresponding, one-to-one changes in brain concentrations of the amino acid.

Several factors appear important in influencing blood and brain tryptophan, including differences in the constituents of the diet, as well as hormonal and other biochemical changes associated with the ingestion, breakdown, and transport of nutrients (Fernstrom and Wurtman, 1972a,b, 1974; Fernstrom *et al.*, 1973, 1975a–c; Madras *et al.*, 1973, 1974; Wurtman and Fernstrom, 1975, 1976; Etienne *et al.*, 1976; Pardridge, 1977; Yuwiler *et al.*, 1977; Woodger *et al.*, 1979; Bloxam *et al.*, 1980). For example, animals fed a diet consisting of carbohydrates and fat, but one that was also deficient in proteins or any other amino acids for a relatively brief, 2-hr period were found to have higher circulating plasma concentrations of tryptophan following the meal and increased brain levels of the amino acid and the neurotransmitter serotonin compared to the fasted control animals (Fernstrom and Wurtman, 1971a). Since no additional tryptophan molecules other than those already sequestered in peripheral or brain tissue, or circulating in the bloodstream, were made available to the animals in this particular experiment, it was hypothesized that changes in the blood hormone insulin, associated with the consumption of the carbohydrate–fat diet, accelerated brain tryptophan uptake from the plasma, thus producing a corresponding enhancement in the rate of synthesis of the neurotransmitter (Fernstrom and Wurtman, 1971a). The possibility that insulin release might play a role in altering brain tryptophan availability was confirmed indirectly in subsequent experiments. When fasted animals with no access to food were injected with subconvulsive doses of insulin, increases in plasma tryptophan and in brain tryptophan and serotonin similar to those seen in animals that consumed the carbohydrate–fat diet were observed (Fernstrom and Wurtman, 1972a,b). Conversely, plasma tryptophan concentrations were found to fall in fasted animals that were injected with the hormone glucagon, which is normally released in response to high circulating levels of insulin (Fernstrom and Wurtman, 1972a,b).

The precise mechanisms that underlie the elevations in plasma tryptophan concentrations observed in animals fed the carbohydrate–fat diet, or in the fasted animals injected with the single dose of insulin, are at present in dispute. This dispute is focused on a discussion of whether brain uptake of tryptophan is correlated with nutrient- or hormone-induced changes in the free, unbound pool of circulating tryptophan molecules in blood or whether changes in brain uptake of the amino acid are determined by changes in the concentrations of tryptophan relative to those of other amino acids that compete with it for an identical brain uptake mechanism. According to one view (see, for example, Wurtman *et al.*, 1980; Fernstrom, 1979a), of all the amino acids known at present, tryptophan is the only one that is elevated by insulin, because the hormone apparently decreases the concentrations of all other amino acids in plasma by enhancing their rate of uptake from the blood into peripheral tissues such as muscle (Luck *et al.*, 1928; Lotspeich, 1949; Wool, 1965; Fernstrom and Wurtman, 1972a,b; Fernstrom *et al.*, 1973).

In addition to the elevations in total plasma tryptophan concentrations caused by the food-related increases in insulin, the hormone also appears to alter the compartmentation between free pools and albumin-bound molecules of the amino acid in blood. It produces this latter effect by decreasing the amount of plasma nonesterified fatty acids normally bound to plasma albumin (McMenamy and Oncley, 1958). Tryptophan and nonesterified fatty acids normally compete for the same binding sites on the albumin molecule; hence, as nonesterified fatty acids become uncoupled from the albumin, these empty binding sites are rapidly occupied by tryptophan molecules, thus increasing the percentage of the amino acid in blood that becomes bound. That serum nonesterified fatty acids compete with tryptophan for blood-albumin-binding sites has also been confirmed in additional studies in which animals fed diets containing progressively larger proportions of fat show positively correlated increases in plasma nonesterified fatty acid and in unbound plasma tryptophan molecules. However, these animals show no change or on occasion, slight reductions in the total circulating pool of both free and albumin-bound molecules of the amino acid (Madras *et al.*, 1974). Interestingly, brain tryptophan and serotonin concentrations were not increased in the animals fed high-fat diets, even though free tryptophan concentrations in the plasma were elevated approximately 2-fold compared to the blood concentrations of the amino acid and the brain tryptophan and serotonin levels seen in fasted control rats (Madras *et al.*, 1974). These latter findings by Madras *et al.* (1974) challenge the notion that increases in free tryptophan concentrations enhance brain serotonin synthesis by increasing uptake of the amino acid. However, these conclusions are controversial (Hutson *et al.*, 1976). Using a different method to measure concentrations of plasma nonesterified fatty acids and free and bound molecules of tryptophan, Hutson *et al.* (1976) claimed that they could not replicate the results of Madras *et al.* (1974) regarding the apparent lack of correlation between free tryptophan concentrations and brain tryptophan levels. These claims and counterclaims regarding changes in free and bound plasma tryptophan may reflect differences in methodologies used by the various investigators to measure the different plasma pools of the amino acid (see, for example, Fernstrom, 1979a; Curzon, 1979).

Perhaps more important in light of this discussion is the fact that Madras *et al.* (1974), repeating the earlier findings of others (Fernstrom and Wurtman, 1972a,b), failed to find a simple one-to-one relationship between acute changes in plasma tryptophan and brain concentrations of the amino acid. It now appears clear that under some conditions, free tryptophan concentrations may influence the brain uptake of the amino acid, whereas under other conditions, the ratio of total plasma tryptophan to other amino acids that compete with it for transport might play a role in its brain uptake. A few examples should clarify this latter point. Animals fed a diet with the milk protein casein as its sole protein source or animals fed a synthetic diet with an amino acid composition identical to that of casein, for a 2-hr period following an overnight fast, had increased plasma concentrations of tryptophan. However, the levels of brain tryptophan and serotonin did not change appreciably after the meal. In contrast, animals fed a different synthetic amino acid diet that contained normal concentrations of tryptophan, but was deficient in the amino acids (tyrosine, phenylalanine, leucine, isoleucine, and valine) that compete with tryptophan for a common transport carrier mechanism, were found to have increased plasma and brain concentrations of tryptophan and increased brain serotonin biosynthesis. Finally, animals fed a synthetic amino acid diet that contained no tryptophan, but did have normal concentrations of the other competing amino acids, showed no change in plasma tryptophan, but were found to have actual reductions in brain tryptophan and serotonin (Fernstrom and Wurtman, 1971a; Fernstrom *et al.*, 1975a,b, 1976; Fernstrom and Faller, 1978). According to these data, then, diet-induced changes in brain tryptophan and serotonin depend on the extent to which plasma levels of tryptophan, and of the other large neutral amino acids that compete with tryptophan for uptake into the brain, change with respect to one another. Manipulations that increase the plasma ratio of tryptophan relative to the competing amino acids tend to elevate brain serotonin, whereas dietary manipulations that decrease the plasma tryptophan/competing amino acid ratio tend to not change or, in some cases, to actually decrease brain serotonin synthesis (Fernstrom and Wurtman, 1972a,b, 1974; Fernstrom *et al.*, 1975a; Colmenares *et al.*, 1975; Fernstrom and Faller, 1978).

The following model regarding diet-induced changes in brain tryptophan and serotonin has emerged: If the quality of the ingested food is such that the competitor amino acids are low or absent, the tryptophan transport into brain is increased as a result of the action of insulin. The release of insulin from the pancreas following the ingestion of food facilitates peripheral-tissue transport of all amino acids except tryptophan; although plasma tryptophan concentrations do not actually change following the ingestion of a carbohydrate–fat, protein-free meal, or after an injection of insulin into fasted animals, the competing large neutral amino acids in plasma fall, thus increasing the tryptophan/competing amino acid ratio, thus favoring subsequent transport of the indoleamino acid from plasma to brain tissue. In contrast, adult animals fed a casein-based protein diet, or other foods that contain the competitor amino acids, have reduced plasma tryptophan/competing amino acid ratios, thus diminishing the total number of tryptophan molecules transported from the plasma to the brain (for reviews, see Wurtman *et al.*, 1980; Fernstrom and Wurtman, 1971b, 1972a,b).

4.3. Nutritional Changes in Brain Serotoninergic Neurotransmission: Meal Effects on Fetal Neurochemistry

The growth of tissues, including the brain, has been divided into three distinct phases (Winick and Noble, 1965): (1) a period of hyperplasia, during which cell number increases at a rapid rate; (2) a period of mixed growth, during which postmitotic cells mature and grow while new nerve cells also continue to emerge as a result of cell division; and (3) a period of hypertrophy, when there is little additional increase in the absolute numbers of cells within a given tissue, but the sizes and weights of postmitotic cells continue to show dramatic increases. These different phases of cell growth have been characterized biochemically. Increases in cell number are generally accompanied by corresponding increases in the synthesis and amount of deoxyribonucleic acid (DNA). The period of mixed growth is characterized by tissue increases in DNA as well as ribonucleic acid (RNA) as the cells increase the synthesis of various proteins required for continued cellular growth. Finally, total tissue protein levels increase dramatically during the period of hypertrophy, whereas DNA concentrations remain relatively constant during this phase of development.

In experiments conducted in our laboratory, we measured changes in total brain weight and in DNA, RNA, and protein levels in the brains of fetal rats during a time period midway in the last trimester of pregnancy. The amount of DNA increased steadily throughout this time period (Fig. 3), a biochemical profile characteristic of tissue observed during the hyperplastic phase of development. Brain weight and RNA and protein content also increased dramatically during this same time period. Interestingly, the amount of weight, RNA, and protein content per brain-cell nucleus remained relatively constant (Fig. 3); hence, total brain-cell development during the last trimester of gestation in the rat is best described as including the mixed-growth phase of hyperplasia and hypertrophy (Winick and Noble, 1965).

Autoradiographic studies using the incorporation of radioactively labeled thymidine into brain neurons in the process of mitosis have shown previously that cells in the raphe nuclei (the loci of brain serotoninergic nerve-cell bodies in adult animals) stop dividing and begin to differentiate between the 11th and 15th days of the 21-day gestation period of the rat (Lauder and Bloom, 1974). These neurons appear capable of synthesizing neurotransmitter molecules shortly after this period of cell division, since serotonin can be detected visually in these cells as early as the 13th day of gestation with histochemical fluorescent methods (Olson and Seiger, 1972) and at approximately the 15th day of gestation with sensitive biochemical assay methods (Howd *et al.*, 1975). Developmental changes in fetal rat brain *levels* of tryptophan and serotonin increase approximately 250 and 400%, respectively, between the 16th and 18th postconception days (Nelson and Lytle, unpublished observations) (Fig. 4). When expressed as the amount of serotonin per gram of brain weight, per protein, or per DNA, however, the concentration of the neurotransmitter is fairly constant from the 17th day of gestation (Nelson and Lytle, unpublished observations). Since other evidence indicates that serotonin neurons have finished mitosis by this time, the fact that changes in serotonin keep pace with these increases in DNA and brain protein synthesis suggests that the growth of fetal rat brain includes both hyperplastic and hypertrophic alterations in total brain cells (the hyperplastic changes appear to occur in other, nonserotoninergic neurons as well as in glial cells). However, even though serotoninergic nerve-cell number does not in-

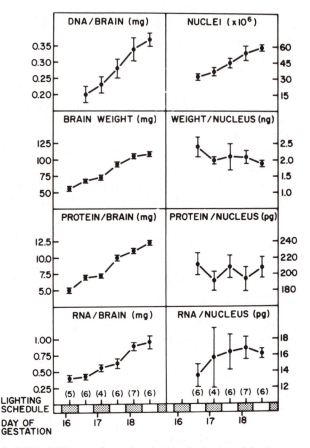

Fig. 3. Changes in DNA, RNA, protein, and weight in the brains of fetal rats between the 16th and 18th postconception days of pregnancy. Gravid female rats were given *ad libitum* access to an 18% casein diet and water, and were exposed to a 12:12 hr lighting schedule [lights on at 0800 hr; stippled (dark period) and clear (light period) horizontal bars at the bottom of the figure indicate whether animals were killed during the dark or light portion of the lighting schedule].

crease sustantially during the last trimester of gestation (Lauder and Bloom, 1974), these cells do show an accelerated growth of their dendritic, axonal and terminal processes during this stage of development (Loizou, 1972; Lauder and Krebs, 1976).

It is interesting to note that although fetal brain serotonin *concentrations* increase approximately 250% during the middle of the last trimester of pregnancy, brain tryptophan remains relatively constant during this time period (top right and bottom left panels of Fig. 4, respectively). At first glance, these data could be taken to mean that precursor substrate availability might not be an important factor for the control of brain serotonin biosynthesis in the fetal animal. However, it should be remembered that it is impossible to accurately determine how much of the total pool of the amino acid in brain might be available for neurotransmitter synthesis in fetal serotoninergic nerves vs. the

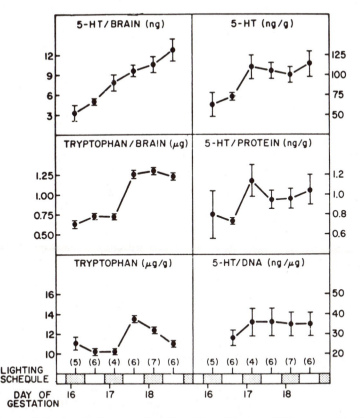

Fig. 4. Changes in fetal rat brain tryptophan (Denckla and Dewey, 1967) or serotonin (Saavedra et al., 1973) between the 16th and 18th postconception days of pregnancy. Gravid female rats were given *ad libitum* access to an 18% casein diet and water, and were exposed to a 12:12 hr lighting schedule [lights on at 0800 hr; stippled (dark period) and clear (light period) horizontal bars at the bottom of the figure indicate whether animals were killed during the dark or light portion of the lighting schedule].

sizes of the tryptophan pools that are available for other biochemical processes, including protein synthesis. In fact, the results of other experiments conducted in our laboratory indicate that fetal brain tryptophan availability does actually rate-limit serotonin synthesis during prenatal development. Injections of tryptophan into gravid female rats, even as early as the 15th postconception day, produce time- and dose-related increases in the concentrations of the amino acid and the neurotransmitter in the fetus (Howd *et al.*, 1975). Hence, if the enzyme tryptophan hydroxylase is indeed saturated with its substrate tryptophan during this stage of development, one would not expect to accelerate prenatal brain serotonin synthesis by producing larger pharmacological increases in the concentrations of the precursor amino acid. How does one account, then, for the normal age-related increases in fetal brain serotonin concentrations seen during the last trimester of pregnancy? Several mechanisms might be important for these developmental changes. The activities of the syn-

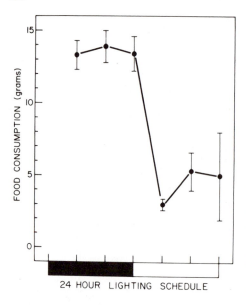

FOOD CONSUMPTION (grams)

24 HOUR LIGHTING SCHEDULE

Fig. 5. Diurnal variation in the amount of food (an 18% casein diet) consumed by gravid female rats at 4 hr intervals on the 17th postconception day of pregnancy. Animals were given *ad libitum* access to the diet and water, and were exposed to a constant 12:12 hr light:dark cycle [lights on at 0800 hr; dark (black horizontal bar) and light (clear bar) phases of the lighting cycle are indicated at the bottom of the figure].

thetic enzymes tryptophan hydroxylase and aromatic L-amino acid decarboxylase increase dramatically during the latter stages of pregnancy (Schmidt and Sanders-Bush, 1971; Deguchi and Barchas, 1972; Lamprecht and Coyle, 1972; Porcher and Heller, 1972), most likely as a result of increased nerve-terminal growth during this stage. In addition, it is also possible that serotoninergic nerves may accelerate their relative abilities to manufacture new synthetic enzyme protein, thus making more tryptophan hydroxylase material available for the synthesis of serotonin molecules from tryptophan during the latter stages of prenatal life. Whatever the reason, the evidence seems consistent with the notion that the relative availability of tryptophan to fetal brain neurons may rate-limit serotonin synthesis. If this latter hypothesis is correct, then it should be possible to show that brain serotonin synthesis in fetal animals is influenced by the quantity or quality of food eaten by their mothers.

As mentioned previously, the adult nongravid rat is a nocturnal animal that consumes nearly 70% of its total daily food intake during the dark portion of a typical 24-hr period (Fitzsimons and LeMagnen, 1969). Similar nocturnal feeding rhythms can also be seen in gravid female animals that are offered *ad libitum* access to an 18% casein protein diet and water (Fig. 5). To determine whether these rhythmic changes in the normal feeding patterns of pregnant rats might influence amino acid availability and brain neurotransmitter synthesis in the brains of their fetuses, different groups of 17-day-postconception pregnant female rats were exposed to a 12:12 hr light–dark cycle (lights on at 08:00 hr) and were killed at 4-hr intervals after being offered *ad libitum* access to an 18% casein protein diet and water. Placental and fetal body and brain concentrations of tryptophan (Fig. 6) were measured at 4-hr intervals. Tryptophan concentrations in the placenta changed rhythmically during the 24-hr period, and peak concentrations of the amino acid were seen during the dark portion of the

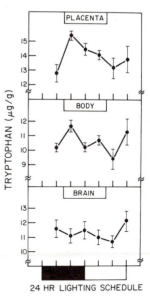

Fig. 6. Rhythmic changes in placental and fetal body or brain tryptophan concentrations in animals obtained from gravid female rats on the 17th day of gestation. Mothers were given *ad libitum* access to an 18% casein diet or water, and were exposed to a constant 12:12 hr light:dark cycle [lights on at 0800 hr; dark (black horizontal bar) and light (clear bar) phases of the lighting cycle are indicated at the bottom of the figure].

lighting cycle (that portion during which time the pregnant mothers consumed approximately 75% of their daily ration of food) (Fig. 5). Rhythmic fluctuations in tryptophan were considerably dampened when measured in the carcasses (Fig. 6, middle) or brains (Fig. 6, bottom) of the fetuses. Additional experiments indicated that the placental tryptophan diurnal rhythm depended on the nocturnal feeding patterns of the pregnant mothers and was not entrained to the lighting cycle *per se* (the peak increases in placental and fetal amino acid concentrations can be made to occur during the light portion of the day–night cycle if the pregnant mothers are offered limited access to an 18% casein diet only during the day).

We also measured the possible extent to which fetal brain serotonin concentrations might change as a function of the nocturnal feeding pattern of the gravid females. No significant rhythmic changes occurred in the concentrations of the fetal brain neurotransmitter over the entire 24-hr period of measurement (Table II). It should be remembered that in this particular experiment, pregnant animals were given *ad libitum* access to an 18% casein diet; this particular diet contains adequate concentrations of proteins to promote normal growth and development. More important, however, casein also contains sufficient concentrations of tryptophan as well as the other large neutral amino acids that compete with tryptophan for transport into the brains of adult animals (Fernstrom, 1979a; Wurtman *et al.*, 1980). Hence, our results indicate clearly that placental concentrations of tryptophan increase significantly as a function of the eating pattern of the pregnant mother, but concentrations of this amino acid do not fluctuate dramatically over the same time period in the bodies or brains of the fetuses. We took these data to mean that the metabolic and endocrinological status of the pregnant mother appears to guarantee an adequate, if not optimal, supply of nutrients to the fetus as long as the mother is provided

Table II. Rhythmic Fluctuations in Fetal Brain Serotonin Concentrations[a]

Time of day (hr)	Fetal rat brain serotonin concentration (ng/g)
Dark portion	
0000	89 ± 9
0400	86 ± 6
0800	83 ± 7
Light portion	
1200	79 ± 8
1600	77 ± 1
2000	100 ± 7

[a] Fetal brain concentrations of serotonin were measured (Saavedra et al., 1973) in animals obtained from 17 day postconception gravid females killed at 4 hr intervals. Pregnant rats had *ad libitum* access to an 18% casein diet and water, and were exposed to a 12:12 hr lighting schedule (lights on at 0800 hr). All values are expressed as the mean ± s.e.m. concentration of serotonin in fetal rat brain (ng/g).

with access to a diet of sufficiently high nutrient quality. The fact that tryptophan concentrations in the placenta and bodies of fetal animals continued to show rhythmic changes correlated with the feeding patterns of the mother, whereas brain concentration of the amino acid did not show these changes, provided indirect support for the notion that the influx of large neutral amino acids (such as tyrosine) might competitively limit the uptake of tryptophan from the maternal plasma to the placenta or from the placenta to the vascular system of the fetus. We directly tested this possibility in another series of experiments (Nelson *et al.*, 1981).

To determine whether foods of different nutrient quality might influence fetal tryptophan availability and, untimately, cause changes in fetal brain serotonin synthesis, we offered pregnant rats *ad libitum* access to an 18% casein diet and water for a 1-week period. Then, on the 17th day of gestation, after an overnight fast of at least 14 hr, different groups of pregnant rats were given access to a carbohydrate–fat, protein-free diet or to the 18% casein diet for a 2-hr period and then were killed. Control groups of pregnant rats were similarly fasted overnight, but received no food for the 2-hr period on the day of the experiment. Tryptophan concentrations were significantly elevated in the bodies and brains of fetuses obtained from pregnant rats allowed to consume the carbohydrate–fat diet for the 2-hr period compared to the concentrations of the amino acid in the control fetuses, the mothers of which were fasted for the same time period (Fig. 7). In contrast, levels of tryptophan were significantly reduced below control values in the bodies and brains of fetuses derived from mothers that consumed the 18% casein diet (Fig. 7). These diet-induced changes in fetal amino acid concentrations were accompanied by corresponding changes in fetal brain serotonin concentrations. The neurotransmitter levels were significantly greater in fetuses the mothers of which ate the carbohydrate–fat diet and were reduced in animals the mothers of which consumed the casein diet (Fig. 7, right). Hence, the results of this experiment show clearly that the type

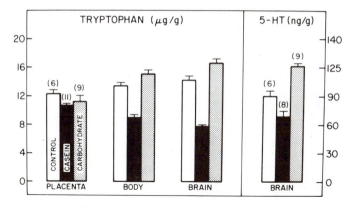

Fig. 7. Effects of acute maternal ingestion of different quality diets on tryptophan and serotonin concentrations in tissues of 17-day-postconception fetal rats. Pregnant rats were fasted overnight on the 16th day of pregnancy, and then were allowed to maintain their fast (control group; white bar), or were given access to an 18% casein protein diet (casein group; black bar) or to a carbo-hydrate–fat, protein-free diet (carbohydrate group; stippled bar). The meal began between 1000 and 1200 hr and the pregnant rats were killed 2 hr after the beginning of the meal. Fetal tissues were removed and assayed for tryptophan (Denckla and Dewey, 1967) or serotonin (Saavedra et al., 1973) concentrations. The numbers in parentheses are the sample sizes for each group. Except for the lack of effect of the carbohydrate diet on changes in placental tryptophan, all other groups differ significantly from fasted control values (p < .02) (Nelson et al., 1981).

of food eaten by pregnant animals, even as a single meal, can influence the synthesis of a brain neurotransmitter in their fetuses.

One factor that may be important for influencing the changes in tryptophan and serotonin in the brains of adult rats involves the secretion of the pancreatic hormone insulin following the ingestion of food (Fernstrom and Wurtman, 1972a,b; Crandall and Fernstrom, 1980). Elevated plasma concentrations of insulin following food consumption in adult animals increase the plasma ratio of tryptophan relative to the other large neutral amino acids, thus effectively facilitating the transport of tryptophan into brain tissue and causing an acceleration in the rate of brain serotonin biosynthesis (Fernstrom and Wurtman, 1972a,b). In the next experiment, we determined whether a similar mechanism might also influence fetal neurochemistry during gestation (Nelson *et al.*, 1981).

On the 17th day of gestation, two groups of pregnant rats, previously fasted for at least 14 hr, were injected with 2 U/kg of regular insulin or with a 0.9% saline control injection. Animals were killed 2 hr later, and fetal tissues were once again examined for possible changes in tryptophan, as well as for changes in fetal brain serotonin concentrations. The insulin treatment significantly elevated tryptophan concentrations above vehicle-injected control values in all fetal tissues examined; similarly, the insulin injection also caused comparable elevations in the concentrations of fetal brain serotonin (Fig. 8). It is interesting to note here that the insulin treatment in the fasted pregnant animals produced changes in fetal neurochemistry that were strikingly similar to those observed in animals the mothers of which consumed the carbohydrate–fat, protein-free

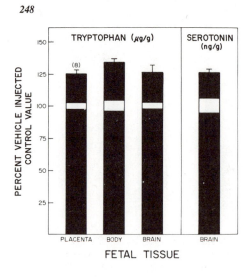

Fig. 8. Effects of insulin on fetal tissue tryptophan and brain serotonin concentrations. After an overnight fast, 17-day-postconception pregnant rats were injected with the vehicle ($N = 10$) or with 2 U/kg of regular insulin (N = 8), and were killed 2 hr later. Data are expressed as percent of control values (mean ± s.e.m. for controls are indicated by the white insert in each bar). All changes for the insulin injected groups are significantly different from control values (p < .01) (Nelson et al., 1981).

diet. Although not proven directly, it is tempting to speculate that the carbohydrate–fat, protein-free diet-induced changes in fetal tryptophan and serotonin concentrations may depend, at least in part, on an insulin-sensitive mechanism. If particular foods eaten by pregnant mammals do not contain large concentrations of the competing neutral amino acids, then the food-associated release of insulin from the pancreas may facilitate tryptophan transport into the placenta and other tissues by effectively increasing the maternal plasma ratio of this amino acid relative to its transport-competitor amino acids. Foods eaten by pregnant animals that contain relatively larger concentrations of the aromatic or branched-chain amino acids or of both (such as are found in the 18% casein diet fed our animals) may decrease the ratio of tryptophan relative to these other transport-competitor amino acids, thus effectively reducing the influx of tryptophan into the fetus and making less of it available for the synthesis of brain serotonin.

Hence, evidence from a wide variety of research conducted with animals indicates that the synthesis of serotonin neurotransmitter molecules appears quite susceptible to nutrient- or hormone-induced changes in the availability of its indoleamino acid precursor, tryptophan. Acute changes in brain tryptophan concentrations associated with alterations in the quality of ingested nutrients cause corresponding changes in the synthesis of the neurotransmitter that may have functional significance. Before considering the possible relevance of the diet-dependent effects on brain neurotransmission, let us briefly review some of the evidence demonstrating changes in the synthesis of the neurotransmitter after chronic alterations in the quantity or quality of ingested nutrients.

4.4. Nutritional Changes in Brain Serotoninergic Neurotransmission: Effects of Malnutrition

As mentioned previously, a wide variety of evidence indicates that malnutrition may severely disrupt the normal development of the structural aspects of the nervous systems of mammals. Similarly, inadequate nutrition incurred early in development may also cause rather striking changes in the ontogenesis of brain neurons utilizing serotonin as their neurotransmitter.

Morphologically, fetal brain serotoninergic nerve cell bodies in the raphe nuclei begin to receive afferent connections during the last trimester of pregnancy, even though the majority of synapse formation in the raphe nuclei does not appear to occur until the early postnatal period (Lauder and Bloom, 1975). Proliferation of the axon terminals of brain cells that contain the indoleamine neurotransmitter is a relatively slow process that begins during the last week of gestation and continues throughout the early postnatal period, reaching adult configuration somewhere between the 6th and 9th postnatal weeks in the rat (Loizou, 1972). Concomitant with maturational changes in the neuronal structure of these cells, serotonin concentrations increase dramatically in the brains of normally fed rats during the latter part of gestation, and this process continues throughout the early postnatal period (Tissari, 1973, 1975; Tissari and Raunu, 1975). At the time of birth, serotonin concentrations are approximately 40% those of the adult rat (Howd *et al.*, 1975). If it can be assumed that wholebrain concentrations of serotonin are dependent to some extent on the morphological maturity of the neurons that contain this compound, then the various neurons that contain this neurotransmitter may be relatively immature during the so-called vulnerable period when malnutrition has been shown to exert its most deleterious and permanent effects on the general structural development of brain cells. It seems clear that during this stage, the capacities of brain and peripheral neurons to synthesize, store, release, and inactivate serotonin are less in the immature altricial animal than in the adult (for reviews, see Tissari, 1973, 1975). Furthermore, some of the mechanisms important for coupling neurotransmitter synthesis and release may also be immature (Bourgoin *et al.*, 1977; Hedner and Lundborg, 1980), thus making neurotransmitter-mediated communication between neurons and other cells even more susceptible to influence by nutritional factors than would otherwise be possible in mature organisms.

Unfortunately, relatively few studies have been carried out to examine this possibility, and the results from these few efforts have not yielded consistent or readily interpretable data. For example, rats that experience protein–calorie malnutrition from the time of birth by being maintained in larger than average litters until weaning, and then are fed 5% of the amount of a laboratory rat chow diet eaten by a control group of animals, show either *no change* (Hernandez, 1973) or *transient decreases* in the whole-brain concentrations of serotonin throughout early postnatal development (Salas *et al.*, 1974). In contrast, the offspring of female rats fed an 8% casein diet for several weeks prior to the time of conception, as well as throughout the periods of gestation and lactation, and that are also maintained on the low-protein diet after weaning, show *increased* concentrations of brain serotonin and 5-HIAA for as long as 300 days of age when compared to control animals maintained on 25% casein diets (Stern *et al.*, 1974, 1975). These latter data are difficult to interpret, especially since malnutrition has been reported to cause transient reductions in the activity of aromatic L-amino acid decarboxylase, the enzyme that catalyzes the conversion of 5-hydroxytryptophan to serotonin (Hernandez, 1973).

The reported malnutrition-induced elevations in brain serotonin concentrations may be caused by several different factors. The uptake process for tryptophan into brain cells appears to have a higher affinity for the amino acid in young compared to older animals (Vahvelainen and Oja, 1972), for example, and it is possible that the state of malnutrition might enhance the rates at which amino acids are transported from the blood into the brain. Along similar lines, malnutrition caused by feeding animals an inadequate protein diet might increase the breakdown of proteins in peripheral tissues, thus liberating tryptophan, which may actually make more of the amino acid available for transport into the brain. Unfortunately, plasma or brain concentrations of tryptophan were not determined in any of these latter studies, and it is difficult to discern the precise mechanisms underlying the malnutrition-induced increases in brain serotonin concentrations.

Measurements of changes in plasma and brain concentrations of tryptophan could be particularly important, especially since normally the brain levels of the amino acid are approximately 2- to 3-fold higher during late gestation and shortly after birth, compared to the amounts of the amino acid typically observed in the brains of adult animals (Tyce *et al.*, 1964; Bourgoin *et al.*, 1974; Howd *et al.*, 1975; Hedner and Lundborg, 1980; Nelson *et al.*, 1981). Hence, brain uptake mechanisms may be relatively more efficient at transporting sufficient quantities of the amino acid into nerve cells at times early in development when changes in brain growth and structure occur at their most rapid rates. Chronic malnutrition incurred early in development and sustained subsequently throughout postnatal life may cause a hyperfunctioning of brain amino acid uptake mechanisms that guarantee sufficient nutrient supplies to maintain the structure and function of this essential organ during times of relative nutrient insufficiency.

Even though malnutrition incurred during the stages of gestation or lactation may actually cause long-term increases in brain serotonin through a mechanism still poorly understood, consistent evidence exists to show that malnutrition incurred later in life appears detrimental for normal synthesis of the indoleaminergic neurotransmitter compound. Some 35 years ago, Zbinden *et al.*, (1958) were able to show that postweanling rats or mice fed a tryptophan-deficient synthetic diet for 2–4 weeks had brain concentrations of serotonin that were roughly half the values of the neurotransmitter measured in the brains of animals fed a tryptophan-adequate, control diet. Shortly thereafter, these basic findings were repeated in a number of other laboratories using either acid-hydrolysate-treated casein or synthetic tryptophan-deficient diets (Gal *et al.*, 1961; Gal and Drewes, 1962; Culley *et al.*, 1963; Thomas and Wysor, 1967). More recently, a number of studies have examined the possible extent to which weanling rats fed diets containing corn as the only source of protein might show changes in plasma and brain tryptophan and brain serotonin concentrations (Fernstrom and Wurtman, 1971c; Fernstrom and Hirsch, 1975; Lytle *et al.*, 1975; Messing *et al.*, 1976; Zambotti *et al.*, 1976). Corn contains inadequate protein to support normal growth and is also deficient in tryptophan. Animals fed the corn-based diet consume only about 10% of the amount of tryptophan normally eaten on a daily basis by rats fed an 18% casein control diet that contains enough protein and tryptophan to guarantee adequate growth and development. Approximately 2 weeks after being placed on corn-based protein diets, rats gradually deplete residual peripheral stores of tryptophan, and plasma and brain tryptophan concentrations fall to approximately 25–30% of those observed in animals fed the casein diet. Interestingly, brain serotonin concentrations fall to approximately 60% of control values and remain depressed for as long as the animals are maintained on the tryptophan-deficient diet. That the reductions in brain serotonin concentrations are due to the inadequate tryptophan content of the diet, and are not the result of nonspecific changes caused by the protein insufficiency of the maize diet, has been shown in a variety of different ways: (1) animals fed a corn-based diet supplemented with tryptophan in an amount equivalent to that found in casein do not show reductions in plasma or brain tryptophan or in brain serotonin (Fernstrom and Wurtman, 1971c; Messing *et al.*, 1976); (2) animals fed a 6% casein diet, which contains inadequate protein but approximately 40% more tryptophan than that consumed by animals maintained on an unsupplemented corn-based diet, have greater amounts of plasma and brain tryptophan and brain serotonin than do corn-diet-fed animals (Fernstrom and Wurtman, 1971c); and (3) the corn-diet-induced reductions in brain serotonin concentrations appear to be due to the reductions in brain tryptophan availability and are not the result of nonspecific reductions in the number of serotonin-containing nerve cells in brain or in the activities of synthetic or degradative enzymes involved in serotonin metabolism (Fernstrom and Hirsch, 1975). Tryptophan-deficient animals chronically fed the corn diet seem to have normal abilities to synthesize and catabolize serotonin when treated systemically with pharmacological doses of tryptophan (Lytle *et al.*, 1975; Fernstrom and Hirsch, 1975). Hence, mammals do not appear capable of synthesizing tryptophan *de novo* and must obtain this indispensable amino acid from the diet. If animals chronically consume foodstuffs that are devoid of, or deficient in, the amino acid, reduced numbers of tryptophan molecules are available to brain serotoninergic neurons, and with time, the synthesis of the brain neurotransmitter is impaired such that its concentrations decline in brain tissue.

Somewhat more rapid reductions in brain tryptophan and serotonin concentrations have recently been observed in rodents fed a maize diet that contains high concentrations of leucine, one of the amino acids that compete with tryptophan for uptake into the brain from blood (Zambotti *et al.*, 1976). Along similar lines, normal casein-based diets that contain a large concentration of leucine have also been found to cause rather rapid reductions in the brain concentrations of tryptophan and serotonin (Krishnaswamy and Raghuram, 1972). The leucine-supplemented dietary reductions in the indoleamino acid and its neurotransmitter in brain could be reversed by supplementing animals with the amino acid isoleucine, as well as by supplementation with nicotinic acid or nicotinamide, two of the principal metabolites of circulating tryptophan that ultimately result following initial catabolism by the enzyme troptophan pyrrolase (Krishnaswamy and Raghuram, 1972). It is possible that brain tryptophan and serotonin concentrations fall in animals fed the leucine-supplemented casein diet because of possible competition between leucine and tryptophan for the same brain transport mechanism. Alternatively, leucine might increase the activity of tryptophan pyrrolase, thus causing fewer molecules of the amino acid to be available for brain transport as a result of its enhanced hepatic catabolism (Rao and Ghafoorunissa, 1972). Along similar lines, supplementation with isoleucine, nicotinic acid, or nicotinamide might reverse the leucine-induced reductions in brain tryptophan and serotonin by inhibiting the activity of tryptophan pyrrolase, thus making greater amounts of the amino acid in plasma available for brain transport (Hankes *et al.*, 1971).

It is interesting to note here that the reductions in brain tryptophan and serotonin observed in postweanling animals fed the corn-based diet or the casein or maize diet containing high concentrations of leucine do not appear to be permanent, but can be reversed within 1–2 weeks by feeding animals diets containing adequate concentrations of tryptophan or other nutrients that interfere with the hepatic catabolism of the indoleamino acid (Krishnaswamy and Raghuram, 1972; Messing *et al.*, 1976; Zambotti *et al.*, 1976). Hence, the diet-induced reductions in brain concentrations of the amino acid and its neurotransmitter appear to represent concurrent, rather than permanent, changes in brain neurochemistry that persist only as long as the organism is exposed to the dietary imbalance. Whether animals fed these same diets might show permanent changes in brain tryptophan and serotonin persisting beyond the immediate period of malnutrition if the nutritional insufficiency is incurred during the vulnerable period of brain growth and development remains to be determined.

A few studies have also examined the possible extent to which the brains of animals chronically fed foods containing supranormal concentrations of tryptophan might show enhanced brain serotonin synthesis. As might be anticipated, Wang *et al.* (1962) and H. Green *et al.* (1962) independently confirmed that brain serotonin concentration could be increased chronically in animals as long as they were maintained on diets supplemented with tryptophan in amounts approximately 10 times greater than those consumed by animals fed a tryptophan-adequate diet. Hence, chronic dietary alterations in postweanling animals that ultimately cause elevations or reductions in brain tryptophan concentration produce corresponding, positively correlated changes in the levels of brain serotonin, apparently by making greater or fewer molecules of the precursor amino acid available to brain nerve cells that synthesize and release serotonin as their neurotransmitter compound.

5. Nutritional Status and Brain Function: A Prospectus

Two major questions have not yet been addressed in this review: First, to what extent are the types of nutrient-induced changes in brain serotoninergic neurotransmission described previously also indicative of the possible effects of nutritional status on other types of central or peripheral neurons utilizing compounds other than serotonin as their neurotransmitter? Second, and perhaps more important, are the types of nutrient-induced alterations in brain and peripheral neurotransmitter biosynthesis also paralleled by changes in the actual functional activity of nerve cells that utilize neurotransmitter compounds known to be sensitive to fluctuations in the quantity or the quality of food ingested?

5.1. Nutritional Changes in Other Brain or Peripheral Neurotransmitter Chemicals or in Both

It seems apparent at this stage of research that potentially all neurotransmitter compounds that are directly or indirectly derived in whole or in part from dietary nutrients may change in response to chronic alterations in nutritional status. Indeed, some evidence indicates that protein or calorie malnutrition or both may influence the synthesis of a rather wide variety of brain and peripheral neurotransmitters (for a review, see Shoemaker and Bloom, 1977). Although not yet studied in great detail, it appears likely that chronic malnutrition may alter the intercellular concentrations of neurotransmitters directly by limiting their total availability to all body cells, including neurons, if the particular neurotransmitter compound is taken up directly into the nerve cell from the surrounding extracellular fluid. On the other hand, chronic changes in nutritional status may indirectly alter the intercellular concentrations of a particular neurotransmitter compound by limiting the relative intercellular availability of precursor substances involved in its synthesis, or possibly by altering other aspects of cellular metabolic function important for the synthesis of enzymes or for the transport, storage, release, or inactivation of the neurotransmitter. Definitive evidence cataloguing changes in the metabolic fate of neurotransmitters that occur during or after periods of chronic nutrient insufficiency is lacking at present; however, it might be anticipated that severe cases of human malnutrition, such as those observed in the states of marasmus or kwashiorkor, should produce rather nonspecific, widespread disruptions in neurotransmitter metabolic competency.

Acute dietary changes in other brain or peripheral neurotransmitter compounds or in both have also been described and appear similar to those documented to influence brain serotonin. Although space does not allow us to describe these changes in great detail, the best available evidence indicates that potentially all neurotransmitter compounds in brain the precursors of which change in direct relationship to their fluctuations in plasma, the synthetic enzymes of which are not normally saturated with respect to their precursor substrates, and the mechanisms for synthesis–release coupling of which are not subject to control by non-nutrient-dependent factors should be susceptible to rather minor alterations in the quantity or quality of food consumed (Wurtman *et al.*, 1980). Moreover, it might be anticipated that neurotransmitter compounds of this type might fluctuate dramatically as a function of the amount or type of food consumed in one meal compared to the next. Thus far, good evidence has been adduced that under some circumstances, the catecholamine neurotransmitters dopamine and norepinephrine, which are ultimately derived from the amino acid precursor compound L-tyrosine, and acetylcholine, the neurotransmitter derived from the precursors choline and acetate, may be subject to nutrient-induced changes in precursor availability (for recent detailed reviews of this evidence, see Wurtman and Fernstrom, 1976; Growdon, 1979; Fernstrom, 1979b; Wurtman *et al.*, 1980) In brief, nutrient-induced elevations in the brain concentrations of tyrosine or choline have been found, under many conditions, to be associated with transient increases in the synthesis of the catecholamine neurotransmitters or acetylcholine, respectively. The possible extent to which many of the other 20–25 putative neurotransmitter compounds currently identified are also susceptible to nutrient influences in a fashion comparable to those demonstrated for serotonin, dopamine, norepinephrine, and acetylcholine awaits further investigation.

5.2. Possible Changes in Neurotransmitter Function following Nutrient-Induced Alterations in Neurotransmitter Synthesis

If nutrient-induced alterations in neurotransmitter biosynthesis are to be functionally significant, then such changes must ultimately increase the total amount of neurotransmitter molecules that are released from nerve terminals *and* that interact with receptors localized on target-cell membranes. It has been difficult thus far to obtain direct evidence that nutrient-induced changes in neurotransmitter synthesis are paralleled by corresponding, predictable changes in the release of neurotransmitters, as well as in the physiological activities of target cells containing receptors appropriate for the neurotransmitter of interest. Part of the difficulty in obtaining relevant evidence regarding the functional effects of nutrients on neurotransmission lies in the fact that with but one possible exception (Lane *et al.*, 1978, 1979; Knott *et al.*, 1980), no method exists at present whereby possible nutrient-induced alterations in neurotransmitter release can be measured *in vivo*, in the intact organism. In addition, it is difficult to lay possible nutrient-induced changes in the gross physiology or behavior of an intact organism to changes in a single neurotransmitter system because these types of macroresponses are mediated by the actions of a multiplicity of nerve cells, each of which utilizes a different neurotransmitter compound. Moreover, nutrient-induced alterations in physiology or behavior could result from changes in neurotransmission, but could just as likely be caused by changes in other metabolic pathways and systems also influenced by increases or decreases in a single nutrient.

Despite these difficulties, some recent tantalizing clues about the possible functional importance of nutrient-induced changes in serotoninergic neurotransmission have been forthcoming. For example, some recent work suggests that neurogenesis and subsequent brain development may depend, at least in part, on the normal maturation of brain serotonin-containing neurons. Fetal animals the mothers of which are treated chronically with parachlorophenylalanine, a halogenated amino acid drug that irreversibly inhibits the enzyme tryptophan hydroxylase (Koe and Weissman, 1963) and also decreases brain tryptophan concentration (Messing *et al.*, 1976), show delayed maturational changes in neurons in specific brain regions that ultimately receive afferent connections from serotoninergic nerves (Lauder and Krebs, 1976, 1978). Since our own data show that relatively mild, physiological fluctuations in the quality of foods eaten as single meals also alter fetal brain serotonin biosynthesis, it is possible that the chronic ingestion of nutritionally imbalanced diets might also produce changes in the rates of neurogenesis and maturation in these same fetal brain regions. At the very least, our results suggest that fetal neurochemical development may be influenced just as dramatically by the types of food eaten during pregnancy as it is by the types of drugs and medications consumed during this time period.

In older animals, we (Lytle *et al.*, 1975; Messing *et al.*, 1976) have also obtained evidence that chronic malnutrition, accomplished by feeding experimental animals a tryptophan-deficient, corn-based diet, might produce predictable changes in at least one behavioral response thought to be mediated, at least in part, by brain serotoninergic neurons. Nociception, or alterations in the sensory or response mechanisms, or in both, caused by presentations of noxious stimuli, appears to be influenced by alterations in the activity of serotonin-containing neurons. A rather wide range of pharmacological and surgical studies indicate that analgesia, or a reduction in pain sensitivity or reactivity or both, results when serotoninergic neurotransmission increases; in contrast, reductions in serotoninergic neurotransmission usually cause hyperalgesia, or increases in sensitivity or reactivity or both, to presentations of noxious stimuli (for a review, see Messing and Lytle, 1977). In initial experiments, we (Lytle *et al.*, 1975) observed that corn-diet-fed rats had a much greater sensitivity to a painful electroshock stimulus delivered to the feet than did a group of well-nourished animals. For example, the shock intensity necessary to elicit a pain response was significantly less (approximately half the amperage) than that needed to cause a similar response in the well-nourished group of animals. This behavioral difference between the two groups of animals fed the diets was first observed within 2 weeks after the experimental animals began consuming the corn diet and remained for as long as the animals continued eating this foodstuff. Interestingly, the changes in behavior correlated very well with the diet-induced reductions in the concentrations of brain serotonin in the corn-diet-fed animals. We also found the behavioral and neurochemical changes associated with the corn malnutrition to be reversible. If corn-fed rats were allowed to ingest the nutrient-adequate, casein protein diet fed previously to the control group of animals beginning 10 weeks after they had been eating the corn diet, then the hyperalgesia and alterations in brain serotonin concentrations were returned to normal within a 2-week period. In subsequent experiments we were able to show that these diet-induced changes in behavior and in brain neurochemistry were specifically related to alteration in brain serotonin-containing neurons, and were not the result of nonspecific alterations caused by the malnutrition. Animals fed a corn diet supplemented with amounts of tryptophan comparable to those found in casein did not develop behavioral hyperalgesia and also did not show any changes in brain serotonin concentrations. Along similar lines, the fact that a variety of drugs (tryptophan, parachlorophenylalanine, and valine) administered to the corn-diet-fed experimental animals or the casein-fed control animals produce predictable, corresponding changes in both nociception and brain serotoninergic neurotransmission was used as evidence to show that the dietary alterations in behavior were caused by relatively specific changes in the metabolic function of brain serotonin-containing neurons. Other types of neuroendocrine and behavioral changes in animals have been observed following manipulation of tryptophan availability to brain neurons, although it still remains to be determined whether all or only some of these changes represent direct effects of the precursor compound on brain serotonin-containing nerves (Wurtman *et al.*, 1980).

Although there is emerging evidence that nutrient-induced alterations in serotoninergic neurotransmission may predictably change some features of the physiological and behavior responses of experimental animals, it has been relatively difficult to obtain parallel, incontrovertible data in humans. Nevertheless, some interesting observations have been forthcoming. For example, in some cases of human clinical depression, levels of cerebrospinal fluid 5-HIAA appear to be reduced (Van Praag and Korf, 1974). Interestingly, treatment with rather heroic doses of L-tryptophan has been reported to be efficacious in alleviating some but not all of the symptoms associated with this mood disorder (for a review, see Wirz-Justice, 1977). Along similar lines, L-tryptophan loading has also proven successful for treating some cases of insomnia in humans (Oswald *et al.*, 1966; Wyatt *et al.*, 1970; Hartmann *et al.*, 1971; Griffiths *et al.*, 1972). Inasmuch as brain serotoninergic neurons have been shown to be important for the occurrence of normal patterns of sleep and waking in experimental animals (Jouvet, 1968), it is tempting to speculate that the precursor-amino-acid treatment of insomnia is caused by an acceleration in brain serotoninergic neurotransmitter function. Although provocative, these findings need further experimental confirmation in larger numbers of humans, using a variety of different approaches.

Precursor-loading therapies, in which large doses of various compounds involved in the synthesis of a particular brain or peripheral neurotransmitter compound or both are given, also appear to be occasionally successful in the treatment of a number of human clinical disorders, including Alzheimer's disease, tardive dyskinesia, depression and mania, sleep disorders, and other types of physiological or psychiatric maladies (for reviews, see Growdon, 1979; Wurtman *et al.*, 1980). Despite the mixed successes of these approaches, it should not be concluded *ipso facto* that diseases that are responsive to precursor-loading therapies therefore result from nutrient-related disorders. Along similar lines, it is not clear at present that those precursor-loading therapies that are successful in reversing or ameliorating some of the symptoms of a human clinical disease will be effective in enhancing the physiological or behavioral response capabilities of the normal population of undiseased humans.

6. Summary

    This brief review has focused on only one small, but important, area of research aimed at understanding and characterizing the possible importance of nutrient status in altering the neurochemical functioning of brain neurons that contain serotonin as their neurotransmitter compound. In brief, the bulk of the evidence available thus far suggests that the synthesis of this neurotransmitter compound can change dramatically as a function of alterations in nutritional status that increase or decrease the relative availability of the precursor amino acid tryptophan to brain cells. Concurrent, long-term, nutrient-dependent changes in brain serotonin biosynthesis have been consistently observed in animals that chronically consume foodstuffs such as corn-based protein that lack adequate amounts of tryptophan. Inasmuch as a substantial population of humans in Third World countries at present consume corn as a major protein source, the reductions in brain serotonin observed in experimental animals may have relevance in understanding some of the intellectual and sensorimotor deficits reported in human malnutrition. Also interesting are the observations that relatively short-term changes in the synthesis of this neurotransmitter appear to occur on a meal-by-meal basis. Depending on the particular dietary constituents eaten in a single meal, brain serotonin synthesis either increases or decreases predictably in direct relation to the enhancement or diminution, respectively, in brain tryptophan uptake. Recent evidence suggests that the synthesis of other neurotransmitter compounds, such as dopamine, norepinephrine, and acetylcholine, may also be subject to similar nutrient-induced alterations that limit the relative availability of precursor compounds involved in their synthesis (see, for example, the reviews by Wurtman and Fernstrom, 1976; Growdon, 1979; Fernstrom, 1979b; Wurtman *et al.*, 1980). Hence, it is possible that chronic or more acute changes in nutritional status may exert an even more profound influence on brain neurotransmitter biosynthesis than was previously thought possible.

    Although the picture is not yet complete, we are beginning to learn a relatively great deal about the importance of several different populations of brain nerves such as those that contain serotonin as their neurotransmitter, and about their roles in mediating various physiological and behavioral responses under normal and pathophysiological disease states. Although it is still too early to know for sure, it seems plausible to assume at this point that future research efforts will be aimed at determining the possible extent to which physiological and behavioral function might be subtly influenced by nutrition-induced changes in the relative capabilities of brain and peripheral nerves to communicate with one another, as well as with other cells in the body, vis-à-vis the actions of nutrient-derived chemical neurotransmitter compounds.

ACKNOWLEDGMENTS. This research was supported in part by grants to L.D.L. from the NIMH (MH-31134) and from the March of Dimes Birth Defects Foundation.

7. References

Aghajanian, G. K., and Bunney, B. S., 1974, Pre- and post-synaptic feedback mechanisms in central dopaminergic neurons, in: *Frontiers of Neurology and Neuroscience Research* (P. Seeman and G. M. Brown, eds.), pp. 4–11, University of Toronto Press.

Aghajanian, G. K., Rosecrans, J. A., and Seward, M. N., 1966, Serotonin release in the forebrain by stimulation of the midbrain raphe, *Science* **156**:402–403.

Ashcroft, G. W., Eccleston, D., and Crawford, T. B. B., 1965, 5-Hydroxyindole metabolism in rat brain: A study of intermediate metabolism using the technique of tryptophan loading, *J. Neurochem.* **12**:483–503.

Björklund, A., Falck, B., and Stenevi, U., 1971, Classification of monoamine neurons in the rat mesencephalon: Distribution of a new monoamine neuron system, *Brain Res.* **32**:1–17.

Black, I. B., 1975, Increased tyrosine hydroxylase activity in frontal cortex and cerebellum after reserpine, *Brain Res.* **95**:170–176.

Blasberg, R., and Lajtha, A., 1965, Substrate specificity of steady-state amino acid transport in mouse brain slices, *Arch. Biochem. Biophys* **112**:361–377.

Bloxam, D. L., and Curzon, G., 1978, A study of proposed determinants of brain tryptophan concentration in rats after porto-caval anastomosis or sham operation, *J. Neurochem.* **31**:1255–1263.

Bloxam, D. L., Warren, W. H., and White, P. J., 1974, Involvement of the liver in the regulation of tryptophan availability: Possible role in the responses of liver and brain to starvation, *Life Sci.* **15**:1443–1455.

Bloxam, D. L., Tricklebank, M.D., Patel, A. J., and Curzon, G., 1980, Effects of albumin, amino acids, and clofibrate on the uptake of tryptophan by the rat brain, *J. Neurochem.* **34**:43–49.

Bourgoin, S., Faivre-Bauman, A., Benda, P., Glowinski, J., and Hamon, M., 1974, Plasma tryptophan and 5-HT metabolism in the CNS of the newborn rat, *J. Neurochem.* **23**:319–327.

Bourgoin, S., Artaud, F., Enjalbert, A., Hery, G., Glowinski, J., and Hamon, M., 1977, Acute changes in central serotonin metabolism induced by the blockade or stimulation of serotoninergic receptors during ontogenesis in the rat, *J. Pharmacol. Exp. Ther.* **202**:519–531.

Carlsson, A., and Lindqvist, M., 1972, The effect of L-tryptophan and some psychotrophic drugs on the formation of 5-hydroxytryptophan in the mouse brain *in vivo*, *J. Neural Transm.* **33**:23–42.

Carlsson, A., and Lindqvist, M., 1978, Dependence of 5-HT and catecholamine synthesis on precursor amino-acid levels in rat brain, *Naunyn-Schmiedeberg's Arch. Pharmacol.* **303**:157–164.

Colmenares, J. L., Wurtman, R. J., and Fernstrom, J. D., 1975, Effect of ingesting a carbohydrate–fat meal on the levels and synthesis of 5-hydroxyindoles in various regions of the rat central nervous system, *J. Neurochem.* **125**:825–829.

Costa, E., and Meek, J. L., 1974, Regulation of biosynthesis of catecholamines and serotonin in the central nervous system, *Annu. Rev. Pharmacol.* **14**:491–511.

Crandall, E. A., and Fernstrom, J. D., 1980, Acute changes in brain tryptophan and serotonin after carbohydrate or protein ingestion by diabetic rats, *Diabetes* **29**:460–466.

Culley, W. J., Saunders, R. N., Mertz, E. T., and Jolly, D. H., 1963, Effects of a tryptophan deficient diet on brain serotonin and plasma tryptophan levels, *Proc. Soc. Exp. Biol. Med.* **113**:645–648.

Curzon, G., 1965, The biochemistry of depression, in: *Biochemical Aspects of Neurological Disorders*, 2nd Series (J. N. Cumings and M. Kremer, eds.), pp. 243–256, Blackwell, Oxford.

Curzon, G., 1979, Relationships between plasma, CSF and brain tryptophan, *J. Neural Transm. Suppl.* **15**:81–92.

Curzon, G., and Green, A. R., 1968, Effect of hydrocortisone on rat brain 5-hydroxytryptamine, *Life Sci.* **7**:657–663.

Curzon, G., and Knott, P. J., 1974, Effects on plasma and brain tryptophan in the rat of drugs and hormones that influence the concentration of unesterified fatty acid in the plasma, *Br. J. Pharmacol.* **50**:197–204.

Curzon, G., Kantamaneni, B. D., Winch, A., Rojas-Bueno, I., Murray-Lyon, M., and Williams, R., 1973, Plasma and brain tryptophan changes in experimental acute hepatic failure, *J. Neurochem.* **21**:137–146.

Curzon, G., Kantamaneni, B. D., Fernando, J. C., Woods, M. S., and Cavanagh, J. B., 1975, Effects of chronic porto-caval anastomosis on brain tryptophan, tyrosine, and 5-hydroxy-tryptamine, *J. Neurochem.* **24:**1065–1070.

Davis, M., and Sheard, M. H., 1974, Habituation and sensitization of the rat startle response: Effects of raphe lesions, *Physiol. Behav.* **12:**425–431.

Deguchi, T., and Barchas, J., 1972, Regional distribution and developmental change of tryptophan hydroxylase activity in rat brain, *J. Neurochem.* **19:**927–929.

Denckla, W. D., and Dewey, H. R., 1967, The determination of tryptophan in plasma, liver, and urine, *J. Lab. Clin. Med.* **69:**160–169.

Dobbing, J., and Smart, J. L., 1973, Early undernutrition, brain development, and behavior, in: *Ethology and Development* (S. A. Barnett, ed.), pp. 16–36, Lippincott, Philadelphia.

Etienne, P., Young, S. N., and Sourkes, T. L., 1976, Inhibition by albumin of tryptophan uptake by rat brain, *Nature (London)* **262:**144–145.

Fernstrom, J. D., 1979a, Diet-induced changes in plasma amino acid pattern: Effects on the brain uptake of large neutral amino acids, and on brain serotonin synthesis, *J. Neural Transm. Suppl.* **15:**55–67.

Fernstrom, J. D., 1979b, The influence of circadian variations in plasma amino acid concentrations on monoamine synthesis in the brain, in: *Endocrine Rhythms* (D. T. Krieger, ed.), pp. 89–122, Raven Press, New York.

Fernstrom, J. D., and Faller, D. V., 1978, Neutral amino acids in the brain: Changes in response to food ingestion, *J. Neurochem.* **30:**1531–1538.

Fernstrom, J. D., and Hirsch, M. J., 1975, Rapid repletion of brain serotonin in malnourished rats following L-Tryptophan injection, *Life Sci.* **17:**455–464.

Fernstrom, J. D., and Wurtman, R. J., 1971a, Brain serotonin content: Increase following ingestion of a carbohydrate diet, *Science* **174:**1023–1025.

Fernstrom, J. D., and Wurtman, R. J., 1971b, Brain serotonin content: Physiological dependence on plasma tryptophan levels, *Science* **173:**149–152.

Fernstrom, J. D., and Wurtman, R. J., 1971c, Effect of chronic corn consumption on serotonin content of rat brain, *Nature (London) New Biol.* **234:**62–64.

Fernstrom, J. D., and Wurtman, R. J., 1972a, Brain serotonin content: Physiological regulation by plasma neutral amino acids, *Science* **178:**414–416.

Fernstrom, J. D., and Wurtman, R. J., 1972b, Elevation of plasma tryptophan by insulin in the rat, *Metabolism* **21:**337–342.

Fernstrom, J. D., and Wurtman, R. J., 1974, Nutrition and the brain, *Sci. Am.* **230:**84–91.

Fernstrom, J. D., Larin, F., and Wurtman, R. J., 1973, Correlations between brain tryptophan and plasma neutral amino acids following food consumption, *Life Sci.* **13:**517–524.

Fernstrom, J. D., Faller, D. V., and Shabshelowitz, H., 1975a, Acute reduction of brain serotonin and 5-HIAA following food consumption: Correlation with the ratio of serum tryptophan to the sum of competing neutral amino acids, *J. Neural. Transm.* **36:**113–121.

Fernstrom, J. D., Hirsch, M. J., Madras, B. K., and Sudarsky, L., 1975b, Effects of skim milk, whole milk, and light cream on serum tryptophan binding and brain tryptophan concentrations. *J. Nutr.* **105:**1359–1362.

Fernstrom, J. D., Shabshelowitz, H., and Faller, D. V., 1975c, Diazepam increases 5-hydroxyindole concentrations in rat brain and spinal cord, *Life Sci.* **15:**1577–1584.

Fernstrom, J. D., Hirsch, M. J., and Faller, D. V., 1976, Tryptophan concentrations in rat brain: Failure to correlate with serum free tryptophan, or its ratio to the sum of other serum neutral amino acids, *Biochem. J.* **160:**589–595.

Fitzsimons, T. J., and LeMagnen, J., 1969, Eating as a regulatory control of drinking in the rat, *J. Comp. Physiol. Psychol.* **67:**273–283.

Friedman, P. A., Kappelman, A. H., and Kaufman, S., 1972a, Partial purification and character-ization of tryptophan hydroxylase from rabbit hindbrain, *J. Biol. Chem.* **247:**4165–4173.

Friedman, P. A., Kaufman, S., and Kang, E. S., 1972b, Nature of the molecular defect in phen-ylketonuria, *Nature (London)* **240:**157–159.

Fuller, R. W., 1970, Daily variations in liver tryptophan pyrrolase, and tyrosine transaminase in rats fed *ad libitum* or single daily meals, *Proc. Soc. Exp. Biol. Med.* **133:**620–622.

Fuller, R. W., 1972, Selective inhibition of monoamine oxidase, *Adv. Psychopharmacol.* **5:**339–354.

Fuxe, K., Hökfelt, T., and Ungerstedt, U., 1968, Localization of indolealkylamines in CNS, *Adv. Pharmacol.* **6A:**235–251.

Gal, E. M., and Drewes, P. A., 1962, Studies on the metabolism of 5-hydroxytryptamine (serotonin). II. Effect of tryptophan deficiency in rats, *Proc. Soc. Exp. Biol. Med.* **110:**368–371.

Gal, E. M., Drewes, P. A., and Barraclough, C. A., 1961, Effect of reserpine and the metabolism of serotonin in tryptophan-deficiency in rats, *Biochem. Pharmacol.* **8:**32.

Gershon, M. D., 1968, Serotonin and the motility of the gastrointestinal tract, *Gastroenterology* **54:**453–456.

Gessa, G. L., and Tagliamonte, A., 1974, Serum free tryptophan: Control of brain concentrations of tryptophan and of synthesis of 5-hydroxytryptamine, in: *Aromatic Amino Acids in Brain,* CIBA Foundation Symposium 22, pp. 207–216, Elsevier, Amsterdam.

Grahame-Smith, D. G., and Parfitt, A. G., 1970, Tryptophan transport across the synaptosomal membrane, *J. Neurochem.* **17:**1339–1353.

Green, A. R., 1978, The effects of dietary tryptophan and its peripheral metabolism on brain 5-hydroxytryptamine synthesis and function, in: *Essays in Neurochemistry and Neuropharmacology* (M. B. H. Youdim, W. Lovenberg, D. F. Sharman, and J. R. Lagnado, eds.), pp. 103–128, John Wiley, London.

Green, A. R., and Curzon, G., 1968, Decrease of 5-hydroxytryptamine in the brain provoked by hydrocortisone and its prevention by allopurinol, *Nature (London)* **220:**1095–1097.

Green, A. R., and Grahame-Smith, D. G., 1975, 5-Hydroxytryptamine and other indoles in the central nervous system, in: *Handbook of Psychopharmacology,* Vol. 3 (L. L. Iversen, S. D. Iversen, and S. H. Snyder, eds.), pp. 169–245, Plenum Press, New York.

Green, A. R., Sourkes, T. L., and Young, S. N., 1975, Liver and brain tryptophan metabolism following hydrocortisone administration to rats and gerbils, *Br. J. Pharmacol.* **53:**287–292.

Green, H., Greenberg, S. M., Erickson, R. W., Sayer, J. L., and Ellison, T., 1962, Effects of dietary phenylalanine and tryptophan upon rat brain amine levels, *J. Pharmacol. Exp. Ther.* **136:**174–178.

Griffiths, W. J., Lester, B. K., Coulter, J. D., and Williams, H. L., 1972, Tryptophan and sleep in young adults, *Psychophysiology* **9:**345–356.

Growdon, J. H., 1979. Neurotransmitters and the diet: Their use in the treatment of brain disorders, in: *Nutrition and the Brain,* Vol. 3 (R. J. Wurtman and J. J. Wurtman, eds.), pp. 117–181, Raven Press, New York.

Growdon, J. H., and Wurtman, R. J., 1979, Dietary influences on the synthesis of neurotransmitters in the brain, *Nutr. Rev.* **37:**129–136.

Guroff, G., and Udenfriend, S., 1962, Studies on aromatic amino acid uptake by rat brain *in vivo,* *J. Biol. Chem.* **237:**803–806.

Haigler, H., 1981, *Serotonergic Receptors in the Central Nervous System,* Chapman and Hall, London.

Hamon, M., and Glowinski, J., 1974, Regulation of serotonin synthesis, *Life Sci.* **15:**1533–1548.

Hankes, L. V., Leklem, J. E., Brown, R. R., and Mekel, R. C. P. M., 1971, Tryptophan metabolism in patients with pellagra: Problem of vitamin B_6 enzyme activity and feedback control of tryptophan pyrrolase enzyme, *Am. J. Clin. Nutr.* **24:**730–739.

Hartmann, E., Chung, R., and Chien, C. P., 1971, L-Tryptophane and sleep, *Psychopharmacologia* **19:**114–127.

Harvey, J. A., and Simansky, K. J., 1982, The role of serotonin in modulation of nociceptive reflexes, in: *Current Aspects of Neurochemistry and Function* (B. Haber, ed.), pp. 125–151, Plenum Press, New York.

Hedner, T., and Lundborg, P., 1980, Serotoninergic development in the postnatal rat brain, *J. Neural Transm.* **49:**257–279.

Hernandez, R. J., 1973, Developmental pattern of the serotonin synthesizing enzyme in the brain of postnatally malnourished rats, *Experientia* **29:**1487–1488.

Howd, R. A., Nelson, M. F., and Lytle, L. D., 1975, L-Tryptophan and rat fetal brain serotonin, *Life Sci.* **17:**803–812.

Hutson, P. H., Knott, P. J., and Curzon, G., 1976, Control of brain tryptophan concentration in rats on a high fat diet, *Nature (London)* 262:142–143.

Jackson, C. M., and Stewart, C. A., 1920, The effects of inanition in the young upon the ultimate size of the body and of the various organs in the albino rat, *J. Exp. Zool.* 30:97–128.

Jacoby, J., Colmenares, J. L., and Wurtman, R. J., 1975, Failure of decreases of serotonin uptake or monoamine oxidase inhibition to block the acceleration in brain 5-hydroxyindole synthesis that follows food consumption, *J. Neural Transm.* 37:25–32.

Jequier, E., Robinson, D. S., Lovenberg, W., and Sjoerdsma, A., 1969, Further studies on tryptophan hydroxylase in rat brainstem and beef pineal, *Biochem. Pharmacol.* 18:1071–1080.

Jouvet, M., 1968, Insomnia and decrease of cerebral 5-hydroxytryptamine after destruction of the raphe system in the cat, *Adv. Pharmacol.* 6:265–279.

Kaufman, S., 1974, Properties of pterin-dependent aromatic amino acid hydroxylases, in: *Aromatic Amino Acids in Brain* (G. E. W. Wolstenholme and D. W. Fitzsimons, eds.), pp. 85–115, Elsevier/North-Holland, Amsterdam.

Kiely, M., and Sourkes, T. L., 1972, Transport of L-tryptophan into slices of rat cerebral cortex, *J. Neurochem.* 19:2863–2972.

Knott, P. J., and Curzon, G., 1972, Free tryptophan in plasma and brain tryptophan metabolism, *Nature (London)* 239:452–453.

Knott, P. J., Hutson, P. H., and Curzon, G., 1977, Fatty acid and tryptophan changes on disturbing groups of rats and caging them singly, *Pharmacol. Biochem. Behav.* 7:245–252.

Knott, P. J., Hutson, P. H., and Curzon, G., 1980, Behavioral and voltammetric evidence for serotonergic inhibition of caudate dopamine release, *Fed. Proc. Fed. Am. Soc. Exp. Biol.* 39(3):607.

Knox, W. E., 1951, Two mechanisms which increase *in vivo* liver tryptophan peroxidase activity: Specific enzyme adaptation and stimulation of pituitary adrenal system, *Br. J. Exp. Pathol.* 32:462–469.

Knox, W. E., and Mehler, A. H., 1950, The conversion of tryptophan to kynurenine in the liver. I. The coupled tryptophan peroxidase–oxidase system forming kynurenine, *J. Biol. Chem.* 187:419–430.

Koe, B. K., and Weissman, A., 1963, *p*-Chlorophenylalanine, a specific depletor of brain serotonin, *J. Pharmacol. Exp. Ther.* 154:499–516.

Krishnaswamy, K., and Raghuram, T. C., 1972, Effect of leucine and isoleucine on brain serotonin concentration in rats, *Life Sci.* 11:1191–1197.

Lamprecht, F., and Coyle, J. T., 1972, DOPA decarboxylase in the developing rat brain, *Brain Res.* 41:503–506.

Lane, R. F., Hubbard, A. T., and Blaha, C. D., 1978, Brain dopaminergic neurons: *In vivo* electrochemical information concerning storage, metabolism and release processes, *Bioelectrochem. Bioenerget.* 5:506–525.

Lane, R. F., Hubbard, A. T., and Blaha, C. D., 1979, *In vivo* voltammetric monitoring of dopamine release and catabolism in the rat striatum, in: *Catecholamines: Basic and Clinical Frontiers*, Vol. 1, pp. 883–885, Pergamon Press, New York.

Langer, S. Z., 1979, Presynaptic regulation of the release of catecholamines, *Pharmacol. Rev.* 32:337–362.

Lauder, J. M., and Bloom, F. E., 1974, Ontogeny of monoamine neurons in the locus coeruleus, raphe nuclei and substantia nigra of the rat. I. Cell differentiation, *J. Comp. Neurol.* 155:469–482.

Lauder, J. M., and Bloom, F. E., 1975, Ontogeny of monoamine neurons in the locus coeruleus, raphe nucleus, and substantia nigra of the rat. II. Synaptogenesis, *J. Comp. Neurol.* 163:251–264.

Lauder, J. M., and Krebs, H., 1976, Effects of *p*-chlorophenylalanine on time of neuronal origin during embryogenesis in the rat, *Brain Res.* 107:638–644.

Lauder, J. M., and Krebs, H., 1978, Serotonin as a differentiation signal in early neurogenesis, *Dev. Neurosci.* 1:15–30.

Lawson, D. M., and Gala, R. R., 1978, The influence of pharmacological manipulation of serotonergic and dopaminergic mechanisms on plasma prolactin in ovarectomized, estrogen-treated rats, *Endocrinology* 102:973–981.

LeMagnen, J., and Tallon, S., 1966, La périodicité spontanée de la prise d'aliments *ad libitum* du rat blanc, *J. Physiol (Paris)* **58**:323–349.

Logan, W. J., and Snyder, S. H., 1972, High affinity uptake system for glycine, glutamic and asparatic acids in synaptosomes of rat central nervous tissues, *Brain Res.* **42**:413–431.

Loizou, L., 1972, The postnatal ontogeny of monoamine-containing neurons in the central nervous system of the albino rat, *Brain Res.* **40**:395–418.

Lotspeich, W. D., 1949, The role of insulin in the metabolism of amino acids, *J. Biol. Chem.* **179**:175–184.

Lovenberg, W., Weissbach, H., and Udenfriend, S., 1962, Aromatic L-amino acid decarboxylase, *J. Biol. Chem.* **237**:89–93.

Lovenberg, W., Jequier, E., and Sjoerdsma, A., 1968, Tryptophan hydroxylation in mammalian systems, *Adv. Pharmacol.* **6A**:21–25.

Luck, J. M., Morrison, G., and Wilbur, L. F., 1928, The effect of insulin on the amino acid content of blood, *J. Biol. Chem.* **77**:151–172.

Lytle, L. D., Messing, R. B., Fisher, L., and Phebus, L., 1975, Effects of long-term corn consumption on brain serotonin and the response to electric shock, *Science* **190**:692–694.

Madras, B. K., Cohen, E. L., Fernstrom, J. D., Larin, F., Munro, H. N., and Wurtman, R. J., 1973, Dietary carbohydrate increases brain tryptophan and decreases serum-free tryptophan, *Nature (London)* **244**:34–35.

Madras, B. K., Cohen, E. L., Messing, R., Munro, H. N., and Wurtman, R. J., 1974, Relevance of serum-free tryptophan to tissue tryptophan concentrations, *Metabolism* **23**:1107–1116.

Mandell, A. J., 1963, Some determinants of indole excretion in man, *Recent Adv. Biol. Psychiatry* **5**:237–256.

Marchbanks, R. M., 1966, Serotonin binding to nerve ending particles and other preparations from rat brain, *J. Neurochem.* **13**:1481–1493.

Martin, J. D., Durand, D., Gurd, W., Faille, G., Audet, J., and Brazeau, P., 1978, Neuropharmacological regulation of episodic growth hormone and prolactin secretion in the rat, *Endocrinology* **102**:106–113.

Mayer, D. J., and Price, D. D., 1976, Central nervous system mechanisms of analgesia, *Pain* **2**:379–404.

McMenamy, R. H., and Oncley, J. L., 1958, The specific binding of L-tryptophan to serum albumin, *J. Biol. Chem.* **233**:1436–1447.

Meltzer, H. Y., Fang, V. S., Paul, S. M., and Kaluskar, R., 1976, Effects of quipazine on rat plasma prolactin levels, *Life Sci.* **19**:1073–1078.

Messing, R. B., and Lytle, L. D., 1977, Serotonin-containing neurons: Their possible role in pain and analgesia, *Pain* **4**:1–21.

Messing, R. B., Fisher, L. A., Phebus, L., and Lytle, L. D., 1976, Interaction of diet and drugs in the regulation of brain 5-hydroxyindoles and the response to painful electric shock, *Life Sci.* **18**:707–714.

Messing, R. B., Pettibone, D. J., Kaufman, N., and Lytle, L. D., 1978, Behavioral effects of serotonin neurotoxins: An overview, *Ann. N. Y. Acad. Sci.* **305**:480–496.

Meyerson, B., 1964, Central nervous monoamines and hormone induced estrus behaviour in the spayed rat, *Acta Physiol. Scand. Suppl.* **63**:241.

Moir, A. T. B., 1971, Interaction in the cerebral metabolism of the biogenic amines: Effect of intravenous infusion of L-tryptophan on tryptophan and tyrosine in brain and body fluids, *Br. J. Pharmacol.* **43**:724–731.

Moir, A. T. B., and Eccleston, D., 1968, The effects of precursor loading in the cerebral metabolism of 5-hydroxyindoles, *J. Neurochem.* **15**:1093–1108.

Moskowitz, M. A., Chiel, H. J., and Lytle, L. D., 1978, Neurotransmitters and diseases of the nervous system, in: *Current Neurology*, Vol. 1 (H. R. Tyler and D. Dawson, eds.), pp. 390–436, Houghton-Mifflin, Boston.

Munro, H. N., 1964, General aspects of the regulation of protein metabolism by diet and hormones, in: *Mammalian Protein Metabolism*, Vol. 1 (H. N. Munro and J. B. Allison, eds.), p. 442, Academic Press, New York.

Murphy, D. L., Baker, M., Goodwin, F. K., Miller, H., Kotin, J., and Bunney, W. E., 1974, L-Tryptophan in affective disorders: Indoleamine changes and differential clinical effects, *Psychopharmacology* **34**:11–20.

Mustard, J. F., and Packham, M. A., 1970, Factors influencing platelet function, *Pharmacol. Rev.* **22**:97–187.

Nelson, M. F., Picon, H. F., and Lytle, L. D., 1981, Maternal meal quality alters fetal rat brain serotonin concentrations, *Life Sci* **28**:231–237.

Nowak, T. S., and Munro, H. N., 1977, Effects of protein–calorie malnutrition on biochemical aspects of brain development, in: *Nutrition and the Brain,* Vol. 2 (R. J. Wurtman and J. J. Wurtman, eds.), pp. 193–260, Raven Press, New York.

Oldendorf, W. H., 1971, Brain uptake of radiolabelled amino acids, amines, and hexoses after arterial injection, *Am. J. Physiol.* **221**:1529–1639.

Olson, L., and Seiger, A., 1972, Early prenatal ontogeny of central monoamine neurons in the rat: Fluorescence histochemical observations, *Z. Anat. Entwicklungsgesch.* **137**:301–316.

Oswald, I., Ashcroft, G. W., Berger, R. J., Eccleston, D., Evans, J. I., and Thacore, V. R., 1966, Some experiments in the chemistry of normal sleep, *Br. J. Psychiatry* **112**:391–399.

Pardridge, W. M., 1977, Regulation of amino acid availability to the brain, in: *Nutrition and the Brain,* Vol. 1 (R. J. Wurtman and J. J. Wurtman, eds.), pp. 141–204, Raven Press, New York.

Parfitt, A. G., and Grahame-Smith, D. G., 1974, Tryptophan transport across synaptosome membrane, in: *Aromatic Amino Acids in Brain,* CIBA Foundation Symposium 22, pp. 175–192, Elsevier, Amsterdam.

Perez-Cruet, J., Chase, T. N., and Murphy, D. J., 1974, Dietary regulation of brain tryptophan metabolism by plasma ratio of free tryptophan and neutral amino acids in human, *Nature (London)* **248**:693–695.

Peroutka, S. J., Lebowitz, R. M., and Snyder, S. H., 1981, Two distinct central serotonin receptors with different physiological functions, *Science* **212**:827–829.

Porcher, W., and Heller, A., 1972, Regional development of catecholamine biosynthesis in rat brain, *J. Neurochem.* **19**:1917–1930.

Rao, B. S. N., and Ghafoorunissa, H., 1972, Effects of leucine on enzymes of tryptophan metabolism, *Am. J. Clin. Nutr.* **25**:6.

Rapoport, M. I., Feigin, R. D., Bruton, J., and Beisel, W. R., 1966, A circadian rhythm for tryptophan pyrrolase and its circulating substrate, *Science* **153**:1642–1644.

Reis, D. J., Joh, T. H., and Ross, R. A., 1975, Effects of reserpine on activities and amounts of tyrosine hydroxylase and dopamine-beta-hydroxylase in catecholamine neuronal systems in rat brain, *J. Pharmacol. Exp. Ther.* **193**:775–784.

Ross, D. S., Fernstrom, J. D., and Wurtman, R. J., 1973, The role of dietary protein in generating daily rhythms in rat liver tryptophan pyrrolase and tyrosine transaminase, *Metabolism* **22**:1175–1184.

Roth, R. H., Morgenroth, V. H., and Salzman, P. M., 1975, Tyrosine hydroxylase: Allosteric activation induced by stimulation of central noradrenergic neurons, *Naunyn-Schmiedeberg's Arch. Pharmacol.* **289**:327–343.

Rowland, D., Steele, M., and Moltz, H., 1978, Serotoninergic mediation of suckling-induced release of prolactin in the lactating rat, *Neuroendocrinology* **26**:8–14.

Saavedra, J. M., Brownstein, M., and Axelrod, J., 1973, A specific and sensitive enzymatic–isotopic microassay for serotonin in tissue, *J. Pharmacol. Exp. Ther.* **186**:508–515.

Salas, M., Diaz, S., and Nieto, A., 1974, Effects of neonatal food deprivation on cortical spines and dendritic development of the rat, *Brain Res.* **73**:139–144.

Schmidt, M. J., and Sanders-Bush, E., 1971, Tryptophan hydroxylase activity in developing rat brain, *J. Neurochem.* **18**:2549–2551.

Shields, J. P., and Eccleston, D., 1973, Evidence for the synthesis and storage of 5-hydroxytryptamine in two separate pools in the brain, *J. Neurochem.* **20**:881–888.

Shoemaker, W. J., and Bloom, F. E., 1977, Effect of undernutrition on brain morphology, in: *Nutrition and the Brain,* Vol. 2 (R. J. Wurtman and J. J. Wurtman, eds.), pp. 148–192, Raven Press, New York.

Sicuteri, F., 1974, Serotonin and sex in man, *Pharmacol. Res. Commun.* **6**:403–411.

Smythe, G. A., Brandstater, J. F., and Lazarus, L., 1975, Serotoninergic control of rat growth hormone secretion, *Neuroendocrinology* **17**:245–257.

Sourkes, T. L., Missala, K., and Oravec, M., 1970, Decrease of cerebral serotonin and 5-hydroxyindoleacetic acid caused by (-)-alpha-methyl-tryptophan, *J. Neurochem.* **17**:111–115.

Spector, S., Gordon, R., Sjoerdsma, A., and Udenfriend, S., 1967, End-product inhibition of tyrosine hydroxylase as a possible mechanism for regulation of norepinephrine synthesis, *Mol. Pharmacol.* **3**:549–555.

Stern, W. C., Forbes, W. B., Resnick, O., and Morgane, P. J., 1974, Seizure susceptibility and brain amine levels following protein malnutrition during development in the rat, *Brain Res.* **79**:375–384.

Stern, W. C., Miller, M., Forbes, W. B., Morgane, P. J., and Resnick, O., 1975, Ontogeny of the levels of biogenic amines in various parts of the brain and in peripheral tissues in normal and protein malnourished rats, *Exp. Neurol.* **49**:314–326.

Tagliamonte, A., Biggio, G., Vargiu, L., and Gessa, G. I., 1973, Free tryptophan in serum controls brain tryptophan level and serotonin synthesis, *Life Sci.* **12**:277–287.

Thoenen, H., 1974, Trans-synaptic enzyme induction, *Life Sci.* **14**:223–235.

Thomas, R. G., and Wysor, W. G., 1967, Alterations of serotonin metabolism in rats deficient in niacin and tryptophan, *Proc. Soc. Exp. Biol. Med.* **126**:374–377.

Tissari, A. H., 1973, Serotoninergic mechanisms in ontogenesis, in: *Fetal Pharmacology* (L. Boreus, ed.), pp. 237–257, Raven Press, New York.

Tissari, A. H., 1975, Pharmacological and ultrastructural maturation of serotoninergic synapses during ontogeny, *Med. Biol.* **53**:1–14.

Tissari, A. H., and Raunu, E. M., 1975, Subcellular distribution of 5-hydroxytryptamine in rat brain during development: Effect of drugs and fasting, *J. Neurochem.* **24**:1143–1150.

Tyce, G. M., Flock, E. V., and Owen, C. A., 1964, Tryptophan metabolism in the brain of the developing rat, in: *Progress in Brain Research*, Vol. 9, *The Developing Brain* (W. A. Himwich, and H. E. Himwich, eds.), pp. 198–203, Elsevier, Amsterdam.

Ungerstedt, U., 1971, Stereotaxic mapping of the monoamine pathways in the rat brain, *Acta Physiol. Scand.* **367**:1–48.

Vahvelainen, M. L., and Oja, S. S., 1972, Kinetics of influx of phenylalanine, tyrosine, tryptophan, histidine and leucine into slices of brain cortex from adult and 7-day-old rats, *Brain Res.* **40**:477–488.

Van Praag, H. M., and Korf, J., 1974, Serotonin metabolism in depression: Clinical application of the probenecid test, *Int. Pharmacopsychiatry* **9**:35–51.

Wang, H. L., Harwalker, V. H., and Waisman, H. A., 1962, Effect of dietary phenylalanine and tryptophan on brain serotonin, *Arch. Biochem. Biophys.* **96**:181–184.

Weiner, N., 1970, Regulation of norepinephrine synthesis, *Annu. Rev. Pharmacol.* **10**:272–290.

Williams, S. R., 1973, *Nutrition and Diet Therapy*, 2nd ed., C. V. Mosby, St. Louis.

Winick, M., and Noble, A., 1965, Quantitative changes in DNA, RNA and protein during prenatal and postnatal growth in the rat, *Dev. Biol.* **12**:451–466.

Winick, M., and Noble, A., 1966, Cellular response in rats during malnutrition at various ages, *J. Nutr.* **89**:300–306.

Winick, M., and Rosso, P., 1969, The effect of severe early malnutrition on cellular growth in human brain, *Pediatr. Res.* **3**:181–189.

Wirz-Justice, A., 1977, Theoretical and therapeutic potential of indoleamine precursors, *Neuropsychobiology* **3**:199–233.

Woodger, T. L., Sirek, A., and Anderson, G. H., 1979, Diabetes, dietary tryptophan and protein intake regulation in weanling rats, *Am. J. Physiol.* **236**:307–311.

Wool, I. G., 1965, Relation of effects of insulin on amino acid transport and on protein synthesis, *Fed. Proc. Fed. Am. Soc. Exp. Biol.* **24**:1006–1064.

Wurtman, R. J., 1970, Diurnal rhythms in mammalian protein metabolism, in: *Mammalian Protein Metabolism*, Vol. 4 (H. N. Munro, ed.), pp. 445–479, Academic Press, New York.

Wurtman, R. J., and Fernstrom, J. D., 1975, Control of brain monoamine synthesis by diet and plasma amino acids, *Am. J. Clin. Nutr.* **28**:638–647.

Wurtman, R. J., and Fernstrom, J. D., 1976, Control of brain neurotransmitter synthesis by precursor availability and nutritional state, *Biochem. Pharmacol.* **25:**1691–1696.

Wurtman, R. J., Larin, F., Axelrod, J., Shein, H. M., and Rosasco, K., 1968a, Formation of melatonin and 5-hydroxyindoleacetic acid from 14 C-tryptophan by rat pineal glands in organ culture, *Nature (London)* **217:**953–954.

Wurtman, R. J., Rose, C. M., Chou, C., and Larin, F., 1968b, Daily rhythms in the concentration of amino acids in human plasma, *N. Engl. J. Med.* **279:**171–175.

Wurtman, R. J., Hefti, F., and Melamed, E., 1980, Precursor control of neurotransmitter synthesis, *Pharmacol. Rev.* **32:**315–335.

Wyatt, R. J., Engelman, K., Kupfer, D. J., Frma, D. H., Sjoerdsma, A., and Synder, F., 1970, Effects of L-tryptophan (a natural sedative) on human sleep, *Lancet* **1:**842–845.

Young, V. R., Hussein, M. A., Murray, E., and Scrimshaw, N. S., 1969, Tryptophan intake, spacing of meals, and diurnal fluctuations of plasma tryptophan in men, *Am. J. Clin. Nutr.* **22:**1563–1567.

Yuwiler, A. W., Wetterberg, L., and Feller, E., 1971, Relationship between alternate routes of tryptophan metabolism following administration of tryptophan peroxidase inducers or stressors, *J. Neurochem.* **18:**543–549.

Yuwiler, A. W., Oldendorf, W. H., Geller, E., and Braun, L., 1977, Effect of albumin binding and amino acid competition on trytophan uptake into brain, *J. Neurochem.* **28:**1015–1023.

Zambotti, F., Carruba, M., Vicentini, L., and Mantegazza, P., 1976, Selective effect of a maize diet in reducing serum and brain tryptophan contents and blood and brain serotonin levels, *Life Sci.* **17:**1663–1670.

Zbinden, G., Pletscher, A., and Studer, A., 1958, Alimentäre Beeinflüssung der enterochromaffinen Zellen und des 5-Hydroxytryptamin-Gehaltes von Gehirn und Darm, *Z. Gesamte Exp. Med.* **129:**615–620.

II

Behavioral Determinants of Food Intake and Nutritional Status

8

The Role of the Mother–Infant Interaction in Nutritional Disorders

Janina R. Galler, Henry N. Ricciuti, Mary A. Crawford, and L. Thomas Kucharski

1. Introduction

Recent efforts to identify the various pathways or mechanisms through which nutritional variations might influence behavioral development have focused on the role of early mother–infant interaction in causing or contributing to infantile mulnutrition. By the term *mother–infant interaction*, we refer to the pattern of behavior engaged in by the mother (or other primary caregiver) and the infant as they relate to each other over time. It is particularly important to examine both the mother's behavior toward the infant and the infant's behavior toward the mother, since each may influence the response of the other (Bell, 1974).

The mother–infant interaction, and its role in nutritional disorders, is a significant issue requiring special attention for three main reasons. First, early malnutrition tends to be particularly frequent in populations living under generally adverse socioeconomic and environmental conditions. Thus, early malnutrition is typically associated with characteristics of the home and family environment that make normal mothering difficult to maintain. Therefore, the influence of malnutrition as such on behavioral development cannot easily be distinguished from the effects of adverse patterns of early infant care and mother–infant interaction. Rather, it has become increasingly clear that nutritional and environmental factors may interact to influence behavioral development jointly. A number of human and animal studies have shown, for

Janina R. Galler • Division of Psychiatry, Boston University School of Medicine, Boston Massachusetts 02118.   *L. Thomas Kucharski* • Department of Child Psychiatry, Boston University School of Medicine, Boston, Massachusetts 02118.   *Henry N. Ricciuti and Mary A. Crawford* • Department of Human Development and Family Studies, Cornell University, Ithaca, New York 14850.

example, that supportive environments—those that facilitate cognitive development—may substantially ameliorate or prevent the adverse effects of even
severe malnutrition on behavior, whereas unfavorable environments may exacerbate adverse consequences. Moreover, individual children vary greatly in
their ability to overcome the effects of both early malnutrition and adverse
socioeconomic and environmental factors.

Second, it is increasingly clear that unfavorable or dysfunctional patterns
of early infant care or mother–infant interaction may be significantly involved
in the *etiology* of malnutrition. This notion goes beyond the view that such
unfavorable mother–infant interactions are simply *associated* with malnutrition
as common risk factors in socially disadvantaged environments. Some recent
research (discussed in Section 4.1) suggests that adverse features of early infant
care, if identified early in life, may be predictive of an increased risk of subsequent clinical malnutrition, as well as of suboptimal behavioral development.

Third, once the child is malnourished, the mother responds differently to
that child than to a nonmalnourished child, thus influencing the infant's early
environment beyond the influence of the malnutrition as such. It is well known
that the care infants receive may be affected by their own behavior or characteristics. Anecdotal and clinical reports, as well as recent empirical studies,
indicate that young infants who are irritable, sickly, or handicapped may be
responded to less sensitively or favorably by their mothers or other primary
caregivers (Lewis and Rosenblum, 1974).

It seems quite clear that malnutrition and alterations in the mother–infant
interaction are intricately intertwined in nature. Animal models, employed to
control potentially confounding variables in research on early malnutrition,
have therefore not been successful in separating an effect of malnutrition from
that of the mother–infant interaction. While many investigators consider this
a methodological limitation of present animal models, it is our view that to
separate malnutrition and mother–infant interaction experimentally would result in a situation so atypical that the explanatory usefulness of such a model
would be limited.

It is the goal of this chapter to examine critically the role of patterns of
mother–infant interaction in nutritional disorders, in the hope of enhancing our
understanding of the interrelationships between early malnutrition and behavioral development in children. In Section 2, we examine the development of
normative patterns of mother–infant interaction in animals as well as humans.
In Section 3, we summarize the rather limited research data concerning
mother–infant interaction in humans as it relates to "normal" early infant-
feeding practices, including breast feeding. Our attention is then directed in
Section 4 to variations in the mother–infant relationship associated with moderate–severe protein–energy malnutrition as it commonly occurs in poor populations in developing countries; we also consider related animal models. We
then discuss the parallels of nonorganic "failure to thrive" with severe malnutrition in human populations. We conclude the chapter (Section 5) with a
summary of some of the common threads in the literature of both the human

and the animal studies concerning mother–infant interaction as a major factor in the etiology and the developmental consequences of nutritional disorders.

2. Normative Patterns of Mother–Infant Interaction

In Table I, we summarize some of the principal dimensions of parent–infant interaction that according to recent empirical studies with both animals and humans, may be significant in infants' developmental outcomes. The table represents the major conceptual areas defined in many studies—the specific behaviors investigated and the terms used to designate particular types of behaviors vary considerably among different researchers. Table I differentiates several levels of description from the broadest, basic aspects of infant care, including general characteristics of the mother's mode of interacting with her child, to the specific behaviors of mothers and infants observed during their interaction. Finally, the table considers interaction in terms of characteristics of the mother–infant pair as a dyad. These dyadic characteristics are difficult to characterize objectively, yet they are highly significant, particularly since they are relevant to the modulation of changes in the infant from one state, such as feeding or crying, to another, and to the reduction or arousal of tension. (See Section 6 for a discussion of the methods used to measure mother–infant interaction.)

2.1. Rats

Patterns of mother–infant interaction (often referred to as "maternal behavior" in animal studies) have been recognized in a wide variety of animals in both field studies and experimental laboratories. In general, these studies have demonstrated that behavioral patterns in the lactating female serve the needs of the newborn young and are synchronized with the physical and behavioral development of the young. Most of the research on mother–infant interaction in animals has been conducted with rats and, to a lesser extent, with monkeys. Maternal behavior in the rat has been explored throughout the reproductive cycle, from conception through weaning of the young. In the rat, this period covers nearly 2 months; during the first 22 days, the female is pregnant, and during the final 28 days (suckling period), she takes care of her young. Three phases of mother–infant interaction have been identified, and these patterns are affected by endocrine as well as environmental influences: the *initiation* of maternal behavior about 24 hr prior to birth, its *maintenance* during the following 2–3 weeks following parturition, and its *decline* as weaning approaches (Rosenblatt, 1969, 1975). These phases of maternal behavior are synchronized with stages in the physical and behavioral development of the young and represent a dyadic relationship contributed to by both mother and pups. Each phase requires analysis in terms appropriate to the events of that phase, and the separate phases need to be related to one another in a developmental sequence.

Table I. Major Features of Parent–Infant Interaction

A. Broadest level: Basic infant care
 1. Meeting infant's basic physical and emotional needs
 2. Providing appropriate sensory and social stimulation
 3. Guiding the infant's interactions with the environment in accordance with infant's developmental status
B. General characteristics of parents' mode of interaction with or relating to infant (range of interaction)
 1. Acceptance....................ambivalence....................rejection
  Warmth   hostility
  Affection
 2. Involvement with infant ....detachment
                                                                 indifference
 3. Sensitivity ....insensitivity
     (extent to which parent is aware of infant's moment-to-
     moment needs, and meets them appropriately)
 4. Contingent responsiveness......................................unresponsiveness
     (extent to which parent's responses are "tuned in to" or
     contingent upon infant's responses, so that infant develops
     sense of regularity and predictability of parental re-
     sponses, rather than experiencing unpredictable parental
     responses unrelated to infant's own behavior)
 5. Encouragement of ....restrictiveness,
  learning, exploration,   interfering,
  independence   encouraging
                                                                  dependence
C. Specific behaviors of parent and infant (from which some of the general characteristics listed above may be inferred or derived)
 1. Parental behaviors toward infant
     Examples: Looks, smiles, touches, cuddles, talks, rocks, sings, encourages, explains, restricts, discourages, nurses, licks, retrieves, pushes away
 2. Infant behaviors toward parent
     Examples: Smiles, fusses, looks at mother, imitates, seeks help, seeks approval, resists, seeks and maintains contact, avoids proximity and contact, licks, suckles
        (Analyses sometimes involve overall frequency of occur-
        rence of such behaviors in a given time period, or they
        may be separated according to whether these behaviors
        are shown in response to various overtures by baby, or
        as initiating contacts by mother.)
D. Characteristics of mother–infant pair as a dyad
 1. Reciprocity
     Mother and infant engage in reciprocal interchanges, such as dyadic gazing, mutual imitation, reciprocal play, where each influences and is influenced by the other
 2. Synchrony, mutual adaptation, or regulation
     Patterns of sustained interaction in which it is apparent that mother and infant gradually adjust to one another over time in mutually adaptive ways

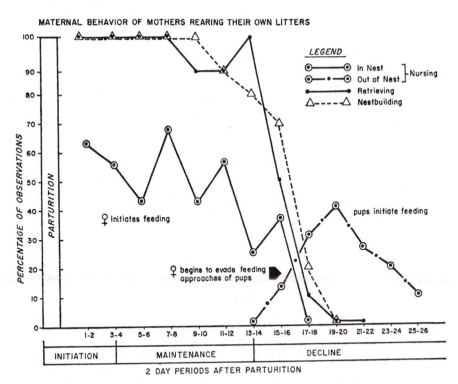

MATERNAL BEHAVIOR OF MOTHERS REARING THEIR OWN LITTERS

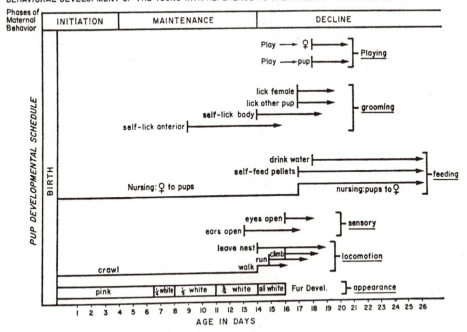

Fig. 1. Summary of observations of maternal behavior (upper graph) and pup development (lower graph) of five litters of laboratory rats during the four-week litter period. Maternal nestbuilding, retrieving, the nursing behavior are shown in relation to changes in the physical and behavioral characteristics of the pups. From Rosenblatt (1969).

As Fig. 1 shows, the *initiation* phase of maternal behavior extends from 24 hr before birth to the 3rd day postpartum. Extensive nursing, retrieving (carrying the young back to the nest when they have strayed from it), licking the pups to stimulate bowel function, and nest-building, which are the principal components of maternal behavior, are characteristic of this phase. The onset of these behaviors occurs immediately after birth and is regulated by the hormonal factors including the rise in estrogen in the mother. Virgin females and males will also develop "maternal" behaviors in the presence of pups, but only if placed in constant contact with foster young for 5 or 6 days (Rosenblatt, 1967). Maternal responses are central to this phase, but they may also be influenced by some features of the pups. The newborn and the placenta possess attractive stimuli that elicit maternal behavior from the female (Denenberg and Whimbey, 1963), yet some features of the newborn (such as movement) appear to inhibit the mother from eating the pup after she has eaten the placenta. In mice and rats, live pups have an immediate effect on the mother, stimulating maternal behavior, while dead pups depress her maternal responsiveness to live pups (Noirot, 1964). The character of maternal behavior in the initiation phase also depends on the behavioral maturity of the pups, as shown by the observation that newborn pups are capable of reviving the maternal behavior of mothers in their 3rd week postpartum, when these behaviors would normally be declining (Nicoll and Meites, 1959, Wiesner and Sheard, 1933). The initiation phase has been established as the "critical period" for the mother–infant relationship, during which secure bonding is established. Interference between the suckling mother and her young at this time results in an immediate decline in maternal behavior and death of the offspring.

During the *maintenance phase* (from the 4th to the 13th or 14th day postpartum), the mother's behavior is stabilized, but she is dependent on the behavior of the young for maintenance of her maternal responsiveness. The pup actively suckles and seeks out the mother during this phase, and in consequence the pup is responsible for the maintenance of maternal behavior during this period. Tests of cage orientation conducted on pups during this phase show their increasing ability over time to find their way back to the nest when displaced from it (Turkewitz, 1966; Galler and Seelig, 1981).

The *decline phase* of the maternal behavior cycle, which begins around the 15th or 16th day and extends until the 21st–28th days, corresponds to the weaning period. Nest-building on the mother's part declines first, followed by a reduction in retrieving several days later and then by a reduction in nursing. This decline of maternal behavior coincides with a period of rapid growth in the pup (Fig. 1). Fur gradually covers the pups' entire bodies, and they show improvement in locomotion, progressing from walking to running and climbing. At around the 16th day, they begin to feed and drink by themselves, thus initiating the process of weaning. These developments coincide with eye-opening and the beginning of hearing. Social interactions among the young and between the mother and the litter also become more varied, involving grooming and play activity. The decline of maternal behavior is therefore associated with the emerging independence of the young.

In summary, the pattern of mother–infant interaction in the rat follows a predictable developmental sequence during the 3–4 weeks following birth. At first, the dyad is primarily dependent on maternal factors, including hormone levels, which heighten the mother's sensitivity to infants after parturition. During the maintenance and decline phases, activities of the pup are necessary to ensure the continuation of maternal behaviors. As pups become progressively more independent, during weaning, the frequency of maternal behavior declines.

2.2. Monkeys

The pattern of mother–infant interaction in the monkey closely parallels that described for the rat. The mother monkey develops greater tolerance for autonomous activity in the infant as the infant becomes more independent; this independence is demonstrated in increasing interest in peer relationships and social explorations. Rhesus monkey infants spend most of their first month of life in close proximity to their mothers. As their locomotor ability improves in the second month of life, they begin to explore their nearby environment for brief periods of time, and by 5–6 months of age, most infants spend relatively more time interacting with peers than with their mothers (Suomi, 1979).

Harlow (1963) has described the relationship between the infant monkey and its mother in terms of an "affectional system" composed of three developmental stages. The first stage, *attachment and protection*, which lasts for about 90 days after birth, is characterized by positive and protective maternal responses. During this stage, mother and infant are in almost constant physical contact. The infant continually explores its mother, with a great deal of time spent in oral contact with the nipples, much of which appears to be nonnutritive suckling. Soon after birth, the infant monkey moves about in close synchrony with the mother. Infants attempt to mouth and eat solid food very early in life in imitation of their mothers' eating behavior, and they actually learn to eat solid food sooner when housed with their mothers than do infants raised separately. During this stage, mothers display a great deal of protective behavior, actively restraining the infant that attempts to leave and rapidly retrieving it when they are separated. This protectiveness is accompanied by almost constant visual orientation by the mother toward her infant. Between 45 and 90 days after birth, these protective behaviors and close attachments wane.

The second stage of development, *ambivalence*, begins approximately 90 days after birth and continues until 6 or 7 months of age. During this stage, positive, protective, maternal behaviors decline and "punishing" behaviors increase. The infant still spends a great deal of time in close physical contact with its mother, especially at night. However, the mother increasingly attempts to avoid the infant and may actually push the infant away when it attempts physical contact. The "punishing" behaviors on the part of the mother appear to promote the gradual emergence of autonomous activity of the infant.

The final stage of development, *separation*, occurs around the time of birth of a new infant. During this time, the first born infant spends a great deal of time in close physical proximity to its mother without actual physical contact. Increasingly, the infant develops more intimate attachment to peers and pre-adolescent play groups. Toward the onset of adolescence, this social play activity rapidly declines, with the young adult monkey expanding into a more diverse pattern of interaction (Suomi, 1983).

In summary, the normative pattern of development of infant monkeys is characterized by three sequential stages. In early life, positive and protective maternal behavior predominates, with infant monkeys spending a majority of their time in direct physical contact with mother. As locomotor skills improve, the infant monkey increasingly explores its environment, using its mother as a secure base. Mothers become generally more tolerant of their infant's brief excursions, and a more ambivalent pattern of responding to the infant emerges. Negative, punishing maternal patterns increase sharply in a manner that propels the infant toward more autonomous functioning. With the birth of the second monkey, the pattern of mother–infant interaction is altered once again. Physical contact is now very limited. Gradually, the firstborn infant's interest in peer relationships outweighs attraction to its mother, and effective separation occurs.

2.3. Humans

The development of human behavior, including mother–infant interaction, is much less biologically determined and more responsive to environmental factors than is the case for animal behavior; the tight linkages found between maternal behavior and the infant's developmental status in the case of the rat and monkey do not exist for humans. It is thus more difficult to chart "normative" patterns of mother–infant interaction in humans as a function of progressive infant development. Nevertheless, the interactions of the human mother and infant are to a considerable degree influenced by such characteristics of the infant as the infant's developmental stage, temperament, or phenotypic characteristics (Lewis and Rosenblum, 1974). There are substantial variations among human mothers, however, in the extent to which their behavior adapts to the infant's characteristics.

Although there are few systematic studies of changes in maternal behavior or in the mother–infant interaction as a function of the infant's development during the first few years of life, certain predictable changes in infants' behavioral repertoire have important implications for mother–infant interaction and pose different challenges for the mother. For example, in the first 3 or 4 months of life, the interactions between mother and infant are characterized by a great deal of mutual looking, smiling, and vocalizing ("mutual play") and by activities that minimize distress in the infant (Stern, 1974). These interactions are consistent with the very young infant's strong visual and auditory attentiveness and responsiveness, particularly to people, and they are very much influenced by the emergence of the infant's social smile and cooing vocalizations at 2–3 months of age. By 3 or 4 months of age, the infant both responds to and initiates maternal behavior, and mother–infant interactions thus involve considerable mutual adaptation and social learning. This pattern is also consistent with the infant's relatively limited motor functions, which restrict its active interaction with objects in the environment.

Starting at around 6 or 7 months of age, a number of dramatic changes take place. Infants are now able to sit up alone, begin to crawl and move about, and are able to use their hands effectively in play with toys or other objects. By 9 or 10 months of age, the infant shows some understanding of familiar words, and by 12 months, he uses several single words, is able to walk independently, can solve simple problems like retrieving a toy hidden behind a screen, and engages in imitative games. Some interesting empirical evidence indicates that systematic changes in maternal behavior are associated with these developmental transitions in the infant. For example, recent studies show that when the infant is 7–8 months old, the mother's speech becomes simpler (Sherrod *et al.*, 1977) and more "conversational," as the infant increasingly understands and interacts with objects and events in the environment (Sherrod *et al.*, 1977; Snow, 1977). A more recent longitudinal study of 14 infants and their mothers observed the mother–infant interaction in the home when the infants were 6, 8, and 12 months of age (Green *et al.*, 1980). Over this period, while the number of mother–infant interactions increased, the proportion of these interactions that were concerned with physical caretaking decreased. Mothers more frequently initiated interactions by responding to their infants' vocalizations and locomotor activites, and they more often played games with their infants, initiated verbal requests, and attempted to prohibit or redirect the infants' activities.

The period from 6 to 12 months of age is characterized by major developmental changes in social and emotional behavior, which have important implications for the mother–infant interaction. By 6 months of age, infants typically recognize familiar people, particularly the mother, and show more pleasure in response to them than to unfamiliar people. Toward the end of the first year, infants develop an attachment, or "focused affectional relationship," to the mother or primary caregiver. At the same time, they become wary or fearful of strangers, particularly when the mother is absent; the infant protests and tries to maintain contact when the mother leaves (particularly if the infant is left with a stranger), uses the mother as a "secure base" from which to explore a new environment, and directs other signs of affection actively toward the mother. This attachment is reciprocal, although the mother's attachment to the infant has been studied less extensively than the infant's attachment to the mother.

From about 10 to 18 months of age, several of the infant's behaviors may conflict and may need to be reconciled. Among them are *attachment*, involving efforts to maintain proximity to the mother; *affiliation*, which inclines the infant to approach and relate to other people; *exploration*, which impels him to investigate new surroundings and objects; and *fear*, which makes him fearful of strangers and strange environments. Much of the interaction between mother and infant during this period involves efforts to help the infant balance these sometimes conflicting behaviors (Ainsworth, 1973).

It is generally acknowledged that a secure attachment in infancy helps to promote normal development by facilitating the infant's exploratory, learning, and problem-solving behaviors and by laying the groundwork for normal social relationships later in life (Ainsworth, 1979). Warmth and affection, and responsiveness from a consistent primary caregiver in a predictable social environment, foster the development of secure attachments. If these are absent, as in the case of severe maternal deprivation in the first 2 years of life, a variety of adverse developmental outcomes can occur. However, in contrast to the animal situation, these adverse outcomes may well be prevented if environmental conditions are later improved. Attachment between mother and infant can be restored in humans even after substantial disruption (Rutter, 1980). Such plasticity is not the case in the ontogeny of mother–infant bonding in animals, where a "critical period" of bonding has been demonstrated.

While maternal behavior toward the infant is indeed influenced by infant characteristics, the response of the mother to the infant's developmental or temperamental characteristics varies considerably. Such variations may be a function of such factors as the mother's socioeconomic level, her knowledge, beliefs, and expectations concerning infant development, her sensitivity in reading infant signals, and her personality. The importance of such variations in maternal adaptation to infant characteristics is illustrated by the finding that the long-term developmental outcomes for fussy, irritable, or "difficult" infants are much more positive if the mother is responsive and achieves "synchrony" with her infant than is the case if she has difficulty in making these adaptations (Thomas and Chess, 1980).

In summary, because of the greater plasticity and flexibility of behavior development in humans, patterns of mother–infant interactions cannot be predicted with the same certainty as in animals. However, patterns of mother–infant interaction are influenced by the infant's developmental or temperamental characteristics, with considerable variation among individual mother–infant pairs. In the first months of life, interactions tend to be focused mainly on mutual looking, smiling, and vocalizing, as well as on the reduction of distress in the infant. In the second half of the first year, interaction patterns increasingly reflect the infant's growing locomotor, language, and intellectual competencies. From 10 to 18 months, much of the interaction between mother and infant centers on the development of a focused attachment relationship and on the infant's concurrent tendencies to explore his environment and to relate socially to others. A secure attachment relationship during this period helps to promote normal psychological development. However, social deprivation at this time may be compensated for later on by secure attachments with primary caregivers under more favorable environmental circumstances.

3. Mother–Infant Interaction and Effectiveness of Early Feeding Practices in Humans

Having discussed normative patterns of mother–infant interaction in rats, monkeys, and humans, we now turn to a review of the rather limited but important literature dealing with human mother–infant interaction as it relates to the effectiveness of feeding practices under generally "normal" conditions.

3.1. Early Infant-Feeding Practices

Although the clinical literature abounds with references to the nutritional consequences of mother–infant interaction during feeding, there has been relatively little study of the relationships between this interaction and the consumption of food and the physical growth of the infant. Prospective studies of maternal characteristics, such as anxiety or acceptance or rejection of pregnancy (e.g., Zemlick and Watson, 1953), have generally failed to show measurable effects on feeding. For example, Shaw *et al.* (1971) found little support for their hypothesis that mothers with anxiety either before or after pregnancy or increased stresses in their life or who are uncertain about their role as mothers would have infants who gain less weight during the first month of life. Instead, anecdotal evidence suggests that the mothers' feeding behaviors are more important than their attitudes and feelings; the infant who gained the most weight in their study had a highly anxious mother, *but* she presented him with 4 oz of formula every 2 hr.

The research of Brody (1956) on mother–infant interaction also indicated that mothers' attitudes were less important than their behavior. Mothers who were undemonstrative during feeding but intent on having their infants suckle had infants who gained the most weight during the first 4 months of life; the infants of mothers who were more warm and affectionate during feeding but less purposeful gained less weight. Infants who gained the least weight had mothers who constantly interrupted them while they were suckling. Finally, in their longitudinal study of mother–infant interaction, Ainsworth and Bell (1969) observed that mothers who discontinued feeding at the first sign of infant disinterest had infants who gained less weight, to the point of being "worrisome" to their pediatrician.

More detailed observations of mother–infant interaction during feeding generally support these findings. Thoman and her colleagues (summarized in Thoman, 1975), using a large sample, examined mother–infant interaction during feeding in the first 3 days of life and found that more experienced (multiparous) mothers spent less time stimulating their infants during feeding than mothers of firstborns, but they were more successful in feeding their infants (i.e., the infant spent more time suckling at the breast or consuming more formula when bottle-fed). Apparently, stimulation of the infant during feeding may in fact disrupt suckling and prolong feeding time (Kaye, 1977; Thoman, 1975). Longitudinal studies indicate that most mothers therefore learn to modify their behavior by providing less stimulation during suckling (Kaye, 1977). By 2 weeks postpartum, feedings go more smoothly for all mothers, and differences between multiparous and primiparous mothers disappear (Dunn and Richards, 1977). Chao (1971) reported that teaching first-time mothers more competent feeding techniques (such as helping newborns to achieve proper sucking grasp, keeping nipples full of milk during feeding, and correctly interpreting the newborn's signs of satiation) was effective in reducing the number of days it took for the infant to regain his birthweight.

Pollitt and colleagues (Pollitt *et al.*, 1978; Pollitt and Wirtz, 1981) examined the correlations between infant weight gain during the first month of life and a large number of maternal and infant behaviors observed twice during the first bottle feeding shortly after birth and during a feeding at 1 month of age. While few of the observed behaviors were significantly related to weight gain, and the data reported are difficult to interpret, the results do underscore the importance of further efforts to identify specific maternal behaviors that may affect feeding, as reflected in weight gain. Pollitt and co-workers also found that weight gain (adjusted for birth weight) at the 1-month feeding was more closely related to the infants' characteristics during feeding than to maternal behaviors. Whether these infant characteristics influenced mothers' behavior during feeding was not addressed directly by these studies, although it is an obvious subject for further study.

Several other studies of mother–infant interaction during feeding show that maternal behavior is sensitive to variations in the infant's state of wakefulness or alertness (Levy, 1958; Osofsky and Danzger, 1974); the characteristics of the infant are likely to have a direct impact on the amount consumed during feedings. Dunn and Richards (1977) found that the rate of sucking was associated with maternal behavior during feeding in the first 10 days of life—a faster rate contributed to smoother feedings. In contrast, the study by Thoman (1975) suggests that infant characteristics *per se* are not responsible for feeding success at least during the first 3 days of life. Whereas the success of feeding by nurses was related to length of labor and type of anesthesia, mothers' feeding success was independent of these factors. This investigation, however, did not directly determine the patterns of mother–infant or nurse–infant interaction.

All the studies discussed thus far were of normal, healthy infants. In abnormal infants, feeding problems may be more prominent. For example, mothers stimulate premature infants more than full-term infants during feeding by encouraging these infants, who are more easily distracted and less attentive, to eat (Di Vitto and Goldberg, 1979; Field, 1977). Also, mothers experienced serious difficulties in feeding infants with cerebral palsy (e.g., Miller, 1976), since this disease impairs the infant's ability to ingest food.

In summary, there are very few empirical studies examining the relationships between mother–infant interaction and feeding. Despite this restricted data base, the studies confirm that when mothers provide an adequate opportunity for the infant to digest food and respond appropriately to infant cues during feeding, their infants grow well. Since the age of the children studied has for the most part been limited to the postpartum period, it may be useful to examine later feeding behaviors. Another area that deserves more attention, though it is beyond the scope of this chapter, is the link between mother–infant interaction and obesity (Olson *et al.*, 1980; Birch *et al.*, 1981).

3.2. Breast Feeding

The question of whether breast-fed babies do better than bottle-fed infants is an issue of long-standing interest that can be discussed only briefly in this review. Nutritionally, breast milk has been considered the most appropriate food for babies because it is more digestible than formula and because it provides natural immunities from infection (Committee on Nutrition, 1980; Gerrard, 1974). Behavioral scientists have been especially interested in the implications of breast feeding for the mother–infant relationship and the subsequent behavioral development of the child. Here, the different consequences of breast vs. bottle feeding are not clear. For example, a comprehensive review by Caldwell (1964) failed to find any behavioral (social or personality) sequelae in children that were consistently associated with feeding method or other childcare practices having nutritional implications, such as age at weaning. Thus, in recent years, the question of different behavioral consequences of breast or bottle feeding has assumed considerably less importance.

Mother–infant interaction during feeding has continued to be a subject of research, however, since this is one of the central activities of early infancy for both mother and infant and since it may be representative of their interactions in other contexts (Brody, 1956; Ainsworth and Bell, 1969). Recent research has shown that there are marked differences in maternal behavior during feeding, depending on the type of feeding involved. Dunn and Richards (1977), for example, found that breast-feeding mothers touch, talk, kiss, and smile at their babies more during feeding than do bottle-feeding mothers; such behaviors were consistent over time for breast-feeding but not for bottle-feeding mothers (Dunn, 1977). These differences in maternal behavior coincide with findings from another study indicating that bottle-feeding mothers view feeding as an emotionally neutral task, while breast-feeding mothers feel that feedings are emotionally gratifying and draw them closer to the infant (Leifer, 1980). Despite differences in their behavior during feeding, mothers do not seem to differ in stimulating their infants to suck (Field, 1977; Kaye, 1977), regardless of feeding method.

It has been suggested that the infant exerts more control over breast feeding than over bottle feeding. While most sucking bouts of bottle-fed infants are ended by maternal removal of the nipple, breast-fed infants are responsible for ending sucking bouts nearly 50% of the time (Dunn and Richards 1977). The implication is that breast-feeding mothers are more responsive to the infant's signals, such as suck rate. Studies of the growth patterns of infants of low and high birth weight indicate that there are important self-regulatory controls within the infant directing growth toward more normal size (Ounsted and Sleigh, 1975). This reversion toward the median (whereby very small infants eat relatively more and grow relatively faster, with the opposite true for large infants) appears to be facilitated by breast feeding, again suggesting greater control over intake for breast-fed infants. This greater self-regulation of intake in breast-fed babies may also be a response to changes in the osmolality of breast milk as the baby suckles (Hall, 1975).

Successful breast feeding beyond the early postnatal period seems to require responsive behavior on the part of the mother, as the physiology of lactation would suggest (Olson *et al.*, 1980). Nursing mothers report that they ignore infant cries less often than bottle-feeding mothers (Freese and Thoman, 1978), but mothers who respond to infant cries by feeding are more likely to continue to breast feed beyond 2 weeks (Bernal, 1972).

Generally speaking, it appears that during feeding, mothers who breast feed are more stimulating, verbal, and responsive to infant signals and show more affection than mothers who bottle feed. Although these differences may or may not be reflected in ongoing mother–infant interactions in settings other than feeding, they involve maternal behaviors associated with more favorable subsequent social and cognitive development in infants (e.g., Beckwith and Cohen, 1980; Clarke-Stewart, 1973). For this reason, and also because most earlier studies of the effects of differential infant feeding were retrospective and therefore limited, some investigators continue to consider the breast vs. bottle distinction important in the psychological outcome of the child and therefore worthy of further research (Ainsworth, 1973; Dunn and Richards, 1977).

Because of the nutritional superiority of breast milk over formula, there has been some recent interest in linking the nutritional advantages of breast feeding to later favorable cognitive consequences. As previous research reviews suggest (Caldwell, 1964), however, it is very difficult to demonstrate causal relationships of this sort, given the broad range of factors capable of influencing cognitive development in children. It is generally recognized that the beneficial properties of breast milk can protect infants born into adverse medical and social circumstances from the risk of developing subsequent malnutrition (Jelliffe and Jelliffe, 1975; Omolulu, 1974). It is much less clear whether breast feeding under these conditions confers a cognitive as well as a nutritional advantage to the infant. A study by Young *et al.*, (1982) followed the growth and behavioral development of over 1000 urban Tunisian infants from 6 to 16 months of age. The data confirm that breast-fed infants in the underprivileged class, where extreme socioeconomic conditions prevail, grow better than do non-breast-fed infants through 12 months of age. In the privileged class, however, growth differences do not show up until 12 months of age, when it appears that non-breast-fed children advance in length and weight beyond their breast-fed peers, perhaps indicating that in the privileged class, older formula-fed infants tend toward obesity. In the behavioral domain, breast feeding was consistently associated with somewhat higher scores on the motor scale of the Bayley Scales of Infant Development from 8 through 16 months of age, although significant differences occurred most frequently among infants in the underprivileged class (fewer significant differences were found in favor of breast-fed infants on the Bayley Mental Scale).

It appears from the study just cited that in populations in which many infants are brought up under adverse environmental conditions, breast feeding may be associated with more favorable mental but especially motor development during the first 18 months of life. It is not clear, however, whether such a relationship would hold in populations where environmental conditions are generally much less severe. Nor is it clear whether the behavioral advantages found for breast-fed infants should be attributed to nutritional factors, to characteristics of mother–infant interaction, or to infant-care practices associated with breast feeding.

A recent study of some heuristic interest but with significant methodological shortcomings was based on the finding that premature infants whose blood tyrosine levels rose above acceptable levels as a result of high-protein feeding subsequently had significantly lower IQ scores and a higher incidence of perceptual dysfunction (Menkes, 1977). Menkes hypothesized that formula feeding, because of its higher protein content, would contribute to subsequent learning disabilities in children. Comparing a group of United States children referred because of learning disorders with a control group referred for other neurological problems, Menkes found that a significantly higher proportion of the children with learning disorders were never breast-fed. However, when comparing formula-fed children in the two groups, there was no difference in the protein content of their formula feedings, thus calling into question the original nutritional hypothesis.

Two recent studies in developed countries, employing careful prospective research designs, found differences in intellectual performance favoring breast-fed children, but the magnitude of these differences was so slight that they must be regarded as having little or no functional significance. Rodgers (1978) followed a cohort of British children representing all legitimate *singletons* born in one week in 1946. By controlling other factors known to be associated with breast feeding and with cognitive development (income, education, birth order, sex of infant, and birth weight), it was found that breast-fed infants performed significantly, but only very slightly, better on Picture Intelligence at 8 years and in nonverbal ability, mathematics, and sentence completion at 15 years.

Edwards and Grossman (1979) followed a large national sample of United States children, representative of noninstitutionalized children aged 6–11 years, born in the mid- to late 1950s. The sample was an advantaged one, being restricted to white children living with both parents. The study controlled for variables in the family background, which might account for the relationship, including whether the mother worked outside the home; on IQ and achievement test scores, breast-fed children attained only very small advantages that had no functional significance.

In summary, the studies available thus far suggest that "successful" breast-feeding mothers may indeed be somewhat more sensitive and responsive to their infants and may have a qualitatively different subjective experience, in the feeding situation. It is not known, however, whether these maternal behaviors and experiences are representative of mother–infant interactions in contexts other than feeding. There is some evidence to suggest that in populations living under extremely adverse environmental conditions, breast feeding leads to physically healthier children and may promote motor development and, to a lesser extent, mental development during the first 18 months or so of life. This is likely to be due in large part to the safer sanitary conditions associated with breast feeding. But there is little or no evidence linking breast feeding to more favorable behavioral development later in childhood.

4. Mother–Infant Interaction and Moderate to Severe Nutritional Disorders

4.1. Human Studies

Disturbances in the mother–infant interaction can contribute significantly to the etiology and the outcome of moderate to severe malnutrition occurring in the first 2–3 years of life. Although the incidence of severe protein–calorie malnutrition is highest in the poorest populations of developing countries, which live under the most adverse socioeconomic and environmental circumstances, clinical malnutrition is not distributed uniformly among the poorest families or even among siblings within the same family. Thus, socioeconomic deprivation alone cannot explain either the occurrence of malnutrition or its outcome. It is therefore likely that there are fundamental disturbances in the home environment of the individual child, including a dysfunctional mother–infant relationship, that increase the risks of malnutrition occurring in that child. We refer to these conditions as being *antecedent* to the onset of malnutrition. Once the child has developed malnutrition, the conditions that led to the episode may persist and result in maladaptive responses, including rejection of the child and apathy on the part of the mother. However, the changes in the mother–infant interaction at the time a child is malnourished (*concurrent* changes) may also be adaptive and may have the effect of protecting the sick child—as in the case of increased physical contact and suckling. These concurrent changes in the mother–child relationship may combine with the physiological effects of malnutrition itself to reduce the child's contact with important features of its environment apart from the mother. This pattern of interference with the child's increasing exploration of the environment results in delays in behavioral and cognitive growth. This can disturb the normal progression of the mother–infant relationship toward autonomy of the child. In addition, a dysfunctional mother–infant relationship can, in and of itself, bring about delays in the development of the child. Figure 2 shows our model of the interrelationships of mother–infant interaction and malnutrition in influencing behavioral outcome.

Perhaps the most persuasive empirical evidence for differences among families that may lead to infantile malnutrition comes from the research of Cravioto and DeLicardie (1972, 1976) in Mexico. This study, which is the only prospective assessment of family characteristics that predict malnutrition, concludes that variations can be identified in specific features of early infant care that are *antecedent* to clinical malnutrition. In a longitudinal study, these investigators followed 334 infants born in 1 year in a poor rural village; 22 of these children (index group) developed severe clinical malnutrition (primarily kwashiorkor) some time in the first 3 years of life. The families of these 22 children were then compared with those of a control group of 22 well-nourished children from the same longitudinal cohort, matched according to birth weight and developmental quotient. The families of the two groups did not differ in terms of socioeconomic features, including literacy, family size, parental age, or education, except that control families listened to radios more often.

Cravioto and DeLicardie used the Caldwell Home Scale (adapted for use in Mexico) to assess the home environment of all 334 children; this technique relies on observation of the home and a structured interview with the mother or primary caregiver. When the data for the 22 malnourished children and their controls were reviewed, the homes in the index children were found to have lower scores in the quantity and quality of social, emotional, and cognitive stimulation and support available to the infant even before the episode of mal-

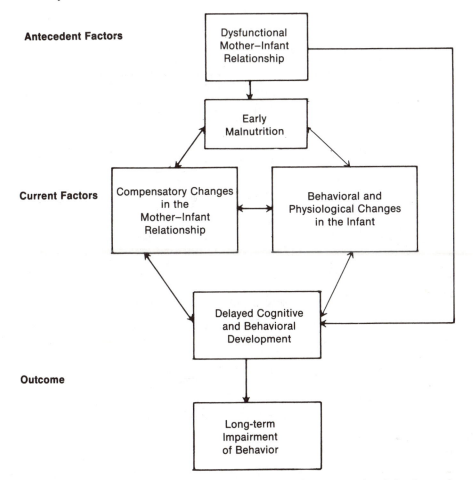

Fig. 2. Schematic diagram showing the possible relationships between mother–infant interaction and malnutrition. Compensatory M–I changes may help to offset deleterious consequences of the behavioral and physiologic changes accompanying malnutrition and, if not adequate, may exacerbate these consequences.

nutrition than did the homes of the control children. Table II shows the mean scores of index and control groups from 6 to 58 months of age. It also shows the distribution of scores on the Home Scale for the index and control groups when the children were 6 months of age, at which time only one child in the index group was malnourished. The difference in mean scores was statistically significant at all ages, and the distribution of scores also differed between the two groups. Additional observations of mother and infant during monthly developmental testing sessions in the first year of life indicated that mothers of the 22 malnourished children were less interested in their childrens' performance, less responsive, less sensitive to the children's needs, less verbally communicative, and less emotionally involved than control mothers. Based on these findings, Cravioto and DeLicardie (1976) concluded (p. 34):

Table II. Relationship between Caldwell Home Scores and the Occurrence of Malnutrition in Young Children[a]

Mean values of total home stimulation for severely malnourished children and controls

	Home stimulation score (mean ± S.D.)		
Chronological age (months)	Severely malnourished children	Control children	*t* Test
6	32.37 ± 2.57	35.35 ± 4.71	3.46[b]
12	36.41 ± 3.44	39.83 ± 4.71	2.47[c]
18	39.68 ± 4.89	42.43 ± 4.01	1.78[c]
24	39.43 ± 4.81	44.16 ± 6.26	2.48[c]
26	39.46 ± 4.98	44.33 ± 5.69	2.51[c]
31	40.66 ± 7.22	45.86 ± 6.68	2.05[c]
34	44.40 ± 5.31	48.20 ± 5.12	2.00[c]
38	44.40 ± 5.31	48.20 ± 5.12	2.00[c]
46	88.53 ± 16.77	101.40 ± 10.62	2.51[c]
52	90.35 ± 11.83	102.80 ± 11.36	2.88[c]
58	97.71 ± 16.98	109.86 ± 6.38	2.52[c]

Proportion of future severely malnourished children and control children showing different total home stimulation scores at 6 months of age

	Proportion of children	
Home score	With future severe protein–calorie malnutrition	Control
27	0.06	—
28	0.06	—
29	0.12	—
30	0.25	—
31	0.31	—
32	0.56	0.12
33	0.62	0.17
34	0.75	0.41
35	0.87	0.65
36	1.0	0.70
37	—	0.76
38	—	0.94
39	—	0.94
40	—	0.94
41	—	1.00

[a] From Cravioto and DeLicardie (1976).
[b] $p < 0.05$.
[c] $p < 0.01$.

A low level of home stimulation and a passive, traditional mother, unaware of the needs of her child, and responding to him in a minimal way as if unable to decide the infant's signals, are two characteristic features of this poor microenvironment that lead to severe clinical malnutrition in children of poor families.

Since Cravioto and DeLicardie's is the only systematic study of conditions in the home antecedent to the development of malnutrition, it is difficult to generalize without additional studies in other cultures. The other studies of mother–infant interaction are concurrent with or follow an episode of malnutrition. In examining the evidence from these studies, it is clear that there are at least two types of alterations in the mother–infant interaction and home environment at these later times. First, the disruptions identified at this time may be a continuation of the earlier patterns of behavior. Second, as seen in other conditions of biological impairment, responses to the malnourished infant or young child may first make their appearance at this time. Often, an increased amount of physical contact and suckling is seen in such cases. In two well-designed and clearly presented studies, Graves demonstrated both patterns of mother–infant interaction. Working in rural West Bengal, India (Graves, 1976), and in Katmandu, Nepal (Graves, 1978), he made 20-min observations of mothers and their infants of 7–8 months of age who had been identified in local health clinics as being either moderately malnourished or well nourished on the basis of anthropometric measures. In both India and Nepal, the malnourished infants showed *reduced* levels of play and spontaneous interaction with their mothers when they were in physical contact ("attachment behaviors"), when compared with the well-nourished infants. Maternal behavior, however, differed in the two cultural settings. In India, mothers of malnourished infants initiated significantly less interaction with their infants as compared with mothers of the well-nourished children; in Nepal, there was no difference between the two groups.

These results provide evidence for cultural differences in patterns of mother–infant interaction under conditions of malnutrition. In both these studies, malnourished infants spent significantly *more time* suckling at the breast and had more physical contact with mother than did comparison children. Chavez and Martinez (1975), working in a Mexican village, support this observation. These investigators provided nutritional supplementation to mothers and infants at risk for developing malnutrition. Not only did the supplemented infants and toddlers show improved physical growth and development when compared to controls, but also the mother–infant interaction was different. Mothers of supplemented children were less overprotective and provided more opportunities for exploration and independence. For example, mothers of supplemented children spent 20% of the time in direct physical contact with their children as compared to 50% for mothers of nonsupplemented children up to 3 years of age. Most of this time was spent in nursing. Such increased contact thus appears to be a common concurrent response to malnutrition.

In their recent study in Jamaica, Grantham-McGregor and her colleagues (Granham-McGregor and Stewart, 1980; Sheffer *et al.*, 1981) did not directly observe mother–infant interaction, but used the Caldwell Home Scale (also used by Cravioto and DeLicardie) to assess home environment. Eighteen infants hospitalized for clinical malnutrition between 6 and 24 months of age were compared with a group of 21 adequately nourished children hospitalized for other reasons. Mothers of both groups of children were visited regularly at home beginning 1 month after the child's release from the hospital and up to 36 months later. The home environments of the two groups in this study were not found to be significantly different on the Home Scale. In contrast, Cravioto and DeLicardie found lower scores for homes of malnourished children up to 58 months of age in their study in Mexico. In addition, Grantham-McGregor and her co-workers also found that mothers of the malnourished children had lower verbal IQ scores than control mothers, whereas the Mexican study did not show this. Grantham-McGregor, in discussing the Mexican and Jamaican studies, suggests that cultural differences in child-rearing patterns may have accounted for the conflicting results. Two additional studies have a bearing on the variable effects of malnutrition on the mother–infant interaction either at the time of the episode or later. One study in Jamaica, with very limited data, is that of Kerr *et al.*, (1978), who reported that severely malnourished children were more likely to come from unstable homes than were well-nourished children. In another study in the United States, Chase and Martin (1970) compared small groups of previously malnourished and well-nourished children generally matched in socioeconomic background, when they had attained 2–7 years of age. They found reduced scores on a home inventory scale, without any difference in maternal IQ.

Thus, the studies of concurrent responses to malnutrition do not present a unified picture. This is due in part to the use of different strategies to examine the home environments of malnourished children. Three of the studies (Chavez and Martinez, 1975; Graves, 1976, 1978) made direct observations of the mother–infant relationship, and the other four (Grantham-McGregor and Stewart, 1980; Sheffer *et al.*, 1981; Cravioto and DeLicardie, 1972, 1976; Kerr *et al.*, 1978; Chase and Martin, 1970) used indirect measures to assess the quality of the home environment. In addition, within each group of studies, there is evidence of culturally determined patterns of mother–infant interaction and variations in the type of home environment. Overall, these studies do not support the general assumption that malnutrition is associated with inadequate mothering. The one prospective study of Cravioto and DeLicardie does reflect this conclusion, but requires replication in other cultural settings, as evidenced by the variable results obtained in studies conducted at the time of the episode of malnutrition or subsequently. In some studies of the concurrent effects of malnutrition, there is evidence that changes in mother–infant interaction actually compensate for the effects of the malnutrition. For example, both Graves and Chavez reported increased physical contact with the mother and increased suckling in malnourished children compared with well-nourished children. Heightened parental attention of this type toward infants with other illnesses, such as cerebral palsy, has also been documented (Boles, 1959). These responses may be a way to compensate for the child's malnutrition and the resultant biological impairment and may thus be protective of the malnourished child at the time of the malnutrition. However, as pointed out earlier, they may also serve to limit the child's further development by fostering excessive dependence on the mother and preventing the child from attaining autonomy. It is possible that these concurrent responses to malnutrition are modified by inadequacies in the home environment, which were apparently present prior to the onset of malnutrition in at least two of the studies described.

The later effects of malnutrition may be modified by changes in the mother–infant interaction prior to or during the acute episode. No study of the long-term consequences of malnutrition has directly measured the mother–infant interaction at the time of malnutrition or earlier. In two studies of children who had suffered from malnutrition many years earlier, extensive data on home environment were obtained on the assumption that these conditions reflected circumstances at earlier points in time. For example, many of the questions asked, such as level of maternal education and reproductive history, relate directly to earlier events. Using a detailed interview, Richardson (1974), in Jamaica, examined home-environment characteristics in a sample of 74 boys, ages 6–10 years, with histories of malnutrition, their siblings, and 74 well-nourished boys. Among other factors, two that distinguished sharply between the index and comparison groups were the "caretaker's level of capability" and the "intellectual stimulation" available to the child.

More recently, Galler *et al.* (1983a–c), in Barbados, identified 129 children aged 5–11 years who had suffered from severe protein–energy malnutrition in the first year of life and their matched controls. Using techniques similar to those used by Richardson, these investigators compared home environments and parental characteristics of the two groups of children and also found that index children came from disadvantaged homes. For instance, the previously malnourished children were read to and told stories less often, had fewer visits and trips providing social contacts with adults, and generally had less interaction with adults in the family. The mothers of the two groups also differed significantly; among other factors, the mothers of previously malnourished children were depressed at the time of the study, had less education, and were less social than mothers of control children. Recent epidemiological studies in England have shown that these factors, especially maternal depression, are most common when children are less than 6 years of age. It is therefore likely in the Barbados study that maternal depression was probably more severe earlier in the life of the child and may even have led to the episode of malnutrition.

In conclusion, the studies of both Richardson and Galler indicate that children with histories of severe malnutrition frequently have mothers with lower levels of competence and that it is a fair assumption that they were even less competent at the time of the episode of malnutrition. As shown in Fig. 2, a dysfunctional mother–child relationship can lead directly to malnutrition (Route A) and thereby produce later behavioral changes. However, such a dysfunctional mother–infant relationship (coupled with inadequate home conditions) can persist and exert a continuing effect on behavioral development of the child (Route B) by limiting opportunities for compensatory factors in the environment.

4.2. Parallels in Nonorganic Failure to Thrive

We turn next to a brief consideration of the role of the mother–infant relationship in the nonorganic syndrome known as failure to thrive (FTT). FTT is defined as severe growth failure in the first few years of life that is not explainable in terms of organic disease or the inadequate food supplies due to extreme poverty or to famine. Unlike most malnutrition, which occurs in developing countries, FTT is thought to be more common in developed countries. FTT occurs most often in families characterized by a high degree of stress, conflict, and disruption, which increase the likelihood of dysfunctional or inadequate mothering. It is more closely linked to deficiencies in the mother–infant relationship than other types of infantile malnutrition. Thus, a dysfunctional mother–child relationship is thought to reduce food intake by the child (Whitten *et al.*, 1969). Many children with FTT are hospitalized, at which time their eating patterns rapidly return to normal and rate of growth improves. Nevertheless, when such children are sent home, the symptoms of FTT recur. This reinforces the notion that the homes of the children are inadequate. Another type of growth failure, ''psychosocial dwarfism,'' is seen in children over the age of 3 whose food intake is apparently adequate. This condition is thought to be due to hormonal disturbances brought about by stress in the home (Krieger and Mellinger, 1971). A review of medical aspects of the FTT syndrome has been undertaken by Brown (1979) in an earlier volume of this series.

Three general groups of factors have been associated with an increased risk of FTT. These include, first, characteristics of the mother; second, characteristics of the child; and third, social, economic, or other environmental stresses confronting the mother and family. Specifically, factors associated with FTT include current or prior emotional disturbances in the mother. Thus, mothers of FTT children are often described as being depressed and as having experienced repeated losses (Evans *et al.*, 1972; Leonard *et al.*, 1966; Fernholt and Provence, 1976; Shapiro *et al.*, 1976). A hostile, insensitive, or rejecting attitude toward the child, or a lack of knowledge about ways to meet the infant's needs (particularly regarding feeding practices), may also contribute. Another important factor contribute to FTT is the mother's personality structure. Polansky *et al.* (1981), in a recent study of child neglect (though not specifically of FTT) in low-income Philadelphia families, suggested that an ''infantile'' or immature personality of the mother may inhibit her capacity to provide adequate physical and psychological care for her child. Fischoff *et al.* (1971) have reported a high prevalence of character disorders in mothers of FTT children.

In the child, characteristics that may contribute to FTT include prematurity or low birth weight and recurrent illness or ''sickliness.'' Many children also have ''difficult'' temperaments. They may be either irritable and passive (Bullard *et al.*, 1967; Elmer, 1960; Gordon, 1979) or overactive and vigorous (Leonard *et al.*, 1966). Family stresses associated with FTT include economic hardship, illness, interpersonal conflict, isolation, and lack of informal supports from relatives, friends, and one's spouse (Evans *et al.*, 1972; Newberger *et al.*, 1977; Kotelchuck and Newberger, 1978, 1983).

Because there are no prospective studies of FTT, it has not been possible to identify with certainty the antecedent factors predisposing to failure to thrive. Nevertheless, the extent of environmental disadvantage has been associated with the severity of the outcome of children with FTT. Thus, children who come from more stable and better-organized homes do better than children who come from less stable homes. Evans *et al.* (1977) studied this relationship in 40 infants and preschool children admitted to Children's Hospital in Pittsburgh and followed these children after discharge. At admission, all were below the 3rd percentile for both height and weight, all the children were reported to have been unplanned or unwanted, all the families were socially isolated with little support from relatives or friends, and all were coping with economic problems. Fathers were present in 27 of the 40 families, but none provided adequate support to their wives or regular fathering to the children. The investigators divided the total sample into three subgroups according to degree of breakdown in parenting. The 14 children who recovered most fully had parents who were able to provide good physical care and maintain good living conditions, even though income was restricted and the mothers were young, immature, rather depressed, and unsure of themselves. Also, their families were small (7 were only children), and only 1 was premature. The children with intermediate levels of recovery had many features in common with the first group, but had experienced more crises and chronic problems. Thus, living conditions were worse, the families were larger, and most of the children received poor physical care. Eleven of the mothers had serious medical problems, and 6 of the children were premature. The 11 children with the least recovery resembled the first in that housing, food supply, and physical care of the children seemed adequate. However, these mothers were emotionally disturbed, were observed to express hostility toward their children, and were difficult to help. These results underscore the contribution of both parental and infant characteristics, as well as environmental stress, to the outcome of children with FTT.

In summary, there are concurrent factors in both children and their mothers that can contribute to the breakdown of the mother–infant relationship and that, in turn, are thought to result in the child's failure to thrive. Family stresses also are seen to play a prominent role in this process. Each of these groups of factors, especially the personality of the mother, has been shown to influence the outcome of the child.

There are parallels between FTT and moderate–severe malnutrition. The breakdown in the mother–child interaction that characterizes FTT in developed regions of the world may also be seen in cases of malnutrition in developing regions where poverty is endemic and food supplies are already limited. Under such extreme conditions of poverty and chronic stress, a dysfunctional mother–child relationship may be even more damaging, resulting in severe malnutrition of the offspring.

4.3. Related Animal Studies

The changes in the mother–infant relationship have been studied in malnourished animals, especially in the laboratory rat. The results of these studies help to clarify certain of the observations made in the human studies just discussed. In particular, as shown in Fig. 2, the animal studies allow a detailed examination of factors concurrent with the condition of malnutrition, namely, the interrelationships of malnutrition during suckling, compensatory changes in the mother–infant relationship, and behavioral changes in the pups. Defining this triad for an animal population allows one to manipulate the early circumstances of infantile malnutrition and permits identification of the impact of each factor on long-term behavioral outcome. Research with rats has not only advantages but also limitations, both of which have been reviewed by Fleischer and Turkewitz in their discussion of the use of animal models in nutritional research (Chapter 2).

In early investigations examining the behavioral effects of malnutrition in rat pups, two techniques were most often employed to produce malnutrition, namely, providing a low-protein diet to the mother rat (Mueller and Cox, 1946) and large-litter rearing—increasing the number of pups suckled by a single mother (Kennedy, 1957). It has been recognized that each of these techniques altered the relationship between the mother and her offspring. In the case of low-protein diets, the mother herself is malnourished, which in and of itself may alter maternal behavior. In the case of large-litter rearing, crowding and competition for a limited number of nipples, among other factors, also modify the early experience of the suckling rat pups.

Using these two techniques to produce malnutrition, a number of investigators looking directly at the mother–pup relationship have confirmed that malnutrition during the suckling period is associated with changes in this relationship, including increased nursing and physical contact between the mother and her young (Table III). When the protein in the diet provided to female rats is restricted during or after pregnancy, reduced milk output by the mother results (Mueller and Cox, 1946; Crnic and Chase, 1978), leading to malnutrition of the young (about a 50% reduction in weight). When the mother's diet is low in protein, the high level of nursing and physical contact continues until the end of the litter period at 21 days, although when a diet adequate in protein is provided, nursing normally declines at an earlier point in time (Massaro *et al.*, 1974; Wiener *et al.*, 1977). Thus, the mother–pup dyad responds to malnutrition as though the pups were younger. Nursing time and physical contact between mothers and pups are also increased when a low-protein diet has been provided over many generations (Galler and Propert, 1981a). Large-litter rearing was first thought to be associated with a decrease in maternal attention to the young (Grota and Ader, 1969). However, Crnic (1980) recently studied mother–pup interaction in large and small litters, using more standard techniques, and reported an increased amount of nursing through day 20 of the litter period. It is clear from these studies that malnutrition during suckling cannot be separated from changes in the mother–infant interaction, at least using the techniques described up to this point.

Because of the close association of malnutrition and changes in the mother–infant interaction in studies of rearing on low-protein diets or with large litters, a number of newer techniques were developed to produce malnutrition in rat pups with the intention of eliminating alterations in the mother–infant relationship. None of these techniques achieved this objective. Galler and Turkewitz (1977) subjected female rats to partial mammectomy before pregnancy and compared their interaction with their pups to that of nonmammectomized and sham-operated controls. Partial mammectomy reduced milk intake by pups, which were about 20% lighter in weight than either of the two groups of control pups. These investigators found that in mammectomized females, nursing was increased through the end of the litter period as compared with both groups of controls. Since the mother was not malnourished, it is clear that the changes in the mother–pup interaction were a response to malnutrition of the young. Slob and Snow (1973) developed the technique of rotating rat pups between lactating and nonlactating females (''aunts''), thereby restricting food intake to those periods spent with the lactating female. When Lynch (1976), Fleischer and Turkewitz (1979), and Crnic (1980) directly examined mother–pup interaction using this technique, they all reported that both lactating and nonlactating females spent more time in the nest with malnourished pups than with control pups, and in the case of lactating females, more time was spent nursing malnourished pups. Thus, investigators have so far been unable to develop techniques that produce malnutrition in the pups without altering the mother–pup relationship. Indeed, it is likely that these two conditions are interdependent and cannot be separated. In addition, in most experimental studies, there is increased nursing and physical contact between the mother and her malnourished young. These changes can be seen as compensating for the reduced food intake of the young by increasing opportunities for suckling and as compensating for other physiological effects of malnutrition, such as reduced body temperature of the pups. This interpretation is supported by the work of Galler and Propert (1982), who reported that among malnourished litters, more weight gain and better survival during the first 28 days of life were associated with high amounts of physical contact between the mother and pups in the first 12 hr after birth. Conversely, mothers that did not have adequate contact with their young as early as the first 3 hr of life had a significantly higher death rate among their offspring.

Increased nursing and physical contact may be present even in rats that have been fed an adequate diet after malnutrition (Galler and Propert, 1981a). Thus, such ''rehabilitated'' mothers and pups show increased nursing, physical contact, and time spent in the nest. Galler and Propert (1981b) also showed that the altered pattern of mother–pup interaction was a result of the nutritional history of the mother, rather than the pups. Thus, in a 2 × 2 crossover design, rehabilitated mothers suckled rehabilitated or well-nourished control pups and control mothers similarly suckled rehabilitated or control pups. Increased nursing and physical contact were observed only in the case of the rehabilitated mothers, and not for the control mothers. The persistence of increased nursing

Table III. Animal Studies of Mother–Pup Interaction and Malnutrition

Reference	Index group	Control group	Results
Levitsky *et al.* (1975)	Well-nourished mothers rearing pups who were malnourished *in utero* (prenatal malnutrition). Pups were born to mothers on low-protein diets (7% casein) provided throughout pregnancy and were cross-fostered to well-nourished mothers at birth.	Compared to well-nourished controls.	Mothers rearing prenatally malnourished pups spent more time in the nest.
Wiener *et al.* (1977)	Females on a low-protein (8%) diet provided *ad libitum* beginning on day 14 of pregnancy and continuing throughout the suckling period (N = 16).	Compared to well-nourished controls (N = 16).	Malnourished mothers spent more time in contact with their litters on days 3–12. Not tested after day 12.
Massaro *et al.* (1974)	Mothers provided with a low-protein (12% casein) diet at parturition and throughout the suckling period (N = 4).	Compared to well-nourished controls (N = 4).	Malnourished mothers spent more time in the nest on day 7 and thereafter.
Galler and Propert (1981a,b)	Intergenerationally malnourished females provided with low-protein (7.5%) casein) diet for 15–19 generations (N = 11).	Compared to well-nourished controls (N = 13).	Malnourished mothers spent more time actively nursing beginning on day 10. They also spent more time in the nest and more time in contact with pups on all days.
	Females with previous intergenerational malnutrition that were nutritionally rehabilitated on an adequate diet (25% casein) for 2–3 generations (N = 15).	See above.	Active nursing was increased in rehabilitated mothers on days 10–20. Time spent in nest and contact with pups were also increased on all days.
Galler and Turkewitz (1977)	Partially mammectomized females (N = 8) who were well nourished.	Compared to nonmammectomized controls (N = 9).	Nursing increased in litters of mammectomized mothers on days 2–20.

Study	Method	Comparison	Findings
Fleischer and Turkewitz (1979)	Rotation of pups between lactating and nonlactating females: group 1 litters (N = 6) spent 16 hr/day with a lactating female and 8 hr with a nonlactating female; group 2 litters (N = 6) spent only 8 hr/day with a lactating female.	Compared to litters (N = 6) rotated between two lactating females to control for effects of rotation without depriving pups of nourishment.	Nursing increased in 8-hr-fed litters as compared to other two groups in 3rd week of lactation.
Lynch (1976)	Rotation of pups between lactating and nonlactating females. Pups (N = 8 litters) spent 12 hr/day with each dam.	Compared to controls who were either rotated between two lactating females (N = 12 litters) or to nonrotated litters (N = 12 litters).	Mothers with malnourished pups spent more time in the nest than other two groups. Similar results for time spent crouching over pups.
Smart and Preece (1973)	Food restriction: females provided with 50% of normal intake through pregnancy and lactation (N = 8).	Compared to controls on an adequate diet (N = 16).	Malnourished mothers spent more time in the nest than controls except as feeding time approached.
Crnic (1980)	Large litters of 16 pups (N = 5).	Compared to litters of 8 pups each (N = 5).	Index mothers spent more time nursing out of the nest; more suckling occurred in the nest on days 10–20.
	Mothers provided with low-protein diet (8% casein) (N = 9).	Compared to well-nourished controls (N = 10).	Some increase in nest attendance among malnourished mothers on days 10–20.
	Mothers provided with restricted access to 40% normal diet (N = 10).	Compared to well-nourished controls (N = 10).	Restricted dams did more nursing outside the nest on days 10–20.
	Pups rotated between lactating and nonlactating females (N = 11 litters). Pups spent 12 hr/day with each dam.	Compared to nonrotated controls (N = 11) that were handled.	Mothers of rotated litters spent more time in the nest on days 1–20.
	Pups separated from mother and kept in incubator for 12 hr/day (N = 11 litters).	Compared to nonseparated controls (N = 5).	Mothers of separated pups spent more time in the nest and more time nursing outside the nest.

in malnourished rats that have been rehabilitated may represent genetic selection of individuals best able to compensate for malnutrition.

In summary, alterations in the mother–pup relationship appear to be an integral aspect of malnutrition of the pups during the litter period. For the most part, these changes appear to compensate for the effects of malnutrition and include increased nursing of pups and physical contact between mothers and pups throughout the litter period. Similar changes were also present in rats with histories of malnutrition that were then rehabilitated with an adequate diet. This evidence suggests that over many generations of malnutrition, there may be selective survival of individuals that show this adaptive response. It is instructive to note the similarity between results of animal studies and certain observations reported in studies of human populations. Increased time spent nursing and increased physical contact have been found in both species when the infant is malnourished. Clearly, the picture in human populations is more complex than in rats, especially since preexisting inadequacies in the mother–infant relationship may contribute to the episode of malnutrition and its outcome.

5. Complexity of the Human Situation

The relationship between the mother and her infant is one of the major socialization influences for the developing child. Under ideal conditions, the mother–infant interaction follows a predictable pattern beginning with the formation of a secure mother–infant bond (attachment). This bond between mother and infant, in turn, helps the child to meet future developmental tasks leading to the attainment of autonomy and independence. Poor bonding results in infants who remain insecure, dependent, and isolated from important developmental influences. A dysfunctional mother–infant relationship may thus directly affect behavioral outcome, but may often express itself as severe malnutrition in the child, particularly in environments in which poverty and limited food supplies are prevalent.

In the human condition, the larger social matrix within which a child lives must be considered in order to understand why a child develops malnutrition in the first place and what mediating factors modify outcome after the episode of malnutrition. Fig. 3 shows the micro- and macroenvironmental conditions that when seriously compromised, put a child at special risk for developing malnutrition and for a poor behavioral outcome. It should be noted that these environmental factors are not specific to malnutrition, but have also been implicated in other childhood disorders stemming from a social origin. The physiological and psychological outcome of the child is therefore determined by multiple levels of influence.

A number of factors are influential in determining the quality of the mother–infant interaction and its interface with nutritional disorders. In addition to the psychological makeup and behavior of the mother, characteristics of the child are important. Children who are of low birth weight, handicapped,

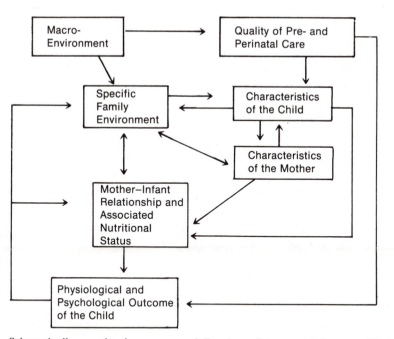

Fig. 3. Schematic diagram showing sources and directions of the many influences affecting the mother–infant relationship. Modified from Ricciuti and Dorman (1983).

or sickly, or who have difficult temperaments adversely affect the mother–infant relationship.

Finally, the specific family environment, including such factors as family size, composition, and stability, may also predispose to and influence the outcome of malnutrition in the child. Each of these three groups of factors is affected by macroenvironmental features such as income, quality of housing, schooling, health care, and other support services. Clearly, under more adverse macroenvironmental conditions, the functioning of the family and the mother–child dyad is compromised.

6. Appendix

In the discussion that follows, we address some of the major methodological issues and common measurement problems that arise in the study of mother–infant interaction. These issues are discussed from the perspective of both human and animal research, since there are many parallels between the two areas of inquiry, which it is useful to be aware of. One of the central problems for developmental psychologists and child psychiatrists studying behavioral development has been how best to characterize variations in the way parents care for or rear their infants, beginning as early as the first days and weeks of life. The focus, of course, is on aspects of infant-rearing and parent–infant interaction that can be assumed to make a difference in the offspring's later psychological development. Thus, for example, investigators may wish to characterize the mother's mode of infant care as warm and affectionate or as cool and detached, as sensitive and responsive to the infant's needs or as insensitive and unresponsive, as optimally or minimally encouraging of learning and exploration, and so on. Their efforts have increasingly been extended in recent years to include the infant-rearing styles of fathers and other primary caregivers, as well as mothers. Also, there has been a pronounced movement toward the identification of specific features of the interactions between parents and infants by means of direct observations of ongoing behavior.

6.1. Major Methods in the Study of Mother–Infant Interaction

Some of the major methodological approaches employed to assess variations in patterns of infant care and parent–infant interaction are described briefly below. Though the first two apply to humans only, direct observation and filming are used in both animal and human studies.

6.1.1. Parental Interviews

Parents may be asked about their typical patterns of infant care and rearing, with the interview format varying from open-ended and unstructured to highly structured and focused on specific questions. Such interviews may lead to qualitative clinical judgments, to overall ratings on specific dimensions of parental behavior, or to quantitative scores based on responses to specific questions.

6.1.2. Child-Rearing Questionnaires

Parents are often asked to record their responses to a large number of specific questions, either by checking one of several specified alternatives or by brief written responses. These questionnaires are often self-administered, but they may be filled in by an interviewer or home visitor in situations where there are literacy or language problems. Such questionnaires may include not only material directly concerned with infant care or child rearing but also questions on various objective features of the home environment that may be relevant in characterizing the child's early experience, e.g., number of people involved in the care of the infant, size of family, amount of space and crowding, and availability of play materials.

6.1.3. Direct Observations of Mother and Infant Together

Many investigators prefer to employ assessments based on systematic observations of the parent and infant together. These observations may be carried out in naturalistic situations or under more structured or controlled conditions—in the laboratory, playroom, or hospital, for example. Assessments may vary from overall ratings or judgments of various dimensions of parent–child interaction to quantitative indices of the frequency of various parent and infant behaviors.

6.1.4. Film or Videotape Recordings of Parent–Child Interaction

With the increasing availability of videotape facilities, many investigators are recording the kinds of observations of mothers and infants just described, so that various measures can be derived from the videotape, rather than from observational notes, written narrative records, or on-the-spot codings of the parent and infant behaviors. The availability of film or videotape has made it possible for investigators to carry out increasingly "fine-grained" analyses of sequential parental and infant behavior. For example, using a tape or film, one could count the number of times a mother responds in various ways to the infant's cues and the number of times she ignores such behavior. One could also determine whether the infant's behavior changed over a 20-min period as a result of the mother's response (and vice versa). In short, one could try to determine how the mother and infant influenced each other or the extent to which they adapted to each other in a synchronous way. Thus, one might begin to identify characteristics of the mother and infant as a dyad, not simply as individuals. In many studies of mothers and infants, separate measurements are made of the frequency of occurrence of various specific maternal and infant behaviors. While providing important information, such measures in themselves may tell us very little about sequential patterns of interaction as discussed above.

6.2. Common Measurement Problems

Each of the general methodological approaches just outlined has obvious advantages and limitations, which cannot be discussed in detail here. One must be aware, however, of several common measurement problems that may arise, regardless of the specific methodological strategy utilized.

6.2.1. Interjudge or Interrater Reliability

Where the measures being derived involve judgments by an interviewer, rater, or coder, one obviously needs to be certain that similar scores can be obtained regardless of the particular individual making the judgments. This problem is typically addressed by determining empirically the degree of agreement between two independent judges. Videotape records are particularly useful, since they can be employed for training judges or raters so as to obtain appropriate levels of inter-judge reliability.

6.2.2. Validity or Representativeness of Parent–Infant Behaviors Observed or Reported

This problem has to do with the question of whether the infant-care practices or interactions observed or reported are representative, or "typical," of what normally goes on between parent and infant. In the case of behaviors reported by parents in interviews or questionnaires, one obviously would like to know whether these reports are accurate reflections of what occurs on a day-to-day basis. Some investigators feel that the most representative (or "valid") measures of infant care or parent–infant interaction are obtained from relatively lengthy observation in natural settings, rather than from much shorter observations in experimental or contrived settings. Because behavior does vary considerably among situations (even in naturalistic settings), one can generally have more confidence in observational measures that are sufficiently long or repeated sufficiently often to sample behavior in a number of different situations, rather than measures based on very brief samples of mother–infant interactions on a single occasion and in a particular setting (e.g., a 10-min videotape of mother feeding newborn infant on the first day after birth).

6.2.3. Short-Term Stability of Measures

This issue is closely related to that of representativeness of observed behaviors. An obvious test for representativeness is to compare the observational measures obtained on two occasions in somewhat different situations (e.g., measures of parent–infant interaction based on observations in the home setting and in a structured laboratory setting, or in a feeding and in a free-play situation). Even if one is interested in patterns of mother–infant interactions in a *particular type* of situation, such as feeding, teaching, or free play, one needs to evaluate empirically the stability of the measures obtained in the same situation on at least two occasions separated by a relatively short time period (e.g., several days to a week). If there is little or no consistency in the behaviors observed on the two occasions, it is unlikely that there are meaningful and consistent relationships between these "unreliable" measures and other variables of interest.

7. References

Ainsworth, M. D. S., 1973, The development of infant–mother attachment, in *Review of Child Development Research* (B. M. Caldwell and H. N. Ricciuti, eds.), Vol. 3, pp. 1–94, University of Chicago Press.

Ainsworth, M. D. S., 1979, Infant–mother attachment, *Am. Psychol.* **34**:932–937.

Ainsworth, M. D. S., and Bell, S. M., 1969, Some contemporary patterns of mother–infant interaction in the feeding situation, in: *Stimulation in Early Infancy* (A. Ambrose, ed.), pp. 133–163, Academic Press, New York.

Beckwith, L., and Cohen, S. E., 1980, Interactions of preterm infants with their caregivers and test performance at age 2, in: *High-Risk Infants and Children: Adult and Peer Interactions* (T. M. Field, S. Goldberg, D. Stern, and A. M. Sostek, eds.), pp. 155–178, Academic Press, New York.

Bell, R. Q., 1974, Contribution of human infants to caregiving and social interaction, in: *Effect of the Infant on Its Caregiver* (M. Lewis and L. A. Rosenblum, eds.), pp. 1–20, John Wiley, New York.

Bernal, J., 1972, Crying during the first 10 days of life and maternal responses, *Dev. Med. Child Neurol.* **14**:362–372.

Birch, L. L., Marlin, D. W., Kramer, L., and Peyer, C., 1981, Mother–child interaction pattern and the degree of fatness in children, *J. Nutr. Educ.* **13**:17–21.

Boles, G., 1959, Personality factors in mothers of cerebral palsied children, *Genet. Psychol. Mongr.* **59**:159–218.

Brody, S., 1956, *Patterns of Mothering*, International Universities Press, New York.

Brown, R. E., 1979, The young child: Failure to thrive, in: *Human Nutrition: A Comprehensive Treatise*, Vol. 2, *Nutrition and Growth* (D. B. Jelliffe and E. F. Patrice Jelliffe, eds.), pp. 171–184, Plenum Press, New York.

Bullard, D., Pivchik, B., and Glaser, M., 1967, Failure to thrive in the neglected child, *Am. J. Orthopsychiatry* **37**:680.

Caldwell, B. M., 1964, The effects of infant care, in: *Review of Child Development Research*, Vol. 1 (M. L. Hoffman and L. W. Hoffman, eds.), pp. 9–87, Russell Sage Foundation, New York.

Chao, Y. M. A., 1971, A comparative study of regain of body weight of newborns during the first ten days of life, *Int. Nurs. Rev.* **18**:15–22.

Chase, H. P., and Martin, H. P., 1970, Undernutrition and child development, *N. Engl. J. Med.* **282**:933–939.

Chavez, A., and Martinez, C., 1975, Nutrition and development of children from poor rural areas. V. Nutrition and behavioral development, *Nutr. Rep. Int.* **11**(6):477–489.

Clarke-Stewart, K. A., 1973, Interactions between mothers and their young children: Characteristics and consequences, *Monogr. Soc. Res. Child Dev.* **38**:153.

Committee on Nutrition, 1980, Encouraging breastfeeding, *Pediatrics* **65**:657–658.

Cravioto, J., and DeLicardie, E., 1972, Environmental correlates of severe clincial malnutrition and language development in survivors from kwashiorkor or marasmus, *Nutrition, the Nervous System and Behavior*, pp. 73–94, PAHO Publication No. 251.

Cravioto, J., and DeLicardie, E. R., 1976, Microenvironmental factors in severe protein–energy malnutrition, in: *Nutrition and Agricultural Development: Significance and Potential for the Tropics* (N. S. Scrimshaw and M. Behar, eds.), pp. 25–35, Plenum Press, New York.

Crnic, L. S., 1980, Models of infantile malnutrition in rats: Effects of maternal behavior, *Dev. Psychobiol.* **13**:615–628.

Crnic, L. S., and Chase, H. P., 1978, Models of infantile undernutrition in rats: Effects on milk, *J. Nutr.* **108**:1755–1760.

Denenberg, V. H., and Whimbey, A. E., 1963, Behavior of adult rats is modified by experiences their mothers had as infants, *Science* **142**:1192–1193.

DiVitto, B., and Goldberg, S., 1979, The effects of newborn medical status on early parent–infant interation, in: *Infants Born at Risk: Behavior and Development* (T. M. Field, A. Sostek, S. Goldberg, and H. H. Shuman, eds.), pp. 311–332, Spectrum Publications, New York.

Dunn, J. B., 1977, Patterns of early interaction: Continuities and consequences, in: *Studies in Mother–Infant Interaction* (H. R. Schaffer, ed.), pp. 457–474, Academic Press, New York.

Dunn, J. B., and Richards, M. P. M., 1977, Observations on the developing relationship between mother and baby in the neonatal period, in: *Studies in Mother–Infant Interaction* (H. R. Schaffer, ed.), pp. 427–455, Academic Press, New York.

Edwards, L. N., and Grossman, M., 1979, The relationship between children's health and intellectual development, in: *Health: What Is It Worth? Measures of Health Benefits* (S. J. Mushkin and D. W. Dunlop, eds.), pp. 273–314, Pergamon Press, New York.

Elmer, E., 1960, Failure to thrive: Role of the mother, *Pediatrics* **25**:717–725.

Evans, S. L., Reinhart, J. B., and Succop, R. A., 1972, Failure to thrive: A study of 45 children and their families, *J. Am. Acad. Child Psychiatry* **11**:440–457.

Fernholt, J., and Provence, S., 1976, Diagnosis and treatment of an infant with psychophysiological vomiting, *Psychoanal. Study Child* **31**:439–459.

Field, T., 1977, Maternal stimulation during infant feeding, *Dev. Psychol.* **13**:539–540.

Fischoff, J., Whitten, C. F., and Petit, M. A., 1971, A psychiatric study of mothers of infants with growth failure secondary to maternal deprivation, *J. Pediat.* **79**:209–215.

Fleischer, S., and Turkewitz, G., 1979, Behavioral effects of rotation between lactating and nonlactating females, *Dev. Psychobiol.* **12**:245–254.

Freese, M. P., and Thoman, E. B., 1978, The assessment of maternal characteristics for the study of mother–infant interactions, *Infant Behav. Dev.* **1**:95–105.

Galler, J. R., and Propert, K. J., 1981a, Maternal behavior following rehabilitation of rats with intergenerational malnutrition. 1. Persistent changes in lactation-related behaviors, *J. Nutr.* **111**:1330–1336.

Galler, J. R., and Propert, K. J., 1981b, Maternal behavior following rehabilitation of rats with intergenerational malnutrition. 2. Contributions of mothers and pups to deficits in lactation-related behaviors, *J. Nutr.* **111**:1337–1342.

Galler, J. R., and Propert, K. J., 1982, Early maternal behaviors predictive of the survival of suckling rats with intergenerational malnutrition, *J. Nutr.* **112**:332–337.

Galler, J. R., and Seelig, C., 1981, Home-orienting behavior in rat pups: The effect of two and three generations of rehabilitation following intergenerational malnutrition, *Dev. Psychobiol.* **14**(6):541–548.

Galler, J. R., and Turkewitz, G., 1977, The use of partial mammectomy to produce undernutrition in the rat, *Biol. Neonate* **31**:260–265.

Galler, J. R., Ramsey, F., Solimano, G., Lowell, W. E., and Mason, E., 1983a, The influence of early malnutrition on subsequent behavioral development. I. Degree of impairment in intellectual performance, *J. Am. Acad. Child Psychiatry* **22**(1):8–15.

Galler, J. R., Ramsey, F., Solimano, G., and Lowell, W. E., 1983b, The influence of early malnutrition on subsequent behavioral development. II. Classroom behavior, *J. Am. Acad. Child Psychiatry* **22**(1):16–22.

Galler, J. R., Ramsey, F., Solimano, G., and Kucharski, L. T., 1984, The influence of early malnutrition on subsequent behavioral development. V) The role of the microenvironment of the household (submitted for publication).

Gerrard, J. W., 1974, Breastfeeding: Second thoughts, *Pediatrics* **54**:757–764.

Gordon, A., 1979, Patterns and determinants of attachment in infants with non-organic failure to thrive syndrome, Ph.D. dissertation, Harvard University School of Education, Cambridge, Massachusetts.

Grantham-McGregor, S. M., and Stewart, M. E., 1980, The relationship between hospitalization, social background, severe protein–energy malnutrition and mental development in young Jamaican children, *Ecol. Food Nutr.* **9**:151–156.

Graves, P. L., 1976, Nutrition, infant behavior and maternal characteristics, a pilot study in West Bengal, India, *Am. J. Clin. Nutr.* **29**:305–319.

Graves, P. L., 1978, Nutritional and infant behavior: A replication study in Katmandu Valley, Nepal, *Am. J. Clin. Nutr.* **31**:541–551.

Green, I. A., Gustafson, G. E., and West, M. J., 1980, Effects of infant development on mother–infant interactions, *Child Dev.* **51**:199–207.

Grota, L. J., and Ader, R., 1969, Continuous recording of maternal behavior patterns in the *Rattus norvegicus*, *Anim. Behav.* **17**:722–729.

Hall, B., 1975, Changing composition of human milk and early development of appetite control, *Lancet* (no. 7910):779–781.

Harlow, H. F., 1963, The maternal affectional system, in: *Determinants of Infant Behavior II* (B. M. Foss, ed.), pp. 3–29, Great Britain, Methuen and Co. Ltd., London.

Jelliffe, D. B., and Jelliffe, E. F. P., 1975, Human milk, nutrition and the world resource crisis, *Science* **188**:557–561.

Kaye, K., 1977, Toward the origin of dialogue, in: *Studies in Mother–Infant Interaction* (H. R. Schaffer, ed.), pp. 89–117, Academic Press, New York.

Kennedy, G. C., 1957, The development with age of hypothalamic restraint upon the appetite of the rat, *J. Endocrinol.* **16**:9–17.

Kerr, M. A., Begues, J. L., and Kerr, D. S., 1978, Psychosocial functioning of mothers of malnourished children, *Pediatrics* **62**:(5):778–784.

Kotelchuck, M., and Newberger, E. H., 1978, Failure to thrive: A controlled study of familial characteristics, The Family Development Study, Children's Hospital, Boston.

Kotelchuck, M., and Newberger, E. H., 1983, Failure to thrive: A controlled study of familial characteristics, *J. Child Psychiatry.* **22**:322–328.

Krieger, I., and Mellinger, R. C., 1971, Pituitary function in the deprivation syndrome, *J. Pediat.* **79:**216–225.

Leifer, M., 1980, *Psychological Effects of Motherhood*, Praeger, New York.

Leonard, M. F., Rhymes, J. P., and Solnit, A. J., 1966, Failure to thrive in infants: A family problem, *Am. J. Dis. Child.* **111:**600–612.

Levitsky, D. A., Massaro, T. F., and Barnes, R., 1975, Maternal malnutrition and the neonatal environment, *Fed. Proc. Fed. Am. Soc. Exp. Biol.* **34:**1583–1586.

Levy, D. M., 1958, *Behavioral Analysis: Analysis of Clinical Observations of Behavior as Applied to Mother–Newborn Relationships*, Charles C. Thomas, Springfield, Illinois.

Lewis, M., and Rosenblum, L. A. (eds.), 1974, *The Effect of the Infant on Its Caregiver*, Wiley-Interscience, New York.

Lynch, A., 1976, Post-natal undernutrition: An alternative method, *Dev. Psychobiol.* **9:**39–48.

Massaro, T. F., Levitsky, D. A., and Barnes, R. M., 1974, Protein malnutrition in the rat: Its effects on maternal behavior and pup development, *Dev. Psychol.* **7:**551–561.

Menkes, J. H., 1977, Early feeding history of children with learning disorders, *Dev. Med. Child Neurol.* **19:**169–171.

Miller, C. J., 1976, Children with feeding problems, *Child: Care, Health, Dev.* **2:**73–76.

Mueller, A. J., and Cox, W. M., Jr., 1946, The effect of changes in diet on the volume and composition of rat milk, *J. Nutr.* **31:**249–259.

Newberger, E. H., Reed, D. B., Daniet, J. H., Hyde, J. N., and Kotelchuck, 1977, Pediatric social illness: Toward an etiologic classification, *Pediatrics* **60:**178–185.

Nicoll, C., and Meites, J., 1959, Prolongation of lactation in the rat by litter replacement, *Proc. Soc. Exp. Biol. Med.* **101:**81–82.

Noirot, E., 1964, Changes in responsiveness to young in the adult mouse: The effect of external stimuli, *J. Comp. Physiol. Psychol.* **57:**97–99.

Olson, C. M., Psiaki, D. L., and Kaplowitz, D., 1980, *Current Knowledge on Breastfeeding: A Review for Medical Practitioners*, Division of Nutritional Sciences, Cornell University, Ithaca, New York.

Omolulu, A., 1974, Nutritional factors in the vulnerability of the African child, in: *The Child at Risk* (E. J. Anthony and C. Koupernik, eds.), Vol. 3, pp. 331–336, Yearbook for the International Association for Child Psychiatry and Allied Sciences, John Wiley and Sons, New York.

Osofsky, J. D., and Danzger, B., 1974, Relationships between neonatal characteristics and mother–infant interaction, *Dev. Psychol.* **10:**124–130.

Ounsted, M., and Sleigh, G., 1975, The infant's self-regulation of food intake and weight gain, *Lancet* No. 7922:1393–1397.

Polansky, N. A., Chalmers, M., Buttenwieser, E., and Williams, D. T., 1981, *Damaged Parents: An Anatomy of Child Neglect*, University of Chicago Press.

Pollitt, E., and Wirtz, S., 1981, Mother–infant feeding interaction and weight gain in the first month of life, *J. Am. Diet. Assoc.* **78:**596–601.

Pollitt, E., Gilmore, M., and Valcarcel, M., 1978, Early mother–infant interaction and somatic growth, *Early Hum. Dev.* **1:**325–336.

Ricciuti, H. N., and Dorman, R., 1983, Interaction of multiple factors contributing to high-risk parenting, in: *A Round Table on Minimizing High-risk Parenting* (*Round Table No. 7*) (R. A. Hoekelman, ed.), pp. 187–211, Harwal, Media, Pennsylvania.

Richardson, S. A., 1974, The background histories of schoolchildren severely malnourished in infancy, in: *Advances in Pediatrics 21* (I. Schulman, ed.), pp. 167–192, Medical Yearbook Publishers, Chicago.

Rodgers, B., 1978, Feeding in infancy and later ability and attainment: A longitudinal study, *Dev. Med. Child Neurol.* **20:**421–426.

Rosenblatt, J. S., 1967, The nonhormonal bases of maternal behavior in the rat, *Science* **156:**1512–1514.

Rosenblatt, J. S., 1969, The development of maternal responsiveness in the rat, *Am. J. Orthopsychiatry* **39:**36–56.

Rosenblatt, J. S., 1975, Prepartum and postpartum regulation of maternal behavior in the rat, *Parent–Infant Interaction*, **33**:17–37.

Rutter, M., 1980, Long term effects of early experiences, *Dev. Med. Child Neurol.* **22**:800–815.

Shapiro, V., Fraiberg, S., and Adelson, E., 1976, Infant–parent psychotherapy on behalf of a child in a critical nutritional state, *Psychoanal. Study Child* **31**:461–491.

Shaw, J. A., Wheeler, P., and Morgan, D. W., 1971, Mother–infant relationship and weight gain in the first month of life, *J. Am. Acad. Child Psychiatry* **9**:428–444.

Sheffer, M. L., Grantham-McGregor, S. M., and Ismail, S. J., 1981, The social intervention of malnourished children compared with that of other children in Jamaica, *J. Biosoc. Sci.* **13**:19–30.

Sherrod, K. B., Friedman, S., Crawley, S., Drake, D., and Devieux, J., 1977, Maternal language to prelinguistic infants: Syntactic aspects, *Child Dev.* **48**:1662–1665.

Slob, A. K., Snow, C. E., and de Natris-Mathot, F., 1973, Absence of behavioral deficits following neonatal undernutrition in the rat, *Dev. Psychobiol.* **6**:177–186.

Smart, J. L., and Preece, J., 1973, Maternal behavior of undernourished mother rats, *Anim. Behav.* **21**:613–619.

Snow, C. K., 1977, The development of conversation between mothers and babies, *J. Child Language* **4**:1–22.

Stern, D. M., 1974, Mother and infant at play: The dyadic interaction involving facial, vocal, and gaze behaviors, in: *The Effect of the Infant on its Caregiver* (M. Lewis and L. A. Rosenblum, eds.), pp. 187–213, Wiley-Interscience, New York.

Suomi, S. J., 1979, Peers, play and primary prevention in primates, in: *Primary Prevention of Psychopathology*, Vol. 3 (J. Rolf, ed.), pp. University of New England Press, Hanover, New Hampshire.

Suomi, S. J., 1983, Social development in rhesus monkeys: Consideration of individual differences, in: *The Behavior of Human Infants* (A. Oliverio and M. Zappella, eds.), pp. 71–92, Plenum Press, New York.

Thoman, E. B., 1975, Development of synchrony in mother–infant interaction in feeding and other situations, *Fed. Proc. Fed. Am. Soc. Exp. Biol.* **34**:1587–1592.

Thomas, A., and Chess, S., 1980, *The Dynamics of Psychological Development*, Brunner/Mazel, New York.

Turkewitz, G., 1966, The development of spatial orientation in relation to the effective environment in rats, Ph.D. dissertation, New York University (unpublished).

Whitten, C. F., Pettit, M. G., and Fischoff, J., 1969, Evidence that growth failure from maternal deprivation is secondary to understanding, *J. Am. Med. Assoc.* **209**:1675–1682.

Wiener, S. G., Fitzpatrick, K. M., Levin, R., Smotherman, W. P., and Levine, S., 1977, Alterations in the maternal behavior of rats rearing malnourished offspring, *Dev. Psyochobiol.* **10**:243–254.

Wiesner, B., and Sheard, N., 1933, *Maternal behavior in the Rat*, Oliver and Boyd, London.

Young, H. B., Buckley, A. E., Hamza, B., and Mandarano, C., 1982, Milk and lactation: Some social and developmental correlates among 1000 infants, *Pediatrics* **69**:169–175.

Zemlick, M. J., and Watson, R. I., 1953, Maternal attitudes of acceptance and rejection during and after pregnancy, *Am. J. Orthopsychiatry* **23**:570–584.

Anorexia Nervosa and Bulimia: Biological, Psychological, and Sociocultural Aspects

Arnold E. Andersen

1. Introduction

Anorexia nervosa is an eating disorder characterized by severe weight loss from self-induced starvation and fear of fatness, with loss of menstrual periods in women or decreased sexual drive in men. It occurs frequently but not exclusively in young women. Its origins are obscure despite the presence of many theories. Anorexia nervosa often presents to the medical profession in disguised form and manifests to the clinician an extreme physiological state usually occurring as a result of serious medical disease. Anorexia nervosa has been the subject of many recent medical and psychological investigations and appears to be increasing in incidence. It presents a very contemporary challenge to medical scientists because of its relationship to societal trends on one hand and its obvious biological and psychological components on the other hand. Its companion disorder, bulimia, is characterized by episodes of compulsive overeating followed by attempts to avoid the caloric consequences of the binge by self-induced vomiting, laxatives, diuretics, or strenous exercise. These two disorders share many biological and psychopathological features and may evolve from one into the other, principally from anorexia nervosa into bulimia. While anorexia nervosa is of ancient origin, bulimia as a separate, diagnosable disorder has only recently been clinically defined.

Arnold E. Andersen • Department of Psychiatry and Behavioral Sciences, The Johns Hopkins University School of Medicine, Baltimore, Maryland 21205.

2. *History*

The first case of anorexia nervosa is lost in antiquity. The first clear case report is commonly attributed to Morton in 1694. Morton (1694) described his case of nervous consumption:

> A nervous atrophy, or consumption is a wasting of the body without any remarkable fever, cough, or shortness of breath; but it is attended with a want of appetite, and a bad digestion, upon which there follows a languishing weakness of nature, and a falling away of the flesh every day more and more.

The first contemporary descriptions of anorexia nervosa were presented in the 1870s by Sir William Gull and by Dr. Charles Lasègue. These almost simultaneous descriptions painted in rich clinical detail the disorder of anorexia nervosa much as we know it at present. Gull (1874) described his case as follows:

> In an address on medicine delivered at Oxford in the Autumn of 1868, I referred to a peculiar form of disease occurring mostly in young women and characterized by extreme emaciation . . . . The subjects of this affection are mostly of the female sex and chiefly between the ages of 16 and 23. I have occasionally seen it in males at the same age . . . . The condition was one of simple starvation . . . . The want of appetite is, I believe, due to a morbid mental state . . . . We might call the state hysterical, . . . I prefer however the more general term "nervosa" since the disease occurs in males as well as in females and is probably rather central than peripheral.

Lasègue (1873) made the following comments:

> La répugnance à s'alimenter suit sa marche lentement progressive. Les repas se réduisent de plus en plus, ... Que dire, en effet? Que la malade ne peut vivre avec une quantité d'aliments dont un enfant en bas âge ne s'accommoderait pas. La malade répond que sa nourriture lui suffit et au delà, elle n'a ni changé ni maigri, on ne l'a jamais vue se refuser à une tâche ou à une fatigue; elle sait mieux que personne ce qu'il lui faut et d'ailleurs il lui serait impossible de tolérer une alimentation plus abondante.

The descriptions by Gull and Lasègue remain the foundation of our modern knowledge of anorexia nervosa. In the early 20th century, the syndrome of pituitary necrosis was delineated and patients with this disorder recognized and excluded from the diagnosis of anorexia nervosa. The rest of the 20th century presents a series of attempts to explain and treat anorexia nervosa from fundamentally different perspectives (Lucas, 1981).

The first decades of the 20th century were characterized by intensive psychoanalytic efforts to understand anorexia nervosa as a disorder growing out of faulty psychosexual development. The motif that remains from this era is the idea that anorexia nervosa is related to a "fear of oral impregnation." Seen out of context from its psychodynamic origins, this archaic phrase now seems of historical rather than of clinical significance. It did represent, however, a prolonged, serious, and energetic effort to understand psychiatric symptoms on the basis of the psychodynamic theories of Freud and his followers.

In the United States, at least, anorexia nervosa was given little clinical attention in the formal literature until the 1950s. Hilde Bruch became a lifelong student of this disorder and contributed many seminal ideas concerning origin and treatment (Bruch, 1973). In the late 1950s and early 1960s, Russell, Crisp, Theander, and several others undertook empirical and longitudinal studies concerning the natural history and the treatment of anorexia nervosa. This disorder continued to be of interest in Great Britain during the time it was lost sight of in the United States (Ryle, 1936). Medical and, subsequently, public interest in anorexia nervosa increased exponentially in the 1970s. Students of this disorder often approached it, however, much like the blind man approached the elephant in the mythical tale. Investigators with specific methodologies in the behavioral, sociocultural, psychodynamic, and pharmacological areas began to view the disorder only in terms of their special focus. The first International Conference on Anorexia Nervosa was held in 1976 and provided a basis for much of the current research by attempting to look at unifying concepts of this disorder (Vigersky, 1977).

The presence of multiple concepts of origin and multiple strategies for treatment is characteristic of a science in its embryonic state. It was not until 1972, for example, that clear interinstitutional criteria were published, allowing for comparison of research studies (Feighner *et al.*, 1972). The last several years have seen an explosion of interest in anorexia nervosa from biological, psychological, and sociocultural perspectives.

The term *anorexia nervosa* derives from the Greek term ἡ ὄρεξις, which means "without appetite." The nomenclature is fundamentally erroneous, because these patients have no primary loss of appetite, but rather a fear of the consequences of caloric ingestion: increased weight. The name, however, has traditional status and is probably so embedded in the clinical descriptive literature that a more accurate term is not practical. The French often refer to this disorder as *l'anorexie hystérique*. The Germans, as is their fashion, describe the disorder with great verbal precision, calling it *Pubertätsmagersucht*.

The term *bulimia*, also of questionable linguistic accuracy, derives from the Greek words for "ox" and "hunger." It had previously been used in a limited fashion, for many years, to describe clinical syndromes of extraordinary appetite usually related to neurological or endocrine disorders. Beginning in the 1970s, however, bulimia has gained nosological status as a separate psychiatric disorder. Bruch (1973) describes symptoms of bulimia existing throughout the ancient world, but these were largely the hedonic experiences of a ruling class inconvenienced by the presence of an engorged stomach while desiring still more food. The *Diagnostic and Statistical Manual of Mental Disorders* (1980) (DSM III) has given bulimia a formal definition as well as anorexia nervosa.

3. Diagnosis

The first clear research criteria for anorexia nervosa were described by Feighner *et al.* (1972). They are as follows:

I. Age of onset less than 25.
II. Accompanying weight loss of at least 25% of original body weight.
III. A distorted, implacable attitude toward eating, food, or weight that overrides hunger, admonitions, reassurance, and threats.
   A. Denial of illness with a failure to recognize nutritional needs.
   B. Apparent enjoyment in losing weight, with overt manifestation that food refusal is a pleasurable indulgence.
   C. A desired body image of extreme thinness, with overt evidence that it is rewarding to the patient to achieve and maintain this state.
   D. Unusual hoarding or handling of food.
IV. No known medical illness that could account for the disorder and weight loss.
V. No other known psychiatric disorder, particularly primary affective disorders, schizophrenia, and obsessive–compulsive and phobic neuroses (the assumption is made that even though it may appear phobic or obsessional, food refusal alone is not sufficient to qualify for obsessive–compulsive or phobic disease).
VI. At least two of the following manifestations:
   A. Amenorrhea
   B. Lanugo
   C. Bradycardia (persistent resting pulse of 60 or less)
   D. Periods of overactivity
   E. Episodes of bulimia
   F. Vomiting (may be self-induced)

The *Diagnostic and Statistical Manual of Mental Disorders* (1980) gives the criteria for anorexia nervosa and bulimia presented in Table I.

These diagnoses are fundamentally syndrome descriptions. They prescind from an etiological concept and instead attempt to provide reliable and valid specific criteria for diagnosis. Especially problematical is the term "bulimia," which according to the DSM III requires the exclusion of anorexia nervosa, yet chronic anorexia nervosa may develop into normal-weight bulimia. The terminology for a patient meeting criteria for anorexia nervosa but employing self-induced vomiting is also unclear. Russell (1979) refers to bulimic patients with a history of anorexia nervosa as having bulimia nervosa.

A major change in the diagnostic approach in the 1970s was the gradual transition from the medical-exclusion approach to the method of eliciting the positive features of the disorder on mental-state examination. In the early 1970s, there was often a delay of several years from the onset of symptomatology to accurate diagnosis. Clinicians in nonpsychiatric specialties, especially, frequently backed into the diagnosis only after excluding all conceivable medical causes of weight loss. In studies at the National Institutes of Health on 32 anorectic patients, no individual diagnosed as having anorexia nervosa by meeting the Feighner criteria was later found to have a medical or other psychiatric cause for her weight loss (Vigersky *et al.*, 1976). These criteria, while empirical and syndromic, are nonetheless valid and reliable criteria.

Cases meeting research criteria must be distinguished from those that require clinical attention even though they do not formally meet these criteria. For example, an individual who has not quite met the 25% weight loss requirement for diagnosis of anorexia nervosa may nonetheless have all the essential clinical features and be urgently in need of treatment, especially if weight loss is rapid. The syndrome approach to diagnosis illustrates the primitiveness of our understanding of these eating disorders, which remain to be understood

Table I. Diagnostic Criteria for Anorexia Nervosa and Bulimia[a]

Anorexia nervosa

A. Intense fear of becoming obese, which does not diminish as weight loss progresses.
B. Disturbance of body image, e.g., claiming to "feel fat" even when emaciated.
C. Weight loss of at least 25% of original body weight or, if under 18 years of age, weight loss from original body weight plus projected weight gain expected from growth charts may be combined to make the 25%.
D. Refusal to maintain body weight over a minimal normal weight for age and height.
E. No known physical illness that would account for the weight loss.

Bulimia

A. Recurrent episodes of binge eating (rapid consumption of a large amount of food in a discrete period of time, usually less than 2 hr).
B. At least three of the following:
  1. Consumption of a high-caloric, easily ingested food during a binge.
  2. Inconspicuous eating during a binge.
  3. Termination of such eating episodes by abdominal pain, sleep, social interruption, or self-induced vomiting.
  4. Repeated attempts to lose weight by severely restrictive diets, self-induced vomiting, or use of cathartics or diuretics.
  5. Frequent weight fluctuations greater than ten pounds due to alternating binges and fasts.
C. Awareness that the eating pattern is abnormal and fear of not being able to stop eating voluntarily.
D. Depressed mood and self-deprecating thoughts following eating binges.
E. The bulimic episodes are not due to anorexia nervosa or any known physical disorder.

[a] From American Psychiatric Association *Diagnostic and Statistical Manual of Mental Disorders* Third Edition, Washington, D.C., APA, 1980.

at a more fundamental level of specific abnormalities of biology or psychology. It must be remembered that the purpose of diagnosis is to prescribe a treatment and to give a prognosis.

Virtually every aspect of every diagnostic criterion for anorexia nervosa has come under attack, and there is lively controversy concerning diagnostic concepts (Rollins and Piazza, 1978).

4. Presentation to Medical Specialists

Anorexia nervosa frequently presents to medical specialists in disguised form. Gynecologists are presented with patients who complain of secondary amenorrhea or occasionally primary amenorrhea resulting from undisclosed anorexia nervosa. Gastroenterologists are consulted because of the concern that an emaciated patient suffers from malabsorption or other gastrointestinal disorders. Neurologists may focus on examinations to eliminate the possibility of lesions in the hypothalamus causing weight loss. Endocrinologists worry about hypoadrenalism and hyperthyroidism when faced with weight loss in an active young person without obvious cause. As medical specialists are increasingly incorporating a mental-state examination into their practice, and understanding the features of anorexia nervosa, they often make early identification of this disorder. Dentists may be the first to spot bulimia because of its effect of eroding dental enamel.

5. Differential Diagnosis

Table II notes the differential diagnosis that may be entertained in patients having anorexia nervosa. Serious consideration of these possibilities is usually not necessary when the patient has the classic psychopathological features of anorexia nervosa and when the crucial diagnostic features of these other entities are lacking.

6. Animal Models

There is no adequate animal model for anorexia nervosa even though a number have been presented. The syndrome of hypophagia secondary to ablation of the lateral hypothalamic nucleus was touted as a model for anorexia nervosa for a number of years. Stricker and Andersen (1980) examined this model and found that except for sharing the feature of weight decrease, these two disorders are fundamentally different. Since the psychopathological requirement for diagnosis is crucial, there can obviously be no adequate animal model. These authors noted global motivational deficits in animals with lesions in the lateral hypothalamic nucleus. In contrast, anorectic patients show only few motivational changes except those related to their psychopathology, unless starvation is extreme. Hypothalamic damage disrupts nonspecific contributions to feeding behavior, by damaging dopaminergic and other fibers passing through the ventral diencephalon. Voluntary behavior in general is disrupted, including feeding, but not limited to it.

Mrosovsky and Sherry (1980) have critically examined models of anorexia in animals. The unifying point in many of these anorexias, such as that occurring during hibernation or incubation, is a lowering of the "set-point" for body fat. Eating very little in the presence of adequate food also occurs during molting and during defense of the animals' territory, illustrating that anorexia can occur when feeding competes with another temporarily more important activity. These authors note that the human anorexia nervosa is nonadaptive, whereas the purpose of animal anorexias is to ensure the survival of the animal or the offspring in the long run. Rolls (1981) has recently reviewed the central nervous mechanisms related to feeding and appetite.

Bernstein (1982) has studied the physiological mechanisms related to cancer anorexia. Several factors contribute to cancer anorexia, including imprecisely understood effects of the tumor, response to medications, and learned aversion. Learned food aversion may occur by pairing foods with the discomfort of either the tumor or its treatment.

Bulimia also lacks a satisfactory animal model. Spontaneous vomiting in animals occurs at times in the setting of abundant food with subsequent en-

Table II. Disorders of Appetite, Eating, and Weight Change

I. Syndromes of weight loss
  A. Psychogenic causes
    1. Anorexia nervosa (primary anorexia)
    2. Secondary anorexia
      a. "Endogenous" depression
      b. Schizophrenia with paranoid delusions regarding food
      c. Some obsessional states, especially psychogenic dysphagia
      d. Globus "hystericus"
      e. Reactive depression and anxiety
      f. Amphetamine or other stimulant medications
  B. Medical causes: weight loss with *decreased* appetite
    1. Neurological
      a. Hypothalamic disorders, especially destruction of lateral nucleus
      b. Change in state of consciousness
      c. Diffuse brain disease (e.g., infections, malignancy)
      d. Dementia
    2. Endocrinological
      a. Addison's disease
    3. Gastrointestinal diseases
      a. Constrictive lesions
      b. Chronic diseases, including regional enteritis, and irritable bowel syndrome
    4. Chronic debilitating diseases
      a. Remote effect on appetite of malignancy anywhere in the body
      b. Metastatic cancer, especially to brain, liver, abdomen
      c. Tuberculosis and other chronic infections
      d. Chronic poisoning, especially heavy metals
      e. Chronic pain of any origin
      f. Chronic fever of any origin
      g. Nausea of any origin, including hyperemesis gravidarum
      h. Other debilitating disorders, including chronic cardiac and renal disease
    5. Iatrogenic causes
      a, Cancer treatment, including medication and radiation
      b. Stimulant medication for hyperactive children or narcolepsy
      c. Side effect of many medications such as propanolamine
      d. After surgery
  C. Medical causes: weight loss with *increased* appetite
    1. Hyperthyroidism (spontaneous or iatrogenic)
    2. Malabsorption syndromes
    3. Diabetes mellitus with proteinuria
II. Syndromes of weight gain
  A. Psychogenic causes
    1. Bulimia
    2. Reactive depression and anxiety
    3. Drug use: marijuana
    4. Iatrogenic causes
      a. Phenothiazines
      b. Tricyclic antidepressants
      c. Lithium
  B. Medical causes
    1. Neurological
      a. Destruction of ventromedial nucleus of hypothalamus
      b. Encephalitis lethargica
      c. Increased intraventricular pressure
    2. Genetic–congenital
      a. Prader–Willi syndrome

(Continued)

Table II. (Continued)

II. Syndromes of weight gain (Continued)
B. Medical causes (Continued)
b. Idiopathic adiposogenital dystrophy
3. Endocrinological
a. Hypothyroidism
b. Pituitary tumors, especially chromophobe adenoma
c. Cushing's syndrome
d. Insulinoma
4. Iatrogenic causes
a. Corticosteroids
b. Birth control pills
c. Cyproheptadine
d. Multiple small feeding regimens for ulcers
C. Multifactorial origin: ordinary obesity
III. Other syndromes of abnormal eating behavior with variable effect on weight
A. Pica
a. Deficiency syndromes
b. Temporal lobe lesions (Kluver–Bucy syndrome)
c. Cultural traditions (starch or clay eating)
d. Chronic schizophrenia with psychosocial dilapidation leading to ingestion of unusual food sources
B. Rumination
C. Anosmia and hypoguesia or dysguesia

gorgement, but the fundamental feature of fear of weight gain is lacking, and the use of large amounts of food to cope with emotional distress is not apparent.

7. Epidemiology of Anorexia Nervosa

Population studies of the eating disorders have led to a description of some common epidemiological factors, but these factors remain probabilistic in relation to any given individual. The most consistent finding has been an increase in the reported cases of anorexia nervosa. The increased incidence of anorexia nervosa is related certainly to increased accuracy of diagnosis and most likely also to an actual increased occurrence of the disorder. The highest estimate is 2.9% in Johannesburg schoolgirls (Ballot et al., 1981). Crisp et al. (1976) estimate that about 0.5% of precollege young women have severe anorexia nervosa. In contrast, few cases have been reported from Third World countries (Buhrich, 1981). The incidence of binge eating and vomiting is only now being studied as a result of recent increased awareness of the widespread nature of this disorder. Halmi et al. (1981) noted in a survey of 355 college students that 19% of the females and 5% of the males experienced all the major symptoms of bulimia as defined by the DSM III. While this figure is extraordinarily high and remains to be confirmed, its order of magnitude correlates with multiple anecdotal reports from college health services.

The reported sex distribution of anorexia nervosa varies from study to study, but all confirm that 85–95% of cases occur in females. Partly this is a definitional issue, since certain investigators define anorexia nervosa only in the presence of amenorrhea, thereby excluding males from typical primary cases. This arbitrary definition fails to take into account the essential similarity of psychopathology in both sexes. Gull (1874) noted male cases in 1874. The concordance rate among identical twins for anorexia nervosa is about 50%, but estimates vary (Hall, 1982). The lack of either 100% concordance or random population concordance suggests that multiple factors are involved, in addition to genetics. Cantwell *et al.* (1977) have noted increased occurrence of affective disorders in families of patients with anorexia nervosa, while Hudson *et al.* (1982) find similarities between bulimic and affective-disorder patients. Halmi *et al.* (1978b) have documented a "weight instability" in parents of anorectic patients.

All studies of anorexia nervosa that have addressed the issue of social class have noted the increased occurrence with increasing socioeconomic status. Multiple etiological implications have been drawn from this fact, but the reasons for the socioeconomic correlation have not been precisely understood. In the developed industrialized Western countries, there is a continuing dialectic among social classes, with increased status attributed to slimness in the upper and middle classes. In contrast, in developing countries where malnutrition is severe, increased social class is often marked by increased weight and slimness is not accorded special recognition.

A number of vulnerable subgroups within the population have been identified. Garner and Garfinkel (1980) identified professional dance and modeling students as being expecially prone to anorexia nervosa. Schwartz *et al.* (1982) postulated, along with other recent authors, that sociocultural forces and psychopathological symptoms interact to increase the probability of anorexia nervosa. Other subgroups with increased vulnerability for anorexia nervosa or bulimia include jockeys, gymnasts, wrestlers, and entertainers, all of whom have enormous pressures toward slimness as a professional requirement. Wrestlers in particular are required to lose 10–20 lb in many cases to reach an artificially low weight class, a feat that may be accomplished with the aid of self-induced vomiting.

8. Natural History of the Eating Disorders

The following is a composite natural history of anorexia nervosa. Typically, the patient is an adolescent young woman who weighs five to ten pounds above her peers at puberty. In temperament, she is perfectionistic and self-critical, manifesting some repugnance toward sexual issues (King, 1963). Her family is usually of middle- or upper-middle-class status and is often undergoing change or discord. A diet is usually the first event in the onset of the disorder, but weight loss from medical illness, a romantic disappointment, or other factors may initiate the weight loss. Not infrequently, a young woman begins along with several peers to achieve weight reduction. As weight is decreased, the desired body weight is set lower and lower. In many but certainly not all patients, perceptual distortion occurs. Patients may feel their bodies to be larger than they actually are and simultaneously set their nutritional needs to lower and lower caloric levels. There is increasing mental preoccupation with food and calories, with the individual often acquiring a precise but flawed view of the caloric content of foods. The focus in the anorectic's mind regarding nutrition is not on balance and appropriateness, but on avoidance of fatness. In the classic anorectic patient, the means of weight loss is limited to food restriction. A subgroup has historically been recognized that maintain their severely lowered weight by self-induced vomiting or by using medications, principally diuretics and laxatives.

There is anecdotal evidence that a larger number of young women go through a milder anorectic period and then presumably improve spontaneously (Button and Whitehouse, 1981). Numerous studies have cited the high percentage of adolescents in Western societies who believe they are overweight and would like to diet (Nylander, 1971). The long-term natural history of anorexia nervosa is imprecisely understood, but students of this disorder such as Theander (1982) estimate a lifetime increase in mortality, compared to the general population, of about 18%. When a patient with anorexia nervosa expires from her illness, it is usually due to starvation in the acute phase and from suicide or medical complications of starvation in the chronic phase.

The natural history of bulimia is even less well understood. Bulimia may occur as an acute phase of anorexia nervosa, as a chronic sequel to anorexia nervosa, or as a disorder that is never associated with lowered weight (Crisp, 1981–1982). Andersen (1983) has noted the high percentage of severe bulimics in a treated hospital population who initially were anorectics. The bulimic patient experiences a repetitive cycle of compulsion–overeating–guilt–vomiting–temporary relief–demoralization–abstinence–compulsion–overeating–etc. When death occurs acutely in a bulimic patient, it is usually due to cardiac arrhythmia associated with hypokalemia, whereas death in the chronic bulimic is often by suicide.

Anorexia nervosa and bulimia are probably best seen as disorders that share a spectrum of similar features with migration in both directions, but principally from anorexia nervosa to bulimia. Table III compares similar and contrasting features of anorexia nervosa and bulimia.

9. Behavior

Eating is one of the "motivated behaviors" and shares a number of features with related activities such as drinking, sexuality, and drug use. These behaviors may be helpfully viewed as having an appetitive (goal-seeking) and consummatory phase. It is important to keep the biological roots of eating behavior in mind because of the tendency to view anorexia in symbolic terms. In fact, much of the behavior of anorectic patients can be explained by an understanding of the common features of all motivated behaviors.

Table III. Comparison of Anorexia Nervosa (Food-Restricting) and Bulimia

Anorexia nervosa (food-restricting type)	Bulimia
Similar features	
Fear of fatness	
Pursuit of weight loss	
Fear of loss of control of eating	
Variable degree of distortion of body size	
Family history of affective disorders increased	
Contrasting features	
Food intake severely restricted	Control of intake lost and actual binges occur
Less vomiting, diuretic, or laxative abuse	Vomiting (self-induced), abuse of laxatives or diuretics or both
Younger	Older
More obsessional, perfectionistic features	More histrionic, antisocial features with loss of impulse control
Denies hunger	Experiences hunger
Severe weight loss	Less severe but variable weight loss
More introverted	Usually more extroverted
Eating behavior source of pride (ego-syntonic)	Eating behavior shameful (ego-alien)
Less sexually active	More sexually active
Amenorrhea or loss of sexual drive	Variable amenorrhea and change in sexual drive
Death from starvation acutely; chronically from starvation or suicide	Death from hypokalemia acutely; chronically from suicide
"Model" child	May have behavioral abnormalities

Many of the behaviors of classic food-restricting anorectic patients grow out of the swirling interaction of two drives in conflict: the body's relentless search for food resulting from starvation and the measures taken because of the fear of fatness from the ingestion of food. Other contributions to the behavior of anorectic patients come from the nonspecific influences of illness ("illness behavior" or "sick role behavior") and membership in adolescent peer groups.

Many of the behaviors resulting from starvation in anorexia nervosa parallel those described in Ancel Keys *et al.* (1950) in their classic studies on human starvation. Individuals who are starved relentlessly scan the environment in search of food. These individuals are physically active in both goal-directed and compulsive activity. They are often immersed in details of preparation of what food is available. Food in front of a starving person may be cut into tiny pieces with purposeful prolongation of mealtime. Sleep may be interrupted by hunger (Crisp *et al.*, 1970). Patients similarly may wake up from sleep many times and consume an apple or a vegetable and then return to sleep for a brief time, the nutrition being inadequate to provide more than a brief cessation from the drive to satisfy the starved state. The starved individual, however, will eat normally when presented with food, but the anorectic patient, in contrast, continues to restrict food in the presence of adequate or abundant nutrition. At a dining table attractively set for the entire family, a patient will consume very small amounts. Interestingly, there is usually an adequate amount of protein eaten and often a vitamin pill is taken daily, preventing the classic manifestation of either vitamin deficiencies or protein malnutrition. Patients will often present their elaborately prepared foods with élan, encouraging others to eat heartily of their products, which are high in lipids and sugars, while themselves abstaining.

While starvation is the source of much of the "foraging" behavior, the other source of excess activity in these patients is strenuous exercise in the pursuit of slimness and compensation for any calories ingested. The overall activity of the patient tends to increase with decreasing weight until a point of physical exhaustion is reached, at which point there is a rapid decrease in activity, with a final, listless state resulting. While these behaviors are typical, they are not diagnostic requirements.

The social behavior of these patients is manifested by increased seclusiveness from the family and from friends. Social patterns become more immature, with childlike dependence on parents. Meals are often taken alone rather than in groups. There is much gazing in the mirror at one's starved habitus with frequent weighings. These descriptions are a composite of many patients and in any given individual may not be manifested. They are certainly not pathognomic.

In contrast, the behavior of bulimic patients is characterized by the rapid ingestion of large amounts of food followed by vomiting or the use of laxatives or diuretics. The natural history of bulimic patients may begin with the food-restricting anorectic state or may start at normal or excessive weight levels. There is some controversy over whether bulimic patients share the central psychopathological feature of anorexia nervosa, the fear of fatness. The amounts ingested can be truly enormous. The most common context for a binge is when the individual is alone, with food available, in the parents' home or a private room elsewhere, after work or school, in the evenings. To induce vomiting, patients may insert their fingers into their throat, leaving scars on the knuckles (Russell, 1979), or induce a Valsalva maneuver with or without pressure on the abdomen.

The quantity of laxatives consumed may be heroic, up to a hundred or more tablets a day of the common preparations. Diuretics are ingested in sufficient amounts to reduce potassium to less than 2 meq/liter in some patients. The social life of a patient with bulimia is carefully orchestrated to ensure exits after meals. Many patients with bulimia hold demanding jobs; the time and effort required to fit in eight to ten binges and purgations a day become feats of ingenuity, since their symptoms are often unknown to peers or families.

As noted previously, patients with anorexia nervosa are only superficially similar to animals having lateral hypothalamic lesions. Patients with bulimia similarly are only tangentially like those with ventromedial hypothalamic lesions producing hypothalamic hyperphagia.

A silent camera viewing the behaviors of anorectic patients would show scenes worthy of Greek drama: an individual who is pitifully thin is surrounded by food and entreated by the family to eat, but responds by pushing away the plate of food and clenching the jaws. Bulimic patients viewed from the perspective of an observing camera would be seen to have an intricate social pattern, allowing for the intermittent, rapid ingestion of large amounts of food followed by secretive disposal. Both groups of patients may frequently consult cookbooks and fashion magazines, although these behaviors also have a high rate of occurrence in the general population.

10. Biology of Anorexia Nervosa and Bulimia

The biological consequences of anorexia nervosa reflect primarily the degree of starvation present and, to a lesser extent, the method chosen to induce weight loss, and secondary medical complications. Although early cessation of menstrual periods before weight loss has occurred is sometimes offered as a unique biological marker of anorexia nervosa, there is evidence supporting the hypothesis that biological changes occur in proportion to the degree of starvation (Vigersky, 1977). Most of the biological changes resulting from self-induced starvation can be subsumed under the phrase, a "protective, hypometabolic" effect (Boyar and Bradlow, 1977).

Virtually every aspect of bodily function is affected in anorexia nervosa. The general systemic effects of starvation include a loss of muscle mass and diminished subcutaneous lipid layers, as well as decreased lipid around organs. In some patients, there is pronounced activity despite cachexia, but eventually a state of inanition and weakness supervenes. Vital signs are decreased. Bradycardia is widely reported and occurs for several reasons, including (1) lowered circulating norepinephrine from the starved state (Young and Landsberg, 1977), (2) a decreased triiodothyronine (Cahill, 1981), and (3) the effect of vigorous exercise as seen in trained athletes.

Luck and Wakeling (1980) have noted a consistently low core temperature with altered thresholds for thermoregulatory sweating and vasodilatation in patients with anorexia nervosa. Myers *et al* (1981) found anemia, leukopenia, and some thrombocytopenia in a substantial percentage of patients. Despite these findings, Golla *et al*. (1981) reported that anorectic patients have fewer infections than comparably starved controls. Other patients with common forms of protein and calorie malnutrition demonstrated depressed cell-mediated immunity. Anorectic patients had normal T-lymphocyte populations, unimpaired proliferative lymphocyte responsiveness, and, in general, a relatively intact cell-mediated immune system.

Luton *et al*. (1981) described "atrophie cérébrale réversible," a report confirmed by Sein *et al*. (1981). The EEG, where changed at all, is altered slightly without lateralizing or focal features.

Gastric emptying may be abnormal. Holt *et al*. (1981) compared 10 female patients having anorexia nervosa with 12 healthy volunteers and found a slower rate of emptying.

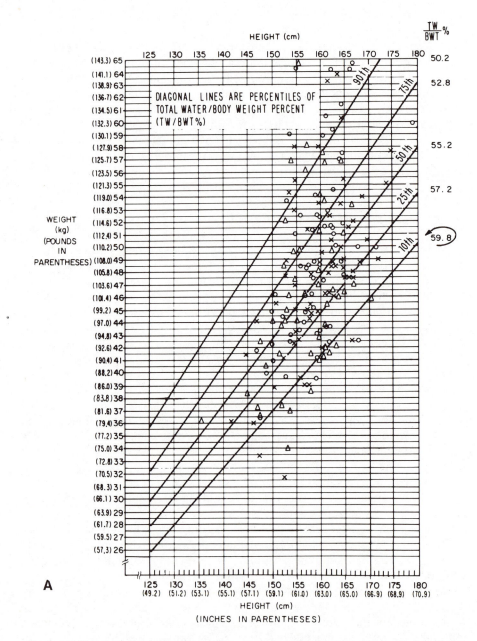

Fig. 1. (A) The minimal weight necessary for a particular height for onset of menstrual cycles is indicated on the weight scale by the 10th percentile diagonal line of total water/body weight percent, 59.8%, as it crosses the vertical height lines. Height growth of girls must be completed, or approaching completion. For example, a 15-year-old girl whose completed height is 160 cm (63 inches) should weigh at least 41.4 kg (91 lb) before menstrual cycles can be expected to start. Symbols are the height and weight at menarche of each of the 181 girls of the Berkeley Guidance Study (○), Child Research Council Study (×), and Harvard School of Public Health Study (△) (Frisch and McArthur 1974). (B) The minimal weight necessary for a particular height for restoration of menstrual cycles is indicated on the weight scale by the 10th percentile diagonal line of total water/body weight percent, 56.1%, as it crosses the vertical height line. For example, a 20-year-

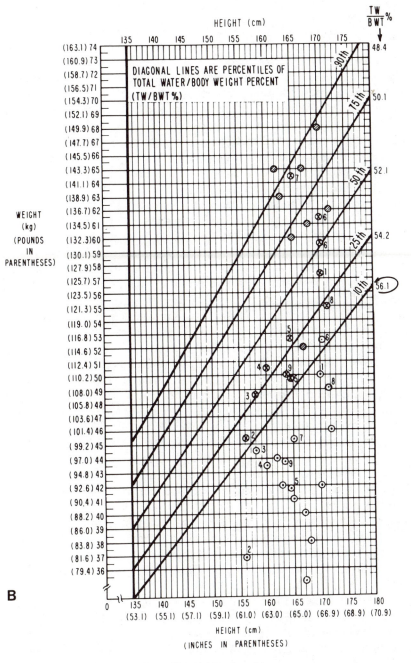

Fig. 1. (*Continued.*)

old woman whose height is 160 cm should weigh at least 46.3 kg (102 lb) before menstrual cycles would be expected to resume. (⊙1⁻⁹) Weights while amenorrheic of patients of; (⊗1⁻⁹) weights at resumption of regular cycles. When two weights are given for a patient, the lower weight is at the first resumed cycle. (⊖) Weights before occurrence of amenorrhea of subjects cited by Lundberg *et al.* (1972); ⊙ weights while amenorrheic.

Andrews (1982) reported on dental erosion due to anorexia nervosa with bulimia, while Walsh *et al.* (1981–1982) described swollen parotid glands from bulimia, a finding widely confirmed.

The most energetic work on biological changes in anorexia nervosa has focused on the endocrine system, especially reproductive endocrinology. Clinicians have been troubled by the lack of return of menses in patients who have otherwise made a good recovery. Russell (1977b) argues the possibility of enduring hypothalamic dysfunction in these patients. Katz and Weiner (1981) summarize well the aberrant reproductive endocrinology in these patients: low plasma and urinary levels of gonadotropins and estrogens, absence of monthly cycling, immature, age-inappropriate circadian patterns of luteinizing hormone (LH) secretion, and deficient LH feedback responses to clomiphene citrate and ethinyl estradiol. They suggest four major reasons for endocrinological abnormalities in patients in addition to low weight: (1) inadequate nutrition, (2) psychological abnormalities, (3) effect of other symptoms of the disorder such as insomnia or excess physical activity, and (4) possible biological predisposition. Beumont *et al.* (1978b) and Palmer *et al.* (1975) have both noted decreased sensitivity to LH-releasing hormone (LHRH) in the starved stages of anorexia nervosa. Frisch and McArthur (1974) have contributed to the understanding of the relationship between body weight and the onset or restoration of menses. Their helpful nomograms are reproduced in Fig. 1.

Male patients have been considered in some studies either to have atypical forms of anorexia nervosa or not to meet strict criteria. This approach appears inadequate, and there clearly are male patients who have all the essential features of anorexia nervosa. Since amenorrhea cannot be used as a diagnostic criterion for these patients, there has been interest in elucidating the reproductive endocrinology of males with anorexia nervosa. Andersen *et al.* (1982) and Crisp *et al.* (1982) documented decreased testosterone in the starved state of anorexia nervosa, with improvement proportional to weight gain. Decreased testosterone in male anorectics correlates with their decreased sexual interest and potency during illness.

Figure 2 demonstrates the changes in plasma testosterone resulting from increased weight in male patients receiving treatment.

The thyroid has several responses to starvation, including: (1) decrease in T_4 to low-normal levels, (2) subclinical T_3, and (3) an associated increase in an ineffectively iodinated molecule, "reverse T_3."

Table IV summarizes the pituitary–end organ functioning in 32 anorectic patients. Most later studies have duplicated or extended details of these findings. Growth hormone is elevated, a presumed teleological response to the starved state, but the exact mechanism behind this elevation is unclear. Basal prolactin is normal. Vigersky *et al.* (1977) describe the hypothalamic abnormalities of secondary amenorrhea associated with simple weight loss.

Gerner and Gwirtsman (1981) found abnormalities in the dexamethasone suppression test (DST) in anorectic patients. They do not suppress adrenal activity after a standard test dose of dexamethasone to the same extent that normal controls do. There are several interpretations of this fact, including the

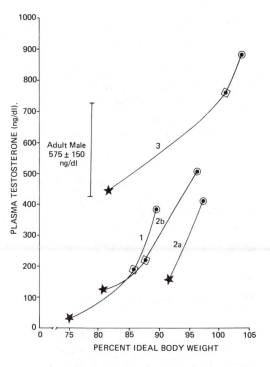

Fig. 2. Changes in plasma testosterone levels in three male patients with anorexia nervosa before, during, and after treatment. (2b) Repeat admission after recurrence of illness. From Andersen *et al.* (1982).

possibility that these patients have depressive illness, a disorder known to be associated frequently with an abnormal DST. It is also possible, however, that starvation affects adrenal functioning in a variety of ways and the abnormal DST is nonspecific rather than related to affective disorder. Hudson *et al.* (1982) found similarities between depressed patients in family history and response

Table IV. Summary of Pituitary–End Organ Function[a]

Measure[b]	Normal[c]	Anorexia nervosa[c]
Basal GH (ng/ml)	10.4 ± 12.2	48.1 ± 80.7[d]
Basal prolactin (ng/ml)	16.2 ± 5.5	19.4 ± 14.4
Max. prolactin p̄ TRH (ng/ml)	82.0 ± 47.0	62.0 ± 37.0
Basal fT$_4$ (ng/ml)	1.5 ± 0.2	1.5 ± 0.3
Basal T$_3$ (ng/100 ml)	156 ± 30	78 ± 34[e]
TSH increment p̄ TRH (μU/ml)	8.6 ± 2.0	9.0 ± 4.0
Diurnal cortisol (μg/ml) 8am/5pm	—	23.0 ± 6.6/16.0 ± 6.0
Basal LH (mIU/ml)	7.9 ± 4.4	3.9 ± 2.5[d]
Basal FSH (mIU/ml)	15.3 ± 2.7	5.6 ± 5.2[d]
LH response to LHRH (mIU/ml/min)	2814 ± 386	2728 ± 2554
FSH response to LHRH (mIU/ml/min)	3143 ± 304	2432 ± 1811

[a] From Vigersky *et al.* (1976).
[b] (GH) Growth hormone; (TRH) thyrotropin-releasing hormone; (T$_4$) 1-thyronine; (T$_3$) triiodothyronine; (TSH) thyroid-stimulating hormone; (LH) luteinizing hormone; (FSH) follicle-stimulating hormone; (LHRH) LH-releasing hormone.
[c] Values are means ±1 S.D.
[d] $P < 0.05$.
[e] $P < 0.001$.

to the DST with ten bulimic patients and suggested that bulimia may be a *forme fruste* of affective disorder. Similar results were reported in anorexia nervosa (Gwirtsman and Gerner, 1981).

Kaye *et al.* (1982) found increased levels of cerebrospinal fluid (CSF) opioid activity in severely underweight anorectic patients. They suggest that increased CSF opioid activity may be a compensatory response to weight loss or may be etiologically associated with anorexia nervosa. A variety of suggestions have been made regarding possible alterations in catecholamine metabolism in anorexia nervosa, but results have been variable (Halmi *et al.*, 1978a; Van Loon, 1980; Abraham *et al.*, 1981).

Figure 3 illustrates some of the medical signs of anorexia nervosa, while Fig. 4 notes the numerous effects of bulimia on the gastrointestinal tract. Since the syndrome of bulimia has only recently been described and accepted as a separate psychiatric entity, many studies are now in progress regarding the physiological effects of self-induced vomiting and of laxative or diuretic abuse.

11. Psychological Aspects of Anorexia Nervosa

The challenge in psychological studies on anorexia nervosa is to distinguish empirical data from theoretical concepts. There is a dynamic interaction between the preillness psychological vulnerabilities and the resulting eating disorder, and then in a circular interaction, there are psychological consequences of the eating disorders.

11.1. Preillness Characteristics

Morton, Gull, and Lasègue all described preillness characteristics of patients who later developed anorexia nervosa. The picture has not changed substantially since that time. The prototype patient is a sensitive, perfectionistic, somewhat emotionally immature young woman who enters puberty with ambivalence or repugnance toward sexuality and a sense of ineffectiveness toward the maturational demands appropriate to her age. Bruch (1981) describes feelings of ineffectiveness and a lack of ability to identify internal feelings as fundamental precursors of anorexia nervosa. King (1963) described one of the first empirical contrasts between anorectic patients and those suffering from weight loss from other causes. He noted consistent differences between these groups, especially the presence of perfectionism, athleticism, and sex disgust in the primary anorectic group. Smart *et al.* have described personality characteristics of patients with anorexia nervosa, finding them to be neurotic, introverted, of average intelligence, but with obsessional features.

One of the major problems in psychodynamic psychiatry is the issue of symptom choice. Many more young people in our society have the preillness characteristics of anorectic patients than actually develop it. Which ones eventually go on to develop the illness is at present a probabilistic prediction from

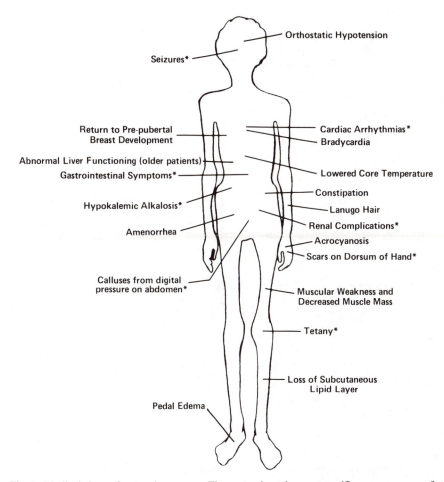

Fig. 3. Medical signs of anorexia nervosa. The systemic and organ-specific consequences of star-vation are noted. (∗) Problems especially prominent in bulimic patients.

large population studies. It is impossible to predict whether a given individual will develop the disease.

11.2. Precipitating Factors

The most common precipitating factor for anorexia nervosa is the decision to go on a diet. This decision, however, has its own antecedents; it usually grows out of a sense that one's current weight is excessive and that fundamental problems such as a sense of ineffectiveness or low self-esteem will be improved with weight loss. There is often competition with peers for who can lose weight

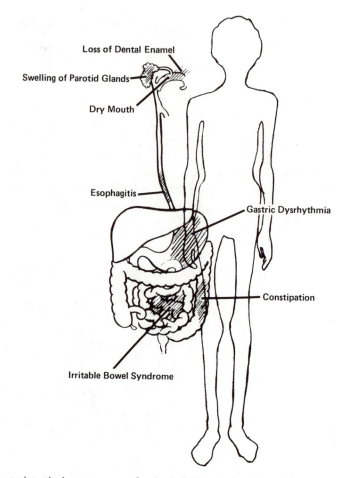

Fig. 4. Gastrointestinal consequences for the bulimic subtype of anorexia nervosa. The gastrointestinal tract is affected at all levels by the bulimic syndrome of self-induced vomiting and diuretic/laxative abuse.

most quickly. Occasionally, the diet is prompted by a comment from a family member either that the patient is becoming overweight or that the patient is showing signs of sexual development. Occasionally, the illness begins in the setting of a romantic disappointment such as terminating a dating situation or breaking an engagement. Interestingly, it may begin with illness that causes weight loss, and only after weight has been lowered do the symptoms of pursuit of thinness and fear of fatness emerge. Crisp (1980), in his book *Anorexia Nervosa: Let Me Be*, develops a plausible interweaving of factual data and psychodynamic hypothesis to explain the predisposition and onset, as well as the sustaining features, of anorexia nervosa.

Beumont *et al.* (1978a) identified a number of experiential and psychological factors related to the psychogenesis of anorexia nervosa, including dependence–independence struggles, sexual challenges, concerns about obesity, and other nonspecific stresses. Beumont *et al.* (1981) later studied the psychosexual histories of adolescents and young women with anorexia nervosa and found that sexual challenges were reported to be a major precipitating factor, but the effect of illness on actual sexual behavior was variable.

The preillness and precipitating factors for bulimia are less well understood. When bulimia results from inadequately treated anorexia nervosa, then the preillness and precipitating characteristics are similar to those of anorexia nervosa. The onset of bulimia is usually dated precisely by patients and related to either feelings of intense hunger at lower than normal weight or feelings of emotional distress that are relieved by gorging. The binge is followed by guilt, feelings of loss of control, and a fear of fatness, leading to panicked attempts to avoid the caloric consequences of the food. Bruch (1978), in her more recent book, *The Golden Cage*, also presents an integrated view of the preconditions and onset of anorexia nervosa, including its bulimic variant.

11.3. Psychological Characteristics of Acute Illness

The psychological aspects of the severely starved anorectic patient can be divided into those that are related to anorexia nervosa itself and those consequent to the starvation process. Since anorexia nervosa is a syndrome (collection of symptoms) rather than a scientifically understood disorder, even its fundamental psychopathological characteristics remain in some dispute. A variety of central psychopathological features have been described, including "fear of fatness" by Russell, "pursuit of thinness" by Bruch, and "weight phobia" by Crisp. In many ways, these features are all different facets of the same prism. What is agreed in general is that anorexia nervosa is a separate entity rather than a *forme fruste* of other disorders.

Reviewing charts of patients treated 20 years ago often reveals a diagnosis of schizophrenia was made in patients having classic features of anorexia nervosa. More accurate diagnosis may be one reason for the increasing incidence of anorexia nervosa. Beumont (1970) has stated that anorexia nervosa "breeds true." This means that the basic form of the anorectic illness remains constant over time, even though its severity may vary. When patients improve and then relapse, they have the same clinical picture as when they first became ill, rather than the appearance of some underlying other disorder.

Another intriguing psychopathological feature of anorexia nervosa that is frequently but not universally found is perceptual distortion. This distortion of one's body size in the direction of overestimation is not found in all patients, nor is it unique to anorexia nervosa. Nonetheless, it alters the perception of the anorectic's body so substantially that it is measurable by several empirical techniques. Recently, however, there has been a critical reappraisal of the meaning of this symptom, and the more recent consensus is that it is a widespread finding in nonanorectic individuals as well (Button *et al.*, 1977; Halmi *et al.*, 1977; Garner and Garfinkel, 1981–1982; Hsu, 1982).

Where there is a marked distortion of body image, Garfinkel *et al.* (1979) have noted it to be a stable finding. Some authors feel that its presence indicates a poor prognosis, but findings vary concerning these data. Druss and Silverman (1979) found that ballerinas were especially prone to feel they were overweight, even when they were very thin.

There are many nonspecific psychological changes in anorexia nervosa, one of the most prominent being depression. Eckert *et al.* (1982) studied depression in anorectic patients and found them to be less depressed than depressed neurotics and about equally depressed as anxious neurotics. Weight gain was correlated with a decrease in depression, as several other investigators have noted.

The preoccupation with food, the general apathy, and the anhedonia are relatively nonspecific and associated with starved states. There is some variation in the psychological features of the acute illness, according to the personality of the patient—whether it is predominantly obsessional or self-dramatizing.

Depression is reported by many patients with anorexia nervosa, but it must be remembered that depressive symptoms are like nonspecific medical symptoms such as fever: diagnosis must be made in association with more specific symptoms. Few patients with anorexia nervosa meet strict criteria for diagnosis of affective disorder; rather, their depression often has an apathetic, anhedonic quality and improves considerably with weight gain alone. Sours (1981) discusses depression in anorexia nervosa from a more psychoanalytic viewpoint. Because of the increased incidence of depression in the family history of anorectic patients, some clinicians have advocated treatment with antidepressants for all patients, but this practice is not widely accepted.

Studies on the psychological aspects of women with bulimia were described by Allerdissen *et al.* (1981). They found that patients with bulimia nervosa had an external locus of control, experienced less sexual pleasure than a control group, felt they would be happier if they were thinner, and were more depressed than controls. They tended to blame themselves, however, rather than others. The most common general feature of patients who are acutely ill with bulimia is that in addition to any effects of weight loss, they feel guilty and demoralized, having experienced actual, rather than feared, loss of control regarding their eating. Pyle *et al.* (1981) noted that bulimic patients shared with anorectics the preoccupation with food and an exaggerated fear of becoming obese. Johnson and Larson (1982) confirmed with "Experience Sample Methods" that bulimic patients experience more dysphoric and fluctuating moods than normal controls.

A completely different view of the "eating disorders" is taken by Rau and Green (1975), who feel that compulsive eating has a neuropsychological basis. They theorize that there is an epileptiform subcortical cerebral dysrhythmia that is primarily responsible for the compulsive eating of bulimia.

11.4. Chronic Illness

Patients with chronic anorexia nervosa have been much better studied than chronic bulimics. Chronic anorectics shows an increased mortality compared to controls, which Theander (1982) noted to be about 18%. They are much more likely than controls to commit suicide or to die of the effects of starvation. Patients who become ill during their adolescence exhibit an arrest of psychological and social maturation, leaving them relatively immature compared to nonafflicted individuals of the same age. Life is seldom joyful for chronically ill anorectics, and the pervasive need to control weight and to be attentive to details of food and calories rules their lives. Few of them are happily married with children.

11.5. Outcome

There is much controversy regarding the outcome of anorexia nervosa, since its exact incidence is not well known. Almost all surveys represent selection and methodological biases of various kinds. There is probably a substantial number of patients who never complain of their illness and live a relatively normal, although underweight, life-span. The most consistent reports of outcome derive from large clinical study programs that specialize in treating referred patients with anorexia nervosa and bulimia. The best outcome occurs in younger patients with good preillness personality and stable family background who have lost less weight than the more severely ill and who receive prompt treatment. The classic food restricters appear to do somewhat better than the bulimic patients. Halmi *et al.* (1979) correlated weight gain and outcome with pretreatment variables.

11.6. Psychological Testing in the Eating Disorders

There is great need for empirical psychological studies of patients with anorexia nervosa and bulimia. One of the first scales to quantify anorectic behavior was devised by Slade (1973), and a more specific "Eating Attitudes Test" that is valid and reliable has been introduced by Garner and Garfinkel (1979). Nonspecific psychological changes in anorectic patients have been measured by Hsu and Crisp (1980), who correlated their acute findings with long-term outcome. Ben-Tovim *et al.* (1979) compared anorectic patients with neurotic and depressive patients and found depressive and obsessional symptoms to be common in the anorectics. Wilbur and Colligan (1981), using the Minnesota Multiphasic Personality Inventory (MMPI) to develop a profile of the anorectic patient, noted that anorectic patients were in more turmoil and psychological distress than medical-patient peers. Strober (1981) has used the MMPI to investigate personality and pathogenetic features of anorexia nervosa.

12. Treatment of Anorexia Nervosa

During the last several years, there has been a clear trend toward a multidisciplinary or combined treatment of anorexia nervosa. Most large studies report a variety of integrated procedures, usually beginning with nutritional rehabilitation and proceeding to some kind of counseling or psychological work with the individual and, to a variable degree, with the family or in a group therapy setting. Russell and Crisp have reported in the past on integrated programs, and these, while much more specific and comprehensive, derive from Gull, who advocated a sequence of increased weight followed by "moral treatment," an older name for psychotherapy. Since Russell and Crisp have had decades of experience, their studies will be discussed in the context of long-term follow-up.

The two most generally successful approaches toward the initial nutritional rehabilitation have been milieu therapy with supportive nursing care and a strict behavioral therapy approach. Hedblom *et al.* (1981) have described in detail a stagewise approach toward treatment in which each stage of treatment has specific functions assigned to every member of the treatment team. The stages of treatment are: (1) nutritional rehabilitation, (2) individualized psychotherapy, (3) maintenance, and (4) follow-up. Much of the success of a treatment program depends on the ability of the nursing staff to encourage safe weight gain and the development of lifelong patterns of normal eating. Close cooperation with a nutritionist is essential. In the early stages of treatment of starved patients, it is unwise to let the patient choose the amount or kind of food; rather, food is presented as "medicine" with much psychological support and encouragement given to eat. Only 1 patient of 140 treated over 6 years in the author's program (Hedblom *et al.*, 1981) required nasogastric feeding. The following basic information was given to each patient on admission, with appropriate variation made according to the age and degree of psychological and medical sophistication of the patient:

> You have an illness which has changed your perception about your body's size and about your body's need for nutrition. We are going to ask you to trust us to prescribe food like medicine for you. This is difficult because you are afraid that you may lose control and eat too much and become overweight. Our goal is never to make anyone overweight and we won't let you do this. But we will ask you to accept our judgment about your nutritional needs and to eat everything that we prescribe for you. Since this is a difficult experience and you will probably be anxious during part of the program, we will have nurses with you much of the time. This is in order to give you the necessary support and encouragement.
>
> We do not know where this illness comes from. It seems to have many factors including some vulnerabilities of personality, such as perfectionism, some family stresses, perhaps some biochemical changes in your body, and usually going on a diet. Whatever the cause, we don't consider anybody to blame. We think people are often blamed for this problem, but we don't feel parents are to blame or patients are to blame.
>
> We would like you to be honest with us and trust us. We know that there is a temptation to dispose of your food or to vomit or to avoid the effects of eating, but we want to concentrate not on gaining weight but on losing starvation. We think that your hands and feet will be less cold, you will have more energy, less restlessness, and clearer thinking when you go through this program. Our goal is to make you happier and healthier. You may experience some discomfort as you go through the nutritional rehabilitation, but we will try to make you as comfortable as possible.

Figure 5 summarizes the dietary analysis of three patients with classic food-restricting anorexia nervosa whose diet was studied by Suelita Delara,

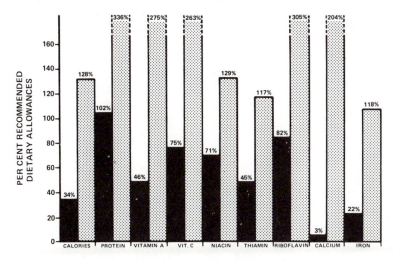

Fig. 5. Adequacy of diet before and after treatment of three food-restricting anorectic patients. Before treatment, total calories are markedly lowered, as are calcium and iron. (■) Pretreatment values; (▨) values obtained at the end of nutritional rehabilitation. Data analysis and figure by Suelita Delara, M.S., R.D., nutritionist, Johns Hopkins Hospital.

M.S., R.D., nutritionist at Johns Hopkins Hospital, before and after treatment. Patients were severely lowered in calories, calcium, and iron before treatment, with moderate decrease in some vitamins. At the end of their nutritional rehabilitation, they were receiving abundant supplies.

Pertschuk *et al* (1981) reported a very high morbidity in 11 anorectic patients receiving total parenteral nutrition (TPN). Of these 11 patients, 1 died, 1 had a pneumothorax, and 7 had significant but less severe side effects. It has been the experience of Andersen and colleagues that TPN is unnecessary even with severely starved patients. Even with the use of moderate feedings of regular food, patients may experience medical problems during nutritional rehabilitation. These are schematically indicated in Fig. 6.

There is considerable controversy in the literature concerning both theoretical and practical aspects of therapy. Bruch (1974) is strongly opposed to behavior modification as a mainstay of treatment for a disorder that she sees as being psychodynamic in origin. Agras *et al.* (1974) have described methods of behavioral therapy, as have Halmi *et al.* (1975), Garfinkel and Garner (1977), and others.

A number of monosymptomatic treatment approaches have been tried with varying degrees of success. Amitriptyline has been utilized by Moore (1977) and by Needleman and Waber (1976) with optimistic results but lack of strict controls. Cyproheptadine has been noted by Vigersky and Loriaux (1977) to show a trend, below statistical significance, toward weight increase. Cyproheptadine has also been used in a controlled trial by Goldberg *et al.* (1979), who found it effective in a subgroup of anorectics. Previous use of chlorpromazine for both weight gain and sedation is decreasing.

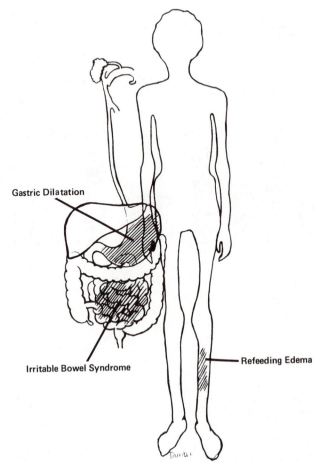

Fig. 6. Medical problems associated with nutritional rehabilitation in patients with anorexia nervosa. The most serious problem is gastric dilatation, but refeeding edema and irritable bowel syndrome are more common and respond to conservative measures.

Treatment centers that have a large experience consistently see numerous unsuccessful approaches in their referral populations. It is not uncommon to receive patients who have experienced bruised jaws from attempts to forcibly feed them, gastric disorders or occasionally gastric ruptures, multiple electro-convulsive therapy (ECT), high doses of phenothiazines, and a variety of other unsuccessful therapeutic modalities.

Some recent studies of interest are the following. Becker *et al.* (1981) treated 38 patients with inpatient psychoanalytic psychotherapy. Treatment lasted about 3 months, and a picture of the good- and poor-outcome patients was developed. The poorest outcome was associated with older onset, longer illness, delay before treatment, lower weight at time of admission, and to some extent poor insight and alcohol abuse. Fifty-six percent had a good to satisfactory outcome. Dally (1981) in Great Britain, Ploog *et al.* (1981) in Germany,

Pierloot *et al.* (1982) in Belgium, Porot (1981) in Paris, and Bassoe and Eskeland (1982) in Norway describe roughly similar programs. In the United States, Halmi (1982), Hedblom *et al.* (1981), and Lucas *et al.* (1976) have developed approaches that integrate nutritional and psychological or behavioral modes of treatment. Garner *et al.* (1981–1982) have described detailed approaches to the psychological and behavioral (Garner and Bemis, 1982) aspects of treatment.

Longer-term studies of outcome are gradually being reported. Theander (1982) has noted a lifetime increased mortality compared to controls of about 18% in anorectic patients. Hsu (1980) critically reviewed the outcome results of anorexia nervosa over the last 25 years, noting the multiple methodological difficulties and errors in many studies. Bemis (1978) reviewed a wide variety of treatment approaches, noting problems inherent in outcome studies. There is general agreement that multiple measures of outcome must be used, since the disorder affects psychological, biological, and social aspects of the individual. Also, there is a wide variety of patients having these disorders, and treatment programs focusing on adolescents, especially younger adolescents exclusively, have a much better outcome independent of the method of treatment because of the relatively good prognosis for this age group. Some of the methodological errors noted by Hsu include the following: (1) lack of rigorous definition of the syndrome, (2) failure to use a direct method of follow-up, (3) failure to trace patients, (4) short duration of follow-up, and (5) inadequate outcome criteria and incomplete information at follow-up. Garfinkel *et al.* (1973) have described an operant conditioning technique for anorexia nervosa. Halmi (1977) has also described a behavioral therapy approach.

Morgan and Russell (1975) noted the following features indicative of poor outcome: late age of onset, long duration of illness, previous admissions to hospital, disturbed relationship with family, and premorbid personality difficulties. Their study reflects long clinical experience.

Hsu *et al.* (1979) followed 100 females for up to 8 years after first presentation. Outcome was good in 48, intermediate in 30, and poor in 20 (2 had died). The poor-outcome features were longer duration of illness, older age of onset, lower weight during illness, presence of bulimia, anxiety when eating with others, poor childhood socialization, and poor parental relationship. Korkina and Marilov (1980) noted the favorable course and prognosis of 17 girls with prepubertal anorexia nervosa having hysterical features. This report bears on the evaluation of authors such as Minuchin *et al.* (1978), whose favorable outcome appears to be partially due to the good prognosis expected of early adolescent and preadolescent patients.

A generally useful maxim is to think of a good outcome of anorexia nervosa as being inversely proportional to some additive or multiplicative equation of premorbid personality disorder, family psychopathology, and chronicity of illness. Death from acute anorexia nervosa with vigorous treatment emphasizing early nutritional rehabilitation has decreased from previous reports of 10–15% to around 1–2% or less.

Treatment approaches to bulimia, similarly, have been either monosymptomatic or integrated. There are enthusiastic unsupported reports that bulimia will respond to antiseizure medications (Green and Rau, 1974; Wermuth *et al.*, 1977). These studies ignore a number of features such as the transition from anorexia nervosa to bulimia and the large percentage of individuals of college age having bulimia, a figure far beyond that of known seizure disorders.

Long and Cordle (1982) describe a sequence of behavioral, dietary, cognitive, and resocialization methods that lead to increased weight and decreased bingeing and vomiting. Fairburn (1981) described essentially the same combination of cognitive psychotherapy and behavioral strategies toward this disorder. Since the natural history of bulimia is largely unknown and controls are usually lacking, these studies need to be evaluated over a longer perspective.

It is the author's experience that the following are helpful components of adequate treatment programs:

1. A rigorous approach to diagnosis with clearly stated criteria. Rollins and Piazza (1978) have critically reappraised the multiple approaches to diagnosis and noted within the same group of 30 successive patients a variation from 23% positive diagnosis to 90% positive diagnosis for anorexia nervosa, according to which diagnostic criteria have been used.

2. An organized approach toward nutritional rehabilitation and normalization of the eating pattern. It is clear that TPN and nasogastric feedings are generally not necessary for most patients. Drugs given to stimulate appetite are based on a fundamental error that anorexia nervosa is associated with decreased appetite. Except in the extreme stages of cachexia, patients with anorexia nervosa are fearful of fatness or in pursuit of thinness, or both, rather than lacking in appetite. The first rule of medicine is, "Above all, do no harm" (*Primum non nocere*). Severe scoldings, the attribution of blame, forced feeding against clenched jaws, multiple ECT, or high doses of medication are inappropriate. There is much controversy as to whether a selective leukotomy is ever acceptable. Russell (1977a) has described a successful approach to nutritional rehabilitation and general principles of treatment.

3. It is necessary to have an overall strategy for the treatment of the psychological distress and social disabilities of these patients. Some symptoms, such as depression, usually decrease steadily as weight increases, without any direct treatment of the affective symptoms, but once this has been accomplished, some form of counseling or psychotherapy is necessary for the underlying conflicts and psychological distress. Here the theories exceed the facts, and one's basic beliefs regarding the nature of this disorder lead to the particular type of treatment chosen. The concepts of Bruch (1978) and Crisp (1980) that anorexia nervosa is a coping response to the maturational crises of early adolescence in an individual who is psychologically unprepared, existing in a family that collaborates by lack of adequate preparation, are general enough, yet helpful enough, to devise programs of psychological treatment that improve understanding of the patient's vulnerabilities, increase self-esteem, and help the person to deal more directly with the Ericksonian stages of separation and individuation. These concepts are less useful with patients of older onset, although with some retrospective maneuvering, they can be fitted into

the same theory. It is probably better to accept the fact that no single theory of the origin of anorexia nervosa is adequate.

4. A long-term view of the disorder is necessary. Where inpatient therapy is indicated, it is only part of a long-term strategy usually spanning several years.

5. The family must be integrally involved in treatment. Approaches to aspects of family therapy are described by Kalucy *et al.* (1977), Minuchin *et al.* (1978), and Yager (1982).

13. Summary

Anorexia nervosa and its companion disorder bulimia are syndromes involving abnormalities of weight and eating that remain basically obscure regarding definite origins. The clinical picture of anorexia nervosa has been accurately described for several centuries, and diagnostic criteria exist that are valid and reliable concerning identification of patients having this disorder. Bulimia has been more recently described as a separate psychiatric disorder. There is probably a large undisclosed population with mild versions of both these disorders who never come to medical or psychological attention because of spontaneous remission, the mildness of the disorder, or lack of motivation to change.

The essential feature of anorexia nervosa is self-induced starvation in an individual having a fear of fatness and a loss of periods or a loss of sexual interest. Predisposing factors include social class, pubertal status, perfectionistic and self-critical qualities of temperament, a variety of abnormalities in family functioning, and an event to initiate weight loss, often dieting but occasionally other causes. Whatever the constellation of causes that launch the disorder, it has a life of its own once it is firmly established.

The best outcome occurs in individuals with early onset, rapid diagnosis and treatment, with relatively better preillness personality and family functioning. The biological effects of anorexia are largely those of starvation plus the effects of exercise, vomiting, or drug use. Patients otherwise successfully treated may still have persisting disorders in reproductive endocrinology for a variety of reasons. There is growing evidence that successful treatment involves combined approaches to nutritional rehabilitation and treatment of the psychological and social dysfunction.

The companion disorder bulimia is more of a description than a diagnosis, since its long-term consequences, especially of the milder versions, are unknown. Bulimia is an occasional outcome of anorexia nervosa, but may be entered into without lowering of weight. In addition to the biological effects of any weight loss the bulimic patient may have suffered, the patient also experiences significant changes in each component of the gastrointestinal tract.

Anorexia nervosa and bulimia are intriguing disorders because they represent abnormalities of eating and weight in the setting of apparently normal functioning of brain controls for eating and weight and a lack of identified biological markers for the disorder. They both fit the "biopsychosocial" model of medical disease suggested by Engel (1980). Treatment must address itself to each of these aspects. There are many ironies connected with this disorder, including the conviction of patients that they are increased in body size despite evidence to the contrary and their abstinence from food even to the point of death in the presence of abundant food. Sociocultural influences make some subpopulations especially vulnerable to eating disorders. While research has increased exponentially during the past decade, elucidation of the true underlying causes of these disorders and their definitive treatment depends on future studies.

ACKNOWLEDGMENTS. The author gratefully acknowledges the very helpful comments about the current state of research in the eating disorders by Herbert Weiner, M.D., Joseph Silverman, M.D., David Garner, Ph.D., and Professor Gerald Russell. This work would not be possible without the expert assistance of Angela Mickalide, Frances Partlow, Linda Ryan, and Kathleen Demuth. Partial support for the preparation of this article was provided by a generous grant from Alvin Margolius, Jr., M.D.

14. References

Abraham, S. F., Beumont, P. J. V., and Cobbin, D. M., 1981, Catecholamine metabolism and body weight in anorexia nervosa, *Br. J. Psychiatry* **138**:244–247.

Agras, W. S., Barlow, D. H., Chapin, H. N., Abel, G. G., and Leitenberg, H., 1974, Behavior modification of anorexia nervosa, *Arch. Gen. Psychiatry* **30**:279–286.

Allerdissen, R., Florin, I., and Rost, W., 1981, Psychological characteristics of women with bulimia nervosa (bulimarexia), *Behav. Anal. Modif.* **4**:314–317.

Andersen, A. E., 1983, Anorexia nervosa and bulimia: A spectrum of eating disorders, *J. Adolesc. Health Care* **4**:15–21.

Andersen, A. E., Wirth, J. B., and Strahlman, E. R., 1982, Reversible weight-related increase in plasma testosterone during treatment of male and female patients with anorexia nervosa, *Int. J. Eat. Dis.* **1**(2):74–84.

Andrews, F. F. H., 1982, Dental erosion due to anorexia nervosa with bulimia, *Br. Dent. J.* **152**:89–90.

Ballot, N. S., Delaney, N. E., Erskine, P. J., Langridge, P. J., Smit, K., Van Niekerk, M. S., Winters, Z. E., and Wright, N. C., 1981, Anorexia nervosa—prevalence study, *S. Afr. Med. J.* **59**:992–993.

Bassoe, H. H., and Eskeland, I., 1982, A prospective study of 133 patients with anorexia nervosa: Treatment and outcome, *Acta Psychiatr. Scand.* **65**:127–133.

Becker, H., Korner, P., and Stoffler, A., 1981, Psychodynamics and therapeutic aspects of anorexia nervosa, *Psychother. Psychosom.* **36**:8–16.

Bemis, K., 1978, Current approaches to the etiology and treatment of anorexia nervosa, *Psychol. Bull.* **85**:593–617.

Ben-Tovim, D. I., Marilov, V., and Crisp, A. H., 1979, Personality and mental state (P.S.E.) within anorexia nervosa, *J. Psychosom. Res.* **23**:321–325.

Bernstein, I. L., 1982, Physiological and psychological mechanisms of cancer anorexia, *Cancer Res.* **42**:715–720.

Beumont, P. J. V., 1970, Anorexia nervosa: A review, *S. Afr. Med. J.* **1970**:911–915.

Beumont, P. J. V., Abraham, S. F., Argall, W. J., George, G. C. W., and Glaun, D. E., 1978a, The onset of anorexia nervosa, *Aust. N. Z. J. Psychiatry* **12**:145–149.

Beumont, P. J. V., Abraham, S. F., Argall, W. J., and Turtle, J. R., 1978b, Plasma gonadotrophins and LHRH infusions in anorexia nervosa, *Aust. N. Z. J. Med.* **8:**509–514.

Beumont, P. J. V., Abraham, S. F., and Simson, K. G., 1981, The psychosexual histories of adolescent girls and young women with anorexia nervosa, *Psychol. Med.* **11:**131–140.

Boyar, R. M., and Bradlow, H. I., 1977, Studies of testosterone metabolism in anorexia nervosa, in: *Anorexia Nervosa* (R. A. Vigersky, ed.), pp. 271–277, Raven Press, New York.

Bruch, H., 1973, *Eating Disorders: Obesity, Anorexia Nervosa, and the Person Within*, Basic Books, New York.

Bruch, H., 1974, Perils of behavior modification in treatment of anorexia nervosa, *J. Am. Med. Assoc.* **230:**1419–1422.

Bruch, H., 1978, *The Golden Cage: The Enigma of Anorexia Nervosa*, Random House, New York.

Bruch, H., 1981, Developmental considerations of anorexia nervosa and obesity, *Can. J. Psychiatry* **26:**212–217.

Buhrich, N., 1981, Frequency of presentation of anorexia nervosa in Malaysia, *Aust. N. Z. J. Psychiatry* **15:**153–155.

Button, E. J., and Whitehouse, A., 1981, Subclinical anorexia nervosa, *Psychol. Med.* **11:**509–516.

Button, E. J., Fransella, F., and Slade, P. D., 1977, A reappraisal of body perception disturbance in anorexia nervosa, *Psychol. Med.* **7:**235–243.

Cahill, G. F., 1981, Role of T$_3$ in fasted man, *Life Sci.* **28:**1721–1726.

Cantwell, D. P., Sturzenberger, S., Burroughs, J., Salkin, B., and Green, J. K., 1977, Anorexia nervosa: An affective disorder?, *Arch. Gen. Psychiatry* **34:**1087–1093.

Crisp, A. H., 1980, *Anorexia Nervosa: Let Me Be*, Grune and Stratton, New York.

Crisp, A. H., 1981–1982, Anorexia nervosa at normal body weight!—the abnormal normal weight control syndrome, *Int. J. Psychiatry Med.* **11:**203–233.

Crisp, A. H., Stonehill, E., and Fenton, G. W., 1970, An aspect of the biological basis of the mind–body apparatus: The relationship between sleep, nutritional state and mood in disorders of weight, *Psychother. Psychosom.* **18:**161–175.

Crisp, A. H., Palmer, R. L., and Kalucy, R. S., 1976, How common is anorexia nervosa? A prevalence study, *Br. J. Psychiatry* **128:**549–554.

Crisp, A. H., Hsu, L. K. G., Chen, C. N., and Wheeler, M., 1982, Reproductive hormone profiles in male anorexia nervosa before, during and after restoration of body weight to normal: A study of twelve patients, *Int. J. Eat. Dis.* **1:**3–19.

Dally, P., 1981, Treatment of anorexia nervosa, *Br. J. Hosp. Med.* **25:**434–440.

Diagnostic and Statistical Manual of Mental Disorders, 3rd ed., 1980, American Psychiatric Association, Washington, D.C.

Druss, R. G., and Silverman, J. A., 1979, Body image and perfectionism of ballerinas: Comparison and contrast with anorexia nervosa, *Gen. Hosp. Psychiatry* **1979:**115–121.

Eckert, E. D., Goldberg, S. C., Halmi, K. A., Casper, R. C., and Davis, J. M., 1982, Depression in anorexia nervosa, *Psychol. Med.* **12:**115–122.

Engel, G. L., 1980, The clinical application of the biopsychosocial model, *Am. J. Psychiatry* **137:**535–544.

Fairburn, C., 1981, A cognitive behavioural approach to the treatment of bulimia, *Psychol. Med.* **11:**707–711.

Feighner, J. P., Robius, E., Guze, S. B., Woodruff, R. A., Jr., Winokur, G., and Munoz, R., 1972, Diagnostic criteria for use in psychiatric research, *Arch. Gen. Psychiatry* **26:**57–63.

Frisch, R. E., and McArthur, J. W., 1974, Menstrual cycles: Fatness as a determinant of minimum weight necessary for their maintenance or onset, *Science* **185:**949–951.

Garfinkel, P. E., and Garner, D. M., 1977, The role of behavior modification in the treatment of anorexia nervosa, *J. Pediatr. Psychol.* **2:**113–121.

Garfinkel, P. E., Kline, S. A., and Stancer, H. C., 1973, Treatment of anorexia nervosa using operant conditioning techniques, *J. Nerv. Ment. Dis.* **157:**428–433.

Garfinkel, P. E., Moldofsky, H., and Garner, D. M., 1979, The stability of perceptual disturbances in anorexia nervosa, *Psychol. Med.* **9:**703–708.

Garner, D. M., and Bemis, K. M., 1982, A cognitive–behavioral approach to anorexia nervosa, *Cognit. Ther. Res.* **6**:123–150.

Garner, D. M., and Garfinkel, P. E., 1979, The eating attitudes test: An index of the symptoms of anorexia nervosa, *Psychol. Med.* **9**:273–279.

Garner, D. M., and Garfinkel, P. E., 1980, Socio-cultural factors in the development of anorexia nervosa, *Psychol. Med.* **10**:647–656.

Garner, D. M., and Garfinkel, P. E., 1981–1982, Body image in anorexia nervosa: Measurement, theory and clinical implications, *Int. J. Psychiatry Med.* **11**:263–284.

Garner, D. M., Garfinkel, P. E., and Bemis, K. M., 1981–1982, A multidimensional psychotherapy for anorexia nervosa, *Int. J. Eat. Dis.* **1**(2):3–73.

Gerner, R. H., and Gwirtsman, H. E., 1981, Abnormalities of dexamethasone suppression test and urinary MHPG in anorexia nervosa, *Am. J. Psychiatry* **138**:650–653.

Goldberg, S. C., Halmi, K. A., Eckert, E. D., Casper, R. C., and Davis, J. M., 1979, Cyproheptadine in anorexia nervosa, *Br. J. Psychiatry* **134**:67–70.

Golla, J. A., Larson, L. A., Anderson, C. F., Lucas, A. R., Wilson, W. R., and Tomasi, T. B., 1981, An immunological assessment of patients with anorexia nervosa, *Am. J. Clin. Nutr.* **34**:2756–2762.

Green, R. S., and Rau, J. H., 1974, Treatment of compulsive eating disturbances with anticonvulsant medication, *Am. J. Psychiatry* **131**:428–431.

Gull, W. W., 1874, Anorexia nervosa, *Lancet* **1868**:22–28.

Gwirtsman, H. E., and Gerner, R. H., 1981, Neurochemical abnormalities in anorexia nervosa: Similarities to affective disorders, *Biol. Psychiatry* **16**:991–995.

Hall, A., 1982, Anorexia nervosa in twins, Presented at the European Conference on Psychosomatic Research, Satellite Conference on Anorexia Nervosa and Bulimia, Noordwijkerhout, The Netherlands.

Halmi, K. A., 1977, Effectiveness of behavioral therapy in anorexia nervosa, *Psychiatry Dig.* **1977**:19–24.

Halmi, K. A., 1982, The diagnosis and treatment of anorexia nervosa, in: *Eating, Sleeping, and Sexuality: Treatment of Disorders in Basic Life Functions* (M. Zales, ed.) pp. 43–58, Brunner/Mazel, New York.

Halmi, K. A., Powers, P., and Cunningham, S., 1975, Treatment of anorexia nervosa with behavior modification, *Arch. Gen. Psychiatry* **32**:93–96.

Halmi, K. A., Goldberg, S. C., and Cunningham, S., 1977, Perceptual distortion of body image in adolescent girls: Distortion of body image in adolescence, *Psychol. Med.* **7**:253–257.

Halmi, K. A., Dekirmenjian, H., Davis, J. M., Casper, R., and Goldberg, S., 1978a, Catecholamine metabolism in anorexia nervosa, *Arch. Gen. Psychiatry* **35**:458–460.

Halmi, K. A., Strauss, A. L., and Goldberg, S., 1978b, An investigation of weights in the parents of anorexia nervosa patients, *J. Nerv. Ment. Dis.* **166**:358–361.

Halmi, K. A., Goldberg, S. C., Casper, R. C., Eckert, E. D., and Davis, J. M., 1979. Pretreatment predictors of outcome in anorexia nervosa, *Br. J. Psychiatry* **134**:71–78.

Halmi, K. A., Falk, J. R., and Schwartz, E., 1981, Binge-eating and vomiting: A survey of a college population, *Psychol. Med.* **11**:697–706.

Hedblom, J. E., Hubbard, F. A., and Andersen, A. E., 1981, Anorexia nervosa: A multidisciplinary treatment program for patient and family, *Soc. Work Health Care* **7**:67–86.

Holt, S., Ford, M. J., Grant, S., and Heading, R. C., 1981, Abnormal gastric emptying in primary anorexia nervosa, *Br. J. Psychiatry* **139**:550–552.

Hsu, L. K. G., 1980, Outcome of anorexia nervosa, *Arch. Gen. Psychiatry* **37**:1041–1046.

Hsu, L. K. G., 1982, Is there a disturbance in body image in anorexia nervosa?, *J. Nerv. Ment. Dis.* **170**:305–307.

Hsu, L. K. G., and Crisp, A. H., 1980, The Crown–Crisp experiential index (CCEI) profile in anorexia nervosa, *Br. J. Psychiatry* **136**:567–573.

Hsu, L. K. G., Crisp, A. H., and Harding, B., 1979, Outcome of anorexia nervosa, *Lancet* **8107**:61–65.

Hudson, J. I., Laffer, P. S., and Pope, H. G., 1982, Bulimia related to affective disorder by family history and response to the dexamethasone suppression test, *Am. J. Psychiatry* **139**:685–687.

Johnson, C., and Larson, R., 1982, Bulimia: An analysis of moods and behavior, *Psychosom. Med.* **44**:341–351.

Kalucy, R. S., Crisp, A. H., and Harding, B., 1977, A study of 56 families with anorexia nervosa, *Br. J. Med. Psychol.* **50**:381–395.

Katz, J. L., and Weiner, H., 1981, The aberrant reproductive endocrinology of anorexia nervosa, in: *Brain, Behavior, and Bodily Disease* (H. Weiner, M. A. Hofer, and A. J. Stunkard, eds.), pp. 165–180, Raven Press, New York.

Kaye, W. H., Pickar, D., Naber, D., and Ebert, M. H., 1982, Cerebrospinal fluid opioid activity in anorexia nervosa, *Am. J. Psychiatry* **139**:643–645.

Keys, A., Brozek, J., Henschel, A., Mickelsen, O., and Taylor, H. L., 1950, *The Biology of Human Starvation*, Vol. II, University of Minnesota Press, Minneapolis.

King, A., 1963, Primary and secondary anorexia nervosa syndromes, *Br. J. Psychiatry* **109**:470–479.

Korkina, M. V., and Marilov, V. V., 1980, Prepubertal nervous anorexia, *Zh. Nevropatol. Psikhiatr. im. S. S. Korsakova* **81**:1536–1540.

Lasègue, P. D., 1873, Memoires originaux, *Arch. Gen. Med.* **1873**:385–403.

Long, C. G., and Cordle, C. J., 1982, Psychological treatment of binge eating and self-induced vomiting, *Br. J. Med. Psychol.* **55**:139–145.

Lucas, A. R., 1981, Toward the understanding of anorexia nervosa as a disease entity, *Mayo Clin. Proc.* **56**:254–264.

Lucas, A. R., Duncan, J. W., and Piens, V., 1976, The treatment of anorexia nervosa, *Am. J. Psychiatry* **133**:1034–1038.

Luck, P., and Wakeling, A., 1980, Altered thresholds for thermoregulatory sweating and vasodilatation in anorexia nervosa, *Br. Med. J.* **281**:906–908.

Lundberg, D. D., Wålinder, J., Werner, I., and Wide, L., Effects of thyrotropin-releasing hormone on plasma levels of TSH, FSH, LH, and GH in anorexia nervosa,

Luton, B., Guilhaume, B., Marre, M., Fredy, D., and Bricaire, H., 1981, Abnormal cerebral CT scans in anorexia nervosa, *Nouv. Presse Med.* **10**:1071–1072.

Minuchen, S., Rosman, B. L., and Baker, L., 1978, *Psychosomatic Families: Anorexia Nervosa in Context*, Harvard University Press, Cambridge, Massachusetts.

Moore, D. C., 1977, Amitriptyline therapy in anorexia nervosa, *Am. J. Psychiatry* **134**:1303–1304.

Morgan, H. G., and Russell, G. F. M., 1975, Value of family background and clinical features as predictors of long-term outcome in anorexia nervosa: Four-year follow-up study of 41 patients, *Psychol. Med.* **5**:355–371.

Morton, R., 1694, *Phthisiologia: Or a Treatise of Consumptions Wherein the Difference, Nature, Causes, Signs, and Cure of All Sorts of Consumptions are Explained*, Sam. Smith and Benj, Walford, London.

Mrosovsky, N., and Sherry, D. F., 1980, Animal anorexias, *Science* **207**:837–842.

Myers, T. J., Perkerson, M. D., Witter, B. A., and Granville, N. B., 1981, Hematologic findings in anorexia nervosa, *Conn. Med.* **45**:14–17.

Needleman, H. K., and Waber, D., 1976, Amitriptyline therapy in patients with anorexia nervosa, *Lancet* **1976**:580.

Nylander, I., 1971, The feeling of being fat and dieting in a school population, *Acta Socio-Med. Scand.* **1**:17–26.

Palmer, R. L., Crisp, A. H., Mackinnon, P. C. B., Franklin, M., Bonnar, J., and Wheeler, M., 1975, Pituitary sensitivity to 50 mg LH/FSH-RH in subjects with anorexia nervosa in acute and recovery stages, *Br. Med. J.* **1975**:179–182.

Pertschuk, M. J., Forster, J., Buzby, G., and Mullen, J. L., 1981, The treatment of anorexia nervosa with total parenteral nutrition, *Biol. Psychiatry* **16**:539–550.

Pierloot, R., Vandereycken, W., and Verhaest, S., 1982, An inpatient treatment program for anorexia nervosa patients, *Acta Psychiatr. Scand.* **66**:1–8.

Ploog, D., Fichter, M., Doerr, P., and Pirke, K. M., 1981, Anorexia nervosa: Neurobiology, psychosomatic aspects and behavior therapy, *Internist* **22**:7–23.

Porot, D., 1981, L'anorexie mentale: Protestation et dépression, *Semin. Hop. Paris* **57**:1837–1840.

Pyle, R. L., Mitchell, J. E., and Eckert, E. D., 1981, Bulimia: A report of 34 cases, *J. Clin. Psychiatry* **42**:61–64.

Rau, J. H., and Green, R. A., 1975, Compulsive eating: A neuropsychologic approach to certain eating disorders, *Compr. Psychiatry* **16**:223–231.

Rollins, N., and Piazza, E., 1978, Diagnosis of anorexia nervosa, *Am. Acad. Child Psychiatry* **1978**:126–137.

Rolls, E. T., 1981, Central nervous mechanisms related to feeding and appetite, *Br. Med. Bull.* **37**:131–134.

Russell, G. F. M., 1977a, General management of anorexia nervosa and difficulties in assessing the efficacy of treatment, in: *Anorexia Nervosa* (R. A. Vigersky, ed.), pp. 277–289, Raven Press, New York.

Russell, G. F. M., 1977b, The present status of anorexia nervosa (letter), *Psychol. Med.* **7**:363–367.

Russell, G. F. M., 1979, Bulimia nervosa: An ominous variant of anorexia nervosa, *Psychol. Med.* **9**:429–488.

Ryle, J. A., 1936, Anorexia nervosa, *Lancet* **1936**:893–899.

Schwartz, D. M., Thompson, M. G., and Johnson, C. L., 1982, Anorexia nervosa and bulimia: The socio-cultural context, *Int. J. Eat. Dis.* **1**(3):20–36.

Sein, P., Searson, D., Nicol, A. R., and Hall, K., 1981, Anorexia nervosa and pseudo-atrophy of the brain, *Br. J. Psychiatry* **139**:257–258.

Slade, P. D., 1973, A short anorexic behaviour scale, *Br. J. Psychiatry* **122**:83–85.

Smart, D. E., Beumont, P. J. V., and George, G. C. W., 1976, Some personality characteristics of patients with anorexia nervosa, *J. Psychiatry* **128**:57–60.

Sours, J. A., 1981, Depression and the anorexia nervosa syndrome, *Psychiatr. Clin. North Am.* **4**:145–157.

Stricker, E. M., and Andersen, A. E., 1980, The lateral hypothalamic syndrome: Comparison with the syndrome of anorexia nervosa, *Life Sci.* **26**:1927–1934.

Strober, M., 1981, The relation of personality characteristics to body image disturbances in juvenile anorexia nervosa: A multivariate analysis, *Psychosom. Med.* **43**:323–330.

Theander, S., 1982, Research on outcome and prognosis of anorexia nervosa, and some results from a Swedish long-term study, Presented at the European Conference on Psychosomatic Research, Satellite Conference on Anorexia Nervosa and Bulimia, Noordwijkerhout, The Netherlands.

Van Loon, G. R., 1980, Abnormal catecholamine mechanisms in hypothalamic–pituitary disorders, *Metabolism* **29**:1198–1202.

Vigersky, R. A., 1977, *Anorexia Nervosa*, Raven Press, New York.

Vigersky, R. A., and Loriaux, D. L., 1977, The effect of cyproheptadine in anorexia nervosa, in: *Anorexia Nervosa* (R. A. Vigersky, ed.), pp. 349–357, Raven Press, New York.

Vigersky, R. A., Loriaux, D. L., Andersen, A. E., and Lipsett, M. B., 1976, Anorexia nervosa: Behavioural and hypothalamic aspects, *Clin. Endocrinol. Metab.* **5**:517–535.

Vigersky, R. A., Andersen, A. E., Thompson, R. H., and Loriaux, D. L., 1977, Hypothalamic dysfunction in secondary amenorrhea associated with simple weight loss, *N. Engl. J. Med.* **297**:1141–1145.

Walsh, B. T., Croft, C. B., and Katz, J. L., 1981–1982, Anorexia nervosa and salivary gland enlargement, *Int. J. Psychiatry Med.* **11**:255–261.

Wermuth, B. M., Davis, K. L., Hollister, L. E., and Stunkard, A. J., 1977, Phenytoin treatment of the binge-eating syndrome, *Am. J. Psychiatry* **134**:1249–1253.

Wilbur, C. J., and Colligan, R. C., 1981, Psychologic and behavioral correlates of anorexia nervosa, *Dev. Behav. Pediatr.* **2**:89–92.

Yager, J., 1982, Family issues in the pathogenesis of anorexia nervosa, *Psychosom. Med.* **44**:43–60.

Young, J. B., and Landsberg, L., 1977, Suppression of sympathetic nervous system during fasting, *Science* **196**:1473–1475.

10

Obesity: Possible Psychological and Metabolic Determinants

Robin B. Kanarek, Nilla Orthen-Gambill, Robin Marks-Kaufman, and Jean Mayer

1. Introduction

Obesity is one of the major health problems in today's society, with estimates that from 20 to 60 million Americans are substantially above their ideal body weight (Bray, 1976). Although a layperson's conception of obesity is primarily based on visual assessment of body weight, obesity is more properly defined as an excessive accumulation of body fat. While several factors, such as age and physical activity, can affect adipose tissue mass, Bray (1976) has suggested that a fat content in excess of 20% in males and 28% in females could be used as a working guideline for defining obesity.

Obesity has been associated with an increased risk for diabetes, hypertension, cardiovascular problems, and, in general, shorter life expectancy (Mayer, 1968; Bray, 1976). The great concern about obesity and its deleterious sequelae is clearly evidenced by the large number of recent publications in both the popular and the scientific press. In recent years, various aspects of obesity have been reviewed in numerous scientific texts (e.g., Bray, 1976, 1979; Stunkard, 1980; Björntorp *et al*, 1981; Cioffi *et al.*, 1981; Silverstone, 1982).

In addition to facing a multitude of potential physiological complications, the obese suffer from the psychological stigma placed on corpulence. Direct empirical evidence indicates that overweight individuals are indeed the pariahs of the Western world. The obese face discrimination in almost all facets of their lives, ranging from difficulties with admission to the college of their choice to later advancement in the job market (Mayer, 1968). The societal bias against

Robin B. Kanarek, Nilla Orthen-Gambill, Robin Marks-Kaufman • Department of Psychology, Tufts University, Medford, Massachusetts 02155. *Jean Mayer* • President, Tufts University, Medford, Massachusetts 02155.

the obese begins early in life, since even children find obesity unattractive. For example, when asked to rate pictures of normal weight, physically handicapped, and obese children, both adults and children rated the obese as least desirable (Richardson *et al.*, 1961; Goodman *et al.*, 1963). In these studies, the obese were judged to be responsible for their own condition, their primary problem being simply a lack of self-control. A presumed lack of self-control has produced a picture of the obese as indiscriminate gluttons. In fact, the word obesity itself is derived from the Latin words *ob* (over) and *edere* (to eat), and thus clearly implies that overeating is the primary culprit in producing obesity.

2. A Psychological Perspective on the Determinants of Obesity

The presumed lack of self-control in obese individuals became the main focus of research on obesity during the 1970s. Schachter and colleagues (Schachter, 1971; Schachter and Rodin, 1974) at Columbia University provided one possible explanation for obese individuals' lack of self-control. These investigators hypothesized that the eating behavior of obese individuals was determined by external food-related cues such as the sight, smell, taste, and immediate availability of food, rather than internal physiological cues for hunger and satiety. During the subsequent ten years, this hypothesis served as the theoretical framework for most investigations of human feeding behavior. Since this hypothesis provided the major impetus for experiments on the psychological basis of obesity in the 1970s, a review of Schachter's original work and the research it generated is warranted.

2.1. Do Obese Humans Behave Like Brain-Damaged Rats?

Taking a very creative approach to the study of obesity, Schachter and colleagues began by comparing the behavior of obese humans to that of animals with obesity thrust upon them by experimental destruction of a portion of the central nervous system, the ventromedial nucleus of the hypothalamus (VMH). Basically, Schacter noted that obese humans seemed to share many of the characteristics of VMH-lesioned rats, which might suggest that certain obese persons could have functional, if not organic, lesions of the VMH. Bilateral destruction of the VMH in rats and other mammalian species leads to a syndrome classically characterized by voracious eating and its subsequent correlate, obesity (e.g., Hetherington and Ranson, 1942; Brooks *et al.*, 1946; Mayer *et al.*, 1955; Mayer and Thomas, 1967; Sclafani, 1976; Powley, 1977). Examination of the behavioral components of this syndrome resulted in a picture of the VMH-lesioned animals as gluttonous, lazy, finicky, hyperemotional, and unable to regulate caloric intake.

The basis for the preceding picture of the VMH-lesioned animal originally came from the observation that although these animals ate more than their neurologically intact controls under free-feeding conditions, they actually were less motivated to work for food. VMH-lesioned rats ate less than controls when required to lift a heavy lid, cross an electrified grid, or expend effort pressing a lever to obtain food (N. E. Miller *et al.*, 1950; Teitelbaum, 1957).

"Finickiness" was added to the classic portrayal of the VMH-lesioned rat because experiments demonstrated that food acceptance in these obese animals was highly dependent on the sensory qualities of their foodstuffs. When the palatability of their diet was decreased by the addition of quinine, lesioned animals drastically reduced food intake to levels below those of normal animals fed similarly adulterated diets (e.g., N. E. Miller *et al.*, 1950; Teitelbaum, 1955; Graff and Stellar, 1962). Conversely, when large quantities of sugar or fat, substances apparently preferred by rats as well as by some humans, were included in the diet, lesioned animals increased food intake to a much greater degree than nonlesioned animals (e.g., Strominger *et al.*, 1953; Teitelbaum, 1955; Graff and Stellar, 1962; Corbit and Stellar, 1964; Carlisle and Stellar, 1969).

While VMH-lesioned rats were pictured as highly sensitive to external food-related cues, they simultaneously were depicted as hyposensitive to internal physiological cues associated with hunger and satiety. In contrast to lean animals, which increased food intake when the diet was diluted with nonnutritive material and reduced food intake when the caloric density of the diet was increased by the addition of fat, VMH-lesioned animals decreased food intake when given diluted diets and increased intake when given high-fat diets (e.g., Kennedy, 1950; Strominger *et al.*, 1953; Corbit and Stellar, 1964; Carlisle and Stellar, 1969).

Experiments demonstrating that obese animals were more responsive to painful stimuli (e.g., Grossman, 1966, 1972) and handling (Singh, 1969) than lean animals added hyperemotionality to the behavioral repertoire of VMH-lesioned animals. Thus, at the time Schachter and colleagues initiated their work, the prevailing picture of the VMH-lesioned rat was that of a highly emotional animal more responsive to the external trappings of its food world than to the internal physiological correlates of hunger and satiety.

2.2. The Search for Behavioral Determinants of Human Obesity

In the early 1970s, only limited empirical information was available on the feeding behavior of obese humans. Work by Stunkard and Koch (1964), however, suggested that like VMH-lesioned rats, obese humans might be less sensitive than their nonobese counterparts to the physiological correlates of hunger. Stunkard and Koch (1964) examined the extent to which stomach contractions, presumed to be a physiological correlate of hunger, coincided with subjective sensations of hunger in lean and obese individuals. Stomach motility, measured by a gastric balloon, and the subjective sensation of hunger, measured by a simple "Yes" or "No" answer to the question "Do you feel hungry?," closely coincided in normal-weight subjects. In contrast, verbal self-reports of hunger did not coincide with stomach motility in overweight subjects.*

2.2.1. Obese Eating Style

With the observation of Stunkard and Koch (1964) and the prevailing view of the obese VMH-lesioned rat as background, Schachter and colleagues initiated a research program to methodically compare the feeding behavior of obese and lean individuals. The notion of an "obese eating style" in humans was of particular interest. The existence of an "obese eating style" characterized by a rapid rate of food intake with large bites and short meal duration was first promoted by Ferster *et al.* (1962). It was hypothesized that obesity was simply a learning disorder that could be corrected by behavioral techniques. If obese individuals could be taught to eat like their nonobese compatriots, they would lose weight. On the basis of this hypothesis, techniques for decreasing bite size, reducing rate of food intake, and prolonging meal duration became pivotal components of behavior-modification programs aimed at the treatment of obesity (e.g., Stuart, 1967; Stuart and Davis, 1972; Brownell and Stunkard, 1978, 1980; LeBow, 1981).

The concept of an "obese eating style" provided a further parallel between the eating behavior of obese humans and that of VMH-lesioned rats. Previous research had shown that VMH-lesioned rats ate larger but less frequent meals than their lean controls (Thomas and Mayer, 1968; Balagura and Devenport, 1970). The assumption that the feeding patterns of both obese humans and brain-damaged rats could be portrayed as consisting of large, albeit a few, daily meals fit nicely with Schachter's hypothesis that the eating behavior of the obese was under external rather than internal control. Within this theoretical framework, obese humans and rats ate only a few meals per day because it required a particularly salient food cue for them to initiate eating, but ate voraciously at each meal because there could be no more compelling stimulus to eat than food immediately before them (Schachter and Rodin, 1974).

Direct evidence for an "obese eating style" in humans, however, was proven to be very illusive. The original assertion of Ferster *et al.* (1962) that obesity was a consequence of maladaptive eating habits was based on clinical impressions, rather than experimental data. Similarly, in his landmark work on the behavioral treatment of overeating, Stuart (1967) provided no data to indicate that the eating patterns of his obese patients were indeed different

---

*Subsequent experiments by Stunkard and Fox (1971) employing more sophisticated measurement techniques and longer time periods failed to support these original observations. In contrast to previous studies using short-term (4-hr) measurements, 24-hr studies of normal-weight individuals revealed no relationship between gastric motility and hunger sensations (Stunkard and Fox, 1971). Additionally, in subsequent 4-hr experiments employing a less intrusive gastric catheter, rather than a gastric balloon, Stunkard and Fox (1971) observed that only a minority of either normal-weight or obese subjects actually associated gastric motility with hunger and that considerable variability existed within subjects from one test session to the next.

from those of lean individuals. However, by describing a variety of techniques aimed at modifying the eating habits of these patients, Stuart's report implied the existence of a distinct "obese eating style." The original assumption of deviant eating patterns in the obese, thus, was formulated without any experimental basis.

During the last 15 years, experiments have been conducted to empirically compare the feeding patterns of obese and nonobese individuals. These experiments are summarized in Tables I and II. To facilitate the interpretation of this wealth of information, Table I summarizes research comparing meal size or duration or both and Table II summarizes data from studies comparing rates of food intake in lean and obese individuals.

2.2.1a. Meal Size. As shown in Table I, there is some evidence to suggest that the obese do consume larger meals than the lean. For example, both Dodd *et al.* (1976) and Gates *et al.* (1975) reported that in a "natural setting" with food freely available (i.e., a fast-food restaurant or college cafeteria), obese individuals ordered larger meals than their normal-weight counterparts. The majority of experiments comparing meal size or duration or both in lean and obese individuals, however, have failed to demonstrate differences in meal size as a function of body weight. Indeed, in some instances, obese subjects have actually been found to consume smaller meals than normal-weight subjects (Bellisle and LeMagnen, 1981).

2.2.1b. Rate of Food Intake. As previously mentioned, a rapid rate of food consumption has been considered an integral part of the "obese eating style." While some investigators have observed that the obese do indeed eat faster than their lean friends (Gaul *et al.*, 1975; Hill and McCutcheon, 1975; Wagner and Hewitt, 1975; Dodd *et al.*, 1976; LeBow *et al.*, 1977; Marston *et al.*, 1977), other research has not provided unanimous support for this presumed difference between obese and lean individuals (Jordan, 1975; Mahoney, 1975; Warner and Balagura, 1975; Shisslak, 1978).

While the evidence for an "obese eating style" in adults is weak, stronger data exist to suggest that obese children may, in fact, have a characteristic eating style. As shown in Table III, obese children do appear to eat faster than their normal-weight friends. It has been assumed that a rapid rate of food intake is a contributing factor to the development of childhood obesity and, further, that slowing the rate of food intake may help the obese child to slim down (Epstein *et al.*, 1976, 1978; Brownell and Stunkard, 1980). In support of these assumptions, Epstein *et al.* (1976) found that a reduction in food intake could be achieved by teaching children to decrease their rate of eating by simply putting down their eating utensils between each bite. Unfortunately, the observed reduction in food intake was not sufficient to prevent weight gain in the obese children. Thus, there is no direct evidence to support the expectation that a decrease in the rate of eating will lead to a reduction in body weight.

Careful consideration of the available evidence thus suggests that while there may be a population of individuals for whom obesity is associated with maladaptive eating habits, this is not universally the case.

Table I. Comparison of Meal Size or Duration or Both of Lean and Obese Individuals

Reference	Subjects	Food	Deprivation	Location	Design	Results
Nisbett (1968b)	Male undergrads 95 Overwt. ($+15$–48%) 83 Norm.-wt. (-1 to 9%) 82 Underwt. (-20–7%)	Sandwich	4 Hr	Laboratory	Subjects told they were participating in a "monitoring" experiment. Given one or three sandwiches after experiment with "dozens" more in the refrigerator.	Overwt. subjects ate less than norm.-wt. subjects when given one sandwich and ate more than norm.-wt. and underwt. subjects when given three sandwiches.
Hill and McCutcheon (1975)	Male undergrads 7 Obese ($+25\%$) 7 Norm.-wt.	Dinner meal	High (18 hr) Low (1.5 hr)	Laboratory	Subjects told they were participating in a food-preference experiment. Given meals with high- or low-preference food items. Videotaped meals.	No differences in meal size between obese and nonobese subjects. Obese ate less of low-preference foods and more of high-preference foods than nonobese.
Dodd et al. (1976)	Adult women 26 Obese 42 Norm.-wt.	Fast-food items	Unknown	Fast-food restaurant	Subjects observed while eating lunch in fast-food restaurant. Unmatched sample of first 51 customers.	Obese ordered larger meals. However, no differences in meal duration between obese and norm.-wt. Larger percentage (96%) of obese than of normals (76%) finished meal.
Mahoney (1975) Expt. 1	14 Male and 6 Female undergrads	Sandwich and apple turnover	Unknown	Laboratory	Subjects ate experimental meal in solitude. Eating responses monitored via closed-circuit TV. Correlated body weight with meal parameters.	No significant correlations between body weight and meal duration.
Expt. 2	18 Males	Hamburger, Fr. fries, soft	Unknown	Fast-food restaurant	Naturalistic field experiment. Obese and	No significant correlations between body weight and

Study	Subjects	Food	Duration	Setting	Procedure	Results
		drink			lean subjects observed eating lunch. Weight and height obtained by postobservation questionnaire.	meal duration.
Warner and Balagura (1975) Expt. 1	Male and female undergrads 32 Obese (+ 20%) 53 Norm.-wt.	Hot dogs	Min. 2 hr	Laboratory	Subjects told they were participating in a taste test. Unaware that food intake was being monitored.	Obese females ate for longer than norm.-wt. females. Obese males ate for shorter time than norm.-wt. males.
Expt. 2	10 Obese males 10 Obese females 10 Norm.-wt. males 10 Norm.-wt. females	"Finger foods," sandwiches, vegetables, dessert	Unknown	College or high school cafeteria	Subjects observed while eating lunch in school cafeteria. Only subjects eating alone observed.	Obese subjects took longer to eat than norm.-wt. subjects. (No data on amount of food chosen or exact amount consumed.)
Gates et al. (1975)	720 Subjects divided according to sex, age, height, and weight	Luncheon foods	Unknown	College cafeteria	Observers recorded total servings of food selected and type of food chosen (i.e., high-calorie/low-nutrient or high-nutrient).	Obese individuals selected more servings of food than norm.-wt. individuals. Obese chose greater percentage of high-calorie/low-nutrient foods than norm.-wt.
Adams et al. (1978)	Women 21 Obese 20 Norm.-wt. 20 Thin	Sandwiches, fruit, dessert, candy	3–4 Hr	Hospital staff lounge	Subjects told purpose of study was to assess the effect of hunger and satiety on "thinking speed." Visual scanning test given preceding and following 30-min meal.	No significant differences in types of foods or number of calories consumed among individuals in the three weight groups.

(Continued)

Table I. (Continued)

Reference	Subjects	Food	Deprivation	Location	Design	Results
Krantz (1979)	Undergrads and faculty 101 Obese (+20%) 96 Norm.-wt.	Luncheon foods	Unknown	University cafeteria	Observed the amount of food chosen by obese and norm.-wt. individuals who ate alone or with others.	Obese individuals ate more when eating alone than when eating with others. Norm.-wt. individuals ate less when alone than with others.
Coll *et al.* (1979)	5291 Individuals divided by trained observer into three weight groups: norm.-wt., overwt., and obese	Variety of normally available foods	Unknown	9 Public eating places: restaurants + snack sites	Observed what food items each individual selected.	No differences in amount of food selected by nonobese and obese with exception that at fast-food restaurant, obese ate slightly more than nonobese.
D. L. Kaplan (1980) Expt. 1	Male undergrads 6 Obese (+20%) 6 Norm.-wt.	Sandwiches	Unknown	Laboratory	Observed subjects for several days eating lunch in laboratory.	No differences in meal size or duration as a function of body weight.
Expt. 2	Male undergrads 6 Obese 6 Norm.-wt.	Sandwiches	Unknown	Laboratory	Subjects told purpose of study was to examine the effect of various rates of ingestion on the taste of food. Subjects asked to consume food at fast, slow, or normal rate.	Irrespective of rate of food ingestion, no differences in meal size or duration between lean and obese subjects. All subjects ate more when consuming food at fast rate.
Bellisle and LeMagnen (1981)	6 Obese females (+25–70%) 7 Norm.-wt. females 3 Norm.-wt. males	Open-face sandwiches	4–5 Hr	Laboratory	Subjects given either one of five types of sandwiches or all five types together. Tested each subject twice under each of the six conditions.	Obese subjects ate smaller meals than nonobese subjects.

Table II. Comparison of Rate of Food Intake in Lean and Obese Individuals

Reference	Subjects	Food	Deprivation	Location	Design	Results
Gaul et al. (1975)	25 Obese females 25 Obese males 25 Norm.-wt. females 25 Norm.-wt. males	Hamburger, Fr. fries, soft drink	Unknown	Fast-food restaurant	Measured number of bites, number of chews per bite, number of sec spent chewing, and number of sips of soft drink.	Obese subjects took more bites per 5 min, performed fewer chews per bite, and spent less time chewing than norm.-wt. subjects.
Hill and McCutcheon (1975)	Male undergrads 7 Obese (+25%) 7 Norm.-wt.	Dinner meal	High (18 hr) Low (1.5 hr)	Laboratory	Subjects told they were participating in a food-preference experiment. Given meals with high- or low-preference food items. Tested under high or low food deprivation.	Obese individuals ate faster than nonobese individuals in all conditions.
Jordan (1975)	24 Obese subjects 28 Lean subjects	Hamburger, Fr. fries, soft drink	Unknown	Fast-food restaurant	Observed lean and obese individuals consuming standard meal in fast-food restaurant.	No difference in calories consumed per min as a function of body weight.
Dodd et al. (1976)	Adult women 26 Obese 42 Norm.-wt.	Fast-food items	Unknown	Fast-food restaurant	Subjects observed while eating lunch in fast-food restaurant. Analyzed data on (1) unmatched sample of first 51 customers and (2) matched 20 obese and 20 norm.-wt. customers on meal size and content.	When obese and norm.-wt. subjects matched for meal size and content, no difference in eating rate (g/min), bite size, or bites; with unmatched sample, obese ate at faster rate and took larger bites. No differences in total bites or bites per minute.

(Continued)

Table II. (*Continued*)

Reference	Subjects	Food	Deprivation	Location	Design	Results
Mahoney (1975) Expt. 1	14 Male and 6 Female undergrads	Sandwich and apple turnover	Unknown	Laboratory	Subjects ate standard meal in solitude on two separate days. During second session, subjects asked to count the number of bites they took.	No significant correlations between bites per minute and body weight. Counting bites did not modify results.
Expt. 2	18 Males	Hamburger, Fr. fries, soft drink	Unknown	Fast-food restaurant	Obese and lean subjects observed eating lunch. Weight and height obtained by postobservation questionnaire.	No significant correlations between eating rate and body weight.
Wagner and Hewitt (1975)	Hospital patients 7 Obese males (+30%) 7 Obese females (+30%) 9 Lean males 7 Lean females	Hospital diet	Unknown	Hospital	Experimenter observed patients during lunch and dinner on two successive days.	Obese individuals spent less time consuming meals and chewed each mouthful for a shorter time than the nonobese.
Warner and Balagura (1975) Expt. 1	Male and female undergrads 32 Obese (+20%) 53 Norm.-wt.	Hot dogs	Min. 2 hr	Laboratory	Subjects told they were participating in a taste-test experiment. Bites per min and total number of bites recorded.	No differences between obese and lean for either bites per min or total number of bites.

Expt. 2	10 Obese males 10 Obese females 10 Norm.-wt. males 10 Norm.-wt. females	"Finger foods," sandwiches, vegetables, dessert	Unknown	College or high school cafeteria	Subjects observed while eating lunch in school cafeteria. Only subjects eating alone observed. Bites per min and total number of bites recorded.	No differences between obese and lean for either bites per min or total number of bites.
Marston *et al.* (1977)	40 Adult males and females Obese Norm.-wt.	Dinner meal	Unknown	Cafeteria	Observed obese and lean patrons eating dinner in public cafeteria. Measured frequency of bites and chews in 2- to 3-min periods.	Obese subjects took more bites per 3-min period than lean subjects. No differences between groups in chews per bite.
LeBow *et al.* (1977)	34 Overwt. 37 Norm.-wt.	Hamburger, Fr. fries, soft drink	Unknown	Fast-food restaurant	Trained experimenters observed overwt. and norm.-wt. customers consuming standard meal in fast-food restaurant.	Overwt. individuals consumed meals faster than norm.-wt. Overwt. took fewer bites and chews and spent less time pausing than norm.-wt. individuals.
Shisslak (1978)	96 Adult males Obese Overwt. Norm.-wt.	Pizza, salad	Unknown	Field study	Trained observers recorded total number of bites, number of chews per bite, and total time spent chewing.	No differences in total number of bites or chews per bite as a function of body weight. Overwt. people spent most time chewing, followed by obese and then norm.-wt.

(Continued)

Table II. (Continued)

Reference	Subjects	Food	Deprivation	Location	Design	Results
Rosenthal and Marx (1978)	Female students 11 Overwt. 13 Norm.-wt. 16 Successful dieters 8 Unsuccessful dieters	Luncheon meal	3 Hr	Laboratory	Subjects told they were participating in a taste-preference experiment. Some subjects previously participated in weight-reduction program. Meal size maintained relatively constant among subjects. Recorded meal duration, number of bites, number of chews per bite, and number of sips.	No difference in eating style of obese and nonobese. No differences in meal duration, number of bites, number of chews, or number of sips as a function of body weight. Subjects who had participated in weight-reduction program had longer meal durations than those not in program.
Adams *et al.* (1978)	Women 21 Obese 20 Norm.-wt. 20 Thin	Sandwiches, fruit, dessert, candy	3–4 Hr	Hospital staff lounge	Subjects told purpose of study was to assess the effect of hunger and satiety on "thinking speed." Visual scanning test given preceding and following 30-min meal. Quartile eating time, active eating time, mouthfuls, and chews recorded.	No differences between obese and norm.-wt. Obese women spent less time actively eating than thin women and less time chewing. All three groups decreased rates of biting, chewing, and drinking over time.

Study	Subjects	Food	Duration	Setting	Procedure	Results
D. L. Kaplan (1980) Expt. 1	Male undergrads 6 Obese (+20%) 6 Norm.-wt.	Sandwiches	Unknown	Laboratory	Observed subjects for several days eating lunch in laboratory. Recorded total bites, bites per sandwich, bites per min and percentage of meal consumed during first and second half of meal.	No differences in total bites, bites per sandwich, or bites per min. Indication that obese may consume smaller proportion of meal during first half of meal than leans do.
Expt. 2	Male undergrads 6 Obese (+20%) 6 Norm.-wt.	Sandwiches	Unknown	Laboratory	Subjects told purpose of study was to examine the effect of various rates of ingestion on the taste of food. Subjects asked to consume food at fast, slow, or normal rate.	No differences in number of bites, bites per sandwich, or bites per min as a function of body weight. Obese consumed smaller percentage of meal during first half of meal than did norm.-wt. subjects.
Bellisle and LeMagnen (1981)	6 Obese females (+25–70%) 7 Norm.-wt. females 3 Norm.-wt. males	Open-face sandwiches	4–5 hr	Laboratory	Subjects given either one of five types of sandwiches or all five types together. Tested each subject twice under each of the six conditions.	When drinking episodes included in rate measurements, no differences between lean and obese in rate of eating. When drinking episodes excluded, obese ate faster than lean regardless of type of meal.

Table III. Comparison of Meal Patterns of Lean and Obese Children

Reference	Subjects	Food	Location	Design	Results
Epstein et al. (1976)	7-Year-old children: 2 Obese females, 1 Obese male (+20%), 1 Norm.-wt. female, 2 Norm.-wt. males	Luncheon meal	School cafeteria	Using time-sampling technique, observed bite rate during lunch meal. Assessed effect of decreasing bite rate (putting utensils down between bites) on meal size. Determined food preferences.	No differences in bite rate between lean and obese children. Decreasing bite rate led to a reduction in food intake in all children. Obese showed greater preference for milk and less preference for bread than lean children.
Drabman et al. (1977)	120 Elementary-school children	Luncheon meal	School cafeteria	2 (Male–female) × 2 (obese–norm.-wt.) × 2 (black–white) design. Time-sampling observations (every other 30-sec period for 5 min) of bites, chews, sips, and talking.	Obese children took more bites per 30-sec interval, chewed less per interval, and performed fewer chews per bite than norm.-wt. children.
Marston et al. (1977)	8 Pairs of lean and obese children matched for sex and age (6–14 yr)	Luncheon meal	School cafeteria	Frequency of bites and number of chews per bite observed during 3-min period. Also, recorded total meal duration and food left on plate.	No differences in total meal time as function of body weight. Obese children ate faster and left less food on plate than lean children.
Drabman et al. (1979)	60 Children divided by age (1.2–2, 3–4, 5–6 yr), sex, and body weight	Luncheon meal	Nursery and preschool cafeteria	Time-sampling technique used. Children observed for 30 sec, then not observed for 10 sec. Recorded number of bites, chews, sips, and talks with neighbor.	Obese children took more bites per 30 sec and averaged fewer chews per bite than nonobese children. Differences in eating patterns between obese and nonobese children as early as 1.5–2 years of age.

Study	Subjects	Food	Location	Method	Results
Waxman and Stunkard (1980)	4 Pairs of brothers: One obese and One nonobese in each pair	Dinner meals Luncheon meals	Home School cafeteria	Observer assessed caloric intake and meal duration during family dinners and school lunches. Nonobese brothers served as controls for dinner and nonobese peers as controls for school lunch.	Obese boys ate faster than their nonobese brothers at dinner and faster than their nonobese peers at lunch.
Geller *et al.* (1981) Expt. 1	24 Obese 24 Norm.-wt. children	Cookie	School lounge	Observed number of bites and the total time to consume a cookie.	Obese children consumed cookie faster and took more bites per min than nonobese children.
Expt. 2	4 Obese 4 Norm.-wt. children	Luncheon meal	School cafeteria	Time-sampling technique used. Children observed for 10 out of 15 sec. Recorded meal duration, number of bites, number of sips, and percentage of food items completed. Observed each child during three lunch meals.	Meal duration and frequency of bites or sips did not differ as a function of body weight. However, obese children did eat greater percentage of food than lean. Shorter interbite interval observed in obese than in lean.

2.2.2. Externality and Food Intake

One of the basic claims of Schachter's theory is that the eating behavior of obese individuals is triggered mainly by salient external food-related cues. In attempts to prove this claim, many studies have directly manipulated the properties of experimental food stimuli. Since earlier research had indicated that diet palatability was a particularly salient cue for determining food intake in the obese VMH-lesioned rat, initial experiments examined the effect of diet palatability on food intake in the obese human.

2.2.2a. Diet Palatability. As can be seen in Table IV, obese individuals do indeed appear to be more sensitive than lean controls to *diet palatability*. Obese individuals eat more good-tasting food (Nisbett, 1968a; Price and Grinker, 1973) and less bad-tasting food (Decke, 1971) than nonobese controls. For example, Nisbett (1968a) found that when subjects were offered expensive, creamy French vanilla ice cream or inexpensive quinine-adulterated ice cream, obese subjects consumed more of the highly palatable ice cream than lean controls. While Nisbett did not observe that obese individuals consumed less quinine-adulterated ice cream than lean subjects, Decke (1971) did find obese subjects to be more responsive than lean counterparts to both good and bad tastes. Obese subjects consumed more good-tasting and less quinine-adulterated milk shakes than lean subjects, who showed no differences in consumption as a function of taste manipulations. It should be noted that the "bad-tasting" milk shake used by Decke contained 0.4 g quinine/quart, while the "bad-tasting" ice cream used by Nisbett (1968a) contained 2.5 g quinine/quart. Nisbett's "bad-tasting" stimulus was thus much more unpalatable than the one used by Decke, which might partly explain the observed differences in results.

2.2.2b. Food-Cue Salience. In addition to being highly responsive to diet palatability, the obese also appear to be very sensitive to empirical changes in *food-cue salience*. As Table IV shows, investigators have used both visual and cognitive manipulations of cue salience. The *visual* salience of food cues has been experimentally manipulated in a number of ways. For example, Ross (1974) presented subjects with either highly illuminated (high salience) or dimly illuminated (low salience) cashew nuts, while Johnson (1974) gave subjects sandwiches that were wrapped in either transparent plastic wrap (high salience) or nontransparent paper (low salience). In both studies, obese subjects ate significantly more under conditions of high than low salience, while lean subjects did not change their eating behavior as a function of food-cue salience. Despite the clear differences between obese and lean subjects found in these experiments, other studies (Levitz, 1975) have not demonstrated differential responsivity of obese and lean individuals to manipulations of visual cue salience.

To manipulate cue salience *cognitively*, researchers have asked subjects to think about (Ross, 1974) or listen to tapes about (Rodin *et al.*, 1977) food-related stimuli (high cognitive salience) or nonfood stimuli (low cognitive salience). In these experiments, obese individuals were again found to eat more under conditions of high than low cognitive salience, while lean persons were

Table IV. Externality and Food Intake

Reference	Subjects	Food	Externality manipulation	Design	Results
Taste manipulations					
Nisbett (1968a)	Undergrads 56 Overwt. (+15–70%) 56 Norm.-wt. (+0–11%) 56 Underwt. (−25–2%)	Ice cream	Good ice cream: expensive, creamy French vanilla Bad ice cream: cheap vanilla with 2.5 g quinine/quart	Examined effects of good- and bad-tasting ice cream on amounts eaten by overwt., norm.-wt., and underwt. subjects.	Found significant relationship between weight and amount eaten of good ice cream; i.e., the heavier the subjects, the more good ice cream they ate. No significant differences seen with bad ice cream. All three groups ate similar, quite small amounts of bad ice cream.
Decke (1971)	Prison volunteers 5 Obese 9 Norm.-wt.	Milk shakes	Good milk shake: regular milk + vanilla ice cream Bad milk shake: condensed milk + water + vanilla ice cream + 0.4 g quinine/quart	Assessed effects of good- and bad-tasting milk shakes on amounts consumed by obese and norm.-wt. subjects.	Obese consumed significantly more good- than bad-tasting milk shake, while lean subjects showed no differences in consumption as a function of taste. In contrast to the previous study, obese consumed significantly less than lean of the bad milk shake. Suggests that obese are more responsive than lean subjects to both good and bad taste.
Price and Grinker (1973)	20 Obese (+122.3%) from obesity program (14 females, 6 males) 20 Norm.-wt. (−3.52%, 14 females, 6 males)	Crackers	Five types of crackers, rated on 5-point palatability scale	Tested effects of palatability on eating behavior of obese and lean subjects.	Obese were more responsive than lean to palatability factors. Obese ate more "best-liked" crackers than lean, but similar amounts of "least-liked" crackers as lean subjects.

(Continued)

Table IV. (Continued)

Reference	Subjects	Food	Externality manipulation	Design	Results
Rodin et al. (1977)	Adult women at wt.-loss clinic 16 Obese 8 Norm.-wt.	Ice milk	High palatability: ice milk rated 9 on 9-point scale Low palatability: ice milk rated 3 on 9-point scale	Expt. 4: Examined whether responsiveness to taste manipulations changes after weight loss.	Obese were more responsive than lean to taste manipulations before weight loss and became even more responsive after weight loss. Obese showed increased consumption of palatable food, not increased rejection of untasty food.
Cue-salience manipulations					
Nisbett (1968b)	69 male undergrads Obese Norm.-wt. Underwt.	Roast beef sandwiches	High-quantity food cues: three sandwiches in front of subject Low-quantity: one sandwich in front of subject In both cases, told more sandwiches were available in refrigerator.	Examined effects of quantity of food on eating behavior of obese, norm.-wt., and underwt. subjects.	Given three sandwiches, obese ate significantly more than norm.-wt. or underwt. subjects. Given one sandwich, obese ate as little as underwt. subjects, which was significantly less than amounts eaten by norm.-wt. subjects.
Levitz (1975)	Cafeteria customers rated as severely overwt. (N = 425), norm.-wt. (N = 2385)	Low-cal.: fruit or gelatin High-cal.: cakes and pies	Availability manipulation: desserts placed in front or back part of shelves Caloric-density manipulation: desserts classified as low-cal. (75 kcal) or high-cal. (350 kcal) Base line: all desserts equally available	"Naturalistic" study in hospital cafeteria. Observed dessert choices made by obese and norm.-wt. subjects as a function of two external cues: availability and caloric density.	Overall, a significantly greater proportion of obese subjects (56%) ate dessert than norm.-wt. subjects (43%). Lean persons consistently selected the desserts that were most available (in front part of shelves), regardless of caloric value. Obese subjects showed an interaction between availability and caloric density. Compared to base line, obese ate more low-

Reference	Subjects	Food	Conditions	Purpose	Results
Ross (1974)	Male undergrads 60 Obese (+15–57%) 60 Norm.-wt. (−8 to +10%)	Cashews	High visual salience: nuts clearly illuminated High cognitive salience: asked to think about nuts Low visual salience: nuts dimly illuminated Low cognitive salience: asked to think about nonfood stimuli	Measured effects of both visual and cognitive food-cue manipulations on the eating behavior of obese and norm.-wt. subjects.	calorie desserts when these were most available. But they still selected low-calorie foods even when these were less available. Obese ate twice as much in high as in low visual-salience condition, while norm.-wt. subjects ate similar amounts in both conditions. Obese ate more than lean in high, but less than lean in low, visual-salience condition. Obese were also more affected than lean by the cognitive-salience manipulation.
Rodin et al. (1977)	Adult women at wt.-loss clinic 9 Obese (>+60%) 8 Moderately obese (<+40%)	Sandwich quarters	High visual salience: platter of sandwiches directly in front of subject High cognitive salience: listened to tape about food-relevant stimuli Low visual salience: platter of sandwiches out of subject's direct line of vision Low cognitive salience: listened to tape about nonfood stimuli	Expt. 3: Examined effects of manipulating both visual and cognitive salience of food cues on food intake in obese and moderately obese subjects, before and after wt. loss.	Before wt. loss: In high-salience conditions, obese ate more than moderately obese. In low-salience conditions, moderately obese ate significantly less than obese. Moderately obese showed a greater drop in intake from high to low salience than obese, indicating that moderately obese are more responsive to differences in cue prominence. After wt. loss: Responsiveness to differences in cue salience did not change after wt. loss. Suggests that responsiveness to external cues is unaffected by wt. loss.

(Continued)

Table IV. (Continued)

Reference	Subjects	Food	Externality manipulation	Design	Results
Effort manipulations					
Schachter and Friedman (1974)	Male undergrads 40 Obese (+15–64%) 40 Norm.-wt. (−13 to +9%)	Almonds	Effort manipulation: shelled vs. unshelled nuts Cue-prominence manipulation: nuts in transparent cellophane or opaque paper bag	Examined effects of work and cue-prominence manipulations on eating behavior of obese and lean subjects.	Work manipulation had major effect on obese and almost no effect on lean. When nuts had shells, 1 of 20 obese ate; when nuts had no shells, 19 of 20 obese ate. Cue-prominence manipulation had no significant effect.
Singh and Sikes (1974)	Undergrads Obese: 30 males (+23%), 30 females (+34%) Lean: 30 males (+2%), 34 females (+5%)	Hershey kisses or cashew nuts	Manipulated both effort and familiarity with effort: "Familiar effort": wrapped Hershey kisses "Unfamiliar effort": wrapped cashews (Low effort: unwrapped chocolates or nuts)	Measured eating behavior of obese and norm.-wt. subjects, after giving them unwrapped or wrapped chocolates (familiar effort) or cashews (unfamiliar effort).	Obese and lean ate same amount of chocolates in both wrapped and unwrapped conditions. However, wrapping did affect cashew consumption in obese: they ate more unwrapped than wrapped cashews. Obese also ate more unwrapped cashews than lean subjects, who were unaffected by wrapping manipulation. Suggests that obese subjects are unwilling to work for food only when the effort involved is unfamiliar.
Johnson (1974)	University students Obese: 31 females, 9 males (+37%) Lean: 29 females, 11 males (−0.2%)	Sandwiches	Cue prominence manipulated. Sandwiches wrapped in either transparent wrap (high prominence) or nontransparent white paper (low prominence)	Examined effects of cue prominence on food-directed performance in obese and lean subjects.	Performance of obese subjects changed as a function of cue prominence; i.e., they worked significantly harder for food when food cues were highly prominent. In contrast, the performance of lean subjects remained relatively stable under different conditions of cue prominence.

Study	Subjects	Food	Manipulation	Purpose	Results
Rodin (1977)	Undergrads Obese (>+60%) Overwt. (+15–40%) Norm.-wt. (−10 to +10%)	Milk shakes	Effort manipulation: Hard: milk shake drunk through narrow straw Easy: milk shake drunk through wide straw	Looked at effect of effort on food intake of obese and lean subjects.	Overwt. group was most affected by effort manipulation. These subjects were significantly more deterred by narrow straw than either obese or norm.-wt. subjects.
Time manipulations					
Schachter and Gross (1968)	Male undergrads 22 Obese (+32%) 24 Norm-wt. (+3%)	Crackers	Fast condition: clock runs at twice normal speed. Slow condition: clock runs at half normal speed.	Measured eating behavior of obese and norm.-wt. subjects who were led to believe that time was later or earlier than real time.	Obese in fast condition ate almost twice as much as obese in slow condition. Time manipulation had opposite effect on lean subjects: they ate much more in slow-time than fast-time condition (several noted they did not want to spoil their dinner).
Goldman et al. (1968)	Air France personnel Overwt. (+0.1–30%) Not overwt. (0 to −22%)	Not used	"Natural" time manipulation: time-zone changes experienced by long-distance east–west travelers	Examined effects of time-zone changes on eating behavior. Plotted relationship between body wt. and complaints about adjusting to new eating time.	Obese appear to adjust to local eating time more easily than lean colleagues. Only 11.9% of overwt. group complained, compared to 25.2% in lean group.
Social manipulations					
Nisbett and Storms (1974)	Male undergrads Overwt. (>+15%) Norm.-wt. (−5 to +5%) Underwt. (−7%)	Crackers	Social facilitation: model ate large amount. Social suppression: model ate only one cracker. Alone: no model present.	Looked at effects of social cues on eating behavior. Measured eating behavior of overwt., norm.-wt., and underwt. subjects in presence or absence of peer model who ate large or minimal amount.	Obese were not more affected than other groups by social manipulation, since all three groups showed similar patterns of food intake. Subjects in all groups ate most in social-facilitation condition. Obese and norm.-wt. subjects ate the least in social-suppression condition.

(Continued)

Table IV. (Continued)

Reference	Subjects	Food	Externality manipulation	Design	Results
Krantz (1979)	Cafeteria customers 101 Obese 96 Norm.-wt.	Typical cafeteria food	Observational study with two conditions: subject had lunch alone or with others	"Naturalistic" experiment on social cues. Measured amount of cafeteria food bought by obese and lean patrons when they had lunch alone or with others.	Obese customers bought more food when alone than with company. Nonobese bought less food when alone than in the company of others. Suggests that social factors inhibit eating in obese. In contrast, social factors seem to have facilitative effect on lean eating behavior.
Conger *et al.* (1980)	114 Male and female undergrads, both obese (+15%) and lean (−6 to +6%)	Crackers	Modeling manipulations: 1. Model present/absent 2. Model's food intake: High-eat (20 crackers) Low-eat (1 cracker) No-eat	Tested importance of social factors in feeding behavior. Looked at effects of same- or opposite-sex model on eating behavior of obese and lean males and females.	Obese and lean were similarly affected by models. Presence of noneating model inhibited food intake in all groups. Presence of model who ate had disinhibitory effect in all groups; i.e., subjects' consumption increased with increases in the model's consumption. Obese females most inhibited by model, since their food intake was as inhibited by low-eat as by no-eat model.

relatively unaffected by the experimental manipulation. Taken together, the results of studies manipulating food-cue salience generally suggest that obese individuals may indeed be more responsive than lean controls to variations in food-cue prominence.

2.2.2c. Manipulations of Effort. One of the original observations about the VMH-lesioned rat was that while this animal ate voraciously when food was easily obtainable, it was unwilling to "earn its daily bread." To determine whether the behavior of obese humans is analogous to the behavior of the obese rat, investigators examined the effects of *effort* (work) on the eating behavior of obese and lean individuals. The results generally suggest that obese individuals, like obese rats, decrease food intake when the effort required to obtain food is increased. For example, Schachter and Friedman (1974) offered subjects nuts with shells (high effort) or without shells (low effort), while Rodin (1977) asked subjects to drink milk shakes through very narrow (high effort) or wide (low effort) straws. In both these studies, obese subjects ate significantly more in the low- than in the high-effort condition. In contrast, the eating behavior of lean individuals was found to be relatively unaffected by changes in the effort required to obtain food. However, certain qualifications may be necessary in describing the effects of effort on the eating behavior of obese individuals. As can be seen in Table IV, Singh and Sikes (1974) found that obese individuals were less willing than lean controls to work for food only when the effort involved was unfamiliar. The obese ate as many wrapped Hershey kisses as lean subjects, but fewer wrapped cashews, presumably because unwrapping cashews was not a "familiar effort."

2.2.2d. Manipulations of Time. While many studies on externality have directly manipulated experimental food stimuli, the externality concept has also been tested by varying factors that are only indirectly tied to food itself. For example, experiments using *time* manipulations have shown that the eating behavior of the obese is triggered more by external time cues than by an "internal" clock. Schachter and Gross (1968) manipulated the passage of time so that subjects believed it was about half an hour before or half an hour after normal dinner time. Obese subjects ate much more when they thought it was past dinner time than when they thought it was still before dinner time. In contrast, the time manipulation had the reverse effect in lean individuals, who ate more when they thought it was before dinner time than when they thought dinner time was past. Other investigators (Goldman *et al.*, 1968) looked at the more "natural" time manipulation experienced by internal airline personnel who travel between different time zones. Obese airline personnel (who are quite rare) apparently have less trouble than their lean colleagues in adjusting to local eating times. Taken together, the results of both laboratory and "naturalistic" studies suggest that obese individuals eat whenever external time cues indicate meal time, rather than eating in response to internal hunger cues.

2.2.2e. Manipulations of Social Factors. In addition to time manipulations, the effects of other external cues that are not directly tied to food intake have been experimentally assessed. As can be seen in Table IV, investigators have examined the effects of *social* cues on the eating behavior of obese and lean subjects. To manipulate social cues, researchers have used peer models who ate either a lot or a little or did not eat at all in the subject's presence (Nisbett and Storms, 1974; Conger *et al.*, 1980). In general, the results show that peer models did not differentially affect the obese and lean subjects. The presence of a noneating model had an inhibitory effect on the eating behavior of all subjects, while the presence of a model who ate was associated with an increase in food intake in all weight categories.

   2.2.2f. Summary. Over the last decade, the externality concept has become so broad that several investigators have used it to explain differences between obese and lean individuals in situations that have nothing to do with food intake. In fact, externality has been used as a general personality characteristic that presumably could affect a person's behavior in any situation. For example, the externality concept has been extended to explain individual differences in thinking behavior (Pliner, 1973), compliance and incidental learning (Rodin and Slochower, 1974), time perception (Rodin, 1975), distractibility (Herman *et al.*, 1978), and field dependence–independence (McArthur and Burstein, 1975).

   While the externality theory has provided a very useful framework for research, some of the enthusiasm for the concept is now waning. The only cases that offer relatively clear support for the theory are studies using direct manipulations of food stimuli (e.g., manipulations of *taste-* and *food- cue salience*). However, much of the research on externality fails to show definite differences between obese and lean individuals. For example, studies using manipulations of *social* cues show that obese and lean persons are similarly affected by these variables. As previously pointed out (Rodin, 1980, 1981; Rodin and Marcus, 1982), a dichotomous concept of externality–internality might be overly simplistic. All obese individuals are not overresponsive to external cues (e.g., Rodin *et al.*, 1977). Further, externality is not exclusively a characteristic of the obese, since people of normal weight can also be highly sensitive to external cues (e.g., Ross *et al.*, 1971; Nisbett and Storms, 1974; Conger *et al.*, 1980). Finally, a person who is highly responsive to external cues is not necessarily insensitive to internal cues.

2.2.3. Responsiveness to Internal Physiological Cues

   2.2.3a. Effects of Preloads on Subsequent Food Intake. Within Schachter's theoretical framework, the obese suffered not only from a heightened sensitivity to external food-related cues but also from a lack of responsiveness to internal physiological cues associated with states of nutritional surfeits and deficiets. Initial experiments (Schachter *et al*, 1968) lent credence to the notion that obese individuals were less responsive than their lean colleagues to internal physiological sensations associated with hunger. In these experiments, obese and normal-weight subjects, who had not eaten for several hours, were either fed sandwiches or kept in a food-deprived condition prior to entering an experimental eating situation. Although subjects thought they were participating in an experiment on taste sensitivity, in reality, the dependent variable was simply the amount of food (crackers) subjects consumed during the experimental session. Normal-weight subjects ate fewer crackers when their stomachs were full following the sandwich preload than when their stomachs were empty. In contrast, obese subjects ate similar amounts of food whether their stomachs were full or empty.

While the study discussed above demonstrated a clear difference between obese and lean subjects, results of subsequent experiments have not unanimously upheld the hypothesized dichotomy in internal responsiveness between obese and normal-weight individuals (see Table V). To explain some of the conflicting findings, it has been proposed that obese subjects may be responsive to liquid-food but not to solid-food preloads (Pliner, 1973). In support of this proposal, obese subjects ate fewer sandwiches following a high-calorie liquid preload than following a low-calorie liquid preload, but did not modify food intake in response to a solid-food preload. In comparison to the obese, normal-weight subjects consumed fewer sandwiches when given either the high-calorie liquid or solid preload than when given the low-calorie preloads. While these results suggest that differences in the types of preloads employed is one reason for the discrepancies among the experiments shown in Table V, these data clearly do not explain all the inconsistencies observed in these studies.

Careful perusal of the experiments in Table V indicates that in many instances, both obese and lean individuals are insensitive to internal cues presumed to be associated with hunger and satiety. Indeed, it has been found that a subject's belief about the caloric value of a preload may be a more powerful determinant of subsequent food intake than the actual caloric value of the preload (S. C. Wooley, 1972; Nisbett and Storms, 1974). Support for this idea comes from an experiment in which sandwich intake was measured in obese and nonobese subjects following liquid preloads containing either 200 or 600 kcals (S. C. Wooley, 1972). In addition, in each preload condition, the sensory qualities of the preload were manipulated to lead subjects to believe that they were receiving either a high or a low-calorie drink. The actual caloric value of the preload had no effect on subsequent food intake. Obese and nonobese subjects ate the same number of sandwiches following the 200-kcal preload as following the 600-kcal preload. In contrast, the perceived caloric value of the preload significantly influenced subsequent food intake. Subjects in both weight categories ate less when they believed they had consumed a high-calorie preload than when they believed they had consumed a low-calorie preload. Thus, cognitive belief about the caloric content of a preload may be more important than the actual caloric content of the food in determining subsequent intake.

2.2.3b. Effects of Caloric Dilution on Food Intake. The responsiveness of lean and obese individuals to the putative physiological cues associated with hunger and satiety has also been examined by manipulating the caloric density of the diet (Table VI). The impetus for employing this technique came from work by Van Itallie and colleagues (Hashim and Van Itallie, 1965; Campbell *et al.*, 1971), who examined the effects of covert variations in the caloric density of a liquid diet on food intake. Normal-weight subjects adjusted food intake appropriately in response to modifications in the nutritive value of the diet (i.e.,

Table V. Effects of Food Preloads on Food Intake of Lean and Obese Individuals

Reference	Subjects	Preload	Meal	Time between preload and meal	Design	Results
Schachter et al. (1968)	Male undergrads 43 Obese (+14–74%) 48 Norm.-wt. (−8% to +9%)	Sandwich or no sandwich	Crackers	Immediate	Subjects told they were participating in an experiment examining interaction of sensations. Measured food intake with some subjects given preload and others given no preload.	Norm.-wt. individuals ate less after preload than after no preload. Obese individuals did not vary food intake as a function of preload condition.
Nisbett (1968a)	Male undergrads 56 Overwt. (+15–70%) 56 Norm.-wt. (0 to +11%) 56 Underwt. (−25–2%)	Sandwich or no sandwich	Ice cream	0–15 Min	Subjects told they were participating in an experiment examining the effect of hunger on the ability to concentrate. All subjects skipped meal preceding experiment. Half the subjects given sandwiches preceding ice cream, half not given sandwiches.	Only underwt. subjects decreased ice cream consumption as a function of the preload. Both obese and norm.-wt. subjects consumed only slightly less after preload than after no preload.
S. C. Wooley (1972)	Undergrads 8 Obese males 8 Obese females (+21–51%) 8 Norm.-wt. males 8 Norm.-wt. females	Liquid diet: High-calorie (600 kcal) Low-calorie (200 kcal)	Sandwiches	20 Min	Subjects told they were participating in an experiment examining the effects of high- and low-calorie foods on appetite. All subjects given one of four preloads on four different days: real high-calorie–apparent high-calorie;	Manipulation of actual caloric value had no effect on subsequent food intake. Manipulation of apparent caloric intake of preloads, however, did influence food intake, with both obese and norm.-wt. subjects consuming

Study	Subjects	Preload	Food	Time	Procedure	Results
					real high-calorie–apparent low-calorie; real low-calorie–apparent low-calorie; real low-calorie–apparent high-calorie.	less when they believed they were given high-calorie preload.
Pliner (1973)	Male undergrads 48 Obese (+15%) 48 Norm.-wt.	Liquid diet: High-calorie (600 kcal) Low-calorie (200 kcal) or Cake: High-calorie (600 kcal) Low-calorie (200 kcal)	Sandwiches	1 Hr	Subjects told they were participating in an experiment examining the effect of a vitamin on ability to concentrate. Told preload necessary for proper absorption of vitamin. Subjects food-deprived for 12–18 hr prior to preload.	Normal subjects adjusted caloric intake as a function of both solid and liquid preload (i.e., they ate less following 600-kcal than 200-kcal preload). Obese adjusted intake with liquid but not with solid preload.
Price and Grinker (1973)	20 Obese male and female patients in obesity clinic (+55–211%) 20 Norm.-wt. male and female volunteers	Sandwich and Coke or no preload	Crackers	Immediate	Subjects told they were taking part in an experiment on effects of hospitalization on taste. Subjects food-deprived for 5.5–7.5 hr prior to experiment.	Obese consumed more of preload and crackers than normals. No difference in amount eaten as function of preload for either normals or obese.
Nisbett and Storms (1974) Expt. 2	Male undergrads Overwt. (+15%) Norm.-wt. (+5%) Underwt. (−7%)	Liquid diet: High-calorie: Nutrament, 750 kcal. No-calorie: Diet Pepsi	Sandwiches	2 Hr	Subjects told they were participating in an experiment examining effect of a vitamin on memory. Subjects given either high-calorie or noncaloric preload. Within each preload condition, subjects were told preload had either 750 cal or no calories.	Subjects in all three weight groups ate fewer sandwiches when given caloric preload than when given noncaloric preload. Subjects also ate fewer sandwiches when told they had received caloric preload than when told they had received noncaloric preload. No significant differences as a function of weight observed.

(Continued)

Table V. (Continued)

Reference	Subjects	Preload	Meal	Time between preload and meal	Design	Results
Expt. 3	Male undergrads Overwt. (+15%) Norm.-wt. (+5%)	Liquid-diet preload (750-kcal) or no preload	Sandwiches	10 Min	Subjects told they were participating in an experiment examining the effect of blood sugar level on memory. Subjects given either placebo pill or liquid-diet preload to "increase blood glucose levels."	Food intake of both obese and norm.-wt. subjects less after liquid-diet preload than after placebo pill.
Tom and Rucker (1975)	Undergrads Obese (+15%) Norm.-wt.	Sandwiches or no sandwich	Crackers	15 Min	Subjects told they were participating in an experiment examining consumer preferences. Subjects given preload or no preload and then presented with slide-rating task followed by preference test for five types of crackers.	Norm.-wt. subjects consumed fewer crackers following sandwich preload than following no preload. Obese subjects consumed slightly more crackers following preload than following no preload.
Hill and McCutcheon (1975)	Male undergrads 7 Obese (+25%) 7 Norm.-wt.	Liquid-diet preload (1000 kcal) or no preload	Dinner meal	1.5 Hr	Within-subject design. Subjects ate one meal a week in lab. Told they were participating in an experiment on how food-preference ratings were affected by eating meals under varying conditions of hunger.	Both obese and norm.-wt. subjects ate smaller meals after liquid preload than after no preload.

Study	Subjects	Preload	Test food	Delay	Procedure	Results
Herman and Mack (1975)	Female undergrads 12 Obese 45 Norm.-wt. Subjects rated as "restrained" or "unre-strained" eaters	0, 1, or 2 milk shakes	Ice cream	Immediate	Subjects told they were participating in an experiment examining the influence of one "sensory experience" (taste) on subsequent experience in the same sensory modality. Subjects given 0, 1, or 2 milk shakes to "taste" prior to ice cream "taste" test.	Consumption following preload did vary as a function of body weight. Unrestrained subjects ate less following milk shake preloads than following no preload. Restrained subjects ate more after milk shake preloads than after no preload.
Hibscher and Herman (1977)	Male undergrads 26 Obese ($X = +27\%$) 28 Norm.-wt. ($X = +6\%$) 32 Underwt. ($X = -9\%$) Using "restraint scale," subjects divided into 44 dieters and 42 nondieters	0 or 2 milk shakes	Ice cream	Immediate	Subjects told they were participating in a taste experiment investigating the influence of prior taste perception on subsequent taste rating of palatable ice cream.	No difference in ice cream intake following different preloads among different weight groups. Dieters, however, consumed somewhat more after 2-milk-shake preload than after 0-milk-shake preload. Nondieters ate less after 2-milk-shake than after 0-milk-shake preload.
Jordan (1975)	2 Obese females 1 Obese male	Liquid diet	Liquid diet	1 Hr	Subjects given Metracal preload equal to 30 or 60% of base-line intake of the liquid diet.	Obese subjects reduced subsequent intake of test meal in response to preload.
Spencer and Fremouw (1979)	Female undergrads 20 Overwt. 20 Norm.-wt. 20 Underwt.	Liquid diet	Ice cream	10 Min	Subjects told they were participating in a taste-test experiment. Subjects rated as "restrained" or "unrestrained" eaters. All	No difference in ice cream intake between subjects who believed given high- or low-calorie preload in any weight group. Unrestrained eaters

(Continued)

Table V. (Continued)

Reference	Subjects	Preload	Meal	Time between preload and meal	Design	Results
					subjects given 500-kcal preload; half the subjects in each weight group told they were given high-calorie preload and half told they were given low-calorie preload.	ate less ice cream when told given high-calorie preload than low-calorie preload. Restrained eaters ate more ice cream when told given high-calorie preload than when told given low-calorie preload.
Durrant (1980)	14 Obese (X BW = 96 kg) 6 Lean (X BW = 67 kg)	Liquid diet	Dinner meal	—	On two days, subjects given 2.51-MJ preload before each meal. On next two days, subjects given 1.26-MJ or 3.77-MJ preload that was indistinguishable from 2.51-MJ preload.	Both obese and lean ate less after 3.77-MJ preload than after 1.26-MJ preload. However, compensation for preload was not precise.
Doerman and Kronenberger (1981)	Obese (+18–61%) Norm.-wt. (+10%) Norm.-wt. with history of obesity and maintained weight loss	Bogus methyl-cellulose tablets	Peanuts	30 Min	Subjects told experiment was to assess gastric distension on gustatory perception. Told methylcellulose tablets expanded either at 1:1 ratio (low expansion) or 1:20 ratio (high expansion). Actually, no difference in expansion.	Obese ate more when believed in "low-expansion" condition than in "high-expansion" condition. No difference in intake between low- and high-expansion condition for norm.-wt. or below-set-point norm.-wt. subjects.

Table VI. *Comparison of Responses of Lean and Obese Individuals to Modifications in the Caloric Density of Their Diet*

Reference	Subjects	Food	Design	Results
Campbell *et al.* (1971)	5 Norm.-wt. males 4 "Grossly" obese females 2 "Grossly" obese male adolescents	Liquid diet	Subjects, maintained on metabolic ward, obtained all food from food-dispensing apparatus that allowed for covert manipulation of nutritive density of diet. Nutritive density varied from 0.5 to 1.5 kcal/ml over 4- to 6-week period.	Norm.-wt. subjects, after slight initial decrease in weight, maintained weight on liquid diet and made regulatory adjustments in volume of diet consumed as function of changes in caloric density. Obese adults lost weight while on liquid diet and did not adjust to changes in caloric density.
O. W. Wooley *et al.* (1972)	2 Obese males 5 Obese females 3 Norm.-wt. males 4 Norm.-wt. females	Liquid diet	Subjects replaced one meal a day with liquid-diet meal of disguised caloric content for 5–10 base-line and 14–21 experimental days. On experimental days, received either a high (1.07 kcal/ml) or low (0.57 kcal/ml) meal. Subjects rated hunger and guessed caloric value of meal.	Subjects displayed very little ability to identify meals as high- or low-calorie and reported hunger more in accordance with initial belief about actual caloric value than about actual caloric value. No differences between obese and lean.
O. W. Wooley (1972)	6 Obese males 5 Norm.-wt. males	Liquid diet	Subjects given only liquid diet to eat for 15 days. Following 5-day base-line 5-day base-line period, subjects given high- or low-calorie diet for 5 days and switched to opposite diet for 5 days.	Subjects showed incomplete adjustment for caloric content. Volume of intake less when food was high-calorie than low-calorie, but absolute caloric intake was greater with high-calorie diet than with low-calorie diet. No differences in total food intake.

(Continued)

Table VI. (Continued)

Reference	Subjects	Food	Design	Results
Spiegel (1973) (Expt. 2)	13 Norm.-wt. males 2 Norm.-wt. females	Liquid diet: Metracal: Standard (1.0 kcal/ml) Diluted (0.5 kcal/ml)	Subjects maintained on liquid diets for up to 21 days. Each subject given both standard and diluted diet.	Six subjects failed to compensate for caloric dilution by increasing food intake and lost weight. Six subjects compensated for caloric dilution by eating more and maintained stable body weight. Compensation for dilution not immediate, but took 3–4 days.
Porikos (1981)	5 "Moderately" obese males 8 Norm.-wt. males	Typical American diet	Obese and norm.-wt. subjects kept on metabolic ward. Base-line measures with normal diet followed by 12-day experimental period in which aspartame replaced sugar in diet (decreased calories by 25%) and then sugar returned for 6 days.	Obese ate 20% more than norm.-wt. in base-line and experimental periods. However, no differences between two groups in response to caloric dilution. Both groups consumed fewer calories (−15%) when given aspartame diluted diet.
Glueck et al. (1982)	3 Obese males 7 Obese females	Conventional hospital diet	Total daily caloric intake measured in obese subjects when sucrose polyester (SPE), a nonabsorbable fat, replaced conventional fats in diet.	Obese did not detect or compensate for covert caloric dilution of fat calories with SPE. Mean daily caloric intake decreased by 23% during dilution of diet with SPE.

they increased intake when the caloric density of the diet was decreased and decreased intake when the density of the diet was increased). Obese subjects, on the other hand, severely reduced food intake when the liquid diet was substituted for a conventional hospital diet. Furthermore, they did not modify food intake in response to variations in the caloric density of the diet and lost substantial amounts of weight across the course of the experiment. The failure of these obese individuals to adapt to alterations in the caloric density of their diet strengthened the notion that food intake in the obese is not under tight physiological control.

Unfortunately, results of subsequent experiments suggest that neither lean nor obese individuals are extremely accurate in regulating food intake in response to variations in the caloric density of their diet (e.g., O. W. Wooley, 1972; O. W. Wooley et al., 1972; Spiegel, 1973). For example, O. W. Wooley (1972) reported that although both obese and normal-weight subjects decreased food intake when a high-calorie liquid was substituted for a low-calorie diet, they did not regulate caloric intake perfectly. Individuals in both weight groups took in more calories when given a high-calorie diet than when given a low-calorie diet. Another experiment by O. W. Wooley et al., (1972) provides additional evidence that both the obese and the nonobese are relatively insensitive to internal physiological cues. When given high- or low-calorie liquid meals, neither lean nor obese subjects were able to identify the caloric content of the meal.

One criticism of the preceding experiments, all of which employed a rather bland-tasting liquid diet, is that this eating situation is extremely artificial (Porikos, 1981; Porikos et al., 1977, 1982), which makes it difficult to generalize to the everyday eating habits of lean and obese individuals. The recent development of palatable, calorically dilute food analogues has made it possible to examine the effects of caloric dilutions using a typical American diet. For example, Porikos and co-workers (Porikos, 1981; Porikos et al, 1977, 1982) have used aspartame, a low-calorie sugar substitute approximately 200 times as sweet as sucrose, as a caloric dilutent. Following an initial base-line period, aspartame was covertly substituted for sucrose in a variety of foods, such as puddings, cakes, and soft drinks. The substitution of aspartame for sucrose reduced the caloric content of these foods by approximately 25%. Neither obese nor lean subjects modified absolute food intake from base-line conditions when given aspartame-containing foods. The failure to increase food intake in response to caloric dilution of their diet obviously resulted in a reduction in total caloric intake. The effects of caloric dilution of "everyday" foods on energy intake have very recently been examined in experiments using sucrose polyester (SPE), a nonabsorbable fatty acid with physical properties similar to those of common dietary fats (Glueck et al., 1982). As for aspartame, when conventional fat was replaced with SPE, obese subjects failed to increase food intake to compensate for the reduction in the caloric density of their diet. The preceding results thus suggest that neither obese nor lean individuals can accurately detect covert caloric dilution of their normal diet.

2.2.4. Effects of Anxiety on Food Intake

While some investigators have attempted to explain obesity in terms of an "obese eating style" or overresponsiveness to external cues, others have stressed the role of underlying emotional factors in the development of obesity.

2.2.4a. Psychosomatic Hypothesis. According to the psychosomatic view of obesity, overeating is an attempt to reduce anxiety and other negative emotions (Bruch, 1961; H. I. Kaplan and H. S. Kaplan, 1957). The inappropriate feeding response is thought to originate in early childhood and is a result of incorrect learning experiences. Because of parental misinterpretations of signals from the child, indiscriminate feeding is used in response to any stress signals from the child. An association thus develops between eating and emotional turmoil, and this early learning experience can result in conceptual and perceptual disturbances (Bruch, 1961). These disturbances involve an inability to correctly identify bodily sensations such as hunger or satiety. Eating occurs as a general response to various emotional states, rather than as a response to a physiological need state such as hunger. The experimental predictions of the psychosomatic position are clear: obese subjects are expected to eat more under conditions of high than low anxiety, and eating behavior is expected to have an anxiety-reducing effect.

2.2.4b. Externality Hypothesis. The psychosomatic view generates predictions about obese eating behavior that are in sharp contrast to Schachter's externality theory. As mentioned before, the externality hypothesis holds that obese eating behavior is triggered mainly by attractive external food cues and is relatively independent of internal physiological cues related to hunger. According to Schachter, fear or anxiety should inhibit hunger, since it reduces gastric motility (Carlson, 1916) and increases blood sugar levels (Cannon, 1915). Since the obese are presumably insensitive to internal cues, Schachter predicted that fear should not affect food intake in obese individuals. On the other hand, fear should decrease feeding in the lean, since it presumably inhibits internal hunger-related signals. In support of his hypothesis, Schachter found that normal-weight subjects decreased their food intake in response to a high-fear manipulation, while obese subjects actually ate slightly more in the high-fear than in the low-fear condition. These results were taken to indicate that the obese were insensitive to the internal hunger-inhibiting cues presumably caused by the fear manipulation.

As can be seen in Table VII, there is some support for both the psychosomatic and the externality view, although the evidence does not clearly support either view. In support of the psychosomatic theory, McKenna (1972) found that obese subjects ate significantly more under high- than under low-anxiety conditions. Additionally, Robinson *et al.* (1980) showed that food intake had an anxiety-reducing effect in obese subjects. Conversely, in support of the externality hypothesis, Abramson and Wunderlich (1972) found that manipulations of fear and anxiety had no effect on the eating behavior of obese subjects. Further, Reznick and Balch (1977) found that a larger number of obese subjects actually ate in a low- than in a high-anxiety condition.

Table VII. Anxiety and Obesity

Reference	Subjects	Food	Emotional manipulation	Design	Results
McKenna (1972)	Male undergrads Obese (+27%) Norm.-wt. (+8%)	Attractive and unattractive cookies	High-anxiety: subjects led to believe they would undergo painful physiological measures. Low-anxiety: subjects made comfortable.	Looked at the effects of anxiety and food-cue valence on the intake of cookies, in obese and lean subjects. Also assessed anxiety reduction after eating.	The obese ate significantly more in high- than in low-anxiety condition. In contrast, lean subjects ate less in high- than in low-anxiety condition. Supports psychosomatic view. Manipulation of food-cue valence had no significant effect—the obese and lean were not differentially responsive to attractive cookies.
Slochower et al. (1980)	Female undergrads 23 Obese (+15–50%) 17 Norm.-wt. (−8% to +8%)	Candy	"Real-life anxiety": final exam period—often felt to be uncontrollable Low anxiety: 3 weeks after exam period	Investigated effects of "real-life stress" on eating. Measured eating behavior and mood of obese and lean students (1) during final exam period and (2) 3 weeks later.	Obese ate significantly more during exams than later. Their eating was positively correlated with anxiety and negatively correlated with sense of control over anxiety. Eating behavior of lean students did not vary as a function of anxiety; i.e., was same during exams and later.
Robinson et al. (1980)	People at wt.-loss clinic 152 Obese 79 Norm.-wt.	"Instant breakfast" in whole milk	Anxiety not manipulated. Psychological scales used to assess psychological state of subjects.	Assessed anxiety-reducing effects of eating. Obese and lean filled out several psychological scales before and after liquid meal.	Following meal, obese showed decrease in anxiety, as indicated by significant changes in pre- and postmeal scores on five anxiety-related questions. No antianxiety effect seen in lean.

(Continued)

Table VII. (*Continued*)

Reference	Subjects	Food	Emotional manipulation	Design	Results
Schachter *et al.* (1968)	43 Obese (+14–74%) 48 Norm.-wt. (−8% to +9%)	Crackers	High fear: strong shock anticipated. Low fear: mild tingle anticipated.	Examined effects of fear on cracker consumption in obese and lean subjects.	Obese ate slightly more in high-fear than in low-fear condition, while lean ate more in low-fear than in high-fear condition. Questionnaire showed that fear decreased slightly after food intake in both groups. No significant between-group differences seen.
Abramson and Wunderlich (1972)	Male undergrads 33 Obese (+15%) 33 Lean (< +10%)	Crackers	Objective fear: threat of shock. Interpersonal anxiety: subjects told they would have problems in relationships. Control condition: subjects did paper-and-pencil tasks.	Wanted to separately assess effects of objective fear and interpersonal anxiety on food intake in obese and lean subjects.	No support for psychosomatic view, because obese ate about the same amounts in all conditions. Could not compare to lean, since they did not respond to anxiety manipulation.
Reznick and Balch (1977)	Undergrads 46 Males, 18 Females 32 Obese; (+17–95%)	Hershey kisses, wrapped (hi-cost) Hershey kisses, unwrapped (lo-cost)	High-anxiety: given hard IQ test and threat of shock. Low anxiety: made to feel comfortable about experiment.	Looked at effects of anxiety and response cost on eating behavior of obese and lean individuals. Results refer to number of subjects who ate in each condition.	More obese ate in low-anxiety, low-response-cost than in high-anxiety, high-response-cost conditions. None of the manipulations affected the number of lean who ate. Results support Schachter.

Study	Food	Subjects	Manipulations	Purpose	Results
Slochower (1976)	Cashews	Male undergrads 40 Obese (+16–40%) 40 Norm.-wt. (−9 to +9%)	Given false heart-rate feedback (very fast): Labeled arousal: given reason for heart rate. Unlabeled arousal: given no explanation. Control condition: low arousal (slow heart rate).	Looked at effects of labeled and unlabeled arousal on the eating behavior of obese and lean subjects.	Aroused obese ate significantly more when arousal was "free-floating" (unlabeled) than when its source was known (labeled). Food intake of obese was not affected by labeling in low-arousal condition. The lean were not significantly affected by any of the manipulations. Obese also showed anxiety reduction after eating.
Slochower and Kaplan (1980)	Cashews	Male undergrads 62 Obese (+15–40%) 67 Norm.-wt. (−9 to +9%)	Same labeled/unlabeled arousal as above. Also: Perceived control: told how to normalize fast heart rate, or given no helpful information.	Examined effects of both labeled and unlabeled anxiety, and perceived control over anxiety, on eating behavior in obese and lean persons.	Obese ate most in unlabeled, uncontrollable anxiety condition and showed marked decrease in food intake when they could both label and control anxiety. Eating also had marked anxiety-reducing effect in the obese. Food intake of controls was not affected by experimental manipulations. Their eating was positively correlated with hunger and negatively correlated with anxiety.

2.2.4c. New Modified Psychosomatic Hypothesis. Recent research (Slochower, 1976; Slochower *et al.*, 1980) now offers an interesting possibility for clarifying at least some of the conflicting results of these earlier experiments. It has been suggested (Slochower, 1976) that a crucial distinction should be made between *controllable* and *uncontrollable* emotional arousal. Studies comparing the effects of diffuse, uncontrollable anxiety and labeled, controllable anxiety show interesting differences. In one study (Slochower and Kaplan, 1980), the anxiety manipulation consisted of making subjects believe that they had a very fast heart rate. Some subjects were then told why their heart rate was so high (labeled), while others were given no reason (unlabeled). Some subjects were also told how they could return their heart rate to normal (controllable), while others were given no information (uncontrollable). The results indicated that obese subjects showed the highest food intake when the anxiety was both unlabeled and uncontrollable. In contrast to the obese group, the feeding behavior of the lean subjects was unaffected by the experimental manipulations. Interestingly, the obese showed marked decreases in eating when the anxiety was labeled and controllable. Other studies (Slochower, 1976; Slochower *et al.*, 1980) have also shown that while obese subjects increase food intake in response to unlabeled anxiety, no increase in eating is seen with labeled anxiety. It thus appears that a qualification might be necessary in describing emotional factors and overeating. The obese might overeat only in response to free-floating, uncontrollable anxiety, but behave like lean counterparts when a label and a sense of control over the anxiety are provided.

2.2.5. Body-Weight Set-Point: Restrained and Unrestrained Eating Behavior

While many investigators have used the concept of external vs. internal control to explain differences in the eating behavior of obese and lean individuals, other researchers offer alternative explanations. As mentioned before, the externality theory might be overly simplistic, since externality is not exclusively a characteristic of the obese. Many normal-weight individuals are also highly responsive to external cues, while many obese are not. Furthermore, obese and lean individuals do not differ very dramatically in their responsiveness to internal cues. To explain some of these discrepancies, an alternative theory of eating behavior has been proposed by Nisbett (1972). He argues that the eating behavior of obese as well as lean individuals is governed by a biologically determined set-point for body weight. The body-weight set-point is thought to be directly related to the number of fat cells (adipocytes) in the organism (Nisbett, 1972). Because of genetic factors and differences in early feeding experiences, there can be large individual variations in the number of adipocytes in the body. Thus, there can also be wide individual differences in body-weight set-points. According to the set-point theory of body weight, obesity simply reflects a higher-than-average set-point that is presumably due to a higher-than-average number of fat cells.

Because of cultural and social pressures to be thin, most obese individuals do not feel comfortable remaining at their high biological set-point and attempt to lose weight by restraining their eating behavior. But attempts to reach a body weight that is below the biological set-point can actually create a state of relative deprivation. According to some investigators (e.g., Nisbett, 1972; Herman and Mack, 1975; Herman and Polivy, 1975, 1980), it is the degree of deprivation (relative to the biological set-point) rather than the absolute body weight that is the crucial factor in explaining individual differences in eating behavior. Many nonobese individuals may also be below their biological set-point because of constant dieting, so people in a state of relative deprivation can be found in all weight categories.

While the set-point concept is theoretically intriguing, it is difficult to evaluate the concept directly, since it is not clear how set-point can be empirically determined. Being below the set-point presumably creates a state of relative deprivation, and attempts have been made to indirectly measure the state of deprivation through the use of a restraint scale (see Herman and Polivy, 1980). The restraint scale is a questionnaire about eating habits, weight fluctuations, and thoughts about food. The questionnaire is constructed in such a way that people who are constantly on a diet and worry about food and eating receive a high score. A high score on the restraint scale thus indicates a "restrained" eater, while a low score indicates an "unrestrained" eater. Since restraint or lack of restraint is thought to be the crucial determinant of individual differences in eating behavior, proponents of the concept have evaluated the effects of experimentally manipulating or removing restraint. A general prediction is that the feeding behavior of unrestrained eaters (who are presumably at or near their set-point) would not be greatly affected by restraint manipulations. In contrast, restrained eaters would be expected to increase eating in response to the experimental removal of restraint, since these individuals are presumably below their set-point and thus in a state of relative deprivation. These predictions were confirmed in a study comparing the eating behavior of subjects who were matched for body weight, but differed on the restraint dimension (Herman and Mack, 1975). The study was a covert "taste experiment," consisting of two phases. In the first phase (preload), subjects were required to consume no, one, or two milk shakes. In the second phase, subjects were asked to taste and rate three kinds of ice cream. All experimental sessions took place after normal meal times, so any eating that took place during the experiment could be considered "excessive." At the end of the experimental session, subjects filled out an "eating habits questionnaire" that was in fact the restraint scale. The restrained and unrestrained experimental groups contained subgroups of both lean and obese subjects. The results demonstrated that restrained subjects, regardless of body weight, consumed more ice cream after preloads than after no preload. In contrast, the unrestrained subjects decreased their ice cream consumption as a function of the size of the preload: i.e., they consumed less ice cream after preloads than after no preload. The data for restrained eaters were taken to indicate that the preloads removed or disinhibited restraint, which resulted in subsequent increases in eating. It is as though the restrained eaters told themselves that after consuming the milk shakes, their diet was ruined anyway, so they might as well "go all the way" and eat until totally satisfied.

It is of interest to note that restrained eaters in the no-preload condition subsequently ate less ice cream than their unrestrained counterparts, presumably because the self-control of the restrained eaters had not been released and they were still successfully restraining their eating behavior. In summary, the results suggest that restraint (which presumably is a reflection of deprivation) may be a better predictor of eating behavior than body weight or obesity.

After receiving experimental support for the idea that a release of inhibition could increase food intake in restrained eaters, proponents of the restraint concept wanted to further examine the effects of disinhibition on eating behavior. Since alcohol is traditionally thought to disinhibit behavior, researchers argued that it might differentially affect the eating behavior of restrained and unrestrained eaters (Polivy and Herman, 1976a,b). Indeed, experiments showed that when subjects were given alcohol as a preload, and were cognitively aware that they were consuming alcohol, subsequent ice cream consumption differed for restrained and unrestrained subjects. As with milk shake preloads, alcohol preloads actually increased the ice cream consumption of the restrained eaters, while it decreased consumption by unrestrained eaters. The results were taken to indicate that alcohol acted as a releaser of inhibition in the restrained eaters.

The disinhibition concept has been further extended to explain the effects of emotional factors on eating behavior. Studies have evaluated the effects of both anxiety (Herman and Polivy, 1975) and depression (Polivy and Herman, 1976c; Frost *et al.*, 1982) on the eating behavior of restrained and unrestrained individuals. Both anxiety and depression increased eating in restrained eaters and decreased food intake in unrestrained individuals. Emotional stress is seen as simply another disinhibitor that releases the self-control of restrained eaters and thus leads to increased food intake. This concept can be contrasted with the psychosomatic view, which sees stress-induced overeating as a way to cope with and reduce emotional distress.

In summary, the set-point theory of body weight and the concept of restrained–unrestrained eating propose that individual differences in eating behavior can best be explained by variations in relative deprivation levels, rather than body-weight differences. The relative deprivation level can be measured by a restraint scale that presumably differentiates between restrained and unrestrained eaters. Investigators simply use the median restraint score as the dividing line between restrained and unrestrained eaters. Any factor that can disinhibit restraint is expected to increase food intake in restrained individuals, because they are presumably below their set-point and in a state of relative deprivation. While very interesting, these concepts are difficult to evaluate critically, since both set-point and restraint are theoretical constructs that can be measured only indirectly. Further, it is not clear that restrained and unrestrained eaters are two mutually exclusive groups. Even proponents of the theory (Herman and Mack, 1975) suggest that the restraint measure may be a continuous variable, so it seems unwarranted to use the median restraint score as the dividing line between restrained and unrestrained eaters.

2.2.6. Use of Salivation as a Dependent Measure in Obesity Research

It is clear from the different studies cited above that food intake is the most commonly used dependent measure in experiments on obesity. But the use of food intake as a measure of hunger can confound hunger with voluntary cognitive control (S. C. Wooley and O. W. Wooley, 1973). A person obviously can be hungry and still not eat if he or she decides to exercise self-control. An attempt to find a measure of appetite that would be independent of the subject's conscious control led investigators to the use of the salivary response as a dependent measure. Salivation to the sight or thought of food was seen as an objective, involuntary measure of hunger or appetite. Preliminary research (S. C. Wooley and O. W. Wooley, 1973) showed that salivation increased with food deprivation and was correlated with subjective hunger ratings. Salivation thus seemed to reflect the internal physiological need state of the organism. Since obese persons were thought to be insensitive to internal cues, it was predicted that their salivary response might not vary as a function of their internal state of hunger or satiety. To test this hypothesis, investigators typically have used the preload method. In salivation studies, subjects are first given preloads of varying caloric density to produce different internal states, and then the salivary response to the sight or thought of food is measured. In addition to preloads, different periods of food deprivation have also been used to manipulate the subject's internal state of hunger. As Table VIII shows, there is some evidence (O. W. Wooley *et al.*, 1975) that the salivary response of obese subjects does not vary as a function of internal state. The obese showed the same salivary response following a high- and a low-calorie preload, while normal-weight individuals showed a much lower salivary response after the high- than after the low-calorie preload. On the other hand, there is also evidence (Durrant and Royston, 1979) that the obese do show differences in salivation as a function of the caloric density of the preload, which suggests that the obese are sensitive to internal cues. As with other lines of research on obesity, salivation experiments unfortunately do not show definite, clear-cut differences between obese and lean subjects. As Table VIII suggests, factors such as the length of the experiment, the degree of obesity, or the wish to lose weight might all affect the outcome of the experiment, but these variables are difficult to control.

2.2.7. General Problems with Research on Psychological Factors and Obesity

Taken together, the large number of different studies on psychological factors and human obesity unfortunately paint a rather disappointing picture. Despite fascinating theoretical formulations, researchers have been able to find very few, if any, undisputable differences between obese and lean individuals. While there are many possible explanations for the empirical discrepancies, one of the most basic problems is that obesity itself is not clearly defined. Typically, experimenters simply describe obese subjects as having a body

Table VIII. Salivation as a Dependent Measure in Obesity Research

Reference	Subjects	Food	Location	Design	Results
S. C. Wooley and O. W. Wooley (1973)	Employees in med. school 2 Lean (1 male, 1 female) 2 Obese (male +15%, female +35%)	"Typical lunch foods" Palatable: pizza and banana splits Unpalatable: bad pizza	"Comfortable" nonlab setting, at lunch time	Measured salivation to the thought and sight of food, before and during exposure to familiar foods. Also looked at effects of mild deprivation and palatability.	Salivation increased with both the thought and the sight of food, in both obese and lean persons. Greatest salivary response seen after deprivation, followed by response just before normal meal time, with lowest salivation response just after eating. Salivation responses correlated with hunger ratings ($r = 0.75$) and food appeal ratings ($r = 0.46$).
O. W. Wooley et al. (1975)	Male prisoners 11 Obese (+34%, age: 24.5) 11 Norm.-wt. (age: 21.1)	Preload: High-calorie (900 kcal) or Low-calorie (450 kcal) liquid meal Test stimulus: chocolate cake or fish sandwich	Prison laboratory	Looked at effects of high- and low-calorie preloads on appetite 1 hr later. Salivation used as "involuntary" measure of appetite. Compared baseline (no food) salivation to salivation during presentation of palatable food.	Obese showed no differences in salivation after high- and low-cal preloads. In both conditions, salivation was significantly greater than base line. Salivation response of lean was significantly lower after high-cal than after low-cal preloads. Hunger ratings were unaffected by the different preloads in both

Reference	Subjects	Stimulus/food	Setting	Procedure	Results/comments
Durrant and Royston (1979)	Expt. 1: 18 Obese female pts., (age: 33, wt.: +95 kg) Expt. 2: 14 Obese female pts. (age: 40, wt.: +99 kg)	Expt. 1 Preload: Hi-cal (300 kcal) or lo-cal (100 kcal) soup or milk shake Test stimulus: Sandwich or mousse Expt. 2 Preload: 200-kcal soup and 1 g methylcellulose Same test food	Weight-loss unit in hospital	Expt. 1: Examined effects of high- and low-cal preloads on salivation to sight of meal 1 hr later. Also used hunger ratings. Expt. 2: Looked at effects of methylcellulose, in addition to preloads.	groups of subjects. Lean rated food higher after low- than after high-cal preload. Salivation, as well as hunger ratings, were significantly greater after low- than after high-cal preloads. Methylcellulose had no significant effect on salivation. Results suggest that obese are sensitive to differences in caloric content of preload (i.e., sensitive to internal cues). Contradicts Wooley and co-workers.
Durrant and Royston (1980)	Expt. 1: 8 Obese female pts. (age: 42, wt.: +102 kg) Expt. 2: 16 Obese female pts. (age: 34, wt.: +93 kg)	Expt. 1 800 kcal/day of varied foods. Expt. 2 Daily intake (kcal): 1000 (wk 1), 500 (wk 2), 1000 (wk 3)	Weight-loss unit in hospital	Expt. 1: Examined long-term effects (3 wk) of 800 kcal/ day on salivation, hunger ratings, and estimates of energy intake. Expt. 2: Same measures, but varied daily kcal: 1000, wk 1; 500, wk 2; 1000, wk 3.	Expt. 1: Salivation measures and hunger ratings declined over 3-wk period on low-cal regime (opposite to short-term effects seen in previous study). Estimates of caloric intake were unchanged. Expt. 2: Salivation related to caloric intake: highest wk 1, lowest wk 2, intermediate wk 3.

(Continued)

Table VIII. (Continued)

Reference	Subjects	Food	Location	Design	Results
Guy-Grand and Goga (1981)	14 Obese females (ages: 15–49, wt.: +20–156%; 7 dynamic, 7 static obese) 10 Norm.-wt. females (ages: 19–27)	Food considered palatable by each subject	Obesity clinic	Salivation to the sight of palatable food after 16-hr fast measured in nonobese, dynamic obese, and static obese females.	Presentation of palatable food increased salivation by 75% in dynamic obese, which was significantly greater than the 19.1% in nonobese and 26.7% in static obese. Suggests that if set-point is taken into consideration, obese persons show normal control mechanisms for food intake.
Klajner *et al.* (1981)	Female undergrads Expt. 1 Lean: 12 Dieters 12 Nondieters Obese: 10 Dieters 10 Nondieters Expt. 2 Lean: 10 Dieters 10 Nondieters Obese: 10 Dieters 6 Nondieters	Expt. 1: Hot pizza Expt. 2: Palatable: Freshly made chocolate chip cookies Unpalatable: Green cookies	Laboratory	Expt. 1: Measure anticipatory salivation to palatable food in obese and nonobese dieters and nondieters. Expt. 2: Same design, but food palatability was manipulated.	Expt. 1: Regardless of weight, dieters salivated significantly more than nondieters. Suggests that dieting rather than body weight is best predictor of anticipatory salivation. Expt. 2: Results with palatable food same as Expt. 1; i.e., dieters in both weight groups salivated more than nondieters. No differences between any groups found with unpalatable food.

weight that is a certain percentage above the Metropolitan Life Insurance Company (1959) tables, which are nothing but statistical averages for large numbers of people. The percentage overweight of obese subjects in different studies, however, varies anywhere from $+15\%$ (even lower in some cases) to several hundred percent above the "normal average." A person who is 15% above ideal weight is obviously quite different from someone who is 150% above the "normal average," and yet they are often treated as the same. Some studies do differentiate between "overweight" ($< +40\%$) and "obese" ($> +60\%$) subjects (e.g., Rodin, 1977; Rodin *et al.*, 1977), and they have in fact found differences between these two subgroups of subjects. So it is not surprising that comparisons of studies in which the "obese" experimental groups were not equally overweight, yield conflicting conclusions. All individuals who are simply above an arbitrary statistical average obviously do not form a homogeneous group. Similarly, individuals who are grouped together simply because they are "nonobese" can also be very heterogeneous. Some studies have subdivided the "nonobese" into groups of "normal weight" and "underweight" subjects (e.g., Nisbett, 1968b; Nisbett and Storms, 1974) and actually found them to respond differentially. It thus seems crucial to use consistent objective criteria in choosing "obese" and "nonobese" subject groups.

The differences in experimental definitions of obesity are only one of the problems that make comparisons and definite conclusions difficult. The generalizability of any possibly important findings is also often limited, since the majority of studies on obesity employ college students (mostly males) as subjects. Many young people gain weight when entering college, possibly because of a new irregular life-style that might include various "social eating binges," like midnight visits to the pizza parlor or the "all-night" bakery. A basically healthy young undergraduate who is overweight mainly because of overindulgence is obviously very different from, for example, an obese middle-aged person who has been forced to exist on cheap high-starch foods for years because of economic necessity. While most experiments have used a highly selective undergraduate subject population, some studies do involve other subject groups, such as prison volunteers (Decke, 1971), clients at weight-loss clinics (Rodin *et al.*, 1977), or airline personnel (Goldman *et al.*, 1968).

To be able to precisely measure eating behavior and standardize other experimental variables, most investigators study feeding behavior in the laboratory. Unfortunately, the laboratory is a very artificial feeding situation that might not reveal the typical eating behavior of an individual. Subjects might feel self-conscious about eating, particularly if they are overweight, so the feeding behavior observed in the laboratory might have little to do with the way subjects normally eat. Some obesity studies have been done in weight-loss clinics (e.g., Robinson *et al.*, 1980), which might result in very misleading interpretations about the eating behavior of obese persons. Obese individuals are obviously in weight-loss clinics to lose weight, so one would hardly expect them to eat freely, but rather to be particularly concerned about controlling their food intake and "doing the right thing."

In addition to being artificial, laboratory studies on eating behavior also are typically based on one single session with each subject. Since the eating behavior of most people fluctuates from day to day, information from only one laboratory session probably does not give an accurate picture of a person's eating behavior.

It is obviously quite difficult to reconcile the problems discussed above, since the switch from one type of experimental paradigm to another is often simply a trade-off between different sets of problems. For example, "naturalistic" studies can examine "real-life" eating behavior, but these experiments often involve very imprecise measures of food intake. Most data are based on visual assessment of various aspects of feeding behavior, which often requires subjective decisions by the observer. Laboratory studies, on the other hand, can use more precise measures of food intake, but these studies are quite artificial and might not give a realistic picture of everyday feeding behavior.

Even if some of the aforementioned methodological problems could be solved, investigators still might fail to find clear-cut differences in the behaviors of obese and lean individuals. Since obesity has multiple determinants, and the obese thus form a very heterogeneous group, certain experimental findings might apply only to specific subgroups of overweight persons. While some obese individuals might indeed be particularly sensitive to external cues or eat in response to emotional stress, many obese persons may not eat any more or any differently than normal-weight individuals. Rather than accusing the obese of being weak and overindulgent, and assuming that the answer to their condition lies in their behavior, we should perhaps also look for nonpsychological determinants of obesity. In fact, there is now accumulating suggestive evidence, as will be seen in Section 3, that some cases of obesity may be closely related to metabolic disturbances, rather than simply being a reflection of overindulgence.

3. Metabolic Obesity: The New Direction for the 1980s?

3.1. A Hypothesized Role for Brown Adipose Tissue in Diet-Induced Thermogenesis in Obesity

If the work of Schachter and colleagues served as the cornerstone for the study of obesity during the 1970s, experiments initiated in 1979 by Rothwell and Stock may provide the foundation for research during the 1980s (for a review, see Rothwell *et al.*, 1982). These experiments have led to the hypothesis that individual differences in energy utilization may play an important role in the determination of body weight.

3.1.1. Overnutrition and Thermogenesis in Experimental Animals

Interest in nutritional obesity led Rothwell and Stock to examine energy regulation in rats offered a selection of palatable food such as cookies, marshmallows, potato chips, ham, and salami (Sclafani, 1978). Although these "cafeteria-fed" rats outgained control animals maintained on a standard laboratory diet, they did not gain as much weight as would be predicted from their excess caloric intake. Thus, "cafeteria-fed" rats were less efficient at retaining dietary energy than rats given a single nutritionally complete diet.

Since the laws of thermodynamics have yet to be disproved, the excess calories that the "cafeteria-fed" rats did not store must have been expended as energy. While there are several ways in which these surplus calories could be expended (i.e., an increase in basal metabolic rate or an increase in physical activity), results of a number of experiments now indicate that these calories most probably were dissipated as heat (e.g., Rothwell and Stock, 1979; Himms-Hagen *et al.*, 1981; Teague *et al.*, 1981). Recent work has led to the hypothesis that the increase in metabolic heat production associated with overnutrition, or diet-induced thermogenesis (DIT), is mediated via brown adipose tissue (BAT) (e.g., Himms-Hagen *et al.*, 1981; Rothwell and Stock, 1979, 1981a).

BAT is a very specialized organ located primarily in the interscapular and paraspinal regions. BAT gets its characteristic brown color from its many mitochondria, which contain a high concentration of iron-containing cytochrome pigments (Dawkins and Hull, 1965; Smith and Horowitz, 1969). The abundance of mitochondria, the major sites of heat production in the cell, make BAT well suited for a role in thermogenesis. Initial support for the hypothesis that BAT may be important in DIT came from the observation that "cafeteria-fed" rats had significantly more interscapular BAT than controls given a single diet (Rothwell and Stock, 1979). Subsequent experiments have shown that BAT is not only larger, but also more metabolically active in "cafeteria-fed" animals than in controls (e.g., Brooks *et al.*, 1980; Himms-Hagen *et al.*, 1981; Rothwell and Stock, 1981a,b; Tulp, 1981).

Increases in BAT weight are also observed in other situations in which animals are given the opportunity to overeat. Rats given access to a sugar solution in addition to standard laboratory diet consumed approximately 15–20% more calories per day than animals given only the standard diet (Kanarek and Hirsch, 1977; Kanarek and Marks-Kaufman, 1979, 1981; Teague *et al.*, 1981; Kanarek and Orthen-Gambill, 1982). In this situation, as in cafeteria feeding, animals do not always gain as much weight as might be predicted from their excess caloric intake. It was recently found that animals with access to a sugar solution had significantly more BAT than animals not given sugar (Teague *et al.*, 1981; Kanarek and Orthen-Gambill, 1982). These findings are particularly interesting in light of work demonstrating that (1) sucrose feeding leads to an increase in sympathetic nervous system activity (Young and Landsberg, 1977; Landsberg and Young, 1981, 1982) and (2) sympathetic activity mediates increased thermogenesis in BAT (Foster and Frydman, 1978).

3.1.2. A Defect in Brown Adipose Tissue in Genetically Obese Animals

While the "cafeteria-fed" rat is less efficient than its lean control in turning energy consumed into weight gain, the genetically obese (*ob/ob*) mouse appears to be more efficient at storing caloric intake as body weight than its lean control. Although overeating is characteristic of the *ob/ob* mouse, hyperphagia is not necessary for the increased deposition of body fat. When the food intake of the obese mouse is restricted to the level of lean littermates, obesity still develops (Alonso and Maren, 1955). The increased metabolic efficiency of the *ob/ob* mouse may be coupled with an impairment in its thermoregulatory abilities. The first indication of this impairment came from early studies in our laboratory that demonstrated that *ob/ob* mice were more sensitive to cold stress than their lean littermates (Mayer and Barnett, 1954; Davis and Mayer, 1954). Later research established that this impairment in thermoregulation preceded the development of obesity (Trayhurn *et al.*, 1977). A defect in thermoregulatory abilities was apparent as early as 12 days of age, a time prior to the development of either hyperphagia or obesity (Rath and Thenen, 1979).

It now appears that the defect in thermoregulation in the *ob/ob* mouse is related to an abnormality in BAT (e.g., Himms-Hagen and Desautels, 1978; Hogan and Himms-Hagen, 1980; Thurlby and Trayhurn, 1980). Work by Himms-Hagen and Desautels (1978) demonstrated that although the total amount of interscapular BAT and the number of mitochondria in the tissue were normal in *ob/ob* mice, the mitochondria themselves were not normal. This abnormality was associated with a decreased heat production in BAT. Additionally, regional blood flow to BAT was impaired in *ob/ob* mice when thermogenesis was induced by norepinephrine administration (Thurlby and Trayhurn, 1980). This defect in BAT may be an important contributory factor to the development of obesity in *ob/ob* mice. If *ob/ob* mice are unable to dissipate excess energy as heat, they presumably store this energy in adipose tissue and thereby become obese.

Impairments in BAT may not be limited to the *ob/ob* mouse, but may also be associated with other animal models of obesity. Interestingly, the VMH-lesioned rat, previously thought to be an indiscriminate glutton, may have a defect in BAT. As for the *ob/ob* mouse, hyperphagia is not necessary for the development of obesity in VMH-lesioned animals. Pair-feeding VMH-lesioned rats with lean controls does not eliminate all of either the excess weight gain or the adiposity seen in the VMH-lesioned animals. Research by Seydoux *et al.* (1981) has demonstrated that the metabolic capacity of BAT is severely depressed following destruction of the VMH. Analogous abnormalities in BAT have also been observed in other types of genetically obese rodents including the Zucker (*fa/fa*) rat and the diabetic (*db/db*) mouse (Trayhurn and Fuller, 1980; Rothwell *et al.*, 1982; Seydoux *et al.*, 1981; Levin *et al.*, 1982).

3.1.3. Is There a Role for Diet-Induced Thermogenesis in Body-Weight Regulation in Humans?

These intriguing data on obese animals have promoted renewed interest in the role of thermogenesis in the regulation of body weight in humans. More than 80 years ago, Neumann (1902) suggested that excess caloric intake may be dissipated as heat. For a 3-year period, Neumann carefully monitored both his daily caloric intake and his body weight. During the first year of this endeavor, he was able to maintain a stable body weight while consuming 1766 kcal a day. Neumann increased his caloric intake to 2199 kcal a day for the second year and then to 2403 kcal a day for the third year. Despite these elevations in caloric intake, Neumann gained only minimal amounts of weight. Neumann hypothesized that the excess calories he consumed, but did not store, were being directly oxidized to heat. Neumann coined the term *luxusconsumption* to describe the heat production resulting from overnutrition. While some researchers came to Neumann's defense (Gulick, 1922; D. S. Miller and Mumford, 1967; D. S. Miller *et al.*, 1967), most others dismissed the concept of "luxusconsumption."

Despite the general lack of interest in the role of thermogenesis in the regulation of body weight, there were other indications that caloric intake might not be a good predictor of body weight. Large variations exist in individuals' abilities to utilize calories. This becomes evident from the fact that two individuals with similar body weights and levels of activity may require very different amounts of food to maintain stable body weights (Rose and Williams, 1961). For example, Rose and Williams (1961) found that pairs of individuals specifically matched to have similar body weights and levels of activity did not have similar caloric intakes. In fact, in some pairs, the daily caloric intake of one individual was twice that of the other. Further evidence of the variability in individuals' abilities to use calories comes from studies conducted by Sims *et al.* (1973) at the University of Vermont. In studies attempting to produce experimental obesity in humans, these researchers were surprised to observe substantial differences in the ability of individuals to gain weight when overfed. Consuming approximately twice their normal daily caloric intake, some individuals easily gained weight and rapidly attained the desired level of obesity. In contrast, others, eating just as much, had difficulty gaining weight and became obese only after an extended period of overfeeding. Some individuals, in fact, never did achieve their obese goal. The preceding observations make it obvious that weight gain is not a necessary consequence of gastronomic indiscretion.

Conversely, some unfortunate individuals appear to eat only modest amounts and yet gain weight. For example, when dietary records of obese and lean individuals were compared, the obese were found to consume no more than the nonobese (M. L. Johnson *et al.*, 1956; Stefanik *et al.*, 1959; Bullen *et al.*, 1964).

Recent research suggests that a defect in thermoregulatory abilities may explain why some individuals gain weight without excessive caloric intake. Jung *et al.* (1979) compared the thermogenic response to norepinephrine in lean, obese, and formerly obese women who were maintaining stable body weights by restricting food intake. Obese women, who gained weight readily and had a familial history of obesity, exhibited less of a thermogenic response to norepinephrine than normal weight women with no history of obesity. Since the formerly obese women also displayed a decreased response to norepinephrine, it appears that a defect in thermogenesis is not the direct consequence of obesity. Complementary results were obtained by Pittet *et al.* (1967), who examined the thermogenic response of lean and obese subjects to oral glucose loads, and Shetty *et al.* (1981), who measured the thermogenic response after a liquid meal in lean, obese, and formerly obese women. In both cases, obese individuals displayed a reduced thermogenic response. In addition, recent work has offered preliminary evidence that the reduced thermogenic response of the obese may be due to a defect in BAT. Contaldo *et al.* (1981) measured oxygen consumption and skin temperature over the interscapular region, an area presumed to be a site of BAT, prior to and after the administration of the sympathomimetic agent Aleudin, in lean and overweight males. After Aleudin administration, lean individuals displayed an increase in both oxygen consumption and surface temperature over the interscapular area. In contrast, obese subjects displayed no increase in oxygen consumption and only minor elevations in intrascapular skin temperature after the administration of Aleudin.

3.2. Involvement of the Metabolic Sodium–Potassium Pump in Obesity

A defect in BAT may not be the only reason for the increased energy retention displayed by some obese individuals. Modifications in total cellular energy expenditure may also play a role in energy retention in the obese. One of the major energy-using processes in the body is the active transport of sodium (Na^+) and potassium (K^+) across the cell membrane. This process may account for up to half the basal metabolic rate. The active transport of Na^+ out of the cell and K^+ into the cell is accomplished by the membrane-bound sodium–potassium pump. The energy that drives this pump is provided by the energy storehouse of the body, adenosine triphosphate (ATP). The activity of the sodium pump can be determined by measuring levels of the membrane-bound enzyme sodium potassium-dependent adenosine triphosphatase (Na^+-K^+-ATPase). It has recently been hypothesized that a deficit in sodium–potassium pump activity and concomitant reduction in ATP utilization may contribute to the increased energy retention in some forms of obesity (DeLuise *et al.*, 1980; Romsos, 1981). In support of this hypothesis, DeLuise *et al.* (1980) reported that relative to a group of lean individuals, obese individuals had a 22% reduction in the number of sodium–potassium pump units in red blood cells. Moreover, the number of pump units was found to be inversely related to the degree of obesity. While DeLuise and colleagues did not establish whether a deficiency in Na^+-K^+-ATPase is an important factor in the etiology of obesity or only a secondary consequence of increased adiposity, they provided some suggestions that the former may be the case. First, they found that weight loss was not associated with modifications in Na^+-K^+-ATPase activity. More specifically, no changes in Na^+-K^+-ATPase activity were observed following dietary restrictions in obese subjects who had lost up to 35% of their body weight. Second, two extremely obese subjects actually displayed greater Na^+-K^+-ATPase levels than those of normal weight controls.

While DeLuise and colleagues presented an intriguing hypothesis, subsequent studies have cast doubt on the importance of the sodium–potassium pump in modulating body weight. Increased, rather than decreased, Na^+-K^+-ATPase activity has been found in obese individuals when this enzyme was measured in other tissues (Bray *et al.*, 1981) or when techniques different from those used by DeLuise *et al.* (1980) were used. These conflicting findings make it obvious that caution must be exercised in drawing conclusions about the significance of the sodium–potassium pump to the maintenance of body weight.

4. Conclusions

Taken together, the preceding information suggests that physiological factors may be of great importance in some cases of human obesity. The large number of experiments conducted during the last few years on the role of energy utilization in the determination of body weight makes evident the enthusiasm researchers hold for these new concepts. However, as has been frequently stressed, obesity is not a unitary disorder, but rather the result of many disparate factors (Mayer, 1968). Therefore, continued research on both the physiological and the psychological aspects of obesity is necessary for a more complete understanding of this major health problem.

5. References

Abramson, E. E., and Wunderlich, R., 1972, Anxiety, fear, and eating: A test of the psychosomatic concept of obesity, *J. Abnorm. Psychol.* **79**:317–321.

Adams, N., Ferguson, J., Stunkard, A. J., and Agras, S., 1978, The eating behavior of obese and nonobese women, *Behav. Res. Ther.* **16**:225–232.

Alonso, L. G., and Maren, T. H., 1955, Effect of food restriction on body composition of hereditary obese mice, *Am. J. Physiol.* **183**:284–290.

Balaguara, S., and Devenport, L. D., 1970, Feeding patterns of normal and ventromedial hypothalamic lesioned male and female rats, *J. Comp. Physiol. Psychol.* **71**:357–364.

Bellisle, F., and LeMagnen, J., 1981, The structure of meals in humans: Eating and drinking patterns in lean and obese subjects, *Physiol. Behav.* **27**:649–658.

Björntorp, P., Cairella, H., and Howard, A. N., 1981, *Recent Advances in Obesity Research III*, John Libbey, London.

Bray, G. A., 1976, *The Obese Patient*, W. B. Saunders, Philadelphia.

Bray, G. A., 1979, *Obesity in America*, NIH Publication No. 79–359, Washington, D. C.

Bray, G. A., Kral, J. G., and Björntorp, P., 1981, Hepatic sodium–postassium-dependent ATPase in obesity, *N. Engl. J. Med.* **304**:1580–1582.

Brooks, C. M., Lockwood, L., and Wiggins, J. L., 1946, A study of the effects of hypothalamic lesions on the eating habits of the albino rat, *Am. J. Physiol.* **147**:735–742.

Brooks, S. L., Rothwell, N. J., Stock, M. J., Goodbody, A. E., and Trayhurn, P., 1980, Increased proton conductance pathway in brown adipose tissue mitochondria of rats exhibiting diet-induced thermogenesis, *Nature*, **286**:274–276.

Brownell, K. D., and Stunkard, A. J., 1978, Behavior therapy and behavior change: Uncertainities in programs for weight control, *Behav. Res. Ther.* **16**:301.

Brownell, K. D., and Stunkard, A. J., 1980, Behavioral treatment for obese children and adolescents, in: *Obesity* (A. J. Stunkard, ed.), pp. 415–437, W. B. Saunders, Philadelphia.

Bruch, H., 1961, Conceptual confusion in eating disorders, *J. Nerv. Ment. Dis.* **133**:46–54.

Bullen, B., Reed, R., and Mayer, J., 1964, Physical activity of obese and nonobese adolescent girls appraised by motion picture sampling, *Am. J. Clin. Nutr.* **14**:211.

Campbell, R. G., Hashim, S. A., and Van Itallie, T. B., 1971, Studies of food-intake regulation in man: Responses to variations in nutritive density in lean and obese subjects, *N. Engl. J. Med.* **285:**1402–1407.

Cannon, W. B., 1915, *Bodily Changes in Pain, Hunger, Fear, and Rage*, 2nd ed., Appleton, New York.

Carlisle, H. J., and Stellar, E., 1969, Caloric regulation and food preferences in normal, hyperphagic, and aphagic rats, *J. Comp. Physiol. Psychol.* **69:**107–114.

Carlson, A. J., 1916, *The Control of Hunger in Health and Disease*, University of Chicago Press.

Cioffi, L. A., James, W. P. T., and Van Itallie, T. B., 1981, *The Body Weight Regulatory System: Normal and Disturbed Mechanisms*, Raven Press, New York.

Coll, M., Meyer, A., and Stunkard, A. J., 1979, Obesity and food choices in public places, *Arch. Gen. Psychiatry* **36:**795–797.

Conger, J. C., Conger, A. J., Constanzo, P. R., Wright, K. L., and Matter, J. A., 1980, The effect of social cues on the eating behavior of obese and normal subjects, *J. Pers.* **48:**258–271.

Contaldo, F., Presta, E., diBlase, G., Scalfi, L., Mancini, M., Maddalena, G., de Divitiis, O., and Rocco, P., 1981, Preliminary evidence for brown fat defect in human obesity, in: *The Body Weight Regulatory System: Normal and Disturbed Mechanisms* (L. A. Cioffi, W. P. T. James, and T. B. Van Itallie, eds.), pp. 143–152, Raven Press, New York.

Corbit, J. D., and Stellar, E., 1964, Palatability, food intake, and obesity in normal and hyperphagic rats, *J. Comp. Physiol. Psychol.* **58:**63–67.

Davis, T. R. A., and Mayer, J., 1954, Imperfect homeothermia in the hereditary obese–hyperglycemic syndrome of mice, *Am. J. Physiol.* **177:**222–226.

Dawkins, M. J. R., and Hull, D., 1965, The production of heat by fat, *Sci. Am.* **213:**62–67.

Decke, E., 1971, Effects of taste on the eating behavior of obese and normal persons, in: *Emotion, Obesity, and Crime* (S. Schachter, ed.), pp. 103–104, Academic Press, New York.

DeLuise, M., Blackburn, G. L., and Flier, J. S., 1980, Reduced activity of the red-cell sodium–potassium pump in human obesity, *N. Engl. J. Med.* **303:**1017–1022.

Dodd, D. K., Birky, H. J., and Stalling, R. B., 1976, Eating behavior of obese and normal-weight females in a natural setting, *Addict. Behav.* **1:**321–325.

Doerman, A. L., and Kronenberger, E. J., 1981, Effects of bogus preloads on the eating behavior of obese, normal, and below set-point normal women, *Psychol. Rep.* **48:**747–750.

Drabman, R. S., Hammer, D., and Jarvis, G. J., 1977, Eating rates of elementary school children, *J. Nutr. Educ.* **9:**80–82.

Drabman, R. S., Cordua, G. D., Hammer, D., Jarvis, G. J., and Horton, W., 1979, Developmental trends in eating rates of normal and overweight preschool children, *Child Dev.* **50:**211–216.

Durrant, M. L., 1980, The effect of changes in preload energy content on energy intake of obese and lean subjects, *Proc. Nutr. Soc.* **39:**87A.

Durrant, M., and Royston, P., 1979, Short-term effects of energy density on salivation, hunger, and appetite in obese subjects, *Int. J. Obesity* **3:**335–347.

Durrant, M., and Royston, P., 1980, The long-term effect of energy intake on salivation, hunger, and appetitie ratings, and estimates of energy intake in obese patients, *Psychosom. Med.* **42:**385–395.

Epstein, L. H., Parker, L., McCoy, J. F., and McGee, C., 1976, Descriptive analysis of eating regulation in obese and non-obese children, *J. Appl. Behav. Anal.* **9:**407–415.

Epstein, L. H., Masek, B., and Marshall, W., 1978, Pre-lunch exercise and lunchtime caloric intake, *Behav. Ther.* **1:**15.

Ferster, C. B., Nurnberger, J. L., and Levitt, E. B., 1962, The control of eating, *J. Math.* **1:**87–109.

Foster, D. O., and Frydman, M. L., 1978, Comparison of microspheres and 86Rb+ as tracers of the distribution of cardiac output in rats indicates invalidity of 86Rb+-based measurements, *Can. J. Physiol. Pharmacol.* **56:**97–109.

Frost, R. O., Goolkasian, G. A., Ely, R. J., and Blanchard, F. A., 1982, Depression, restraint, and eating behavior, *Behav. Res. Ther.* **20:**113–121.

Gates, J. C., Huenemann, R. L., and Brand, R. J., 1975, Food choices of obese and non-obese persons, *J. Am. Diet. Assoc.* **67:**339–343.

Gaul, D. J., Craighead, W. E., and Mahoney, M. J., 1975, Relationship between eating rates and obesity, *J. Consult. Clin. Psychol.* **43**:125.

Geller, S. E., Keane, T. M., and Scheirer, C. J., 1981, Delay of gratification, locus of control, and eating patterns in obese and non-obese children, *Addict. Behav.* **6**:9–14.

Glueck, C. J., Hastings, M. M., Allen, C., Hogg, E., Baehler, L., Gartside, P. S., Phillips, D., Jones, M., Hollenbach, E. J., Braun, B. and Anastasia, J. V., 1982, Sucrose polyester and covert caloric dilution, *Am. J. Clin. Nutr.* **35**:1352–1359.

Goldman, D., Jaffa, M., and Schachter, S., 1968, Yom Kippur, Air France, dormitory food and eating behavior of obese and normal persons, *J. Pers. Soc. Psychol.* **10**:117–123.

Goodman, N., Richardson, S. A., Dornbush, S. M., and Hastorf, A. H., 1963, Variant reaction to physical disability, *Am. Soc. Rev.* **28**:429–435.

Graff, H., and Stellar, E., 1962, Hyperphagia, obesity, and finickiness, *J. Comp. Physiol. Psychol.* **55**:418–424.

Grossman, S. P., 1966, The VMH: A center for affective reactions, satiety, or both, *Physiol. Behav.* **1**:1–10.

Grossman, S. P., 1972, Aggression, avoidance, and the reaction to novel environments in female rats with ventromedial hypothalamic lesions, *J. Comp. Physiol. Psychol.* **78**:274–283.

Gulick, A., 1922, A study of weight reduction in the adult human body during over-nutrition, *Am. J. Physiol.* **60**:371–395.

Guy-Grand, B., and Goga, H., 1981, Conditioned salivation in obese subjects with different weight kinetics, *Appetite* **2**:351–355.

Hashim, S. A., and Van Itallie, T. B., 1965, Studies in normal and obese subjects with a monitored food dispensing device, *Ann. N. Y. Acad. Sci.* **131**:654–661.

Herman, C. P., and Mack, D., 1975, Restrained and unrestrained eating, *J. Pers.* **43**:647–660.

Herman, C. P., and Polivy, J., 1975, Anxiety, restraint, and eating behavior, *J. Abnorm. Psychol.* **84**:666–672.

Herman, C. P., and Polivy, J., 1980, Restrained eating, in: *Obesity* (A. J. Stunkard, ed.), pp. 208–225, W. B. Saunders, Philadelphia.

Herman, C. P., Polivy, J., Pliner, P., and Threlkeld, J., 1978, Distractibility in dieters and non-dieters: An alternative view of "externality," *J. Pers. Soc. Psychol.* **36**:536–548.

Hetherington, A. J., and Ranson, S. W., 1942, The spontaneous activity and food intake of rats with hypothalamic lesions, *Am. J. Physiol.* **136**:609–617.

Hibscher, J. A., and Herman, C. P., 1977, Obesity, dieting, and the expression of "obese" characteristics, *J. Comp. Physiol. Psychol.* **91**:374–380.

Hill, S. W., and McCutcheon, N. B., 1975, Eating responses of obese and non-obese humans during dinner meals, *Psychosom. Med.* **37**:395–401.

Himms-Hagen, J., and Desautels, M., 1978, A mitochondrial defect in brown adipose tissue of the obese (*ob/ob*) mouse: Reduced binding in purine nucleotides and a failure to respond to cold by an increase in binding, *Biochem. Biophys. Res. Commun.* **83**:628–634.

Himms-Hagen, J., Triandafillou, J., and Gwilliam, C., 1981, Brown adipose tissue of cafeteria-fed rats, *Am. J. Physiol.* **241**:E116–E120.

Hogan, S., and Himms-Hagen, J., 1980, Abnormal brown adipose tissue in obese mice (*ob/ob*): Response to acclimation to cold, *Am. J. Physiol.* **239**:E301–E309.

Johnson, M. L., Burke, B. S., and Mayer, J., 1956, Relative importance of inactivity and overeating in the energy balance of obese high school girls, *Am. J. Clin. Nutr.* **4**:37–44.

Johnson, W. G., 1974, Effect of cue prominence and subject weight on human food-directed performance, *J. Pers. Soc. Psychol.* **29**:843–848.

Jordan, H. A., 1975, Physiological control of food intake in man, in: *Obesity in Perspective* (G. A. Bray, ed.), pp. 35–47, DHEW Publication No. (NIH) 75–708.

Jung, R. T., Shetty, P. S., James, W. P. T., Barrand, M. A., and Callingham, B. A., 1979, Reduced thermogenesis in obesity, *Nature (London)* **279**:322–323.

Kanarek, R. B., and Hirsch, E., 1977, Dietary-induced overeating in experimental animals, *Fed. Proc. Fed. Am. Sec. Exp. Biol.* **36**:154–158.

Kanarek, R. B., and Marks-Kaufman, R., 1979, Developmental aspects of sucrose-induced obesity in rats, *Physiol. Behav.* **23**:881–885.

Kanarek, R. B., and Marks-Kaufman, R., 1981, Increased carbohydrate consumption induced by neonatal administration of monosodium glutamate to rats, *Neurobehav. Toxicol. Teratol.* **3:**343–350.

Kanarek, R. B., and Orthen-Gambill, N., 1982, Differential effects of sucrose, fructose, and glucose on carbohydrate-induced obesity in rats, *J. Nutr.* **112:**1546–1554.

Kaplan, D. L., 1980, Eating style of obese an nonobese males, *Psychosom. Med.* **42:**529–538.

Kaplan, H. I., and Kaplan, H. S., 1957, The psychosomatic concept of obesity, *J. Nerv. Ment. Dis.* **125:**181–189.

Kennedy, G. C., 1950, The hypothalamic control of food intake in rats, *Proc. R. Soc., London Ser. B.* **137:**535–549.

Klajner, F., Herman, C. P., Polivy, J., and Chhabra, R., 1981, Human obesity, dieting, and anticipatory salivation to food, *Physiol. Behav.* **27:**195–198.

Krantz, D. S., 1979, A naturalistic study of social influences on meal size among moderately obese and nonobese subjects, *Psychosom. Med.* **41:**19–27.

Landsberg, L., and Young, J. B., 1981, Diet-induced changes in sympathoadrenal activity: Implications for thermogenesis and obesity, *Obesity Metab.* **1:**5–33.

Landsberg, L., and Young, J. B., 1982, Effects of nutritional status on autonomic nervous system function, *Am. J. Clin. Nutr.* **35:**1234–1240.

LeBow, M. D., 1981, *Weight Control: The Behavioral Strategies*, John Wiley, New York.

LeBow, M. D., Goldberg, P. S., and Collins, A., 1977, Eating behavior of overweight and non-overweight persons in the natural environment, *J. Consult. Clin. Psychol.* **45:**1204–1205.

Levin, B. E., Comai, K., O'Brien, R. A., and Sullivan, A. C., 1982, Abnormal brown adipose composition and β-adrenoreceptor binding in obese Zucker rats, *Am. J. Physiol.* **243:**E217–E224.

Levitz, L. S., 1975, The susceptibility of human feeding behavior to external cues, in: *Obesity in Perspective* (G. A. Bray, ed.), pp. 53–60, DHEW Publication No. (NIH) 75-708.

Mahoney, M. J., 1975, The obese eating style: Bites, beliefs, and behavior modification, *Addict. Behav.* **1:**47–53.

Marston, A. R., London, P., and Cooper, L. M., 1976, A note on the eating behavior of children varying in weight, *J. Child Psychol. Psychiatry* **17:**221–224.

Marston, A. R., London, P., Cohen, N., and Cooper, L. M., 1977, *In vivo* observation of the eating behavior of obese and nonobese subjects, *J. Consult. Clin. Psychol.* **45:**335–336.

Mayer, J., 1968, *Overweight, Causes, Costs, and Control*, Prentice-Hall, N. J.

Mayer, J., and Barnett, R. J., 1953, Sensitivity to cold in the hereditary obese–hyperglycemic syndrome of mice, *Yale J. Biol. Med.* **26:**38–45.

Mayer, J., and Thomas, D., 1967, Regulation of food intake and obesity, *Science* **156:**328–337.

Mayer, J., French, R. G., Zighera, C. F., and Barrnett, R. J., 1955, Hypothalamic obesity in the mouse: Production, description, and metabolic characteristics, *Am. J. Physiol.* **182:**75–82.

McArthur, L. Z., and Burstein, B., 1975, Field dependent eating and perception as a function of weight and sex, *J. Pers.* **43:**(3):402–420.

McKenna, R. J., 1972, Some effects of anxiety level and food cues on the eating behavior of obese and normal subjects: A comparison of the Schachterian and psychosomatic conceptions, *J. Pers. Soc. Psychol.* **22**(3):311–319.

Metropolitan Life Insurance Company, 1959, New weight standards for men and women, *Stat. Bull. Metrop. Life Ins. Co.* **40:**1–4.

Miller, D. S., and Mumford, P., 1967, Gluttony. 1. An experimental study of overeating low- or high-protein diets, *Am. J. Clin. Nutr.* **20:**1212–1222.

Miller, D. S., Mumford, P., and Stock, M. J., 1967, Gluttony. 2. Thermogenesis in overeating man, *Am. J. Clin. Nutr.* **20:**1223–1229.

Miller, N. E., Bailey, C. J., and Stevenson, J. A. F., 1950, Decreased "hunger" but increased food intake resulting from hypothalamic lesions, *Science* **112:**256–259.

Neumann, R. O., 1902, Experimentelle Beiträge zur Lehre von dem Täglichen Nahrungsbedarf des Menschen unter besonderer Berücksichtigung der notwendigen Eiweissmenge, *Arch. Hyg.* **45:**1–87.

Nisbett, R. E., 1968a, Taste, deprivation, and weight determinants of eating behavior, *J. Pes. Soc. Psychol.* **10:**107–116.

Nisbett, R. E., 1968b, Determinants of food intake in human obesity, *Science* **159:**1254–1255.

Nisbett, R. E., 1972, Hunger, obesity, and the ventromedial hypothalamus, *Psychol. Rev.* **79:**433–453.

Nisbett, R. E., and Storms, M. D., 1974, Cognitive and social determinants of food intake, in: *Thought and Feeling: Cognitive Alteration of Feeling States* (H. London and R. E. Nisbett, eds.), pp. 190–208, Aldine, Chicago.

Pittet, P., Chappuis, Acheson, A., de Techtermann, F., Jequier, E., 1976, Thermic effect of glucose in obese subjects studied by direct and indirect calorimetry, *Brit. J. Nutr.* **35:**281–292.

Pliner, P. L., 1973, Effect of external cues on the thinking behavior of obese and normal subjects, *J. Abnorm. Psychol.* **82:**233–238.

Polivy, J., and Herman, C. P., 1976a, Effects of alcohol on eating behavior: Influences of mood and perceived intoxication, *J. Abnorm. Psychol.* **85:**601–606.

Polivy, J., and Herman, C. P., 1976b, The effects of alcohol on eating behavior: Disinhibition or sedation?, *Addict. Behav.* **1:**121–125.

Polivy, J., and Herman, C. P., 1976c, Clinical depression and weight change: A complex relation, *J. Abnorm. Psychol.* **85:**338–340.

Porikos, K. P., 1981, Control of food intake in man: Response to covert caloric dilution of a conventional and palatable diet, in: *The Body Weight Regulatory System: Normal and Disturbed Mechanisms* (L. A. Cioffi, W. P. T. James, and T. B. Van Itallie, eds.), pp. 83–87, Raven Press, New York.

Porikos, K. P., Booth, G., and Van Itallie, T. B., 1977, Effect of covert nutritive dilution on the spontaneous food intake of obese individuals: A pilot study, *Am. J. Clin. Nutr.* **30:**1638–1644.

Porikos, K. P., Hesser, M. F., and Van Itallie, T. B., 1982, Caloric regulation in normal weight men maintained on a palatable diet of conventional foods, *Physiol. Behav.* **29:**293–300.

Powley, T. L., 1977, The ventromedial hypothalamic syndrome, satiety, and a cephalic phase hypothesis, *Psychol. Rev.* **1:**89–126.

Price, J., and Grinker, J., 1973, Effects of degree of obesity, food deprivation, and palatability on eating behavior of humans, *J. Comp. Physiol. Psychol.* **85:**265–271.

Rath, E. A., and Thenen, S. W., 1979, Use of tritiated water for measurement of 25-hour milk intake in suckling lean and genetically obese (*ob/ob*) mice, *J. Nutr.* **109:**840–847.

Reznick, H., and Balch, P., 1977, The effect of anxiety and response cost manipulations on the eating behavior of obese and normal-weight subjects, *Addict. Behav.* **2:**219–225.

Richardson, S., Goodman, N., Hastorf, A., and Dornbush, S., 1961, Cultural uniformity and reaction to physical disability, *Am. Soc. Rev.* **26:**241–247.

Robinson, R. G., Folstein, M. F., Simonson, M., and McHugh, P. R., 1980, Differential antianxiety response to caloric intake between normal and obese subjects, *Psychosom. Med.* **42:**415–427.

Rodin, J., 1975, Causes and consequences of time perception differences in overweight and normal weight people, *J. Pers. Soc. Psychol.* **31:**898–904.

Rodin, J., 1977, Implications of responsiveness to sweet taste for obesity, in: *Taste and Development: The Genesis of Sweet Preference* (J. M. Weiffenbach, ed.), pp. 295–308, DHEW Publication (NIH), Bethesda, Maryland.

Rodin, J., 1980, The externality theory today, in: *Obesity* (A. J. Stunkard, ed.), pp. 226–239, W. B. Saunders, Philadelphia.

Rodin, J., 1981, Current status of the internal–external hypothesis for obesity, *Am. Psychol.* **36:**361–372.

Rodin, J., and Marcus, J., 1982, Psychological factors in human feeding, *Pharmacol. Ther.* **16:**447–468.

Rodin, J., and Slochower, J., 1974, Fat chance for a favor: obese–normal differences in compliance and incidental learning, *J. Pers. Soc. Psychol.* **29(4):**554–565.

Rodin, J., Slochower, J., and Fleming, B., 1977, Effects of degree of obesity, age of onset, and weight loss on responsiveness to sensory and external stimuli, *J. Comp. Physiol. Psychol.* **91:**586–597.

Romsos, D. R., 1981, Efficiency of energy retention in genetically obese animals and in dietary-induced thermogenesis, *Fed. Proc. Fed. Am. Soc. Exp. Biol.* **40**:2524–2529.

Rose, G. A., and Williams, R. T., 1961, Metabolic studies on large and small eaters, *Br. J. Nutr.* **15**:1–9.

Rosenthal, B. S., and Marx, R. D., 1978, Differences in eating patterns of successful and unsuccessful dieters, untreated overweight and normal weight individuals, *Addict. Behav.* **3**:129–134.

Ross, L., 1974, Effects of manipulating the salience of food upon consumption by obese and normal eaters, in: *Obese Humans and Rats* (S. Schachter and J. Rodin, ed.), pp. 43–52, Earlbaum/Halsted, Washington, D.C.

Ross, L. D., Pliner, P., Nisbett, P., and Schachter, S., 1971, Patterns of externality and internality in eating behavior of obese and normal college students, in: *Emotion, Obesity and Crime* (S. Schachter, ed.), pp. 118–121, Academic Press, New York.

Rothwell, N. J., and Stock, M. J., 1979, A role for brown adipose tissue in diet-induced thermogenesis, *Nature (London)* **281**:31–35.

Rothwell, N. J., and Stock, M. J., 1981a, Regulation of energy balance, *Annu. Rev. Nutr.* **1**:235–256.

Rothwell, J. J., and Stock, M. J., 1981b, Influence of noradrenaline on blood flow to brown adipose tissue in rats exhibiting diet-incuded thermogenesis, *Pfluegers Arch.* **389**:237–242.

Rothwell, N. J., Stock, M. J., and Stribling, D., 1982, Diet-induced thermogenesis, *Pharmacol. Ther.* **17**:251–268.

Schachter, S., 1971, Some extraordinary facts about obese humans and rats, *Am. Psychol.* **26**:129–144.

Schachter, S., and Friedman, L. N., 1974, The effects of work and cue prominence on eating behavior, in: *Obese Humans and Rats* (S. Schachter and J. Rodin, eds.), pp. 11–14, Erlbaum Associates, Potomac, Maryland.

Schachter, S., and Gross, L., 1968, Manipulated time and eating behavior, *J. Pers. Soc. Psychol.* **10**:98–106.

Schachter, S., and Rodin, J. (eds.), 1974, *Obese Humans and Rats*, Erlbaum Associates, Potomac, Maryland.

Schachter, S., Goldman, R., and Gordon, A., 1968, Effects of fear, food deprivation, and obesity on eating, *J. Pers. Soc. Psychol.* **10**:91–97.

Sclafani, A., 1976, Appetite and hunger in experimental obesity syndromes, in: *Hunger: Basic Mechanisms and Clinical Implications* (D. Novin, W. Wyrwicka, and G. A. Bray, eds.), pp. 281–296, Raven Press, New York.

Sclafani, A., 1978, Dietary obesity, in: *Recent Advances in Obesity Research 2* (G. A. Bray, ed.), pp. 123–132, Newman, London.

Seydoux, J., Rohner-Jeanrenaud, F., Assimacopoulos-Jeannet, F., Jeanrenaud, B., and Girardier, L., 1981, Functional disconnection of brown adipose tissue in hypothalamic obesity in rats, *Pfluegers Arch.* **390**:1–4.

Shetty, P. S., Jung, R. T., James, W. P. T., Barrand, M. A., and Callingham, B. A., 1981, *Clin. Sci.* **60**:519–525.

Shisslak, C. M., 1978, Naturalistic observations of eating patterns in humans: Relationships between obesity and eating style, *Diss. Abstr. Int.* **38**:3416B.

Silverstone, T., 1982, *Drugs and Appetite*, Academic Press, New York.

Sims, E. A. H., Danforth, E., Horton, E. S., Bray, G. A., Glennon, T. A., and Salans, L. B., 1973, Endocrine and metabolic effects of experimental obesity in man, *Recent Prog. Horm. Res.* **29**:457–469.

Singh, D., 1969, Comparison of hyperemotionality caused by lesions in the septal and ventromedial hypothalamic areas in the rat, *Psychon. Sci.* **16**:3–4.

Singh, D., and Sikes, S., 1974, Role of past experience of food-motivated behavior of obese humans, *J. Comp. Physiol. Psychol.* **86**:503–508.

Slochower, J., 1976, Emotional labeling and overeating in obese and normal weight individuals, *Psychosom. Med.* **38**:131–139.

Slochower, J., and Kaplan, S. P., 1980, Anxiety, perceive control, and eating in obese and normal weight persons, *Appetite* **1**:75–83.

Slochower, J., Kaplan, S. P., and Mann, L., 1980, The effects of life stress and weight on mood and eating, *Appetite* **1**:115–125.

Smith, R. E., and Horowitz, B. A., 1969, Brown fat and thermogenesis, *Physiol. Rev.* **49**:330–425.

Spencer, J. A., and Fremouw, S., 1979, Binge eating as a function of restraint and weight classification, *J. Abnorm. Psychol.* **88**:262–267.

Spiegel, T., 1973, Caloric regulation of food intake in man, *J. Comp. Phsiol. Psychol.* **84**:24–37.

Stefanik, P. A., Heald, F. P., and Mayer, J., 1959, Caloric intake in relation to energy output of obese and non-obese adolescent boys, *Am. J. Clin. Nutr.* **7**:55–62.

Strominger, J. L., Brobeck, J. R., and Cort, R. L., 1953, Regulation of food intake in normal rats and in rats with hypothalamic hyperphagia, *Yale J. Biol. Med.* **26**:55–74.

Stuart, R. B., 1967, Behavioral control of overeating, *Behav. Res. Ther.* **5**:357–365.

Stuart, R. B., and Davis, B., 1972, *Slim Chance in a Fat World: Behavioral Control of Obesity*, Research Press, Illinois.

Stunkard, A. J., 1980, *Obesity*, W. B. Saunders, Philadelphia.

Stunkard, A. J., and Fox, S., 1971, The relationship of gastic motility and hunger: A summary of the evidence, *Psychosom. Med.* **33**:123–134.

Stunkard, A. J., and Koch, C., 1964, The interpretation of gastic motility. 1. Apparent bias in the reports of hunger by obese persons, *Arch. Gen. Psychiatry* **11**:74–82.

Teague, R. J., Kanarek, R. B., Bray, G. A., Glick, Z., and Orthen-Gambill, N., 1981, Effect of diet on the weight of brown adipose tissue in rodents, *Life Sci.* **29**:1531–1536.

Teitelbaum, P., 1955, Sensory control of hypothalamic hyperphagia, *J. Comp. Physiol. Psychol.* **48**:156–163.

Teitelbaum, P., 1957, Random and food directed activity in hyperphagic and normal rats, *J. Comp. Physiol. Psychol.* **50**:486–490.

Thomas, D. W., and Mayer, J., 1968, Meal taking and regulation of food intake by normal and hypothalamic hyperphagic rats, *J. Comp. Physiol. Psychol.* **66**:642–653.

Thurlby, P. L., and Trayhurn, P., 1980, The role of thermoregulatory thermogenesis in the development of obesity in genetically obese (*ob/ob*) mice pair-fed with lean siblings, *Br. J. Nutr.* **42**:377–385.

Tom, G., and Rucker, M., 1975, Fat, full and happy: Effects of food deprivation, external cues, and obesity on preference ratings, consumption, and buying intentions, *J. Pers. Soc. Psychol.* **32**:761–766.

Trayhurn, P., and Fuller, L., 1980, The development of obesity in genetically diabetic-obese (*db/db*) mice pair-fed with lean siblings, *Diabetologia* **19**:148–153.

Trayhurn, P., Thurlby, P. L., and James, W. P. T., 1977, Thermogenic defect in pre-obese *ob/ob* mice, *Nature (London)* **266**:60–62.

Tulp, O. L., 1981, The development of brown adipose tissue during experimental overnutrition in rats, *Int. J. Obesity* **5**:579–591.

Wagner, M., and Hewitt, M. I., 1975, Oral satiety in the obese and nonobese, *J. Am. Diet. Assoc.* **67**:344–346.

Warner, K. E., and Balagura, S., 1975, Intrameal eating patterns of obese and nonobese humans, *J. Comp. Physiol. Psychol.* **89**:778–783.

Waxman, M., and Stunkard, A. J., 1980, Caloric intake and expenditure of obese boys, *J. Pediatr.* **96**:187–193.

Wooley, O. W., 1972, Long-term food regulation in the obese and nonobese, *Psychosom. Med.* **33**:436–444.

Wooley, O. W., Wooley, S. C., and Dunham, R. B., 1972, Can calories be perceived and do they affect hunger in obese and nonobese humans?, *J. Comp. Physiol. Psychol.* **80**:250–258.

Wooley, O. W., Wooley, S. C., and Woods, W. A., 1975, Effect of calories on appetite for palatable food in obese and nonobese humans, *J. Comp. Physiol. Psychol.* **89**:619–625.

Wooley, S. C., 1972, Physiologic versus cognitive factors in short term food regulation in the obese and nonobese, *Psychosom. Med.* **34:**62–68.

Wooley, S. C., and Wooley, O. W., 1973, Salivation to the sight and thought of food: A new measure of appetite, *Psychosom. Med.* **35**(2):136–142.

Young, J. B., and Landsberg, L., 1977, Stimulation of the sympathetic nervous system during sucrose feeding, *Nature (London)* **269:**615–617.

11

Psychological Factors Affecting Food Selection

Daisy Lau, Magdalena Krondl, and Patricia Coleman

1. Introduction

Food selection, a multidimensional behavior leading to food intake, the ingestive process, is dependent on access to edible substances and the availability of options. It incorporates the concept of food acceptance and rejection and ultimately results in personal eating patterns. Although food selection overlaps with food preference, the terms are not synonymous; food preference presupposes ranking of foods according to degree of liking or disliking, whereas food selection involves actual choices based on the influence of various exogenous and endogenous factors.

Factors affecting food selection include physiological controls, sensory considerations, and the learning mechanism. Their relative effect differs from one person to another. Flexibility exists, since food selection is a dynamic process that increases in complexity as individuals acquire experience and beliefs about food, increase social activity, and respond to environmental changes. Genetic differences also occur; these affect food choices via physiological factors. Although food-selection determinants operate in an integrated manner, they influence particular foods differently according to individual needs and wants. The role of a food may be to satisfy physiological requirements, maintain health, provide pleasure, or function as a vehicle for socialization.

An understanding of the process of food selection is vital to all health professionals applying nutritional theory. Isolating, analyzing, and integrating the numerous variables involved in food selection is now recognized as an important area of research.

Daisy Lau • Brescia College, University of Western Ontario, London, Ontario N6G 1H2. *Magdalena Krondl and Patricia Coleman* • Department of Nutritional Sciences, Faculty of Medicine, University of Toronto, Toronto, Ontario M5S 1A3, Canada.

The objectives of this chapter are to (1) discuss the role of the physiological, sensory, and cognitive factors that affect food selection; (2) consider the variability and changeability of the factors and consequently the flexibility of food selection; and (3) illustrate the contribution of food perceptions in integrating the various factors and determining their relative significance in consumption behavior.

2. Physiological, Sensory, and Cognitive Factors

Individuals select from their available food resources the items that are most useful, acceptable, and pleasing according to the particular time, internal state, and occasion. This is preceded by rejection of harmful, unknown, or disliked foods. The factors influencing food selection can be arbitrarily classified as physiological, sensory, and cognitive.

In the framework of existing knowledge, the physiological factors encompass at least two important functions. One is to ensure a supply for the body of the two important nutrients, protein and carbohydrate; the other is to protect the body from harmful substances. The sensory factors categorize foods according to degree of liking and disliking. The cognitive factors screen the sensory cues from the point of view of healthfulness and facilitate adaptation of food-related behavior to the social environment.

2.1. Physiological Factors

2.1.1. Nutrient-Related Effects on Food Selection

Food choice can be affected by previous food intake with particular nutrients such as protein and carbohydrate exerting a regulatory effect. The ability of the rat to regulate protein intake at a constant proportion of the dietary energy was reported first by Musten *et al.* (1974) with findings that the animals choose among dietary options to control the proportion of total food consumed as protein. More recent studies (G. H. Anderson, 1979) suggest that separate regulation of nutrients and energy intake exists, although it is believed that the mechanisms interact.

An understanding of the mechanisms related to control of protein intake is emerging with evidence that food selection is influenced by shifts in the plasma tryptophan concentration relative to other neutral amino acids as a result of prior food consumption (G. H. Anderson, 1979). High protein ingestion is followed by a more rapid rise in other neutral amino acids than of tryptophan in the plasma. Competition for attachment to the carrier molecule acting as the means of transport across the blood–brain barrier results in low uptake of tryptophan. Since this amino acid serves as the precursor of serotonin, a neurotransmitter triggering feeding reflexes, decreased stimulation of the sertonergic system leads to a decreased preference for protein in the next meal.

Serotonin is similarly involved in the regulation of carbohydrate intake (Wurtman and Wurtman, 1979). High carbohydrate ingestion results in increased insulin release and a reduction in the plasma level of neutral amino acids. Tryptophan levels remain comparatively high because of binding to plasma albumin (McMenamy and Oncley, 1978). Thus, comparatively more tryptophan reaches the neurons. Increased serotonin synthesis may lead to a decreased preference for carbohydrate.

The desire to consume a nutrient such as protein or carbohydrate is related to its depletion, which in turn depends on the interval between meals. Thus, the regulatory mechanism is presumably manifested as a feeling of satiation caused by the nutrients. Since protein and carbohydrate are contained in particular foods, it could be postulated that these foods would participate in a specific satiating effect.

2.1.2. Food Intolerance

Food aversion, intolerance, and sensitivity affect food selection through food rejection and may be categorized as physiological in nature. Food aversion may develop through conditioning when the substance in a prior experience has been followed by negative internal consequences and illness, even though no disagreeable sensory experience has taken place. Subsequent consumption of the substance decreases. Garcia *et al.* (1974) demonstrated that in rats, gustatory cues can play a major role in the acquisition of conditioned food aversion. This may occur with a single flavor–illness pairing even though the ingestion preceded the illness by several hours (Smith and Roll, 1967). Humans acquire aversions for flavors eaten just prior to an illness despite their awareness that the food was not the cause of their symptoms. Bernstein and Webster (1980) found that adult cancer patients acquired an aversion to a distinctly flavored ice cream consumed before receiving a single chemotherapeutic drug treatment. Conditioned aversions appear to be independent of cognitive processes (Garcia and Brett, 1977).

Food intolerance is a term generally reserved for illness resulting from the ingestion of a substance that has nutritive potential (Herman and Hagler, 1979). Lactose intolerance is the most common. It may be present at birth or occur later in life. It is due to a deficiency of the jejunal brush border enzyme, lactase. Osmotic diarrhea occurs together with fermentation of the ingested lactose by intestinal microorganisms. The gene for maintaining a high lactase level is dominant and not sex-linked (Kretchmer, 1972). The exact mode of the hereditary defect is not well understood (Woteki *et al.*, 1976). In certain ethnic groups, intolerance to milk products may be due to factors other than lactase deficiency, as shown by double-blind studies of tolerance to dairy drinks by adolescents (Haverberg *et al.*, 1980) and elderly people (Rorick and Scrimshaw, 1979).

Food sensitivity, a designation preferred to food allergy, is the term used when considering immunologically mediated adverse reactions to food (Bock, 1980). Proteins or protein fragments cross the mucosal barrier, encounter sensitized cells, and produce a sequence of immunological reactions that result in clinically apparent manifestations. Reaginic reactions can be shown to have identifiable reagins, principally the IgE class of antibodies, provoked by a particular food. Nonreaginic reactions have been less clearly delineated. An example is celiac disease, in which the individual's intestine produces an increased amount of specific antigluten antibody in response to a challenge with gluten.

Despite the distinguishing features, the terms food aversion, intolerance, and sensitivity are often merged and described in general terms by the affected individuals as food disagreement or simply food intolerance.

2.2. Sensory Factors

A variety of sensory-related mechanisms aid the living organism in the search for food. Of the five senses, taste has the greatest impact and ultimately determines whether the food will be swallowed and more will be eaten. Vision and audition serve to establish the initial location of a food (Garcia and Brett, 1977) and to reinforce the gustatory and olfactory discrimination as pleasant or unpleasant. Texture is important with some but not all foods (Amerine *et al.*, 1965).

The taste receptors, end organs of the taste nerves located in the mouth, are activated by a large variety of different compounds and solutions (Amerine *et al.*, 1965). Although under question at various times, it has been usual to consider four basic tastes as identifiable by humans: sweet, sour, salt, and bitter. Other sensations such as astringent and metallic may also be distinguishable, but are difficult to typify verbally (Boudreau *et al.*, 1979). It has been suggested that tastes in combination may not remain distinct but produce a new sensation (Erickson, 1982).

Various taste substances may influence sensitivities to one or more of the basic tastes (Bartoshuk, 1980). These taste modifiers include sodium laurel sulfate, a detergent used in some toothpastes and mouthwashes. It reduces the sweetness effect of sugar and adds bitterness to the sourness of orange juice. The African berry "miracle fruit" induces sweetness in normally sour substances. Sodium content of saliva varies among individuals and may influence taste acuity for salt.

The olfactory system is able to identify a nearly infinite range of volatile substances. Several physiochemical properties of the molecules are recognized by the receptor sites with which they interact at the nasal sensory epithelium (Pager, 1977). Food odor is chemically more complex and stimulates a greater number of receptors than nonfood odors, as shown in studies of fish (Hara and Law, 1972). The activation level appears to be related more to the nutritional meaning of the odors than to their chemical features.

The relative contribution of the sensory- and nutrient-related cues in food selection is indicated from the study on the role of amino acids in food intake by Rogers and Leung (1977). These researchers conclude that although animals make use of the olfactory and gustatory senses to enable them to select a proper diet in a chosen situation, these senses, unlike the nutrient-related cues, are dispensible.

Closely associated with sensory sensation of food, especially taste, is hedonic quality, the experience of pleasantness or unpleasantness. The commonality of the hedonic findings has led to the idea of a biological basis of certain fundamental taste preferences. From the ontogenic studies, starting with the newborn animal and human, it appears that there are certain innate determinants of the responses to taste stimuli (Beauchamp *et al.*, 1982; Steiner, 1977). Bitter is mostly aversive, sugars are mostly preferred. Saltiness and sourness may be pleasant at low levels but unpleasant at higher values. Any generalizations about these sensations produced by food substances and beverages must be made with caution. Bitterness in beer and coffee may become attractive on experience. Stimulus intensity is only one factor influencing hedonic value that may be changed by changes in internal state (Pfaffmann, 1980).

Sensory discrimination between food and nonfood is related to body states. Olfaction is sharper in the state of hunger than in the state of satiation (Pager, 1977). Cabanac (1971) suggests that pleasure or displeasure of a sensation is not stimulus-bound but depends entirely on hunger–satiety signals related to usefulness of the stimulus in relation to the need of the subject. He proposed the term "alliesthesia" to describe a change in sensation of pleasantness depending on internal state. A study by Moskowitz (1977) has shown a limited degree of change in hedonic perception of taste after a satiating glucose load and only for sweet preference; others (Mower *et al.*, 1977; Krondl, 1982) reported that changes in hedonic value depended on internal state for some stimuli more than others, and the response varied among individuals.

2.3. Cognitive Factors

Cognition, the act of knowing, involves the processing of information in the higher brain centers. The steps involved include examination, comprehension, storage, and finally retrieval for evaluation and decision-making purposes (Olson and Sims, 1980). Although it is generally believed that behavior is a function of cognition, research has failed to demonstrate a strong relationship between storage of nutrient information and its application in food selection (Brown *et al.*, 1963; Sims, 1978). Some of the information may have been inadequately understood and recoded into personal meanings before storage, resulting in "believed" or "perceived" knowledge (Reddin and Anderson, 1978; Dugdale *et al.*, 1979). The weak association found between nutrition knowledge and food behavior may also be the result of inappropriate evaluation methods.

Cognitive factors affecting food selection may be more elementary than the acquisition of nutrition knowledge. They may be related simply to the learning of the existence of a previously unknown food, familiarization with its satiating and sensory properties, and determination of its tolerance. The food may then take on specific meanings when the utility becomes associated with healthfulness or when the food is perceived as having a social function. In addition, cognition may play a role in ascribing prestige, monetary, or convenience value to a food.

Health beliefs associated with specific foods were common among ancient cultures and reemerged in North American and other technologically advanced societies in the 1970s as a wave of food faddism. Some understanding of these beliefs has come from anthropological and ethnographical studies (Snow, 1974; Snow and Johnson, 1978).

In Old World societies, the association between food and health is often intertwined with social, cultural, and historical influences. For example, in China, foods were classified along with herbs and medications for the maintenance of health or for treatment of diseases as early as 1000 B.C. Records in the medical classic *Yellow Emperor's Book of Internal Medicine* showed that health, illness, and food were integrated into the dualistic concepts of yin and yang, sometimes loosely referred to as hot and cold (Libassi, 1974). This hot–cold dichotomy is based on the belief that various parts of the body vary in temperature and that foods and diseases can be grouped into hot and cold classes as well. Spices such as ginger are considered to be hot, while watermelon is cold.

Changes in personal income, food costs, and leisure time, along with societal emphasis on prestige or gourmet foods, may produce changes in an individual's dietary patterns. Although these social factors may affect the use of secondary and peripheral foods more than core items, they nevertheless play a major role in mediating an individual's choices within a changing food supply. They will be discussed separately as price, convenience, and prestige.

When income approaches the poverty level and the balancing of the total budget for the needs of housing, clothing, transportation, and food is in question, price considerations may limit the range of foods used. Foods perceived to be too expensive will not be purchased, and the foods selected will be mainly those required for the satisfaction of hunger and maintenance of life. The esthetic qualities of food then become secondary as choice determinants. When several affordable options are available, price considerations may be less important than eating qualities or health aspects of the product (Monroe, 1973).

The relationship of food purchasing to income is expressed in terms of income elasticity, which indicates the percentage change in demand with income changes of 1%. Income elasticity for most foods is greater than zero, implying that consumption increases with increased income. Income elasticities are lowest for tubers, higher for cereals, and progressively higher for pulses, fruits, vegetables, and animal products (Caliendo, 1979).

In their analysis of the national food balance sheets of 85 nations, Perisse *et al.* (1969) expressed the relationship of income and food intake in terms of food contribution to total food energy. With increased national product per capita, they documented increased use of animal fats, animal protein, and sugar and decreased use of complex carbohydrates, vegetable protein, and vegetable fat. These data agree with the elasticity-of-income theory. However, McKenzie (1979) in the United Kingdom reported that the effects of price were not predictably linked to food patterning. Indeed, they could work in a most haphazard fashion, since the impact of higher prices on food budget can be moderated by using high-priced foods in smaller amounts or with reduced frequency.

In some sectors of North American society, the value of time is equated with money. Numerous food items are developed with time-saving features so that more hours can be spared for money-making purposes. Generally, the term convenience denotes anything that saves or simplifies work and adds to one's ease or comfort. More specifically, convenience foods are those that have services incorporated into the basic ingredients to reduce the amount of preparation required in the home (Glicksman, 1971).

Convenience-oriented consumption was identified by W. T. Anderson (1971) as representing a focal point between rising affluence and the increasing significance of time. Thus, it could be hypothesized that people with time constraints would be more inclined to choose foods requiring little manipulation before consumption. As a food-choice motive, convenience was shown to fluctuate in importance with age, sex, income, and education of the household head, as well as with location—rural or urban—of the household (Tinklin *et al.*, 1972). The use of convenient snack foods was reported by Buchanan (1974) to be greater in city homes, with children in the household largely responsible for the regular use of these foods. Most North Americans under 25 years of age readily accept convenience foods, presumably because they became familiar with them early in their lives, are knowledgeable about their time-saving qualities, or perhaps are less knowledgeable about preparation of food from basic ingredients.

Prestige is a measure of the position of a food in the hierarchy relative to society's values. Uniform criteria for the evaluation of prestige do not exist. Jelliffe (1967) categorized prestigious foods as those served to illustrious members of the community for important occasions. The consumption of prestigious foods would not occur on a regular and frequent basis, if defined as those reserved for guests and for special occasions. Cussler and De Give (1970) have proposed that people seem to be motivated by a desire to consume foods that are prestigious rather than those that are good for health. Peer influence, as an external social cue, may induce changes in values and affect the rationale for assigning status value to foods. When people tend to eat foods of the social group to which they aspire, prestige foods can become an important attribute to vertical mobility.

It is evident from the preceding discussion that food intolerance and a high degree of taste acuity for unpleasant tastes such as bitterness may have a limiting effect on food selection. Pleasant tastes will enhance the range of food acceptance. Nutrient regulation has a role in ensuring some balance in selection; knowledge should but may not exert a similar effect. Social factors such as price, convenience, and prestige may upset this balance, depending on the particular foods affected and the extent of influence.

3. Individualization and Changeability in Food Selection

The food-selection factors previously described vary in intensity of manifestation from one individual to another and from one group to another. Although some universal mechanisms are evident, individuality exists in accordance with biological variability. The variability is mainly genetic in nature, while changeability is related to learning and environmental adaptation.

3.1. Heredity and Food Choices

There is evidence of a genetic component in at least three factors leading to individuality in food selection. The first set of evidence relates to a twin study of mature females documenting greater concordance for monozygotic than for dizygotic twins in percentage of total daily kilocalories as protein and carbohydrate as well as for total carbohydrate consumption (Wade *et al.*, 1981). This implies that although the protein and carbohydrate content in relation to food energy in individuals is biologically controlled, there is person-to-person variation in the level of satisfaction from these nutrients.

There is also evidence of differences in food intolerance due to the inherited deficiency of an enzyme such as lactase, a common disorder, or phenylalanine hydroxylase, a rather rare deficiency. Lactose intolerance is most prevalent among Negroes, Orientals, and American Indians and was considered normal among early human populations. Simoons (1973) reviewed various hypotheses that have been advanced to explain ethnic and racial differences in the tolerance of lactose by human adults. He favors the hypothesis that a high prevalence of lactose tolerance probably developed after certain human groups, having large amounts of unprocessed milk and inadequate amounts of alternate sources of protein, may have initiated the postweaning use of lactose-rich dairy products. Through time and by selection, high intestinal lactase levels were maintained throughout life in these particular groups.

The third set of evidence linking heredity to food choices is related to taste factors. Best known is the insensitivity to the bitter taste of phenylthiocarbamide (PTC), a type of antithyroid compound, which is a genetic trait involving a homozygous pair of recessive alleles (Kaplan *et al.*, 1976). About one third of all Americans are nontasters. Caffeine and saccharine taste more bitter to PTC tasters than to nontasters, indicating differences in sensitivity to other compounds as well (Kaplan *et al.*, 1976).

3.2. Gender and Cultural Individuality

Gender differences in taste, specifically preference for higher levels of sweetness by males than by females 9–15 years old, was reported by Desor *et al.* (1977). Conversely, Zucker *et al.* (1972) have demonstrated that in the presence of estrogens, there is a preference for sweet taste and an aversion to bitter.

Intergroup differences in food selection are most pronounced when comparing different cultures. Milk is the best-known example. Apart from its intolerance, which can be genetic, milk is refused for religious reasons. The *ahimsa* concept, noninjury to living creatures, is fostered by religions of Indian origin such as Buddhism. Milk is also considered an "unclean fluid" among both Asian and African nonmilkers (Simoons, 1973). Conversely, prestige ratings of milk by adult Western women is very high (Reaburn *et al.*, 1979).

3.3. Changeability of Food Selection

Changes in food selection are to be expected, since an organism must exhibit adaptive behavior simply to survive. The rate and degree of changes in food choices depend on the factors involved, internal state, stage of life, and state of health.

Learning, which is a form of adaptation to the environment, may begin as early as the fetal stage. A study of the effect of maternal diet on feeding behavior of self-selecting weaning rats has shown that maternal dietary protein concentration during gestation and lactation influenced the quantity of protein later selected by the offspring (Leprohon and Anderson, 1980).

Individual food preference is affected by exposure and learning. In the domain of taste, there are many instances of the significance of early learning specific for time and taste modalities. Rats exposed to quinine during the post-weaning period or in adulthood showed increased quinine ingestion, but exposure while weaning had no effect. In contrast, rats exposed to citric acid while nursing acquired a liking for sour substances. Even garlic flavor, considered unpalatable to rats, was better liked by those exposed to it during lactation and early weaning (Wyrwicka, 1981). Foods unusual for a species, such as bananas for cats, can become accepted if facilitated by the mother (Wyrwicka, 1974).

In humans, the most adaptive stage in establishing food preferences is early in life through associative learning, as shown by Rozin and Schiller (1980) with chili pepper among the Mexican children. Repeated exposure to coffee has led to positive affective changes (Cines and Rozin, 1982). The tamarind fruit, which is extremely sour, is considered pleasant by Indian laborers. This may be due to a lowering of sensitivity to sour (Moskowitz *et al.*, 1975). Learning continues throughout life, although the assimilation of information in old age may require a longer time (Burton and Hennon, 1981; Woodruff and Walsh, 1975; Gounard and Hubicha, 1977).

In contrast to previously described changes that contribute to the use of a variety of foods, others may have a negative effect. They are related to aging and disease. Gustatory changes are due to decreased functionings of taste buds compared to earlier years (Booth *et al.*, 1982). Busse (1978) reports that salt and sweet taste sensitivities decline earlier than bitter and sour sensitivities.

Certain disease states alter taste sensitivity (Henkin *et al.*, 1963; Schelling *et al.*, 1965; Carson and Gormican, 1976; Conger, 1974; DeWys, 1977). Patients with cancer may show a preference for higher concentrations of sucrose and a lower threshold for bitter taste. Hypertensive humans may show decreased sensitivity to salt taste (Contreras, 1978). Some medications cause abnormal taste sensations or produce an unpleasant aftertaste (Rollin, 1978).

Since sensory receptors involve the tongue and other oral structures as well as the nervous system, it would be expected that many nutrient deficiencies would affect their functioning (Gershoff, 1977).

The genetic component appears to be a strong influence on individual-ization of food choice. Learning may be the most important factor in change-ability of food selection. Negative sensory changes are related to aging and disease.

4. Integration of Factors through Food Perceptions

Food selection is a complicated area of study because the food basket consists of a variety of items. Krondl *et al.* (1982) determined the average number of different foods used within 1 year in Western society, specifically among free-living seniors, to be 70 with a range of 20–150 items. Before a food is accepted or rejected, it must undergo consideration of all relevant factors almost simultaneously by the selector and be categorized at each level whether physiological, sensory, or cognitive. Thus, each food will acquire an indivi-dualized set of labels or meanings. In light of earlier discussion, it is assumed the person will rate the food first according to its specific satiation value and label it as to its association with illness, discomfort, or overall tolerance. The memory of flavor and other sensory factors will also be recalled and the degree of experience or acquaintance with the food, familiarity, mentally ascertained. In addition, belief about healthfulness and knowledge of the social factors, prestige, price, and convenience, will be considered.

4.1. Holistic Concept and Perceptions

The range of factors influencing food selection had intrigued Kurt Lewin (1935), a follower of the Gestalt holistic psychological approach, as early as the mid-1930s. It led him to introduce the concept of a multidimensional ap-proach to delineating these factors by estimating the value that people attach to foods (Markin, 1974). He measured consumer satisfaction using four di-mensions: work, taste, health, and expense.

The food meanings of Lewin have much in common with perceptions that, along with learning and motivation, have formed the central conceptual core of psychological science since its beginning. The idea of their use in the study of food selection was developed by the authors because food perceptions offer fairly clear and simple intuitive interpretations.

In simplest terms, to perceive is to observe through the senses. Young (1948) suggested the paradigm for perceiving as:

$$\text{To perceive is} \begin{cases} \text{to see} \\ \text{to hear} \\ \text{to touch} \\ \text{to taste} \\ \text{to smell} \\ \text{to sense internally} \end{cases} \text{some} \begin{cases} \text{thing} \\ \text{event} \\ \text{relation} \end{cases}$$

Perceptions are formed when electrical impulses are processed by the brain

(Goldstein, 1980). The source of the electrical impulses is the sensory receptors—organs that are responsible for the reception of visual, olfactory, gustatory, auditory, and tactile stimuli. These sensory stimulations are translated into organized experience at the higher brain centers.

What is perceived can be considered as a blend of three factors: (1) the external and internal realities, (2) the messages of the stimuli that are conveyed by the nervous system to the integrative centers where thinking and evaluation take place, and (3) the resultant interpretations of the messages with feedback from past experiences.

The senses generally do not operate in isolation, but together contribute to the formation of the perception. As well, perceptions have a close relationship to learning and cognition. They have sensory data at their core, and in turn, they play a central role in the cognitive and thinking process. They enable man to turn raw data into meaning (Bliss, 1970) and create a relatively stable environment out of an otherwise chaotic assortment of sensory impressions (Williams, 1981).

There is fairly general agreement that perceptions involve both internal and external stimuli (Bliss, 1970). Thus, the concept of food perceptions allows for the integration of the internal physiological stimuli as well as the external environmental impacts presenting themselves as physical, verbal, or symbolic stimuli. These signals are then translated into organized experience and stored as meaningful impressions in the higher brain centers. Accordingly, food perceptions can be viewed as mental impressions or pictures of foods formed as the result of visual, olfactory, gustatory, auditory, and tactile experiences. It follows that the comparative analysis of the effect of the different stimuli on food selection is possible because they result from a similar formulation process. When presented with the same food item, each person may focus on what for him is relevant and salient and ignore or treat as background the rest of the stimuli.

4.2. Food Perceptions

The problem of evaluating factors affecting food selection has been confounded by the lack of a common conceptual measurement. To overcome this drawback, Krondl and Lau (1978) proposed a model for the study of food perceptions intended as a basis for food-habit modification. Based on the perceptions approach and using the man–system theory proposed by Brody (1973), methodology was developed to study food selection in various populations (Lau *et al.*, 1979; Reaburn *et al.*, 1979; Krondl *et al.*, 1982). It incorporated the concept of labeling of foods according to perceived meanings. Five-point scales similar in format to Likert scales (Likert, 1932) were developed for each factor. The scales extend from strongly favorable through neutral to strongly unfavorable. ''Don't know'' and ''No answer'' categories were included but treated separately. These latter responses were excluded in the analysis of the food perceptions. Median ratings were determined to judge the relative strength of the perceptions. The perceptions were related to frequency of food use. Ken-

Table I. Perceptions of Beef Liver and Chocolate Bars by 135 Adolescents of British Descent

Perceptions[a]	Beef liver[b]		Kendall's tau[d]	Chocolate bars[c]		Kendall's tau[d]
	N	Median scores		N	Median scores	
1. Satiety	100	3.7	0.40–0.49	133	2.1	0.20–0.29
2. Tolerance	104	1.3	0.40–0.49	121	1.0	0.15–0.19
3. Taste	120	4.6	>0.50	133	1.2	0.30–0.39
4. Familiarity	116	1.8	0.30–0.39	133	1.2	<0.14
5. Healthfulness	118	1.2	<0.14	131	4.0	0.30–0.39
6. Prestige	113	3.2	0.20–0.29	133	1.3	0.15–0.19
7. Price	65	2.3	<0.14	122	2.6	<0.14

[a] The rating scale used was 1–5 for perceptions 1, 3, 4, 5, and 6, and 1–3 for perceptions 2 and 7 (the lower the score, the more positive the rating).
[b] Tried but do not use.
[c] Used two to three times per month.
[d] Coefficients indicating correlation range between frequency of use and perception; values above 0.30 are considered relatively strong.

dall's tau correlations with two-tailed tests of significance were carried out to assess the food perception–food use associations as well as intergroup differences.

The use of this approach allows for food-to-food comparisons of the perceptions and has led to a better understanding of intergroup differences in food-related behavior. In addition, it has enabled elucidation of the routes of change in food-selection patterns. It has provided a data base for speculation on the relative strength of the factors in the ultime selection of a food basket.

4.3. Perceptions and Food Selection

Some results from studies of adolescents incorporating food perceptions are presented to illustrate their use with selected populations and specific foods.

An assessment of perceptions and interfood differences among adolescents of British descent (Krondl, 1981) is shown in Table I. Beef liver and chocolate bars were chosen to represent rarely used and moderately used foods. Thus, the data reflect the personal reality of a specific age and ethnic group. The sample is well defined in terms of age and culture. It is readily apparent that for beef liver, more of the perception ratings are unfavorable than for chocolate bars. The greatest differences are in the perceptions of healthfulness, taste, and satiety. These ratings reflect the prevailing societal view of healthfulness of each food as well as degree of pleasantness of flavor and texture. Some awareness of the physiological effects may have influenced the perception of satiety. No difference was observed for tolerance. Ratings of the perception of price suggest that this group shops for chocolate bars but not beef liver. Prestige ratings were higher for chocolate bars as expected, since this food is associated with socializing with peers and thus has status. Foods eaten at home are usually considered less prestigious.

Table II. Perceptions of Lettuce and Soft Drinks by 57 Adolescent Boys and 78 Adolescent Girls of British Descent

Perceptions[a]	N	Boys' median scale score	N	Girls' median scale score	Kendall's tau[b]
Lettuce[c]					
1. Satiety	54	2.1	75	1.5	0.25*
2. Taste	53	1.4	75	1.2	0.19†
3. Familiarity	53	1.2	75	1.1	0.21†
4. Healthfulness	55	1.3	75	1.1	0.25*
5. Prestige	54	1.6	72	1.3	0.20†
6. Price	43	2.0	61	2.1	N.S.
Soft drinks[c]					
1. Satiety	55	1.6	74	2.1	N.S.
2. Taste	55	1.3	75	1.3	N.S.
3. Familiarity	54	1.3	75	1.3	N.S.
4. Healthfulness	55	3.8	73	3.1	−0.18†
5. Prestige	55	1.3	74	1.2	N.S.
6. Price	49	2.4	70	2.3	N.S.

[a] The rating scale used was 1–5 for perceptions 1, 2, 3, 4, and 5, and 1–3 for perception 6 (the lower the score, the more positive the rating).
[b] Levels of significance: *$p < 0.01$; †$p < 0.05$. (N.S.) No significance. A negative correlation designates a higher rating by boys than by girls; a positive correlation, a higher rating by girls than by boys.
[c] Used several times a week.

The rejection of beef liver is obvious in the association of low use with the unfavorable ratings for the perceptions of satiety, tolerance, and taste, representing the physiological and sensory factors. Chocolate bars have greater acceptance than beef liver. The perceptions of taste and healthfulness have comparable weight in influencing the use of chocolate bars. This may indicate that health belief is contradicting the influence of taste and thus limiting its use to two to three times per month.

The male–female differences in food perceptions for two foods used with similar frequency, lettuce and soft drinks, were examined (George and Krondl, 1983) and are shown in Table II. The greatest differences were in the higher ratings by the girls for the perceptions of satiety and healthfulness for lettuce. Similar but less pronounced were the findings for taste, familiarity, and prestige. No gender difference was found in the perception of price of lettuce. For soft drinks, healthfulness was the only factor showing gender differences; it was rated higher by the boys, indicating their greater recognition that soft drinks are not healthy.

Soft drinks were rated as less healthful than lettuce by both boys and girls. Taste was rated highly favorable for both foods. Since frequency of use was similar for lettuce and soft drinks, the sensory factor would appear to exert a stronger influence than the cognitive factor.

Following immigration, changes occur in both the perceptions and use of foods. The effect of acculturation on perceptions of ice cream, an item typical of North American cuisine, is illustrated in Table III with populations of first-

Table III. Perception of Ice Cream Used Weekly by 38 First-Generation and 16 Second-Generation Chinese Boys

			Scales						
			1	2	3	4	5	Don't know[c]	Kendall's tau b[d]
Perceptions	Generation[a]	N[b]	Distribution (%)						
1. Satiety	1	37	10.8	35.1	24.3	13.5	16.3	—	
	2	16	37.5	12.5	18.8	12.5	18.8	—	N.S.
2. Tolerance	1	37	100.0	—	—	—	—	—	0.22‡
	2	15	93.3	6.7	—	—	—	—	
3. Taste	1	37	67.6	24.3	8.1	—	—	—	−0.28†
	2	16	93.8	6.2	—	—	—	—	
4. Familiarity	1	37	56.8	35.1	8.1	—	—	—	N.S.
	2	16	75.0	18.7	6.2	—	—	—	
5. Healthfulness	1	34	38.2	29.4	26.5	5.9	—	8.1	
	2	15	26.7	60.0	6.7	6.6	—	—	N.S.
6. Prestige	1	37	59.5	21.6	16.2	2.7	—	—	−0.33*
	2	15	93.3	6.7	—	—	—	—	
7. Price	1	37	18.9	67.6	13.5	—	—	—	
	2	15	33.3	60.0	6.7	—	—	—	N.S.

[a] (1) First-generation (born in China); (2) second-generation Chinese (born in Canada).
[b] Excludes "Don't know" responses.
[c] Percentage of "Don't know" responses based on maximum sample size.
[d] Generation differences; level of significance: $*p < 0.025$; $†p < 0.05$; $‡p < 0.01$. (N.S.) No significance. A negative correlation designates a higher rating by second-generation than by first generation; a positive correlation, a higher rating by first-generation than by second-generation.

and second-generation Chinese adolescents (Hrboticky and Krondl, 1981). Most significant were the changes in taste and prestige; they were rated higher by the children of immigrants, whereas tolerance was scored higher by the first-generation Chinese. No differences were found for satiety, familiarity, and price. The positive change in taste corresponding to the length of time of immigration is likely influenced by the fact that the food is used in the company of friends; thus, the peer effect is implied. With increased exposure, ratings of prestige and taste become more favorable. The social influence is likely to be the route for changing food use and consequently the food-patterning of adolescents.

A food basket with representative foods was chosen to rank the relative influence of different factors on food use by adolescents of British descent. The findings are presented in Table IV. The perceptions that represent the sensory and physiological factors, taste, satiety, and tolerance ranked highest, followed by healthfulness. The social factors, prestige and price, as well as familiarity showed the least influence on use. A previous study of low-income homemakers (Reaburn *et al.*, 1979) also showed a weak association between the social factors and food use.

When all factors were simultaneously analyzed using an actual food basket, the results were similar to the fragmented literary evidence showing the effect of various factors on food selection in animal and human studies using test

Table IV. Ranking of Food Perceptions According to Their Relationship with Food-Use Frequency among 135 Adolescents of British Descent

Perception	Median correlation value (Kendall's tau)[a]
1. Taste	0.36
2. Satiety	0.29
3. Tolerance	0.26
4. Healthfulness	0.21
5. Prestige	0.18
6. Familiarity	0.15
7. Price	0.04

[a] Median value in the correlations between the perceptions and frequency of use of 48 foods.

diets. The advantage of the perception methodology lies in its focus on the depth of understanding of the mechanisms involved in the process of food selection. It also enables comparisons of the relative influence of the factors on foods selected by different populations as well as changes in food selection over time.

5. Conclusions

Food selection is influenced by physiological, sensory, and cognitive factors. The physiological and sensory factors contribute to constancy in food selection, in contrast to cognitive processes, which lead to environmental adaptation. Individualization is partially genetic in nature, as specifically demonstrated in the regulation of food sources of protein and carbohydrate, intolerance due to inherited enzyme deficiencies, and some forms of sensitivity to the bitter taste modality. Learning of food preference occurs as early as the perinatal stage and continues throughout life.

The application of the concept of food perceptions in the integration of all the factors is useful in determining the relative strengths of the individual factors as they apply to specific foods as well as the total food basket. It also enables comparisons of the relative influence of the factors on foods selected by different populations as well as changes in food selection over time. The studies of food perceptions illustrated by adolescent populations show the strong influence of sensory and physiological factors and the weak influence of social factors in established food patterns. Nevertheless, the social factors play a role in the changing of the patterning.

6. References

Amerine, M. A., Pangborn, R. M., and Roessler, E. B., 1965, Visual, auditory, tactile, and other senses, *Principles of Sensory Evaluation of Food*, pp. 220–244. Academic Press, New York.

Anderson, G. H., 1979, Control of protein and energy intake: Role of plasma amino acids and brain neurotransmitters, *Can. J. Physiol. Pharmacol.* **57**:1043–1057.

Anderson, W. T., 1971, Identifying the convenience-oriented consumer, *J. Market Res.* **8**:179–182.

Bartoshuk, L. M., 1980, Separate worlds of taste, *Psychol. Today* **14**(4):48–63.

Beauchamp, G. K., Bertino, M., and Moran, M., 1982, Sodium regulation: Sensory aspects, *J. Am. Diet. Assoc.* **80**:40–45.

Bernstein, I. L., and Webster, M. M., 1980, Learned taste aversions in humans, *Physiol. Behav.* **25**(3):363–366.

Bliss, P., 1970, Perception, in: *Marketing Management and the Behavioral Environment*, pp. 68–79, Prentice-Hall, Englewood Cliffs, New Jersey.

Bock, S. A., 1980, Food sensitivity, *Am. J. Dis. Child.* **134**:973–982.

Booth, P., Kohrs, M. B., and Kamath, S., 1982, Taste acuity and aging: A review, *Nutr. Res.* **2**:95–109.

Boudreau, J. C., Oravec, J., Hoang, N. K., and White, T. D., 1979, Taste and the taste of foods, in: *Food Taste Chemistry* (J. C. Boudreau, ed.), pp. 1–32, American Chemical Society, Washington, D.C.

Brody, H., 1973, The systems view of man: Implications of medicine, science, and ethics, *Perspect. Biol. Med.* **17**(1):79–91.

Brown, A. M., McKenzie, J. C., and Yudkin, J., 1963, Knowledge of nutrition amongst housewives in a London suburb, *Nutrition* **17**:16.

Buchanan, R. O., 1974, Two marketing surveys reveal convenience food attitude and usage, *Food Sci. Market* **36**:51–56.

Burton, J. R., and Hennon, C. B., 1981, Consumer education for the elderly, *J. Home Econ.*, Summer, pp. 24–28.

Busse, E. W., 1978, How mind, body and environment influence nutrition in the elderly, *Postgrad. Med.* **63**:119–125.

Cabanac, M., 1971, Physiological role of pleasure, *Science* **173**:1103–1107.

Caliendo, M. A., 1979, Poverty and its relationship to malnutrition: Strategies for increasing the purchasing power of the poor, *Nutrition and the World Food Crisis*, pp. 157–176, Macmillan, New York.

Carson, J. S., and Gormican, A., 1976, Disease–medication relationship in altered taste sensitivity. *J. Am. Diet. Assoc.* **68**:550.

Cines, B. M., and Rozin, P., 1982, Some aspects of the liking for hot coffee and coffee flavour, *Appetite* **3**(1):23–24.

Conger, A. O., 1974, Loss and recovery of taste acuity in patients irradiated in the oral cavity, *Radiat. Res.* **53**:338.

Contreras, R. J., 1978, Salt taste and disease, *Am. J. Clin. Nutr.* **31**(6):1088–1097.

Cussler, M., and De Give, M., 1970, *Twixt the Cup and the Lip*, Consortium Press, New York.

Desor, J. A., Maller, O., and Greene, L. S., 1977, Preference for sweet in humans: Infants, children and adults, in: *Taste and Development* (J. M. Weiffenbach, ed.), pp. 161–172, U.S. Department of Health, Education and Welfare, Bethesda, Maryland.

DeWys, W. D., 1977, Changes in taste sensation in cancer patients: Correlations with calorie intake, in: *The Chemical Senses and Nutrition* (M. R. Kare and O. Maller, eds.), pp. 381–392. Academic Press, New York.

Dugdale, A. E., Chandler, D., and Baghurst, K., 1979, Knowledge and belief in nutrition, *Am. J. Clin. Nutr.* **32**:441–445.

Erickson, R. P., 1982, Studies on the perception of taste: Do primaries exist?, *Physiol. Behav.* **28**:57–62.

Garcia, J., and Brett, L. P., 1977, Conditioned responses to food odor and taste in rats and wild predators, in: *The Chemical Senses and Nutrition* (M. K. Kare and O. Maller, eds.), pp. 277–291, Academic Press, New York.

Garcia, J., Hankins, W. G., and Rusiniak, K. W., 1974, Behavioral regulation of the milieu interne in man and rat, *Science* **185**:824–831.

George, R., and Krondl, M., 1983, Perceptions and food use of adolescent boys and girls, *Nutr. Behav.* **1**(2):115–125.

Gershoff, S. N., 1977, The role of vitamins and minerals in taste, in: *The Chemical Senses and Nutrition* (M. K. Kare and O. Maller, eds.), pp. 201–212, Academic Press, New York.

Glicksman, M., 1971, Fabricated foods, in: *CRC Critical Reviews in Food Technology 2*, Chemical Rubber Company Press, Cleveland, Ohio.

Goldstein, E. B., 1980, *Sensation and Perception*, pp. 2–3, Wadsworth, Belmont, California.

Gounard, B. R., and Hubicha, C. M., 1977, Maximizing learning efficiency in later adulthood: A cognitive problem solutions approach, *Educ. Gerontol.* **2**:424.

Hara, T. J., and Law, Y. M. C., 1972, Adaptation of the olfactory bulbar responses in fish, *Brain Res.* **47**:259.

Haverberg, L., Kwon, P. H., and Scrimshaw, N. S., 1980, Comparative tolerance of adolescents of differing ethnic backgrounds to lactose-containing and lactose-free dairy drinks. I. Initial experience with a double-blind procedure, *Am. J. Clin. Nutr.* **33**:17–26.

Henkin, R. I., Gill, J. R., and Barter, F. C., 1963, Studies on taste thresholds in normal man and in patients with adrenal cortical insufficiency, *J. Clin. Invest.* **42**:727–735.

Herman, R. H., and Hagler, L., 1979, Food intolerance in humans, *West. J. Med.* **130**:95–116.

Hrboticky, N., and Krondl, M., 1981, Food habit acculturation of Chinese adolescents, XIIth International Congress of Nutrition, San Diego, California, Abstract 146.

Jelliffe, D., 1967, Parallel food classifications in developing and industrialized countries, *Am. J. Clin. Nutr.* **20**:279.

Kaplan, A. R., Powell, W. E., Moorhouse, A. E., and Hinko, E. N., 1976, Taste sensitivity and human variation, in: *Human Behavior Genetics* (A. R. Kaplan, ed.), pp. 401–423, Charles C. Thomas, Springfield, Illinois.

Kretchmer, N., 1972, Lactose and lactase, *Sci. Am.* **227**:70–78.

Krondl, M., 1981, Study of adolescent food habits, *Report to Ontario Ministry of Health*, DM. 413, Department of Nutritional Sciences, University of Toronto.

Krondl, M., 1982, The effect of hunger–satiety state on sensory reactions to food, *Report to the Human Nutrition Research Council of Ontario*, Department of Nutritional Sciences, University of Toronto.

Krondl, M., and Lau, D., 1978, Food habit modifications as a public health measure, *Can. J. Public Health* **69**:39–43.

Krondl, M., Lau, D., Yurkiw, M. A., and Coleman, P. H., 1982, Food use and perceived food meanings of the elderly, *J. Am. Diet. Assoc.* **80**:523–529.

Lau, D., Hanada, L., Kaminskyj, O., and Krondl, M., 1979, Predicting food use by measuring attitudes and preference, *Food Prod. Dev.* **13**(5):66–72.

Leprohon, C. E., and Anderson, G. H., 1980, Maternal diet affects feeding behaviour of self-selecting weanling rats, *Physiol. Behav.* **24**:553–559.

Lewin, K., 1935, *A Dynamic Theory of Personality*, McGraw-Hill, New York.

Libassi, P. T., 1974, Two for the cellular seesaw, *Sciences* **14**:15–20.

Likert, R., 1932, A technique for the measurement of attitudes, *Arch. Psychol.* **22**:5.

Markin, R. J., Jr., 1974, *Consumer Behavior*, pp. 51–76, 197–259, Macmillan, New York.

McKenzie, J., 1979, Economic influences on food choice, *Nutr. Food Sci.* **60**:4–7.

McMenamy, R. H., and Oncley, J. L., 1978, The specific binding of L-tryptophan to serum albumin, *J. Biol. Chem.* **233**:1436–1447.

Monroe, K., 1973, Buyers' subjective perception of price, *J. Market Res.* **10**:70.

Moskowitz, H. R., 1977, Intensity and hedonic functions for chemosensory stimuli, in: *The Chemical Senses and Nutrition* (M. R. Kare and O. Maller, eds.), pp. 71–103, Academic Press, New York.

Moskowitz, H. R., Kumaraiah, V., Sharma, K. N., Jacobs, H. L., and Sharma, S. O., 1975, Cross-cultural differences in simple taste preferences, *Science* **180**:1217–1218.

Mower, G. D., Mair, R. G., and Engen, T., 1977, Influence of internal factors on the perceived intensity and pleasantness of gustatory and olfactory stimuli, in: *The Chemical Senses and Nutrition* (M. R. Kare and O. Maller, eds.), pp. 104–123, Academic Press, New York.

Musten, B., Peace, D., and Anderson, G. H., 1974, Food intake regulation in the weanling rat: Self-selection of protein and energy, *J. Nutr.* **104**:563–572.

Olson, J. C., and Sims, L. S., 1980, Assessing nutrition knowledge from an information processing perspective, *J. Nutr. Educ.* **12**(3):157–161.

Pager, J., 1977, Nutritional states, food odor, and olfactory function, in: *The Chemical Senses and Nutrition* (M. R. Kare and O. Maller, eds.), pp. 51–68, Academic Press, New York.

Perisse, J., Lizaret, F., and François, P., 1969, The effect of income on the structure of diet, *FAO Nutr. Newslett.* **7**(3):1–9.

Pfaffmann, C., 1980, Wundt's schema of sensory affect in the light of research on gustatory preferences, *Psychol. Res.* **42**:165–174.

Reaburn, J., Krondl, M., and Lau, D., 1979, Social determinants in food selection, *J. Am. Diet. Assoc.* **74**:637–641.

Reddin, J. E., and Anderson, D. M., 1978, Adolescents and their food: Changes through nutrition education and school lunch, Report No. 2, *Conceptual Framework and Methodology*, Home Economics Department, University of Prince Edward Island, Charlottetown, Prince Edward Island, Canada.

Rogers, J. R., and Leung, P. M. B., 1977, The control of food intake: When and how are amino acids involved? in: *The Chemical Senses and Nutrition* (M. R. Kare and O. Maller, eds.), pp. 213–249, Academic Press, New York.

Rollin, H., 1978, Drug related gustatory disorders, *Ann. Otol. Rhinol. Laryngol.* **87**:37.

Rorick, M. H., and Scrimshaw, N. S., 1979, Comparative tolerance of elderly from differing ethnic backgrounds to lactose-containing and lactose-free dairy drinks: A double blind study, *J. Gerontol.* **34**:191–196.

Rozin, P., and Schiller, D., 1980, The nature and acquisition of a preference for chili pepper by humans, *Motivation Emotion* **4**:77–101.

Schelling, J. L., Tetrault, L., Lasagna, L., and Davies, M., 1965, Abnormal taste threshold in diabetes, *Lancet* **1**:508.

Simoons, F. J., 1973, The determinants of dairying and milk use in the old world: Ecological, physiological and cultural, *Ecol. Food Nutr.* **2**:83–90.

Sims, L. S., 1978, Dietary status of lactating women. 2. Relation of nutritional knowledge and attitudes to nutrient intake, *J. Am. Diet. Assoc.* **73**:147–54.

Smith, J. C., and Roll, D. L., 1967, Trace conditioning with X-rays as the aversive stimulus, *Psychon. Sci.* **9**:11–12.

Snow, F., 1974, Folk medical beliefs and their implications for care of patients—a review based on studies among black Americans, *Ann. Intern. Med.* **81**:82–96.

Snow, L. F., and Johnson, S. M., 1978, Folklore, food, female reproductive cycle, *Ecol. Food Nutr.* **7**:41–49.

Steiner, J. E., 1977, Facial expressions of the neonate infant indicating the hedonics of food-related chemical stimuli, in: *Taste and Development* (J. M. Weiffenbach, ed.), pp. 143–189, U.S. Department of Health, Education and Welfare, Bethesda, Maryland.

Tinklin, G. L., Fogg, N. E., and Wakefield, L. M., 1972, Convenience foods: Factors affecting their use where household diets are poor, *J. Home Econ.* **64**:26–28.

Wade, J., Milner, J., and Krondl, M., 1981, Evidence for a physiological regulation of food selection and nutrient intake in twins, *Am. J. Clin. Nutr.* **34**:143–147.

Williams, S. R., 1981, Cultural, social and psychologic influences on food habits, in: *Nutrition and Diet Therapy*, 4th ed. p. 278, C. V. Mosby, St. Louis.

Woodruff, D. S., and Walsh, D. A., 1975, Research in adult learning: The individual, *Gerontologist* **15**(5):424–430.

Woteki, C. E., Weser, E., and Young, E. A., 1976, Lactose malabsorption in Mexican-American adults, *Am. J. Clin. Nutr.* **29**:19–24.

Wurtman, J. J., and Wurtman, R. J., 1979, Drugs that enhance central serotonergic transmission diminish elective carbohydrate consumption by rats, *Life Sci.* **24**:895–904.

Wyrwicka, W., 1974, Eating bananas in cats for brain stimulation award, *Physiol. Behav.* **12**:1063–1066.

Wyrwicka, W., 1981, *The Development of Food Preferences*, Charles C. Thomas, Springfield, Illinois.

Young, P. T., 1948, Appetite, palatability and feeding habit: A critical review, *Psychol. Bull.* **45:**289–320.

Zucker, J., Wade, J. N., and Ziegler, R., 1972, Sexual and hormonal influences on eating taste preferences and body weight of hamsters, *Physiol. Behav.* **8:**101–111.

12

Sociocultural Aspects of Nutrient Intake and Behavioral Responses to Nutrition

Ellen Messer

1. Introduction

This chapter will consider sociocultural factors that influence nutrient intake and behavioral responses to diet. With the possible exception of modern Western society, no cultural group evaluates the individual foods and combinations it ingests in terms of the scientific categories—energy, fat, protein, vitamins, and minerals. People must therefore have some cultural rules by means of which to classify items as "edible" vs. "inedible," to rank edible items as "preferred" to "less preferred," and to construct nutritionally adequate diets. Yet nutritional patterns also have their symbolic and social dimensions. "You are what you eat" goes the German proverb; "Tell me what you eat and I'll tell you who you are" say the French. Every culture in every age has consumed a distinctive diet that sets its members apart from other cultural groups. Within cultures, membership in particular social strata or status is often marked by distinctive consumption styles. Social relations are idealized and materialized in terms of food exchange; friendship and close social relations are marked by food sharing, enmity or social distance by refusal to give or receive food. Dominant–subordinant relationships may also be marked symbolically by selective food consumption and restrictions. The emotional force of social membership is impressed on participants and initiates by special ritual foods and fasts. Food, then, functions as a cultural code and social currency to create and nourish the sociocultural as well as biological individual. Given available natural or market resources, or both, people acquire, process, and distribute food to maintain themselves as a biological population, but one with a particular social structure and culture.

Ellen Messer • Department of Anthropology, Yale University, New Haven, Connecticut 06520.

Taking a perspective that considers human nutritional systems as a complex interaction of biological and cultural features, one can also begin to describe how cultural patterns and social organization are shaped by nutritional concerns or are themselves the consequences of nutritional deficiencies. The extent to which people's physical and intellectual growth and activities are circumscribed by inadequate nutrition and the extent to which social well-being and community development are stymied by insufficient food resources at the individual, household, and community levels are additional biocultural concerns. Sociocultural patterns of food procurement and rules of food distribution within households and communities can interact with other biological factors such as illness and sociocultural factors such a postweaning residence rules to endanger nutritional and health status during early child development. Moreover, early cognitive stimulation, physical growth, and intellectual development may be affected by poor health and nutritional status. These interrelated factors in turn limit the potential of human populations to change and improve their biological and cultural environment. All are topics of biological as well as cultural concern.

There have been a range of studies, spanning the nutritional and behavioral sciences, on food systems—production, cultural classifications, rules for social distribution, and nutritional patterns of consumption—and their evolution. These studies can be located in annotated bibliographies (C. S. Wilson, 1973, 1979; Freedman, 1981) and in several recent volumes of collected essays on nutritional, cultural, and social dimensions of food habits (Arnott, 1976; Fitzgerald, 1976; Greene, 1977; Jerome *et al.*, 1979; Fenton and Owen, 1981). While many focus on the food habits of a group—usually a cultural community—some authors have emphasized the need to study intracultural variation in human food patterns (diversity within one human population or cultural group) (P. J. Pelto and G. H. Pelto, 1975; G. H. Pelto and Jerome, 1979) as a way of more precisely analyzing food preferences and why and how food habits change.

Communication between nutritionists and social scientists has also been furthered by the publication of a number of cross-disciplinary journals, which bring together recent research and publications on political, sociocultural, and nutritional perspectives on food: *Ecology of Food and Nutrition*, *Medical Anthropology*, *Social Science and Medicine*, *Medical Anthropology Newsletter* (*Medical Anthropology Quarterly*), *The Communicator* (Newsletter of the Committee on Nutritional Anthropology of the American Anthropological Association), *The Digest* (Publication of the University of Pennsylvania Food Group of the Department of Folklore and Folklife), *Food and Nutrition Bulletin* (Publications of the United Nations University), *Nutrition Research*, *Food Policy*, *Appetite*, *Human Ecology*, *Ethnobiology*, and a gastronomic section in *Social Science Information* (see Jerome, 1979b). Reviews of diet and culture, both general (Tannahill, 1973; Farb and Armelagos, 1980) and for specific cultures (Forster and Forster, 1975; Cosman, 1976; Chang, 1977; Braudel, 1981), have also contributed to understanding of the historical relationships among culture, biology, and international relations. Particularly the "Columbian exchange" of animals, plants, and disease between the New World and the Old (Crosby, 1972) and the international trade in sugar (Mintz, 1979–1980) illustrate the tremendous impact of international exploration and commerce on nutrition and sociocultural development.

Food and nutrition have been conceptualized as components of more general investigations of cultural patterns, social organization, social relations, symbol systems, the material bases of society, human ecology, political economy, household economics, and sociocultural history. Nutritional studies that consider "sociocultural factors" have as their aim the understanding of the determinants of nutrient intake, the consequences of deficiencies, and the etiologies of malnutrition syndromes. By contrast, anthropological studies often view human nutrition as a variable dependent on ecological and economic resources, cultural classifications of food, social organization of production and consumption, and relationships among local, regional, state, and international economies. The interactions of ecosystems and symbol systems, biological variables, and cultural factors are analyzed to understand how biological–physical conditions, culturally perceived and used—including resource availability and human health nutritional status—determine the limits to human productivity and creativity in particular environments.

In the following sections, I will review in more detail the approaches, methods, and general findings of anthropological studies of food as an aspect of culture and the aspects of behavior that are related to nutrition (Section 2). Next, I will consider sensory, psychological, and sociocultural factors that enter into decisions about food intake (Section 3). I will conclude with a discussion of biocultural determinants of nutrient intake and human physical and sociocultural functioning at individual, household, and community levels (Section 4).

2. Anthropological Studies

The rules for food production, distribution, and consumption often form the basic organizing principles of society and its elementary social groups. Whether or not anthropologists have focused directly on the food quest and nutrition, this observation has been central to anthropological understandings of other aspects of culture. Early studies of the economics and social organization in nonindustralized societies subsisting on local resources noted how the search for, preparation, and consumption of food provided the primary focus rather than an interval in the day's activities and how, in such contexts, symbolic and emotional values of foods were often used ritually to mark social status, intervals in time, and culturally important environmental resources (Bell, 1931–1932; Raymond Firth, 1929, 1936). Subsequently, studies have emphasized the centrality of social cooperation in the food quest and food-sharing to the structure and evolution of human social organization and culture. Anthropologists of different backgrounds in different historical periods have found that the relationships between people and their "food" resources are basic to their understandings of society, culture, and biocultural evolution, whether taking as their unit of analysis a "society" ("social relations" and "social organization")—the approach of *social anthropology*; a "culture" (the ideas and behavior of a human group)—the approach of *cultural* including *psychological, anthropology*; a "human population" (the biological and cultural attributes of a group of humans interacting with other populations and physical features in their local and regional environments)—the approach of *ecological* and *biocultural anthropology*; or some socioeconomic grouping—an *economic anthropological* approach. The following subsections will discuss the questions, units of analysis, and applications of each of these approaches, some of which were designed to answer practical problems, others for scholarly ends.

2.1. Social Anthropological Studies

British social anthropologists working in pre-World War II colonial Africa provided some of our earliest and most complete studies of the interrelationships among food supply, social organization, and nutrition. Although as social anthropologists they were interested in defining and analyzing the basic units of sociopolitical organization, they found that the study of food and hunger was central to their understandings of the societies, social relations, and changing cultures disrupted by British rule. Richards (1939), in her classic study of the Bemba of Northern Rhodesia, concluded that the reason natives did not work harder (a primary concern for British mining and other economic interests) was not a question of sloth, but of undernutrition. In this society, the men had been drawn away from their roles in local gardening production to the mines, and the women found it difficult to perform the heavy clearing tasks traditionally assumed by males, and their own cultivation and gathering roles. Analyzing the activities throughout the year and the social relations surrounding food, Richards found that during the period of the year when women most needed food energy to sustain clearing and planting of fields, food was in shortest supply. Thus, they were enmeshed in an ongoing cycle of underproduction and undernutrition.

As part of her study, Richards carefully examined all social relations as food relations. She analyzed the emotional qualities assigned to different foods: their desirability in terms of taste and digestibility, their importance in native ceremonial life—for instance, the importance of grains used in beer-brewing—and people's perceptions of the nutritional qualities and physiological effects of different staple grains and relishes eaten with them. She also described the social dimensions of food production, preparation, distribution, and consumption, noting how all kinship relations were marked by prescribed rules of sharing, as well as how these obligations would break down in times of dearth.

Fortes and Fortes (1936), working among the Tallensi of West Africa during the same period, described how natural and ritual seasonal cycles, social organization, and social relations were interrelated in terms of the food supply. They found among the Tallensi the same productive bottleneck as Richards had found among the Bemba—that during the planting time, the season of greatest energy needs, the food supply was exhausted. They concluded that this was in part a problem of underproduction, but also one of allocation of resources, since the Tallensi spent grain liberally in the season immediately following the harvest on beer-brewing and began rationing only later in the season. Both Richards's and the Fortes' studies are examples of an early British trend in interdisciplinary studies of food that called for nutritionists, biochemists, botanists, and anthropologists to work together to understand native diets and their potential for change (Worthington, 1936). They provided initial data that enabled social scientists and administrators to consider food systems and nutritional status, among other variables, to "explain" native behaviors. In part in reponse to their work, Malinowski (1946) included dietary changes as one aspect of social change to be studied by anthropologists. He furthermore suggested that one would have to study the nutritional advice and food being offered by the mines and administrators, in addition to the nutritional potential of native diet, native social structure, and cultural values, to adequately predict directions of dietary change.

Other British social anthropologists, while not focusing on food and nutrition to such a degree, have still found the social units of food production and consumption and seasonal food cycles essential to their understandings of economic groups, community organization, and rituals. Goody (1958), for example, in analyzing the reasons for breakup of polygynous (multiple-wife) consumption units among the LaDagaba of the Gold Coast, relied on native explanations of child-feeding customs. Not so much jealousies or personality incompatibility among co-wives but individual mothers' needs to supervise their children's food from separate rather than common pots caused households that formerly cooked and ate together to divide into mother–child units. Gluckman (1963), in considering the timing of "rituals of rebellion" in African societies, noted that they took place following the harvest and cited the energy and euphoria resulting from the renewed food supply and nutrient intake as a triggering factor for such functional social activities. While these later studies focus less directly on nutrition than earlier studies such as Richards's, they nevertheless continue a tradition of interpreting social organization at least in part in terms of food supply and rules for sharing as indices of social relations.

2.2. Psychological Anthropological Studies

The social psychological approaches to culture that were characteristic of American anthropology of the 1930–1940s, by contrast, focused on how attitudes toward food developed in particular cultures and affected later social relationships, behavior, and personality as part of larger "culture and personality" studies (see, for example, Kardiner, 1945). DuBois (1941, 1944), in her classic study of the Alorese, suggested that the child's early experiences of frustration and neglect when his need for food was not met established the basic insecurity and suspicious distrust that characterize the adult personality, cultural orientations, and social relations. Basic affective bonds between males and females, first between mothers and sons and later between wives and husbands, were considered, on native testimony, to be based on men's reliance on women for food from the earliest age. Conversely, men's roles as providers of meat, which was considered a delicacy, established male prestige. Hunger was also seen as the basic motivation for foraging, thieving, and learning adult skills such as gardening. Yet DuBois also argued that hunger and food anxieties were really social fictions, but so pervasive that people made no effort to increase their food supply, since they felt, helplessly, that such efforts would be subject to natural or cultural deprivations from rats or theft.

Other ethnographers, e.g., Fortune (1932) and Holmberg (1950), demonstrated how food anxieties could be raised to a pitch so beyond real physiological need that they dominated cultural, social, and psychological functioning among the Dobu (Melanesia) and Siriono (Peruvian Amazon), respectively. Along the same lines, Malinowski (1935, 1948) and M. Young (1971) interpreted gardening beliefs and practices among certain Pacific Island groups (Trobriand and Goodenough Islanders, respectively) as efforts to magically control appetite and thereby conserve and extend food supplies in contexts in which both real and fictional shortages induced anxieties about hunger. D. N. Shack (1969) and W. Shack (1971) analyzed the personality characteristics, abstemious eating behaviors but also ritual gorging of the Gurage of Ethiopia as also due to hunger anxieties. But he interpreted such anxieties as social fictions, manufactured from historical memories of warfare, raiding, and want out of line with contemporary food conditions.

More generally, it has been argued that anxieties formed through early food experiences determine whether eating habits are easily amendable to change and whether food is a chief explanation for illness in later life (Whiting and Child, 1953). Mead (1962) suggested that in societies where children are fed lovingly and where food is a source of great pleasure and delight, food habits will later become very intractable. On this dimension, she compared Italian-American immigrants to certain German-American immigrants, whose early experiences with food as one element of a rigid system of discipline allowed their food habits to be changed more easily than those of the lovingly fed Italians.

In general, social psychological analyses of the relationships among food supply, early feeding experiences, emotional responses, and personality development have continued to be important for studies of eating disorders, as well as for other contexts in which there is some desire to change food habits. They also provide important cross-cultural data on cultural and psychological factors in hunger vs. appetite.

As a subgroup within psychological anthropology, more specialized studies on cultural food habits, often called ethnic "foodways," were also carried out in the United States during the 1930s, 1940s, and subsequently. They were meant to serve both scholarly and practical ends (for a review, see Montgomery and Bennett, 1979; also Committee on Food Habits, 1943, 1945; Cussler and De Give, 1952; Mead, 1964; Committee on Food Consumption Patterns, 1981). In addition to improving academic understanding of food as an aspect of culture, these studies were designed to clarify how cultural rules of food classification, dietary structure, and sharing as well as the channels by which food reached households influenced adequate nutrition within cultural communities and how food habits could be changed (in 1941–1943 in anticipation of World War II food rationing). While these studies seem to have had very little influence on political or nutrition policy (Montgomery and Bennett, 1979), some of the studies of the ethnic foodways have served as base-line data for more recent studies of the symbolic and social food repetoire of the same cultural groups [cf. Theophano and Curtis (1981) in Section 2.4]. Freedman (1976) has provided a brief review of the literature on food habits in the development of nutritional anthropology.

2.3. Symbolic Anthropological Studies

Less nutritionally oriented have been those studies within social and cultural anthropology that analyze food prohibitions, prescriptions, and exchange as indices of social identity and social relations. Such studies, often termed "symbolic" or "semiotic," analyze how foods or components of foods and manners of preparation or serving provide a set of elements, or a "code," by which other aspects of social relations, such as male–female relations, or of social status, such as caste or class position, are expressed in ordinary or ritual life. Semiotic approaches to cultural food habits have considered, for example, why certain animal species are classified as "edible" while others are judged "inedible" for everyone, for certain social categories, or for all but certain ritual contexts (Leach, 1964; Douglas, 1957, 1966; Tambiah, 1968). Such food classifications and ritual uses of particular foods or quantities of foods are viewed as encoding information about social categories, ritual concepts of time, and ritual or social events (Douglas, 1966; Turner, 1969; Lévi-Strauss, 1969, 1973, 1978; Leach, 1970; Sperber, 1975), the role of the semioticist (anthropologist or literary critic) being to decipher the cultural meanings of such food usages.

Classic studies in food symbolism have addressed elementary topics such as totemism—the relationship of a group to a particular animal, plant, or other quantity in its environment (Lévi-Strauss, 1963, 1966)—and food taboos (Steiner, 1956). Such analyses are attempts to discern how rules on consuming or not consuming particular species mark social groups and promote social identities in common. The meanings of food-giving and -receiving, i.e, food gifts as preliminaries to other social exchange, but also symbolic and material indicators of social rank, status, and power of respective givers and receivers, have also been addressed (Mauss, 1967; Sahlins, 1972). "Fighting" and "friendship" with food (the types and quantities of foods offered on different occasions) are ways to communicate hostile, amicable, or hierarchical relations in most societies.

As a special case in point, Hindu Indian food classifications and rules of food exchange elaborate principles of social organization, caste hierarchy, and relative caste status position. Probably no other culture has so elaborate a food code, which is also reflected in the close association of particular deities with particular foods and food attributes in Indian mythology (Ferro-Luzzi, 1977; Apte and Katona-Apte, 1981) and rigid regulations surrounding consumption on the human level. The code includes classifications of raw and processed foods according to relative degrees of purity and pollution (Harper, 1964) as well as rankings of persons as relatively pure to polluted depending not only on caste and sexual status, but also on the foods they consume and from whom and with whom they take comestibles. Marriott (1964) has provided an excellent example of how castes manipulate food transactions to improve their relative caste status; i.e., castes that are in the process of improving their socioeconomic position will try to substantiate these gains on the spiritual plane by ceasing to eat meat or other "polluting" foods, refusing to accept foods from lower castes, and trying to force food on those who classify themselves as spiritually higher.

Hindus also vary in the fastidiousness with which they observe food regulations and preparations (Khare, 1970, 1976). Day-to-day food rules seem to lapse particularly as people move away from local communities, that give support and meaning to careful observances (e.g., Katona-Apte, 1976). Localities may furthermore exhibit substantial leeway in interpreting regulations to suit personal or group historical circumstances (Cantlie, 1981).

Other approaches towards tracing the relationship between food and social classifications includes that of O'Laughlin (1974), who has brought together the perspectives of Lévi-Strauss, Douglas, and others in her symbolic–economic analysis, "Why Mbum women do not eat chicken." In her insightful analysis of the prevailing sexual division of labor, male dominance over production (control of land and granaries) and reproduction (allotment of women as wives), she concludes that food restrictions on women (women cannot eat chicken, goat, or the preferred white flour porridge) reify the real social relations in this northwest Chad society. Women are subordinate to men in all things, including diet, and are therefore kept from eating the chickens, which like the women who raise them, are kept for food production and reproduction.

Barthes (1975), in contrast to these other writers, has dealt with food symbolism and its relationship to social classifications in modern state societies. He has, on one hand, considered the various cultural meanings attributed to certain substances, like sugar and coffee, by different national groups such as the French vs. Americans. On the other hand, he has tried to identify certain "tastes" with particular social class status—e.g., lower-class preferences generally for extremely sweet or strong flavors.

Food with its multiple referents can thus be used or interpreted to symbolize (mark) social-group membership, relative social status and social relations, prevailing dominant–subordinate relationships, or relative class or national position. While all the foregoing analyses have been qualitative, in terms of analyzing the meanings of individual symbols or the structures of particular dietary laws relative to other types of cultural rules or social distinctions, Douglas and Gross (1981) (the latter a mathematician) have recently collaborated on a project to establish a more quantitative measure of the "structure" or "structuredness" of cultural food systems, which they term "intricacy." Intricacy is described as a measure of information, i.e., how many lines on a standardized computer program it would take to describe a culture's food rules (specifications for meal components, feasting, food exchange). They envision that such an approach should contribute to cross-cultural analyses of the rigidity vs. flexibility of a culture's food habits. By such an analysis, they hope to link studies of sociocultural rules for consumption (structural and symbolic studies of food and food-based social relations) to nutrititional studies. They plan by such a measure to predict, for example, the incidence of malnutrition on the basis of food-distribution rules under variable resource conditions. As a second example, the "structuredness" of a culture's food habits might be used to predict the potential for acceptance of new foods. It is difficult on the basis of their current descriptions to envision precisely how such measures of "intricacy" will be operationalized and applied in particular cultural and nutritional case studies. More ethnographic examples will have to be studied according to this "intricacy" principle before it can be said to be useful.

2.4. Cognitive and Ethnoscientific Anthropological Studies

Within American anthropology, a further development that has as yet unfulfilled promises for contributing to cross-cultural understandings of food and nutrition has been that of "cognitive anthropology" or "ethnoscience." Studies using this approach analyze particular domains, such as fauna, flora, kinship, or medicine, in terms stemming from the descriptions of the natives themselves. One can then compare and contrast such "native" understandings of their universe and behaviors resulting from such evaluations with "objective," scientific measures of the orders of experience, e.g., native vs. scientific classifications of plants, animals, and illnesses.

Ethnoscientific studies have largely concentrated on the "linguistic" aspect of culture ("linguistic anthropology" or "ethnographic semantics") and have had as their primary aim the elucidation of the structures of plant, animal, social, or illness term taxonomies. Ideally, one should be able to move from such "folk classifications" to actual behaviors and evaluate the practical consequences of these folk classifications and their resulting usages to environment, nutrition, and health. Of particular relevance for understanding relationships between nutrition and behavior are ethnomedical studies, which include analyses of native illness classifications, etiologies, and nosologies. They usually include as well general principles of cosmological structure relating processes within the human body to social and natural processes outside of it. Studies of ethnobiology and ethnoecology provide additional valuable information on food production and consumption. In his ethnobotany of the Hanunóo, Conklin (1954) carefully described, as aspects of understanding the total range of relations between people and plants, the principles for classifying and naming all flora as well as the native's understanding of local plant communities and ecology, the growth, visual, and "taste" characteristics of different rices, their roles in local ecology and diet, and the structure of diet. Messer (1978), as another ethnobotanical example, presented the local taxonomic and functional classifications of plants in Mitla, Oaxaca, Mexico. In addition to analyzing local cultural understandings of plants in terms of systematic nomenclature, she showed how people categorized plants for food and medicine and how classifications in terms of growth habits and life stages of particular species, taste (more generally flavor, texture, and color), and prestige values influenced formation of categories and rankings of "edible" plants. She then observed how such classifications affected consumption.

Unfortunately, there have been few ethnoscientific studies specifically dealing with native food choices and with the dimensions and objective consequences of "folk" nutritional concepts and practices. Potentially, ethnoscientific studies of diet could have informants sort elements into culturally meaningful dietary categories such as "starchy staples" and "relishes," elicit the criteria, including costs, by which individuals and groups choose among them, and evaluate the nutritional consequences of such choices. The study by Theophano and Curtis (1981) of Italian-American food habits, which analyzed patterns of food consumption and inter-/intrahousehold food exchange utilizing categories supplied by the people themselves, illustrates the potential usefulness of this approach both for describing current food habits and for tracing cultural changes. Studies of this type also build on the kinds of structural analyses of food proposed by Douglas (1966).

Another topic for further ethnoscientific–nutritional research is that of the chemical senses. So far, there has been little comparative ethnographic analysis of taste–smell terms or careful analysis of how such conceptual categories organize and direct the selection and arrangement of food. In many cultural systems, taste and smell are identified by the same word, particularly to identify various classes of "edible" vs. "foul" or "disgusting" items. Foods, animals, plants, and medicines can be classified as tasting–smelling sweet, bitter, salt, sour, pungent, or foul in most cultures. Other taste categories found in some cultures include acid, not-sweet, unripe, and tasteless or not-bitter, terms that identify desirable vs. undesirable taste qualities of specimens within classes of items generally judged to be "edible." A few ethnomedical (Colson, 1976), ritual (Kuipers, 1981), and kinship–social organization (Seeger, 1981) studies suggest that in certain languages, taste–smell terms are extended beyond the primary chemical sense referents to label symbolic categories of foods, especially animals, inappropriate for members of certain social (age, sex, ritual) status, as well as other nonfood symbolic categories. Taste–smell, then, is one of the dimensions by which people sort foods into categories that have sociocultural meanings beyond nutrition. However, sociocultural research into the *nutritional* consequences of "taste" classifications remains underdeveloped.

Also lacking are ethnoscientific investigations of nutrition-related diseases like diabetes. The elegant ethnomedical study of Piman shamanism by Bahr *et al.* (1974) offers no insights into cultural conceptions of diabetes, an illness that afflicts large numbers of this and other native American populations, although certain classes of illnesses in this culture are associated with the production of "dirty stuff" that is believed to be generated by excessive consumption of soda pop, ice cream, sweet wines, and other sweet and sticky substances. More promising is the copious literature on "hot–cold" qualities and other symbolic classifications of foods found to be related to native concepts of health and illness. These will be discussed further below.

2.5. Ecological and Biocultural Anthropological Studies

Complementing analyses by ethnoscientists have been studies by cultural materialists (Harris, 1968), ecological anthropologists (Steward, 1955; Vayda and Rappaport, 1968), and nutritional anthropologists (Jerome *et al.*, 1979). These have been more explicitly concerned with understanding the nutritional and environmental determinants and consequences of food habits in scientific terms. From the cultural materialist perspective, environmental conditions and subsistence needs are seen as directly or indirectly shaping cultural practices that maintain the material basis of society. In this view, "culture," including social organization, symbolic, and other aspects of cultural attitudes (toward food), consciously or unconsciously serves ecological and economic ends. The units of analysis are objective measurements, sometimes nutritional, such as quantities of energy (calories), protein, and other nutrients in foods, and sometimes ecological, relating species within a local or regional ecosystem, and sometimes economic, involving financial calculations.

Materialist analyses have been used to "explain" the rationale of eating habits as diverse as the Indian prohibition on eating cows to Aztec cannibalism. Harris (1974) used this approach to find an ecological and economic logic for the Hindu custom of *ahimsa* (respect for animal life, but especially cattle), New Guinea highland peoples' customs of periodic pig feasts, the Hebrew ban on eating pigflesh, and the endemic warfare between Yanamamo groups. He concluded that all were customs that ensured group nutritional survival within their given food environment. Ross (1978) interpreted South American lowland animal food taboos as mechanisms for maintaining the populations of animal species within ecosystems. Harner (1977) explained Aztec cannibalism as an adaptive response to protein deficiency. In each case, however, critics have shown that the materialist explanations are defective. They generally do not incorporate all the available ecological, social, or cultural data into their analyses (see "Comments" in Ross, 1978; Chagnon, 1977; Chagnon and Hames, 1979). As Douglas (1966) commented, whatever the health rationale for the Hebrew dietary laws, to treat Moses as no more than an enlightened public health administrator (the "medical materialist" view) misses the multiple levels of meaning of such food rules in the culture. Harner's critics have shown that there was sufficient protein in the Aztec diet without human flesh; in fact, more energy would have been expended in capturing, processing, and finally eating the parts of a cannibal victim classified as edible than would have been consumed (Ortiz de Montellano, 1982). Thus, some of the cultural materialist arguments have also been refuted in their own "materialist" terms.

Other examples of materialist reasoning can be found in those studies that attribute episodes of spirit possession to nutritional deficiency (e.g., Gussler, 1973; Kehoe and Giletti, 1981) and witch-craft behaviors to the use of hallucinogens (Caporael, 1976; Matossian, 1982). The attempt to link the Salem witchcraft incidents to ergotism, caused by the consumption of ergot fungus from infected rye, is a representative example of the weaknesses of materialist reasoning. The argument is insubstantial and unsubstantiated and rejects more complex explanations that stem from more detailed sociocultural and socio-economic analyses.

Studies focusing on "human ecology," by contrast, have attempted to widen and refine the understanding of social and cultural components of material systems. While still working with such "objective" units as biological species and energy flow, they also investigate cultural understandings of resources and resulting patterns of resource use as basic to understanding the synchronic and evolutionary dynamics of ecosystems and cultures. While human ecological studies take their models from animal ecology, studies of the environmental resources, cultural food classifications, and patterns of food-sharing among hunters and gatherers such as the !Kung San of the Kalahari (Lee and Devore, 1968) have shown how the adaptations and food systems of even the simplest human societies involve a complexity unknown in other primates or lower animals. Relatively precise knowledge about the location and seasonality of upwards of 80 plant resources and their careful classification into preferred and less preferred species on the basis of palatability, digestibility, and availability is a cultural corpus of knowledge (Lee, 1968, 1969) that has to be carefully taught, often within same-sex groups (Draper, 1975; Dahlberg, 1981). People must also learn the names, whereabouts, and habits of many animal species. The organization of hunting tasks, which include formation of cooperative groups, manufacture of tools, and time-planning, are clearly activities that set off human primates from non human (Strum, 1981). Water availability, and the relative resource positions of other social units from which one can expect to give and receive food, other material goods, and friendship, are other aspects of !Kung San environment and comprise part of their resources.

More generally, ecological hunter–gatherer studies have shown how such societies adjust to scarce resources by moving, eating less-preferred foods, limiting activities, resting more, or mobilizing trade networks (Lee and DeVore, 1976; Marshall, 1976; Wilmsen, 1978; Weissner, 1981). Occasionally, they have also been forced to consider the nutritional and cultural implications of the changing political and environmental settings of hunter–gatherer groups. In particular, Wilmsen (1978, 1982) has noted how !Kung San, on a cornmeal, sugar, and milk diet in contrast to "bush" diets, are heavier, seem to show no seasonal weight variations, and have lost seasonal patterns in fertility. Whether looking at hunters and gatherers in the "bush" or "in town," such ecological studies do not take the environment as a "given" but emphasize the sociocultural knowledge and behaviors that allow people to use that environment in ways that preserve their physical and social life.

The study by Rappaport (1968) of ritual in the ecology of the Maring, a highland New Guinea people, marked another watershed in anthropological studies of the relationships between human populations and their environments.

By comparing what he termed "operational" (scientific, objective) and "cognized" (native) environmental perspectives, he tried to show how native ecological understandings and resulting cultural practices influenced the relationships of local populations to land, natural species, and other local groups within a regional ecosystem. Concurrently, he considered how horticulture and food-gathering contributed to adequate nutrition and enabled the populations to reproduce themselves. In contrast to earlier "cultural ecological" studies (Steward, 1955), which had distinguished between the "core" (subsistence) and secondary (ritual elaboration) features, Rappaport considered such cultural and biological components of ecosystems to be not separate but interrelated. He argued that the pig festivals of the Maring, though viewed by them as a ritual cycle, were regulatory mechanisms, periodically adjusting people to land, to their sweet potato–pig resource base, and, through food taboos associated with the pig cycle, to some other potentially edible animal species in the environment. At the individual level, the pig festivals also allowed people to consume high-quality protein periodically when they most needed it and during festivals to display their availability as potential mates. While Rappaport's claims about the ecological triggering of the pig festivals and the nutritional well-being of the Maring have been attacked (McArthur, 1974), the study remains a model of how to study human population adjustments to land, available biological resources, and other human populations from native and scientific perspectives.

Among the other studies employing a human ecological framework have been those examining the cultural and biological evolution of particular cropping and herding systems (Flannery, 1973), the nutritional values of indigenous foods (Robson and Elias, 1978), and traditional diets that selectively use a wider range of foods within local ecosystems than is common under modern cropping systems (Forbes, 1976; Messer, 1976a). Ecologically oriented ethnographies have examined local human populations in their ecological context from the perspective of energy flow—how energy inflow matches energy outflow (expenditures), or how energy passes through the food chain (e.g., Brush, 1977; Thomas, 1973). Several studies have also calculated input–output efficiency of human cultivation systems operating at different levels of technology in terms of energy flow, human energy and fossil fuel energy inputs, and the amounts of energy it takes for particular foods to reach the "tables" of modern consumers (Rappaport, 1971; Pimentel *et al.*, 1973, 1975; Steinhart and Steinhart, 1974; Revelle, 1976; Pimentel and Pimentel, 1979). Efforts to interrelate local, regional, national, and international levels of energy flow through world economic system models are still preliminary. There have been a few attempts to demonstrate how local populations become "hooked" on consumer goods and integrated into the larger cash economy, with accompanying loss of (food) self-sufficiency (e.g., Gross *et al.*, 1979).

Using the social group rather than the human population of an area as their focus of analysis, "human socioecological" studies (usually of hunter–gatherers) have adopted and adapted models from animal ecology to pursue questions such as how the structure of resources in the environment affects the social structure and behavior of the human groups exploiting them (Winter-halder and Smith, 1981). While the archeological or ethnographic data on human social organization or the resources used rarely conform exactly to the predictions of the models, the departures from the models offer ample opportunity for expanding the frameworks within which one interprets foraging behavior (Durham, 1981).

Along biocultural lines, nutritional (physical) anthropologists, beginning with biological models of nutrient intake and behavioral or biological outcome or both, have attempted to trace how culture affects both food intake (food habits) and resultant health and behavior (Jerome *et al.*, 1979). Additional attempts to combine biological and cultural models and data have taken the form of physiological studies of "adaptation" to particular nutritional environments (Damon, 1977; Harrison, 1975; Weiner, 1977), biocultural analyses of human nutrition and functioning at the individual, household, and community levels (Haas and Harrison, 1977), and biochemical and pharmaceutical analyses of the contents, chemical reactions, and possible physiological advantages of certain food and medicinal practices (Etkin, 1979).

Interrelationships among environment, human genetics, behavior, and social organization have also been demonstrated (Greene, 1977). Greene, (1976), employing such a biocultural model (Fig. 1), showed how iodine deficiency produces not only severe human disability in the form of endemic cretinism, but also a range of milder mental–physical impairments, which in local socio-cultural terms may be accepted as "normal." In the iodine-deficient Ecuadorian community he studied, cultural expectations for human performance had been adjusted downward to accommodate moderately impaired individuals. Thus, the whole community operated with lower standards of tasks and achievement than unaffected communities. Mead had earlier asserted that mild iodine deficiency affected cultural patterns by reducing the complexity of behaviors and social organization, particularly in rituals that she described as "slowed down" (and therefore easier to study). On this basis, she chose to study trance and dance patterns in a Balinese village that suffered from endemic goiter (Mead, 1977). The social and behavioral effects of other forms of malnutrition, including the consequences of mild to moderate energy deficiencies, have been discussed in Greene (1977).

An additional area of biocultural research has been the possible nutritional bases of certain behavioral disorders (some of which case studies were considered above as examples of cultural materialism). Relationships between arctic hysteria and calcium deficiency (Wallace and Ackerman, 1960; Foulks, 1972) and between membership in women's possession cults and calcium or niacin deficiencies (Kehoe and Giletti, 1981; Gussler, 1973) have been proposed. In each case, however, investigators need more objective data on nutrient intakes and cultural and clinical descriptions of the supposed nutrient deficiencies to demonstrate convincingly that particular cultures, through environmental lack or food rules or both, predispose certain classes of individuals (e.g., women) to nutrient deprivation and then classify and respond to resultant symptoms, to, in certain cases, "solve" the nutritional problem (Gussler, 1973).

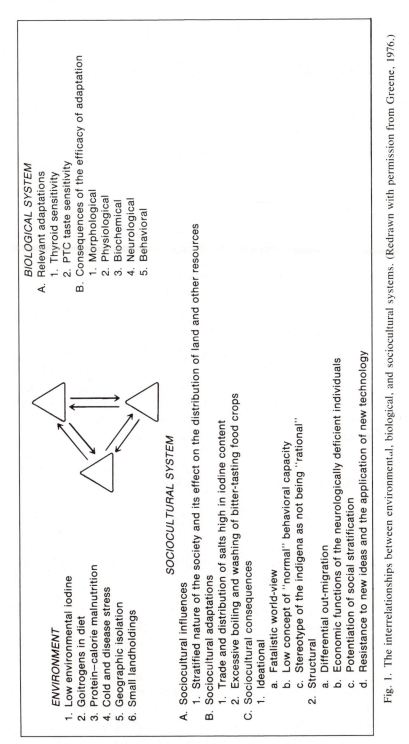

BIOLOGICAL SYSTEM
A. Relevant adaptations
   1. Thyroid sensitivity
   2. PTC taste sensitivity
B. Consequences of the efficacy of adaptation
   1. Morphological
   2. Physiological
   3. Biochemical
   4. Neurological
   5. Behavioral

ENVIRONMENT
1. Low environmental iodine
2. Goitrogens in diet
3. Protein–calorie malnutrition
4. Cold and disease stress
5. Geographic isolation
6. Small landholdings

SOCIOCULTURAL SYSTEM

A. Sociocultural influences
   1. Stratified nature of the society and its effect on the distribution of land and other resources
B. Sociocultural adaptations
   1. Trade and distribution of salts high in iodine content
   2. Excessive boiling and washing of bitter-tasting food crops
C. Sociocultural consequences
   1. Ideational
      a. Fatalistic world-view
      b. Low concept of "normal" behavioral capacity
      c. Stereotype of the indigena as not being "rational"
   2. Structural
      a. Differential out-migration
      b. Economic functions of the neurologically deficient individuals
      c. Potentiation of social stratification
      d. Resistance to new ideas and the application of new technology

Fig. 1. The interrelationships between environmental, biological, and sociocultural systems. (Redrawn with permission from Greene, 1976.)

Bolton (1973, 1981), in seeking a relationship between hypoglycemia and aggression among natives in the Lake Titicaca Basin, argued that nutritional deficiency, in combination with social pathology on the individual level, led to widespread social pathology and undercooperativeness within the group, thereby preventing individuals from ever working together to improve their nutritional environment. Thus, the system was a psychobiologically self-perpetuating one. The argument, however, suffers from the general problem of moving directly from the individual and household level of interaction to the social level of explanation and from the related problems of proceeding from one biological factor to a general explanation of social behavior. Lewellyn (1981), in addition, has suggested that the people Bolton studied were neither homicidally aggressive nor hypoglycemic and has offered the alternative hypothesis that both aggression and low blood sugar were caused by high alcohol consumption in the context of harsh environment and poor food.

Biocultural approaches to nutrition as a factor in illness have been summarized in the medical anthropology text by McElroy and Townshend (1979), a volume that concludes with case studies of the nutritional and health consequences of contact between traditional and modern cultures.

2.6. *Anthropological and Related Studies of Nutrition and Economic Development*

Because international development efforts are concerned with the social, cultural, and physical barriers to economic advance, there has been a concerted effort to understand the impact of international development on diet (Den Hartog and Bornstein-Johannson, 1976), the "nutrition factor" in economic development (Berg, 1973), and the role and nutritional consequences of women's work in food production, preparation and distribution, household maintenance and child care, and cash employment (Cowan, 1978). Political aspects of national policies on food self-sufficiency in the context of other political–economic objectives have been treated by Lappé and Collins (1977) and Franke and Chasin (1980), among others. The nutritional impact of highly mechanized "Green Revolution" agriculture has also been investigated as one aspect of national agricultural, general economic, and nutritional policies. Some analyses have suggested that the economic and nutritional effects have been at least in part negative, because mechanization, being capital-intensive rather than labor-intensive, eliminates land, jobs, and income for the poor (see essays in Winikoff, 1975; Hewitt de Alcantara, 1976). Regional development projects aimed at improving earning capacity and nutritional–health status of the rural poor have also had mixed success, when measured in terms of children's nutritional status. Evaluations of two Mexican regional development projects indicated that people thrust into a cash-cropping scheme may benefit when they can still produce some of their own food and when they have improved health facilities that prevent nutritional losses due to infection (Brown, 1978; Dewey, 1980). Government policies that favor home production of food over cash crops may also prove nutritionally beneficial, although the outcome of such policies depends on a complex of landholding, market, and employment variables, as well as food practices, all of which affect food intake and nutritional status (Marchione, 1977). In general, all the aforecited anthropologists favor policies promoting food self-sufficiency over cash-cropping and over exclusive reliance on market food resources at the local level.

At the household and community levels, the relationship among nutrition, productive behavior, and reproductive behavior has been investigated by economists in a growing literature on women's and children's work and household time allocation (Popkin, 1980; Evenson, 1981; U.S.A.I.D. Women and Food Information Network). Some studies have shown that the extra income of the mother seems to be the main "economic" factor accounting for lower levels of malnutrition within populations (Tripp, 1981). Others have suggested that mother's time away from home and the child, in the absence of adequate supplementary domestic and child-care arrangements, may account for the poorer nutritional status of the child either because the mother has inadequate time to prepare a balanced diet to feed the child (Popkin and Solon, 1976; Popkin, 1980) or because the child suffers hunger, exposure, and illness while the mother works (Kumar, 1977). A related question is intrahousehold distribution of food under different work conditions. Gross and Underwood (1971), in a classic study of energy flow among sisal workers, showed that male wage-earners received preference in the allocation of calories within the household. They were fed first, in sufficient quantity to sustain their work, often at the expense of children and women, who received inadequate calories if total household food supply was insufficient. The impact of woman's work and income in changing the intrahousehold distribution of energy and protein away from a pattern favoring senior male workers to one that better meets the needs of women and children still needs further investigation.

2.7. Summary

The previous subsections have discussed the kinds of models and methods used by anthropologists and social scientists in related disciplines to discuss sociocultural aspects of food production, selection, preparation, and distribution, all of which affect nutritional outcome at the individual, household, and community levels. Given that anthropologists tend to organize their data in sociocultural frameworks, the data on sociocultural factors in food selection and behavioral outcomes of malnutrition addressed by nutritionists in nutritional studies are not always easily accessible. Therefore, in the next section, I will discuss some of these factors by topic.

3. Sociocultural Determinants of Food Intake

Humans accept food items as "edible" or reject them as "inedible" and establish preferences among edible foods on the basis of a number of sensory and cultural characteristics (for current research, see Solms and Hall, 1981).

The term *sensory* incorporates here psychophysical, cognitive, and affective factors that enter into "taste" discriminations and preferences in food selections; *cultural* includes symbolic, social, and economic factors that in interaction with sensory data and preferences, shape food patterns and influence the selection of their individual component foods. People select foods to conform to particular styles of eating, as well as to cultural concepts of what items are edible and preferred. They eat discontinuously; that is, they structure their food intake throughout the day and also over longer periods of time according to cultural convention. In some contexts, they eat whether or not they are physiologically hungry. In others, they abstain from consumption in the presence of both food and hunger for social and symbolic reasons. Furthermore, people may ingest or feed to others items that they know or should recognize as harmful. However, in evaluating food habits—the combinations of cultural and social ideas and behaviors that surround nutrition—we generally assume that food knowledge and practice are "adaptive" overall—that is, that they serve to preserve human life and the cultural and material bases of the human group. I will return to this issue at the conclusion of this section.

3.1. Sensory Characteristics

At a most basic level, food selection is governed by a number of sensory characteristics, such as taste–smell, texture, color (and other visual characteristics), even sound (as in crunchiness), and physiologically perceived characteristics, such as "fillingness' or "burn." The exact dimensions of these characteristics, as well as degrees of acceptance or preference for them, differ among and also within different cultural populations.

3.1.1. Taste

Liking for sweet taste and a rejection of bitter seem to be innate, perhaps as a result of biological "coding" of "safe" vs. "poisonous" foods. Experiences within particular cultures, however, affect how taste qualities are conceived and labeled, how preferences for degrees of sweet, bitter, and other flavors are formed, and how both of these inform intake. Mexicans, for example, enjoy beverages sugared to a degree that is "too sweet" by most American standards, while Mexicans find the bitterness of unsweetened American coffee unpleasant (Messer, 1979). Some African populations seem to be raised from an early age with a preference for sour tastes, like tamarind (Jerome, 1977), which American children might find "puckering." Certain cultures have also "discovered" taste modifiers that chemically alter sensory perception of taste, such as miracle fruit, which makes "sour" taste "sweet" (Bartoshuk *et al.*, 1974). Cultural populations also seem to differ in their preferences and tolerances for salty tastes. It is unclear, however, to what degree these preferences are genetic and to what extent they are environmentally developed (Desor *et al.*, 1975). Cultural experience also determines the acceptibility of and preferences for pungent or "biting" flavors, like chili peppers (P. Rozin and Schiller, 1980). Tastes for more complex flavors—whether strong like the fermented sauces and pastes of oriental cuisines and the ripe cheeses of Western Europe or bland like the slightly fermented staple grain porridges of Africa—are also culturally conditioned and may be judged unpleasant, at least initially, by those unaccustomed to them. In either case, the development of taste tolerances has enabled people to extend the range of nutritious substances in the environment, to select, among other things, microorganisms, spices, and the products they transform and preserve, and to channel the evolution of human populations-in-ecosystems in particular directions (see, for example, Ritenbaugh, 1978). Yet the mechanisms by which some flavors become acceptable, and likes and dislikes are formed, have not been well established (P. Rozin and Fallon, 1981). Flavors of foods, either psychophysical "taste" or culturally extended meanings of "flavor" or both, may also be closely tied to concepts of digestion and health. One example is the classic Chinese belief that the five basic flavors have predictable physiological effects by acting in systematic fashion on particular organs of the body (Veith, 1949).

Flavor (spicing) is also one means by which people mark the cultural identity of their foods (E. Rozin, 1973) and a procedure by which they render unaccustomed foods familiar. Vietnamese refugees in the United States, for example, experimented with American condiments to create an acceptable substitute for their usual spices and used this sauce to transform American foods that were new for them into more familiar comestibles (P. Rozin and Schiller, 1980). Even within the same regional cuisines, different local cultures distinguish themselves by variations in their spicing patterns; e.g., certain Malaysian groups are known for their "sweet" sauces. At the local level, individual households and household groups may further express their distinctiveness by variations in food preferences and flavoring within the same general pattern of a community. Additionally, flavoring patterns, including sweetening, are sometimes interpreted as a class marker. Indulgence in high degrees of sweetness within American culture is interpreted by some as a lower-class (or foreign) phenomenon (see Barthes, 1975). In some countries, such as Guatemala, pungent and highly spiced foods are sometimes associated with "less civilized" elements within the nation. The decreasing use of strong tasting–smelling spices like *Chenopodium ambrosioides (epazote)* and chili peppers among some urban Guatemalans has been interpreted as a way of demonstrating that they are "civilized," since these strong flavors are judged by some as suitable only for "less civilized" palates (Messer, 1979, field notes). In similar fashion, the strong odors of certain ethnic cuisines may be rejected by members of these groups who see them as barriers to socializing in the larger culture. Algerian immigrants in Marseilles were reported to have noted how the aroma of their lunches offended other workers and made them conspicuous and to have arranged to have a food-training course for their wives (Autret, 1962). While bland, less strongly flavored foods seem to characterize the "civilized" Western diet, it should also be noted that such preferences may be in part a result of technological factors that encourage efficiency and uniformity in mass production and processing, more than flavor and other subjective qualities (Amerine, 1966).

Finally, as noted earlier, the taste and smell of foods are in many cultures evaluated jointly. In many cases, they are, for practical purposes, indistinguishable.

3.1.2. Texture

Texture is a second essential element in food selection and preferences. In large part, it shapes what makes foods familiar and may influence acceptance of new foods. In Africa, where the basic staple is porridge, ranging from thick to watery, different groups distinguish themselves by the texture they prefer their staple food to be. When given a new commodity, they will accept it more easily if it can be manipulated in forms that are texturally familiar. Within certain European cultures, groups distinguish themselves by preferences for "thick" or "thin" soups. In rice-eating cultures, donated food in the form of grits or bulgur (wheat) may be more acceptable than grains delivered in the form of flours or meals. The glutinous properties of different varieties of grains such as wheat, maize, and rice may affect their suitability for producing familiar foods such as bread, tortillas, and rice dishes, respectively. Such qualities often affect acceptance of new varieties of foods designed to improve nutrition. For example, the high-lysine maize developed to help correct the lack of protein in Mexican maize diets was largely rejected by those it was intended to benefit because its poor adhesive qualities did not allow it to form a workable tortilla dough. New "miracle" rices have been disparaged by the consumers toward whom they were targeted as having textures that "stick in the throat" (Franke, personal communication); they are less preferred than traditional rice varieties. Even within American culture, whole grain breads, which have positive "health" connotations, are denser and so may be rejected by some who have developed a taste for the lighter, softer white bread.

Certain textural properties are also intrinsic in judgments of qualities like "crispness," "crunchiness," and "freshness," all of which are important in the selection of food items by classes and by individuals within a class. Barthes (1975) has suggested that there exists a general symbolic opposition between "crisp, brisk, sharp" foods and "soft, soothing, sweet" foods in Western cultures. Texture, like flavor, may also be associated with palatability and digestibility. Natives sometimes use these descriptive characteristics in distinguishing between preferred and less-preferred foods, those consumed under normal vs. starvation conditions (e.g., Rosemary Firth, 1966).

3.1.3. Visual Characteristics

Size, shape, and color are all additional sensory properties of foods that influence food selection and preferences. These aspects of visual appearance often encode information about other taste and textural characteristics by which people judge foods to be more or less appetizing, appealing, or valuable for certain purposes. At the most basic "gut" levels, visual inspection may contribute to acceptance vs. rejection of particular items as "food," as in the usual rejection of items that "look" (as well as "smell") rotten or of those that carry symbolic connotations of "abomination," e.g., the rejection of foods that resemble genitalia in cultures where such are judged to be "disgusting." Visual characteristics furthermore often supply the morphological and other criteria for "folk" classifications of foods such as cereal grains, legumes, tubers, and sauces and for additional varietal discriminations within these categories. By means of such classifications, people construct diets that incorporate cultural ideas of the nutritional contents of particular food types as well as their other social and symbolic meanings.

Color, as one visual characteristic of foods, often provides a code by which people label and rank varieties within more general "species" of foods with particular textural or taste properties. Consequently, color, because it also encodes other dimensions of cultural value, may influence food selection more than reputed nutritional worth. Throughout much of Mexico, for instance, white maize is preferred for tortillas, since white tortillas are said to look "cleaner," to have a softer texture, and to taste better than tortillas made of yellow, red, or purple maize (Messer, 1976b). New high-nutrient maize varieties have been rejected because of their yellow color among other culturally "undesirable" characteristics.

Color may also operate as a kind of "folk index" of purity and refinement and affect the "prestige values" of varieties within particular categories of foods. The prestige value of "white" ("refined") or "black" ("coarse," "unrefined") breads has been a distinction basic to European food habits. White "polished" rices in most parts of the world are preferred because they are more processed and refined, quicker cooking, more desirable in taste and texture, and consequently associated with higher cultural status.

Color may also provide a sign of expected quality, including ripeness and wholesomeness, and therefore taste, in fruits and vegetables. For such cosmetic reasons, oranges are colored orange in the United States; other fruits are anticipated to taste "green," "ripe," or "overripe" on the basis of, among other qualities, color. Color may also be perceived as a sign of decay and inedibility, as in blue or green cheese (except if cheese is expected to be moldy), discolored meats, and spotted or brown fruits and vegetables. But colors that are normally "unappetizing" may be not only accepted but also chosen under ceremonial circumstances and at certain times of the year. On Saint Patrick's Day in the Boston area, even the beer and the bagels are dyed green! Thus, colors may carry symbolic meanings beyond ordinary "health" connotations.

Along these symbolic lines, food colorings, like flavorings, may also have customary or ritual values, as in Mexican "green," "yellow," "red," and "black" sauces, which may identify at a glance dishes of particular condiment compositions and flavor and, in certain instances, the special occasions at which they are served. Color thus provides in various cases a code for "safe" as well as appropriate foods.

3.1.4. Perception of Physiological Effect

Individual foods or methods of food preparation are also classified and ranked according to their perceived physiological effects. Physiological sensation of satiety is a first factor. For example, the Bemba of Northern Rhodesia classify the different grains that they prepare as porridges according to "fillingness." They aim toward a "turgid" feeling in the stomach and further rank the different grains (millet higher than maize or sorghum) in terms of the amount of time porridges made from them will supply this feeling, which suppresses hunger pains (Richards, 1939). Hondurans have been reported to prefer maize over sorghum for the same reason—tortillas made of the former are described as more filling (DeWalt, 1982, personal communication). Protein may be an additional factor. Hunting–gathering cultures put a value on "meat" and claim that without it, they are "hungry" no matter how much vegetable food they have ingested (Lee, 1968; Holmberg, 1950). Agricultural populations may also be "hungry for meat," though it is open to question whether it is the actual ingestion of the meat or the anticipation of the feasting, the heightened activity, and the feelings of exhilaration that accompany meat consumption that constitute the "hunger." Others have suggested that the emotional and physiological cravings for meat reported in the ethnographic literature might be more accurately a function of the salt or fat content of meat (Hayden, 1981; Jochim, 1981). Fat and salt, along with other condiments, are recognized as physiologically necessary by most peoples subsisting on a predominantly cereal grain, tuber, or plantain diet. The Bemba, for example, consume grain porridges with a relish of animal or vegetable protein, at a ratio of three parts grain to one part relish. If the female of the household has not had the energy to gather the relish ingredients from the bush, she may not prepare even the grain, since, in her cultural view, one cannot get the porridge down without the relish. One might make a similar case for the consumption of tortillas with chili peppers in Mesoamerica. Without pungent spices, natives claim they have no "appetite." Cultural dietary practices thus take into account the need for spices to help people ingest sufficient quantities of their bulky, starchy staples to meet energy needs. Some spices may furthermore provide other physiological and emotional responses that people learn to classify as desirable; e.g., chili peppers are appreciated for their "burn," while at the same time the hot peppers increase the flow of saliva and intestinal activity, both of which aid in digestion of a maize diet (P. Rozin and Schiller, 1980). Other physiological effects that become incorporated into cultural dietary preferences include (1) the stimulating effects of caffeine in coffee, tea, chocolate, and cola drinks; (2) the sudden rush of energy that people experience after consuming heavily sugared beverages (Messer, 1979); and (3) the appetite-stimulating or dulling effects of certain "quasi-food" medicinals and tonics, like marijuana, coca leaves, *ch'at* (*Catha edulis*), a leaf stimulant chewed commonly in Ethiopia (Simoons, 1960), and alcohol-based substances. Ingredients of bitter taste are often included as elements of herbal brews and tonics prepared to stimulate appetite in many parts of the world, since "bitterness" is recognized as characteristic of certain underlying chemical constituents like tannins, which do have a perceptible physiological effect on gastric secretions (Etkin and Ross, 1982).

Negative short-term physiological effects may, conversely, form the basis of food avoidances. Fisher *et al.* (1977), in their analysis of foods avoided by a group of Caroline Islanders, found that most of the foods that were tabooed (ritually prohibited) were potential allergens such as shellfish and eels. Furthermore, "punishments" for breaking food taboos, whether purposefully or accidentally, corresponded to the symptoms of allergic reaction. They included chastisements such as skin eruptions, skin sores, swelling, and shortness of breath. Within this culture, violations of prohibitions against consumption of the "runts" of species also existed. Consumption of "runts," which would be less likely than allergens to produce such adverse physiological effects, were supposed to be punished by a deformed birth in the future.

Culturally prescribed avoidances such as prohibitions on consuming shellfish and pork may also have their health rationale, although the cultural reasons for avoiding such items may go beyond the simple reasoning that "they will make you sick" (Douglas, 1966). Other adverse physiological reactions culturally encoded as food dislikes may be at least in part genetically based. For example, genetically based lactose intolerance may be at the root of milk avoidances in certain cultures (Simoons, 1973; Harrison, 1975), although the physiological argument still does not explain why certain lactose-intolerant populations, such as the Chinese, do not like cultured milk products that they should be able to digest. Mediterranean populations tested to be lactose-intolerant have adjusted to inherited inability to digest fresh milk by "predigesting" it into processed forms, such as yogurts and cheese, that can be consumed without gastric distress. Thus, physiological and cultural factors must be considered in evaluating the evolution and distribution of food dislikes.

That the potentially harmful properties of certain otherwise apparently edible substances have been discovered and translated into "cultural" dislikes and prohibitions some time in each culture's history is not surprising. Ethnologists observing contemporary populations have noted how new foods are classified as "good" or "bad" for you on the basis of their perceived physiological effect—whether they are easily digested or make people sick. Such idiosyncratic observations for particular foods may be incorporated into an existing cultural symbolic classification such as hot–cold that systematizes health and dietary interactions (see Foster, 1953; Cosminsky, 1977; Messer, 1981a). What is still little understood, however, is how certain otherwise innocuous foods come to be classified as not only unacceptable, but also as actually disgusting. Semiotic approaches to food rules and taboos have cleverly analyzed why certain natural animal and plant categories should be interpreted as "anomalous" and made the target of special cultural attention. But why such foods should produce violent physiological reactions like vomiting if accidentally ingested by those who classify them as "taboo" (Holmberg, 1950) is still a topic that requires further analysis (P. Rozin and Fallon, 1981).

Similarly, psychological and cultural components of food cravings are not well understood. The factors permitting or encouraging consumption of "dangerous" foods—i.e., those that are tainted, toxic, or carcinogenic—are an additional area for investigation. Thus, the sensory characteristics of foods, while an important guide to selecting a diet that is healthy, do not in all cultural cases provide an infallible route to safe and nutritious eating.

3.2. Cultural Symbolic Dimensions

In addition to sensory characteristics, foods may also be classified according to a number of cultural factors that can be constructed from sensory data and other cultural information. Not all these criteria are relevant to food selection in modern "Western" culture, but they are useful for understanding the construction of diets in other cultural contexts.

Hot–cold, wet–dry, male–female, heavy–light, yin–yang, pure–impure, clean–poison, and ripe–unripe, as well as "flavor," "sharpness," "itchiness," and "color," are all dimensions that singly, or in combination, have been used in different cultural contexts to classify food and to direct food intake (see, for example, Reichel-Dolmatoff, 1968; Ahern, 1975; Colson, 1976; Beck, 1969; Messer, 1981a; Manderson, 1981). Such categories are often considered "symbolic" because they may not refer to easily measurable or single "objective" qualities of food or other items, and also because such classifications often bring together a number of different sociocultural domains, such as flora, fauna, medicine, health, ritual, and social relations. They reach their greatest elaboration in Eastern, particularly Indian and Chinese, cultures, where they are part of a complex system of humoral medicine and philosophy. Their meaning and nutritional significance vary according to cultural context, as well as according to individual inclination to "follow the rules."

Hot–cold classifications, as they interrelate domains of food, health, and social relations, are among the most discussed in the literature, in part because they appear to span the Old World and the New and are therefore useful for comparing and contrasting nutritional and medical ideas from different cultures and also for tracing diffusion of foods, medicaments, and medical systems. Hot–cold values, which refer to an intrinsic quality rather than temperature or spiciness, are ideally present within the human body in approximate balance. Too much heat or cold, caused by overconsumption of either hot or cold substances, overexertion (overheating the body), overexposure (to climatic or other heat or chills), an illness agent that is classified as giving rise to heat or cold within the body, or usually a combination of these factors produces an imbalance that is believed to result in illness. For example, an elderly Mexican peasant, "hot" from years of work in the hot sun, avoids overconsumption of chocolate, classified as hot, because he will find it difficult to digest, a cause-and-effect sequence that he describes in terms of "too much heat" (Messer, 1981a). Imbalances are corrected by the principle of opposites: the individual is carefully dosed with foods or medicines, or both, of the opposite quality to restore a healthy balance and is also encouraged to avoid exposure to the offending quality. In the Mexican case just cited, the elderly farmer may correct an imbalance of too much heat with "cooling" herbal teas, e.g., tea of lemon grass, or a diet that avoids fats and spices classified as hot. In some cases, small quantities of the excessive quality (in the direction of imbalance) are introduced to prevent too severe a jolt to the system, which it is believed would further aggravate illness.

Hot–cold classifications of foods, medicines, and illnesses have been reported from many areas of the world (for reviews, see Logan, 1977; Messer, 1981a), but they display great variability in how they are conceived and how they operate in local dietary and health practices. For example, Hong Kong Chinese classify the healthy body state as slightly hot and foods as either hot or cold, although food dimensions may include, in certain instances, a wet–dry coordinate and perhaps represent overall a condensation of the classical Chinese health cosmology built out of yin and yang and the five elements (Anderson and Anderson, 1977). The Indian cosmological system was based on the concept of five humors, although foods, again, for practical purposes, are classified as either hot or cold—cold foods being, in general, those that are easier to digest—and the ideal body state is conceived as slightly cool. Greco-Arabic medicine and food classifications, by contrast, were expressed in terms of hot–cold and wet–dry dyads, with ideal body state more balanced. In the Mexican example above, hot and cold are again the predominant idiom, but in contrast to classical doctrines of the Chinese, Indian, and Greco-Arabic, the classification has little import beyond, in certain cases, regulating diet in relation to health (Messer, 1981a).

Opinions may differ as to qualities to be assigned to particular items even within the same culture. In the absence of a written tradition or authoritative oral sources, there are often no absolute classifications in most popular cultures that use hot–cold as a standard to diagnose and treat illness and to maintain health. People combine inherited knowledge with cause-and-effect reasoning from the basic hot–cold principle of balancing opposites in particular curing contexts and thereby certify the classification of particular items for future reference. In most cultures, the hot–cold values of particular foods and medicines may be judged by assessing a number of sensory factors (e.g., color and taste) and reputed physiological effect, as well as nominal factors (inherited lists of hot and cold foods and remedies), which are either accepted on faith or added to and reconfirmed through experience. In certain cases, the classifications of food involve modification of their reputed initial quality by certain cooking and spicing techniques (Molony, 1975). In other cases, foods may be evaluated as "good" for curing certain conditions that are known to be either hot or cold, so the classification (by the principle of opposites) is arrived at by recognition of physiological effect (for a review, see Messer, 1981a).

The hot–cold classification, however, may have little practical value for determining the majority of foods that people ordinarily eat. While all members of a population may know of the potential existence of such categories, individuals may refer to such categories for the purpose of food selection only under conditions of physiological stress (infancy, pregnancy, illness, old age), rather than for regular selection of dietary items. More significant factors in ordinary food selection are usually flavor preferences and cost. Nevertheless, hot–cold and related categories may be very useful for individuals to diagnose, *ex post facto*, what foods caused particular illnesses, and hot–cold reasoning will then direct what remedies, including corrective diet, are administered. For modern Western health practitioners, it may be necessary to understand such classifications to prescribe a diet that will not go against hot–cold beliefs and practices (Harwood, 1971). For those introducing new foods to infants and children, the perceived hot–cold values (primarily whether, during initial trials, the new foods made infants sick) may affect the acceptance of new foods (Golpadas *et al.*, 1975). In some contexts, new foods may be perceived as "nourishing" and outside the hot–cold evaluation system (Cosminsky, 1977; Messer, 1981a).

Additional folkloric dimensions may also condition the appropriateness of certain foods, particularly for those in a delicate condition, e.g., pregnant, postpartum, and lactating women (e.g., Manderson and Mathews, 1981). In each cultural case, however, individual knowledge and use of hot–cold and other cultural information must be considered to understand whether they have significance for actual dietary constructions and, consequently, nutritional significance.

3.3. Health Factors

Other characteristics that affect cultural and individual food choices are health factors as culturally construed. These include concepts of "safe" or "harmless" foods (foods that do not make one sick), as well as foods that are believed to be positively good for health and well-being, i.e., "nutritious," "vitamin-rich," and "tonic" foods, all of which qualities may alternatively be glossed as "healthful." These categories often subsume perceptions of physiological effect; foods that are good to eat and good for you are also those that seem to elicit good appetite and promote energetic well-being. In children, they furthermore include foods that seem to foster growth and good health. Within the "healthful" category, foods classified as "nourishing" and "vitamin-rich" are often analyzed as "neutral" categories with respect to hot–cold, in the sense that they are considered to be generally beneficial, and can be consumed in quantity without harm (e.g., Cosminsky, 1977; Messer, 1976b). Such neutral foods often incorporate the basic staple of the diet, the nourishing value of which is also evaluated according to "fillingness," i.e., the ability to produce and sustain feelings of satiety. Certain Latin American cultures also judge the food values of comestibles according to their "juiciness" (J. C. Young, 1981). "Dry" foods like cheese and dried fish are considered insufficient to sustain "life force" or "strength" over long periods of time. More generally, the "cooked" food that defines a "meal" as opposed to a "snack" in particular cultures, such as "cooked rice" in the Philippines, may be categorically defined as "nourishing" and therefore basic to health and nutrition.

The reputed healthfulness of particular foods for promoting both physical and spiritual well-being may also be integrally tied to hot–cold, yin–yang, or some other "humoral" or cultural symbolic quality. In Indian cultures, for instance, foods classified as "cool" are generally considered as more healthful than those classified as "hot." Such beliefs are tied to more general cosmological conceptions of the physiological relationships among ingestion, digestion, and health. Health food "faddism" in Western culture is an additional example of how people use food classified as "healthful" to control both physical and emotional well-being. In this case, "natural foods" (and natural food therapies) are seen as a kind of alternative spiritual and physical health-maintenance system. In contrast to Indian or Chinese conceptual systems, Western health food faddists can point to no single classification system relating foods and health or any unified philosophy, and there is great individual variation among participants (Kandel and Pelto, 1979). Needless to add, "folk" and "scientific" ideas of "healthfulness" vary considerably.

People acquire ideas about the healthfulness of foods from diverse sources. Currently, in both Western countries and the Third World, information about the nutritional value of foods is derived from advertising and medical personnel, as well as from accumulated cultural hearsay. Like the hot–cold classifications discussed above, nutritional and "vitamin" categories ordinarily may have little impact on diet, but under conditions of stress, particularly weakness due to illness, people may attempt to eat more "vitamin-rich" food or take vitamin tonics (Messer, 1981a).

Recent attempts to quantify the nutritional knowledge, beliefs, and attitudes that characterize food choices seem to indicate that the nutritional knowledge alone of the food provider is insufficient to predict beneficial choices (for methodologies see Sims, 1978; Worsley, 1980; Caliendo *et al.*, 1976). Even with adequate scientific nutrition knowledge, considerations of flavor and cost seem to take precedence over criteria of healthfulness (Dewalt and Pelto, 1976). Foods that are sweet or fatty or both are still preferred in many areas despite growing evidence and dissemination of nutrition information that high intakes of either are unhealthful. Along these lines, it has been suggested that preferences for both sugar and fat are part of the human evolutionary heritage and will be overcome only by concerted cultural and nutritional strategies aimed at offsetting these basic biological dietary tendencies (see, for example, Neel, 1962; Beller, 1977). Tastes for sugar, fat, and salt may also be culturally ingrained from an early age. As mentioned above, a rush of energy accompanies concentrated sugar consumption; people who find that this sensation is pleasant and motivates them for work, may be reluctant to forego it (c.f., Messer, 1979). Fatty portions, because they are considered tastier and perhaps more filling, are often the meat cuts of choice in developing countries. High salt intakes and preferences for salty foods may also be culturally ingrained from an early age. In some cultures, for example, in certain parts of Mexico, meat, certain other foods, and even water if the body is overheated, are considered to be "unhealthful" (that is, they will make one sick) unless taken with liberal quantities of salt (McCullough, 1973; Messer, 1981a).

In each of these cases, healthful food habits or "tastes" in one environment may prove unhealthful in another. For example, the types and concentrations of calories supplied by sugar and fat may be excessive where people do less physical work. People may also be disinclined to believe that there are negative health effects of particular diets unless they or close relatives have personally experienced the cause-and-effect relationships between particular foods and illness, such as diabetes, which is scientifically associated with high sugar intake. People thus tend to continue to select diet in large part on the basis of taste and appeal, rather than reputed nutritional value and possible larger health consequences, unless presented with the immediate specter of supposed nutrition-related illness.

3.4. Economic Factors

Economic factors also influence how people act on nutrition information. Availability of foods in the environment and the relative cost (in time and money) of acquiring and then processing particular classes of foods are considerations in any cultural survival strategy. Seasonality of wild and cultivated resources and scheduling of food-acquisition activities among different resources have been basic factors in the evolution of diets in both the Old World and the New (Flannery, 1973). With the expansion of the modern market system and cash economies, limitations of seasonality have to some extent been overcome. In their place, limitations on cash income and lack of cultural knowledge about nutritional needs and the nutritional content of foods have become key factors in the construction of diets. Similarly, allocation of time to non-food-related activities influences how much time people can schedule for food acquisition and preparation and consequently affects food selection and nutritional value of diets in modern market economies.

Several recent studies have shown that even with adequate nutrition knowledge, people may not have the economic wherewithal to feed themselves at optimal or even adequate levels. Calloway and Gibbs (1976), in their study of food choices on native American reservations, concluded that inadequate nutrition levels were a result not of lack of nutritional knowledge, but of lack of money. Dewalt and Pelto (1976), by having informants in a Mexican village rate foods according to taste, healthfulness, and economic value, similarly found that people had accurate notions of "nutrition," but chose foods largely according to budgetary considerations.

The need to pay more attention to economic factors in patterns of food intake has been noted in general nutrition surveys as well. Omawale (1980) concluded from a Philippine household survey that malnutrition among pre-school-age children generally was correlated to the ratio of household income to the cost of calories in the diet. Households were likely to eat more expensive calories when they shifted from subsistence agriculture to cash-cropping. He suggested that the income must double to maintain "subsistence" levels of calorie intake. Schuftan (1979) suggested that since the real problem in malnutrition is lack of food-purchasing power, nutrition programs should address malnutrition in economic rather than nutritional terms; i.e., they should measure deficits in the purchasing power of the households and give priority to nutrition programs that generate income and new employment opportunities in food production and food-related services.

Thus, there is no simple way to calculate food selection in strictly "economic" terms without also considering the cultural context.

3.5. Status Factors

Given these cultural classifications, foods are also judged to be more or less appropriate for certain classes of individuals and for certain occasions. Age, sex, physiological, and social (ritual and economic) status enter into the decisions regarding food choices, as do the "prestige" values of certain items. Certain foods, for example, are judged to be especially good or "edible" only for children. They are those observed to be pleasing to children; often they are bland (not overly spicy) and easy to digest. In simple terms, such foods do not make children sick and are therefore believed to foster growth. They may include food preparations not ordinarily consumed by adults, for example, the usual staples of a culture prepared in a teethable form or ground into a consistency acceptable for infants. Special preparations without spices and special gruels may also be part of early-childhood cuisines. Not all cultures, however, provide foods prepared especially for children, a feature that often makes it difficult for very young children to take in sufficient calories of the adult diet.

In approaching adult diet, certain foods may be withheld until the child can chew, others until the child demands them. Such restrictions may have their basis in ideas of indigestibility; for example, whole grains of rice and beans are never served to children in certain cultures because people observe that whole grains pass undigested through the digestive tract into the stools. Or hot–cold classifications may influence the foods deemed healthful for children; e.g., young children in the cultural case of Mexico, among others, are classified as predominantly cool and have their intakes of "cold" foods restricted, particularly if they are ill (Messer, 1981a). Where these limitations are observed [in some parts of Southeast Asia, they seem to be losing force (Manderson, 1981)], nutritional interventions must either try to change these food beliefs and practices or look for alternative sources of vitamin A in the environment that are not classified as "cool" (Van Veen, 1971).

As this example suggests, local beliefs and practices may interfere with or may not provide for children's receiving the optimal nutritional benefit from the local environment. For example, local infant-feeding customs may not provide for a nutritionally well-rounded mixture of locally available grains and legumes or tuber and legume mixtures prepared especially for children. Or legumes may not be specially prepared in a form that is easily digested by children and may therefore, as local cultural wisdom in these cases suggests, not be good for them. Nutrition programs have tried to remedy these problems both by designing special high-nutrient weaning foods and by developing from local ingredients nutritionally balanced mixtures that can be blended at the household or community level (for a review of weaning foods, see Uddin Ahmed *et al.*, 1981). However, in certain cases, cultural factors affecting poor infant and child nutrition go beyond simply developing appropriate foods. They may involve more complex beliefs about and attitudes toward children at the household level, e.g., systematic favoring or neglect of children of one sex or at one position in the birth order. In such cases, rules for intrahousehold distribution of food as well as actual practices must be examined to see whether there is systematic discrimination against, for example, children of female sex, as has been suggested to be true in rural Bangladesh, or whether such practices vary according to other socioeconomic factors (Rizvi, 1981).

Points in the female or male reproductive cycle, physiological change stages, and advanced age are also marked by food prescriptions and proscriptions in many cultures.

Similarly, illnesses, particularly those believed to be related to superhuman contacts—possession by spirits—may result in special food demands and privileges. More generally, culturally recognized "illness" is marked by alteration in eating behavior as part of "social behaviors" in most of the world. Unwillingness to eat and lack of appetite or occasionally overindulgence are signs of illness, restoration of "normal" appetite the sign of renewed health state. Curing procedures often include initial restriction on food intake, based either entirely on physical cues or on a set of culturally prescribed rules, usually followed by a regimen of foods to regain strength while protecting digestion (e.g., Fortes and Fortes, 1936; Colson, 1976). "Psychosomatic" states, including "fright" and "sadness" classified as "illness" with known sets of symptoms in some cultures, may also be marked, on one hand, by perceptible changes in appetite on the part of the patient and, on the other, by special foods and medicines to stimulate return to a normal state, including "healthy" appetite. Such food restrictions, proscriptions, and prescriptions may be of general health benefit for resting digestion or providing for rehydration, restoration of electrolyte balance, or even pharmacological benefit (e.g., Etkin and Ross, 1982). Alternatively, rules restricting fluids and other protein foods, particularly in cases of infant diarrhea, may be of great harm (Escobar, 1982).

Finally, "appropriate foods" may be more generally related to the society's social division of labor and male domination–female subordination roles (e.g., Tuzin, 1972; McKnight, 1973; O'Laughlin, 1974; Sillitoe, 1981). More generally, ritual and economic status may direct food intake and avoidances. As mentioned above, members mark totemic, caste, and religious-group affiliations by sharing food avoidances, festival foods, and ordinary food preparations and consumption in common. Prestige factors also affect food choices, concepts of what is culturally "appropriate" for one of given socioeconomic status (or pretensions) to consume. Thus, in Latin America, ethnographers report an increasing tendency of people to forego "wild" greens in favor of cultivated vegetables because consumption of wild foods is considered to be a sign of poverty (Messer, 1978; DeWalt *et al.*, 1979). The problem of acceptability of low-cost foods for the poor has been commented on as well in the United States, where there are no foods produced especially for "the poor," but even if there were and even if they were highly nutritious, poor people would not willingly accept them (Gershoff, 1971). To increase the acceptability

of foods aimed at the poor, whether new high-yield cereal grains of less preferred taste and texture or other nutritious, cost-effective foods that otherwise
lack high status, their prestige value must also be raised, perhaps by consci-
entious campaigns showing their consumption by prestigious members of the
community or government (Committee on Food Habits, 1943). On the opposite
side of the budgetary spectrum, relatively expensive foods may be consumed
out of proportion to expenses for other food items because they are of high
cultural value, as is the case with carbonated beverages. Thus, the various
dimensions of personal status—biological, social, and economic—can positively or negatively affect nutrient intake; they supply in large part the contexts
in which other symbolic factors in food classification and selection operate and
thereby condition the nutritional impact of these other cultural food rules.

3.6. Dietary Structure

Beyond such food classifications with respect to general edibility, preferences, symbolic values, and appropriateness according to status, people,
unlike other animals, also structure food consumption in terms of dietary patterns. These include ordinary daily rounds of meals and snacks, as well as
annual cycles of feasts and fast days, which in combination with classifications
of individual foods comprise the group's "food habits." How often one eats,
the times of day or night that one eats, the kinds of food one eats in general
and on each eating occasions, and with whom one eats are ways of commu-
nicating information about one's own social and cultural identity and relationships with others. The "food code"—the kinds of foods (for instance, cooked
vs. uncooked) one consumes at particular times of the day—and the individuals
one invites to share each repast are basic to deciphering the cultural rules of
etiquette, identity, and membership in social groups (Douglas, 1972). They
provide an additional key to criteria for food selection and individual nutrition.

Food patterns are constructed by encoding foods according to their proper
place in the diet. Foods may be classified as "meal items," including staples
such as cereal grains, tubers, and plantains; "relishes" (i.e., preparations eaten
along with staples) and "condiments" such as salt, mineral earths, and chili
peppers; or as categories appropriate for particular meals, such as "toast" for
breakfast and "meat and potatoes" for dinner foods in American society
(Lewin, 1943). Alternatively, they may be deemed "snack foods," i.e., those
that are "uncooked," consumed without the major cooked staple, or eaten
outside the regular meal settings. Or certain dishes may be considered as special
foods eaten only at feasts or as famine food consumed only if other more
preferred foods are unavailable.

Studies in the "foodways" of different groups classify foods into "core,"
"secondary core" ("specialty"), and "peripheral" ("alternative") items to
describe the stable vs. more variable elements of the cultural diets (Passin and
Bennett, 1943), or they study meal-planning according to folk categories such
as the Italian-American mode of varying in meals the basic elements "pasta"
and "gravy" (Theophano and Curtis, 1981). What constitutes a "meal," as

opposed to a "snack," is defined in every culture and determines what are "appropriate foods" at different times of the day. In Malaysia, for example, cooked rice defines a "meal"; all food eaten without cooked rice is defined as a "snack." For those recording dietary intake, it is useful to realize that "snack foods" may not be listed in recall surveys, because they may not be considered "real foods"; also, comestibles considered inferior and exploited only in times of scarcity or those ordinarily consumed only by infants or foraging youngsters, not ordinarily by adults, may go unreported even if they are part of dietary intake for those individuals or adults.

How many meals are prepared and their timing during the day are also culturally determined. The number, timing, structure, and contents of meals may be relatively fixed, as in upper-class British society (Douglas, 1972), or the numbers of meals and ingredients may vary throughout the year according to seasonal resources and festivities. Many African horticultural societies adjust downward from two meals to one meal a day and rely on more or less palatable wild foods as harvest stores are depleted (Fortes and Fortes, 1936). Rationing, for the purpose of extending limited resources, is common in most cultures without unlimited food supplies, although the timing of such rationing and the mechanisms by which people compute how much food to allow to daily meals have not been carefully studied. If firewood or other fuel supplies are short or expensive or both, and if the food provider–preparer has other work responsibilities, the number of meals defined as "cooked food" may be restricted. Employment or school schedules also affect how work or meals structure the day in modern industrialized societies. For example, Rotenberg (1981) analyzed how changing work schedules in post-World War II Viennese society caused a shift from five to three meals a day and also the consumption of more convenience (prepared) foods.

Factors of income levels, perceived prestige values, and availability of fresh foods and other ingredients in different cultural contexts also affect continuities in ethnic dietary structure, as does the ongoing presence of a social group that shares such ethnic foodways (Kolasa, 1978; Jerome, 1979a).

3.6.1. Dietary Complexity

In addition to examining these multiple sociocultural determinants of food selection and dietary construction, nutritionists have tried to quantify some of these qualitative aspects of dietary patterns and relate them to nutritional content and adequacy of diets. "Food diversity indices" or "dietary complexity" scores that calculate the numbers of different food items consumed by individuals or households over a given period of time, for example, have been used as a surrogate measure of nutritional adequacy of diets (Schorr *et al.*, 1972; Romero de Gwynn and Sanjur, 1974; Sanjur and Romero, 1975; Caliendo *et al.*, 1976). The measures act as an even more accurate predictor, particularly of infants' and children's nutritional status, when certain protein factors, such as milk, are weighted (Romero de Gwynn and Sanjur, 1974). Food diversity indices might be of potential utility for understanding why, in certain cultural contexts, children undereat and suffer malnutrition despite apparently adequate supplies of the staple grain or tuber; the diet may be too boring (N. L. Wilson and R. H. L. Wilson, 1971; Gershoff *et al.*, 1977). Food diversity indices might also be developed as a tool to monitor how availability of variety influences food choices and therefore levels of consumption of those who overeat. In modern industrialized societies, where people are exposed to extreme amounts and varieties of foods, people need to have new ways to control intake, since the satiety that accompanies habitual exposure to a very monotonous diet is absent (Mead, 1964). Analysis of factors such as novelty, satiety, and complexity can also contribute to studies of food acceptance and preference (Köster, 1981).

Dietary complexity might also be used to roughly screen households for malnutrition within a poor community where increased complexity indicates a diet with greater variety and, consequently, superior nutritional value. DeWalt *et al.* (1979), by contrast, used food complexity scores as one of several ways to compare diets of rich and poor, both qualitatively and nutritionally, in a Mexican community. Particular aspects of dietary complexity, such as demand for a minimum level of variety or for a certain number of "fatty" meal components at different socioeconomic levels, might be worked into linear-programming models of dietary construction (Calavan, 1976). However, for purposes of anticipating nutritional outcome, rules of community food redistribution and intrahousehold distribution of food may be more significant.

3.7. Social Factors: Food Distribution and Sharing

Generally, throughout the world, sharing food is indicative of friendly social relations. Food offerings are the sign of hospitality; giving and accepting food are the introduction to further social intercourse. In most societies, the people to whom one gives food, and the types of food one gives them, are specified by the degree of kinship or other formalized or friendly social relations. Not offering food is a sign that one is not honoring expected social obligations. The patterns of "eating together" may also replicate certain other basic social relations, such as relations between the sexes (see, e.g. Johnson, 1980).

Food-sharing, which has been cited as basic to human society and the evolution of social organization, is one of the means by which people in hunting–gathering and horticultural societies maintain cooperation for other pursuits. Food-sharing rules are also one of the ways in which patchy (uneven) environmental resources may be systematically passed from fortunate to less fortunate groups and individuals (Weissner 1981); competitive food-giving is a mode of assuring that power and prestige accrue to individuals by virtue of their largess (and skill) at managing food supplies (M. Young, 1971). Even in more complex societies, food-sharing has its political and nutritional dimensions, as in the institutionalized food-sharing that goes on during feasting. Diskin (1978) and Greenberg (1981) have suggested that high-quality protein and vitamin-rich foods that are assembled and distributed during the cycles of festivals in Mexican towns may make substantial contributions to the diets of poor participants. Traditional timing, furthermore, seems to have been such that food distribution took place at times of the year when the population was most in need of protein and nutrients.

While cultural practices of food-sharing between individuals or households and within groups would seem to ensure the survival of the group, they apparently operate more consistently in times of plenty than in times of dearth. Various ethnographic cases indicate that under conditions of potential starvation, people avoid their obligations to share by hiding and hoarding food. Since most "hospitality" obligations are triggered when food is visible, during times of famine, food is simply not brought out. People lie about their food supply, saying that they have nothing to offer; later, they eat alone (Richards, 1939). Generosity may be limited to immediate, rather than extended, kin networks (Raymond Firth, 1959). In general, food-sharing obligations shrink; they apply to fewer people (Dirks, 1978). Or food sharing may be eliminated altogether, even among close kin, including parents and children. Turnbull (1972) found that among the drought-plagued Ik, families not only did not share with one another, but also children over 3 years of age were left to fend for themselves. In less desperate situations, where food is feared to be in short supply, people may steal food that is ordinarily forbidden or shared (Turnbull, 1965). However, Raymond Firth (1959) found that food taboos deemed essential to maintaining social identity and the sanctity of authority held, even as theft increased.

More generally, Laughlin and Brady (1978) have suggested how the extent of cooperation in social relations can be quantified under different conditions of resource availability (Fig. 2). If such a scheme could be combined with quantitative measures of the amount of structure in rules for classifying and sharing different types of food (Douglas and Gross, 1981), one might be able to diagram changes in the social units of food production and distribution. This could be a first step to "quantifying" descriptions of social breakdown in nutritional systems, including the kinds of changes in social organization of food-related tasks that accompany moves from rural to urban environments.

Each of the foregoing studies suggests additional ways to conceptualize and quantify the factors in social organization that contribute to nutritional well-being. Such factors are rarely captured by correlations of *single* social factors with nutritional status, since nutritional well-being of individuals is the result of the interaction of numerous factors operating within the household and community, such as the social organization of food production and food preparation, rules for distribution of food within the household, and regulations for eating and exchange with members of other households. Only by considering such social factors, in combination with other cultural and psychological aspects of food habits, and environmental factors impinging on health status, will a more accurate picture of the sources of malnutrition be drawn.

3.8. Eating Behaviors: Hunger, Satiety, and Body Image

In addition to the sensory and cultural aspects of individual foods and the sociocultural considerations in the ways diets are constructed, the food selec-

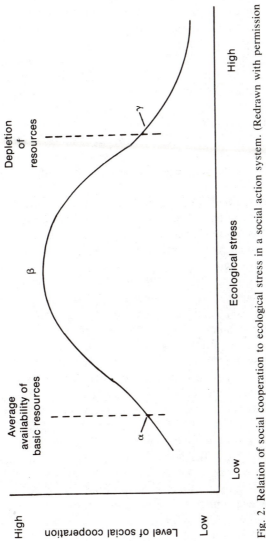

Fig. 2. Relation of social cooperation to ecological stress in a social action system. (Redrawn with permission from Laughlin and Brady, 1978, p. 34.)

tions of humans are also determined by cultural conceptions of appropriate eating behaviors—in particular, by culturally prescribed responses to hunger, satiety, appetites for energy and for specific nutrients, the food demands of infants and children, and particular food cravings. The initiation, duration, and termination of feeding episodes are culturally as well as biologically determined.

Although people in most cultures seem well aware of their needs for energy and other nutrients—they associate lack of physical energy and consciously restrict activity with not having enough to eat and describe seasonal patterns of weight loss or gain in terms of fluctuations in diet (e.g., Richards, 1939; Marshall, 1976; Wilmsen, 1978)—certain investigators have questioned whether it is a physiological change in what food means inside the body or a cultural sense of what constitutes "hunger" that affects perceptions of hunger, what people eat, and consequently the value of nutrients taken in in different resource seasons (e.g., Ogbu, 1973; Pagezy, 1982). Several of the studies already described (e.g., Holmberg, 1950; W. Shack, 1971) indicate that cultural rules and practices, rather than the actual food supply or physiological hunger, often dictate how much people consume of what kinds of foods in particular contexts.

Desirable body image and the cultural desire to control appetite are other factors that may be integrally linked to culturally conditioned fear of hunger, which in turn affects consumption styles and body weight. The spiritual and material ideal of underconsumption may be expressed by a cultural ideal of thinness, the visible sign of moral "success," as among the Gurage (W. Shack, 1971), and women in Indonesia (Poorwo, 1962). United States and European cultures place a high value on thinness as a sign of successfully controlled appetites; fatness is associated with lack of self-control, even immorality (Beller, 1977). Food anxieties in these Western cases, expressed as fears of overconsumption and fatness, can lead to dieting and to several species of eating disorders—anorexia nervosa and bulimia, illnesses specific to middle- and upper-class Western populations and virtually exclusive to women.

In other instances, eating behavior may be conceived as subject to magical or spiritual control, as among the Trobriand and Goodenough Islanders. Malinowski (1935) and M. Young (1971), respectively, studied forms of yam magic that attempted to protect supplies by controlling appetites.

Efforts to overcome the desire to eat by spiritual means are part of the cultures of religious ascetics or of general religious observance in many parts of the world. Fasting as a votive or pentitential act or as a political tactic (the hunger strike) is well known. The symbolic impact of such fasting probably reached its pinnacle as used by Mahatma Gandhi to protest British subjugation of India. In that case, the political message was built on the sentiments surrounding real material dearth, a highly developed religious tradition of fasting to gain spiritual merit, and the pervasive sociocultural symbolism of creating social identity and social relations through differential food acceptance (see, for example, Erikson, 1969).

By contrast, traditional fear of hunger, due to scarcity in one environment, may place a cultural value on eating and even cause overeating if the food environment changes. Change in foods available, along with limitations on physical activity, may contribute to obesity and its health consequences when food is abundant. For example, overconsumption and obesity among Puerto Ricans living in a mainland environment of greater plenty have been blamed, in part, on residual fear of hunger and on warm emotional bonds traditionally associated with feeding (Massara, 1980). Such cultures also tend to see fatness in general as a sign of health, wealth, and well-being. Their conceptual images of what are unacceptably high body weights may fall far above weights recommended for health by medical researchers and physicians (Massara, 1980). Associations of fat or thin body images with particular personality types, and potential fertility and sexual allure (especially for females) have been characteristic of different cultures since ancient times (cf. Beller, 1977).

Finally, particular cultural dietary patterns can influence physiological determinants of hunger and satiety and perhaps specific food cravings. As mentioned above, many groups have tastes or cravings for meat, fat, salt, and certain other flavors, textures, and substances that are culturally developed, even if they are physiologically based. The custom of meeting a pregnant woman's food cravings, for instance, is widespread. Individual cravings for nonfood or quasi-food substances like clays, starch, and paint—an eating behavior known as "pica"—are represented in many parts of the world (e.g., Cooper, 1957; Lackey, 1978). Certain Latin American cultures recognize by the name *antojos* ("desires") otherwise inexplicable skin blemishes or other illnesses (Suárez, 1974), which they believe originate from unmet food cravings. The remedy for them is to feed (or apply locally) the longed-for food. Thus, types of individual food cravings may be culturally recognized and labeled.

In summary, then, ideas of what constitute unhealthful, unattractive, and unwealthy body weights may also differ by culture. All these factors contribute to eating habits that may not be favorable for individual nutrition and health.

3.9. Nutritional and Health Consequences of Dietary Practices

Nutritional consequences of a given diet are a function of how well people recognize nutrient needs, then select, process, and organize consumption of available foods to meet such requirements. Factors affecting the nutritional quality of diet and nutrient intakes of individuals, households, and communities include cultural techniques of food-processing and customs dictating what foods are eaten together, as well as social rules for food distribution within and between households and other sociocultural behaviors that affect the health environment.

In general, the "wisdom" of many traditional diets, which put together a reasonably balanced vegetable protein from cereals and legumes or a healthy mix of calories and protein from a starchy staple plus meat or fish, has been recognized and lauded, particularly in studies of diets subsequently disrupted by economic "development." The numbers of species from which hunter–gathers (e.g., Lee, 1968) and horticulturalists (e.g., Brokensha and Riley, 1980) gained their living, some of the more unusual animal and plant species that have been selectively captured and consumed by different peoples, and the mineral-rich salts that were traditionally exploited by native American groups among others (Kühnlein, 1980) suggest that native peoples, left undisturbed, were usually well adapted nutritionally to their local and regional environments. Traditional food-processing techniques also contributed to the nutritional quality of locally developed diets. Such techniques further extended the quantity and variety of foods available for cultural consumption. Examples are the custom of alkali-processing maize in Latin America, which improved digestibility, enhanced protein quality, and added calcium, and the detoxification of poisonous tubers and cultivation of microorganisms as part of food preparation, spicing, and preservation in South America, Oceania, and Southeast Asia, respectively (e.g., Katz *et al.*, 1974; Rogers, 1965; Yen, 1975; Stanton, 1969). Additional customs, such as the oriental habits of eating sprouted legumes and the consumption of certain fermented products, increased the vitamin B content of such diets. The use of chili peppers high in vitamins A and C and the indigenous Northwest Pacific Coast habit of preserving very sour berries high in vitamin C along with less sour (Charlene Martinsen, 1982, personal communication) are examples of culturally developed taste and dietary techniques for enhancing the quantities of these vitamins in annual consumption. Even the cultural "rules" for hot–cold balancing through spices have been interpreted as nutritionally, metabolically, and hygenically beneficial from medieval through modern times (Arber, 1953). Not only may they enhance vitamin and mineral contents of diets by enforcing variety and restore electrolyte balance after exertion by mandating consumption of water with salt (McCullough, 1973), but also rules for eliminating dangerous "cold" qualities of stored cooked foods may include heat processing useful for preventing growth of bacteria (Erasmus, 1952). As indicated above, other food habits have been noted to be of pharmacological benefit (Etkin, 1979). Additional examples include the possible anthelmintic action of *Chenopodium* and cucurbits consumption, among Mexican and paleo-Indian populations of the Great Basin (Kliks, 1975); the protective role of fibers in such prehistoric diets and the possible significance of naturally occurring dietary antibiotics to human health (Keith and Armelagos, 1983).

In each of these situations, it is necessary to ask what specific cultural food tastes or rules enabled a group to meet its nutrient needs and how information from sensory perceptions and preferences was channeled into rules of food preparation and dietary construction. While in general such dietary adaptations were "successful" in that they enabled populations to survive, there are also abundant examples in history of "maladaptive" strategies; among them are the culturally preferred white polished rice of Far East Asia, known to be associated with thiamine deficiency (beri-beri); nutritionally disadvantageous food restrictions in many cultures on women and children; and sociocultural food habits that contribute to food poisoning, either because foods or parts of foods are initially poisonous, improperly prepared to render them nonpoisonous, or distributed in such a manner that they are tainted when consumed. Over the long run, certain foods—such as large quantities of salted fish and pickled vegetables in the Japanese diet; nitrosamines, consumed in Western European and American diets in the form of spiced, smoked meats and fish; and Western diets high in refined carbohydrates and saturated fats but low in dietary fiber—are dangerous in that they seem to be carcinogenic (Dunn, 1975; Marquandt *et al.*, 1977). Yet, prior to modern scientific nutritional and health analyses, there appeared no external rules discouraging such consumption patterns.

Thus, certain cultural diets seem to militate against health in the short or long run; furthermore, cultural patterns of food intake may not be optimal for all individuals or social categories even where they seem to be of overall benefit.

3.9.1. Infant- and Child-Feeding Practices

One category of nutritional behavior that can negatively affect the youngest members of a society is infant- and child-feeding practices, particularly the widespread shift in developing countries from breast feeding to bottle feeding and "social factors" in child care that lead to inadequate provisioning of the child's needs or to neglect. The causes of pathologies related to poor infant-feeding practices are both sociocultural and biological. Usually, Western culture, particularly the promotional techniques of multinational corporation formula food manufacturers (Grenier, 1975, 1977a,b), is blamed. Alternatively, the decrease in breast feeding has been interpreted bioculturally. Mothers who have trouble initiating breast feeding due to failure of the "let-down" response, particularly where young women have no support networks to teach and encourage them, may give up and bottle feed. Another difficulty, "insufficient milk syndrome" (Gussler and Briesemeister, 1980), refers to premature weaning or early termination of breast feeding due to mothers' perceptions that their infants were hungry; i.e., they had "insufficient milk" to satisfy them. This syndrome has been traced to cultural styles of breast feeding that space nursing at sufficiently long intervals to induce hunger and anxiety in the infant, reduce stimulation of the breast, and perhaps result in further feeding difficulties (Gussler and Briesemeister, 1980). Some have suggested that early introduction of supplementary foods, including sweet infant formulas, may lead to an infant's dissatisfaction with the breast and thus induce such behavioral symptoms of "milk insufficiency" (Grenier *et al.*, 1981).

Also affecting decisions to breast feed are social philosophies, convenience, and the relative costs and "prestige" of breast feeding or not; women may also bend to the preferences of their households in such decisions (G. Pelto, 1981). The effect of medical advice should also be considered (Helsing, 1976).

Feeding customs following or in addition to breast feeding and the mother's use of time—how much time she has to spend with the child—and how well she supervises the child's nutrient intake and health also affect the child's ongoing nutritional status and health (for a review of this literature, see Popkin, 1980). The time immediately following weaning may be important. Physical separation from the mother, as well as inadequate food, may contribute to cultural patterns of failure to thrive (Dean, 1962). The social organization of food-provisioning and child care enter into the equation of whether the child thrives—important issues are who feeds the child, how often, in what quantities, and what types of foods, and who meets the child's emotional needs. How the mother manages her time, her money, and her children are aspects of "mothercraft" that contribute to why some children thrive and others fail to thrive in apparently the "same" environments (Pollitt, 1973; Cravioto and DeLicardie, 1976; Kerr *et al.*, 1978; Franklin and Valdés, 1979; Waldmann, 1976, 1980).

The quantity and quality of parent–child interactions within household units are additional "social factors" affecting the child's nutritional status and well-being, although the relative contributions of the extra stimulation vs. the extra nutrient intake are not clear. For example, intensive observations in a Mexican intervention study indicated that infants who received nutritional supplementation were substantially more active and interactive with their environments (including their mothers and fathers) than their unsupplemented counterparts (Chávez and Martinez, 1979), although the effects of the extra stimulation in addition to the food provided by the program may have contributed to this outcome.

Local understandings of and responses to mild to moderate nutritional disorders and clinical nutritional disease also affect nutritional and behavioral outcome. In many cases, child-feeding customs and food classifications contribute to clinical deficiencies such as protein–energy malnutrition and avitaminosis A (Burgess and Dean, 1962). The general sanitation conditions and particular sources of infection within the household or larger residential and community areas also have impact on the nutritional well-being of the child. Interactions among all economic, sociocultural, and disease factors further condition how the child fares in nutrition, health, and other interrelated aspects of behavioral development (Fonaroff, 1968; Frankenberg and Leeson, 1976; Taha, 1979; McKay, 1980; Mamarbachi *et al.*, 1980; Herrera *et al.*, 1980). While there have been numerous studies attempting to link levels of childhood malnutrition with particular socioeconomic factors in (usually household) environments, simply quantifying the socioeconomic factors within households may not be sufficient to understand the relationships between "social factors" and malnutrition. Rather, it is the way the various resources are organized, how adults manage their time in relation to child-rearing, and how a child learns to fend for himself within such environments that determine the nutritional outcome, measured in physical and functional terms.

3.10. Adaptive Value of Food Classifications, Dietary Structure, and Eating Behaviors

The preceding section has considered how different sociocultural groups classify, prepare, and consume foods to maintain the well-being of individuals, households, and communities.

While the transformation from hunting–gathering to food-producing economies generally reduced the diversity of foodstuffs produced and consumed within any local environment, it has also been argued, in general, that traditional "food habits" in most areas were "adaptive" before they were disrupted by the influences of Western culture; that is, they sustained the local populations that evolved with them and ensured the perpetuation of the natural environment and food resources as well. Although the preceding section has noted how food rules restricting the intakes of women and children, cultural customs that do not optimize use of the environment or that do not arrange consumption throughout the year to ensure maximum energy and nutrient availability, and social customs that encourage the consumption of nonnutritious or poisonous substances all interfere with the well-being and survival of individuals and populations, they face the problem presented by any studies that attempt to assess the "adaptive" values of particular traits, which is that they are often shortsighted. The "adaptive value" of a trait might be to preserve a particular social or cultural pattern that, in turn, better enables that human population, as a group, to survive, although such food habits may still be amenable to nutritionally "rational" change (see e.g. Cassell, 1955). Or traits may simply be neutral, neither positive nor negative in furthering the survival of individuals and the group, and therefore not selected against. These biological vs. cultural aspects of adaptation have been summarized by Durham (1976, 1979), among others.

In addition, the time frame during which traits are evaluated as adaptive is usually too short; as has been indicated, certain tastes, such as those for sugar and fat, may have been adaptive and therefore selected for in one environment or at one period, but prove detrimental in the next. Also, spatial and sociocultural frames of reference are usually too limited. Although most of the discussions of how cultural factors in food selection presented in this section have concentrated on local or regional sociocultural factors that affect how the nutritional environment is created and used, the nutritional adaptability of local populations involves political and economic processes at national and international levels. The articulation of these different levels and the formation and evaluation of food policy will be referred to in Section 4 (see also Berg, 1973; Scrimshaw and Wallerstein, 1982).

4. Nutrient Intake and Function

The potential for human and economic development of larger social units cannot ignore the impact of adequate health and nutrition at the local, household, and individual levels. How does the nutritional intake of individuals, households, and communities affect all aspects of their human performance? Conversely, how does performance of individuals, households, and communities affect their food intake and general health and well-being? What are the interconnections between mild to moderate malnutrition at the individual level and well-being at the household and community levels? Correspondingly, how does community organization of food supplies affect the individual and the household?

A comprehensive interdisciplinary project to address these questions (Calloway *et al.*, 1979) considered the impact of mild to moderate malnutrition on disease and the immune response, physical activity, work performance, reproductive behavior, cognitive performance, and general levels of social functioning as a biocultural system (Fig. 3). Some of the "sociocultural factors" affecting nutrient intake included the general cultural conditions of food production or market availability, cultural rules for food selection, preparation, distribution, and storage, and the social networks that provide food and other kinds of care for individuals of all ages. Nutrient intake was considered for the purposes of this study primarily in terms of energy, or calories.

While the research protocol for this ongoing project proposes to collect and analyze data on disease and activity patterns in primarily biological terms, such measurable outcomes of nutritional status also have their cultural components. For example, cultural perceptions of health and disease and cultural conceptions of the role of diet in maintaining physical and emotional well-being affect nutrient intake, on one hand, and the *social* (behavioral) response to illness, on the other. While the maximum work capacity at a given nutritional or health status can be measured in terms of VO_2 max, the kinds of training tasks the culture provides and the expectations it builds for work performance and energy expenditure are biocultural constructions. Various aspects of reproductive behavior are also affected by sociocultural rules—among them are decision-making regarding pregnancy and childbirth and the food beliefs and practices surrounding the pregnant, postpartum, and lactating woman. Postpartum taboos on sexual activity—for instance, until a child is weaned—affect birth-spacing and thereby influence various dimensions of the mother's and child's health and activities. The overall availability of food within the culture may also affect seasonality of births, an aspect of birth-spacing (Wilmsen, 1982; Lindenbaum, 1976). Cultural constraints on breast feeding have already been discussed above. Inadequate infant-feeding patterns will also affect the child's cognitive and behavioral development, both directly, by limiting brain and physical growth, and "indirectly," by creating a context in which the child may be ill and lethargic much of the time, incapable of interacting with and learning from his environment to the fullest extent. Inadequate nutrition of adults in the child's environment also affects his mental development, since the adults are less energetic and therefore act and interact with him less.

"Social functioning," relative to nutrient intake, however, is more difficult to describe and quantify in biological and psychological terms. It refers to the capacity of an individual to perform sociocultural tasks relative to some cultural or scientific standard—as in the case of mother's ability or inability to complete all housekeeping and feeding tasks, to fully or less adequately care for children and to provide them with a clean, safe, healthful, and stimulating environment—because of a particular level of energy intake, contributing to better to poor nutritional status, which may be further aggravated by illness, apathy, and irritability. The implications of one person's functioning, particularly that

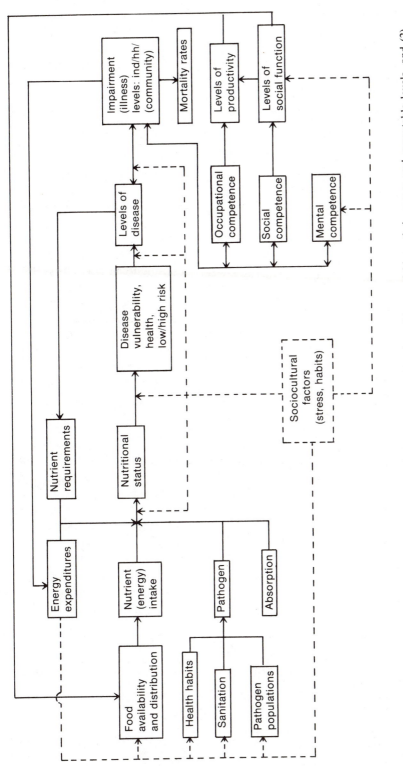

Fig. 3. General research procedure from the systems perspective: (1) identification of key variables—their ranges and acceptable levels, and (2), establishment of systemic interrelationships between variables in (——, biological; ----, sociocultural) terms of energy, information, rates of flow; bottlenecks in e.g. seasonal, social structural terms. ind (individual), hh (household).

of the focal female, for the health and nutritional welfare of the rest of the household, particularly the children, have already been suggested in a number of the economic, nutritional, and anthropological studies cited above. The household organization and the individual woman's abilities, within the general sociocultural "patterns," also make nutritional and behavioral differences, and demand further study (Messer, 1981b). In the case of children, one might look at such behavioral measures as school attendance, learning performance and conduct as these relate to nutrient intake and other aspects of their social and physical environment. In the case of adults, e.g. male workers, additional food intake, particularly where it overcomes deficiencies in energy, protein, and iron, can contribute to the overall quality of their lives; after work, they are better able to participate in other social activities, including community development projects (see summary of studies in Viteri, 1982).

Such studies are attempts to deal with nutritional status or outcome in terms other than more standard anthropometric biofunctional or even cognitive measures, and try to trace how malnutrition perpetuates the cycle of poverty. Although nutritionists and physiologists have contributed to understanding how the human body adjusts to inadequate calorie and protein intake, e.g. by limiting activity, growth, and slowing basal metabolic rate, they also recognize that such so-called "adaptations" are an important part of the problem of undernutrition and social behavior, by which some populations and individuals do not achieve their physical and social potential because of the prevailing availability and distribution of resources (Calloway, 1978; Gopalan, 1978; for policy implications see also essays in Scrimshaw and Wallerstein, 1982). It remains for nutritionists, psychologists, anthropologists, and others—working together—to establish the dimensions and refine the measures of such nutritional and sociocultural interactions at theoretical and practical levels. They can then use such knowledge to contribute to the formation, at international and national levels, of food policies that will provide for more adequate distribution of food resources within communities and households and establish the basis for individuals and communities to achieve their full biological and sociocultural potential.

5. References

Ahern, E., 1975, Sacred and secular medicine in a Taiwan village: A study of cosmological disorders, in: *Medicine in Chinese Culture*, pp. 91–113, U.S. Government Printing Office for Fogarty International Center, N.I.H., Washington, D.C.

Amerine, M. A., 1966, Flavor as a value, in: *Food and Civilization* (S. M. Farber, N. L. Wilson, and R. H. L. Wilson, eds.), pp. 104–120, Charles C. Thomas, Springfield, Illinois.

Anderson, E. N., and Anderson, M. L., 1977, Modern China: South, in: *Food in Chinese Culture: Anthropological and Historical Perspectives* (K. C. Chang, ed.), pp. 317–382, Yale University Press, New Haven, Connecticut.

Apte, M. L., and Katona-Apte, J., 1981, The significance of food in religious ideology and ritual behavior in Marathi myths, in: *Food in Perspective* (A. Fenton and T. N. Owen, eds.), pp. 9–22, John Donal, Edinburgh.

Arber, A., 1953, From medieval herbalism to the birth of modern botany, in: *Science, Medicine, and History: Essays on the Evolution of Scientific Thought and Medical Practice Written in Honor of Charles Singer*, Vol. 1 (E. A. Underwood, ed.), pp. 317–336, Oxford University Press, London.

Arnott, M. (ed.), 1976, *Gastronomy: The Anthropology of Food Habits*, Mouton, The Hague.

Autret, M. M., 1962, Comment on the social psychology of food habits, in: *Malnutrition and Food Habits* (A. Burgess and R. F. A. Dean, eds.), p. 78, Macmillan, New York.

Bahr, D. M., Gregorio, J., Lopez, D. I., and Alvarez, A., 1974, *Piman Shamanism and Staying Sickness*, University of Arizona Press, Tucson.

Barthes, R., 1975, Toward a psychosociology of contemporary food consumption, in: *European Diet from Pre-Industrial to Modern Times* (E. Forster and R. Forster, eds.), pp. 47–59, Harper Torchbooks, New York.

Bartoshuk, L., Gentile, R. L., Moscowitz, H. R., and Meiselman, H. L., 1974, Sweet taste induced by miracle fruit (*Synsepalum dulcificum*), *Physiol. Behav.* 12:449–456.

Beck, B., 1969, Colour and heat in South Indian ritual, *Man* 4:553–572.

Bell, F. L. S., 1931–1932, The place of food in the social life of central Polynesia, *Oceania* 2:117–135.

Beller, A. S., 1977, *Fat and Thin: A Natural History of Obesity*, McGraw-Hill, New York.

Berg, A., 1973, *The Nutrition Factor: Its Role in National Development*, Brookings Institution, Washington, D.C.

Bolton, R., 1973, Aggression and hypoglycemia among the Quolla: A study in psychobiological anthropology, *Ethnology* 12:227–257.

Bolton, R., 1981, Susto, hostility and hypoglycemia, *Ethnology* 20:261–276.

Braudel, F., 1981, *Civilization and Capitalism, 15th–18th Century*, Vol. 1, *The Structures of Everyday Life: The Limits of the Possible* (translated from the French and revised by S. Reynolds), Collins, London.

Brokensha, D., and Riley, 1980, Mbeere knowledge of their vegetation and its relevance for development: A case study from Kenya, in: *Indigenous Knowledge and Development* (D. Brokensha, D. Warren, and O. Werner, eds.), pp. 111–127, University Press of America, Washington, D.C.

Brokensha, D., Warren, D. M., and Werner, O. (eds.), 1980, *Indigenous Knowledge Systems and Development*, University Press of America, Washington, D.C.

Brown, A., 1978, The impact of economic development on health in the Papaloapan, Presented at the Annual Meeting of the American Anthropological Association, Los Angeles.

Brush, S., 1977, *Mountain, Field, and Family: The Economy and Human Ecology of an Andean Valley*, University of Pennsylvania Press, Philadelphia.

Burgess, A., and Dean, R. F. A. (eds.), 1962, *Malnutrition and Food Habits*, Macmillan, New York.

Calavan, M. M., 1976, A model of food selection in northern Thailand, *Ecol. Food. Nutr.* 5:63–74.

Caliendo, M. A., Sanjur, D., Wright, J., and Cummings, G., 1976, Use of PATH analysis as a statistical technique for the analysis of qualitative nutritional data, *Ecol. Food. Nutr.* 5:75–81.

Calloway, D. H., 1978, Discussion, in: *Progress in Human Nutrition*, Vol. 2 (S. Margen and R. Ogar, eds.), pp. 157–158, Avi, Westport, Connecticut.

Calloway, D. H., and Gibbs, J. C., 1976, Food patterns and food assistance programs in the Cocopah Indian community, *Ecol. Food Nutr.* 5:183–196.

Calloway, D., 1979, Précis: Collaborative Research Support Program on Intake and Function, University of California, Berkeley (mimeographed).

Cantlie, A., 1981, The moral significance of food among Assamese Hindus, in: *Culture and Morality: Essays in Honour of Christoph von Furer-Haimendorf* (A. C. Mayer, ed.), pp. 42–62, Oxford University Press, Delhi.

Caporael, L. R., 1976, Ergotism: The Satan loosed in Salem?, *Science* 192:21–26.

Cassell, J., 1955, A comprehensive health program among South African Zulus, in: *Health, Culture, and Community* (B. D. Paul, ed.), pp. 1–41, Russell Sage Foundation, New York.

Chagnon, N., 1977, *Yanamamo: The Fierce People*, 2nd ed., Holt, Rinehart, and Winston, New York.

Chagnon, N., and Hames, R., 1979, Protein deficiency and tribal warfare in Amazonia: New data, *Science* **203**:910–913.

Chang, K. C., (ed.), 1977, *Food in Chinese Culture: Anthropological and Historical Perspectives*, Yale University Press, New Haven, Connecticut.

Chávez, A., and Martinez, C., 1979, *Nutrición y Desarollo Infantil*, Interamericana, Mexico City.

Colson, A. B., 1976, Binary oppositions and the treatment of sickness among Akawaif, in: *Social Anthropology and Medicine* (J. Louden, ed.), pp. 422–499, Academic Press, New York.

Committee on Food Consumption Patterns, Food and Nutrition Board, National Research Council, 1981, *Assessing Changing Food Consumption Patterns*, National Academy Press, Washington, D.C.

Committee on Food Habits, 1943, The Problem of Changing Food Habits, National Research Council Bulletin No. 109, Washington, D.C.

Committee on Food Habits, 1945, Manual for the Study of Food Habits, National Research Council Bulletin No. 111, Washington, D.C.

Conklin, H. C., 1954, The relation of the Hanunóo culture to the plant world, Ph.D. dissertation, Yale University, New Haven, Connecticut.

Cooper, M., 1957, *Pica*, Charles C. Thomas, Springfield, Illinois.

Cosman, M. P., 1976, *Fabulous Feasts: Medieval Cookery and Ceremony*, George Braziller, New York.

Cosminsky, S., 1977, Alimento and fresco: Nutritional concepts and their implications for health care, *Hum. Org.* **36**:203–207.

Cowan, A. B. (ed.), 1978, *The International Conference on Women and Food*, 3 volumes, Consortium for International Development, United States Agency for International Development.

Cravioto, J., and DeLicardie, E. R., 1976, Microenvironmental factors in severe protein-calorie malnutrition, in: *Nutrition and Agricultural Development: Significance and Potential for the Tropics* (N. S. Scrimshaw and M. Béhar, eds.), pp. 25–35, Plenum Press, New York.

Crosby, A., 1972, *The Columbian Exchange: Biological and Cultural Consequences of 1492*, Greenwood Press, Westport, Connecticut.

Cussler, M., and De Give, M. L., 1952. *'Twixt the Cup and the Lip: Psychological and Socio-Cultural Factors Affecting Food Habits*, Twayne, New York.

Dahlberg, F. (ed.), 1981, *Woman the Gatherer*, Yale University Press, New Haven, Connecticut.

Damon, A., 1977, *Human Biology and Ecology*, W. W. Norton, New York.

Dean, R. F. A., 1962, Malnutrition in its setting, in: *Malnutrition and Food Habits* (A. Burgess and R. F. A. Dean, eds.), pp. 19–29, Macmillan, New York.

Den Hartog, A. P., and Bornstein-Johannson, A., 1976, Social science, food and nutrition, in: *Development from Below: Anthropologists and Development Situations* (D. C. Pitt, ed.), pp. 97–123, Mouton, The Hague.

Desor, J. A., Greene, L. S., and Maller, O., 1975, Preference in sweet and salty in 9- to 15-year-old and adult humans, *Science* **190**:686–687.

DeWalt, K. M., and Pelto, G. H., 1976, Food use and household ecology in a Mexican community, in: *Nutrition and Anthropology in Action* (T. Fitzgerald, ed.), pp. 79–93,

DeWalt, K. M., Kelly, P. B., and Pelto, G. H., 1979, Nutritional correlates of economic microdifferentiation in a Highland Mexican community, in: *Nutritional Anthropology* (N. Jerome, R. Kandel, and G. Pelto, eds.), pp. 205–221, Redgrave, Pleasantville, New York.

Dewey, K. G., 1980, The impact of agricultural development on child nutrition in Tabasco, Mexico, *Med. Anthropol.* **4**:55–78.

Dirks, R., 1978, Resource fluctuations and competitive transformations in West Indian slave societies, in: *Extinction and Survival in Human Populations* (C. D. Laughlin and I. A. Brady, eds.), pp. 123–180, Columbia University Press, New York.

Diskin, M., 1978, Discussion, in: Symposium on Mexican Food Systems, 77th Annual Meeting of the American Anthropological Association, Los Angeles.

Douglas, M., 1957, Animals in Lele religious thought, *Africa* **27**:46–58.

Douglas, M., 1966, *Purity and Danger*, Penguin Books, Baltimore.

Douglas, M., 1972, Deciphering a meal, *Daedalus* **101**:61–81.

Douglas, M., and Gross, J., 1981, Food and culture: Measuring the intricacy of rule systems, *Soc. Sci. Inform.* **20**:1–35.

Draper, P., 1975, !Kung women: Contrasts in sexual egalitarianism in foraging and sedentary contexts, in: *Toward an Anthropology of Women* (R. R. Reiter, ed.), pp. 77–109, Monthly Review Press, New York.

DuBois, C., 1941, Food and hunger in Alor, in: *Language, Culture, and Personality: Essays in Memory of Edward Sapir* (L. Speir, A. I. Hollowel, and S. S. Newman, eds.), pp. 272–281, Sapir Memorial Publication Fund, Menasha, Wisconsin.

DuBois, C., 1944, *The People of Alor, A Social–Psychological Study of an East Indian Island*, University of Minnesota Press, Minneapolis.

Dunn, J. E., Jr., 1975, Cancer epidemiology in populations of the United States—with emphasis on Hawaii and California—and Japan, *Cancer Res.* **35**:3240–3245.

Durham, W., 1976, The adaptive significance of cultural behavior, *Hum. Ecol.* **4**:89–121.

Durham, W., 1979, Toward a coevolutionary theory of human biology and culture, in: *Evolutionary Biology and Human Behavior: An Anthropological Perspective* (N. A. Chagnon and W. Irons, eds.), pp. 39–59, Duxbury Press, North Sciutate, Massachusetts.

Durham, W., 1981, Overview: Optimal foraging strategies in human ecology, in: *Optimal foraging Strategies* (B. Winterhalder and E. A. Smith, eds.), pp. 218–231, University of Chicago Press, Chicago.

Erasmus, C., 1952, Changing folk beliefs and the relativity of empirical knowledge, *Southwest. J. Anthropol.* **8**:411–428.

Erikson, E. H., 1969, *Gandhi's Truth: On the Origins of Militant Non-violence*, W. W. Norton, New York.

Escobar, G., 1982, Social and cultural factors affecting infantile diarrhea in Lima, Peru, Yale University Medical School thesis, New Haven, Connecticut.

Etkin, N. L. (ed.), 1979, Biochemical evaluation of indigenous medical practices: Introduction, *Med. Anthropol.* **3**(4).

Etkin, N., and Ross, P. J., 1982, Food as medicine and medicine as food: An adaptive framework for the interpretation of plant utilization among the Hausa of Northern Nigeria, *Soc. Sci. Med.* **16**:1559–1572.

Evenson, R. F., 1981, Food policy and the new home economics, *Food Policy* **3**:180–193.

Farb, P., and Armelagos, G., 1980, *Consuming Passions: The Anthropology of Eating*, Houghton Mifflin, Boston.

Fenton, A., and Owen, T. M. (eds.), 1981, *Food in Perspective*, Proceedings of the 3rd International Conference on Ethnological Food Research, Cardiff, Wales, 1977, Edinburgh: John Donald Publishers, Ltd.

Ferro-Luzzi, G. F., 1977, Ritual as language: The case of South Indian food offerings, *Curr. Anthropol.* **18**:507–514.

Firth, Raymond, 1929, *Primitive Economics of the New Zealand Maori*, George Routledge, London.

Firth, Raymond, 1936, *We, the Tikopia: A Sociological Study of Kinship in Primitive Polynesia*, Allen and Unwin, London.

Firth, Raymond, 1959, *Social Change in Tikopia*, Allen and Unwin, London.

Firth, Rosemary, 1966, *Housekeeping among Malay Peasants*, 2nd ed., London School of Economics Monographs on Social Anthropology No. 7, Humanities Press, New York.

Fischer, J. L., Fischer, A., and Mahony, F., 1977, Totemism and allergy, in: *Culture, Disease, and Healing* (D. Landy, ed.), pp. 154–159, Macmillan, New York.

Fitzgerald, T., 1976, *Nutrition and Anthropology in Action*, Van Gorcum, Assen, The Netherlands.

Flannery, K. V., 1973, The origins of agriculture, *Annu. Rev. Anthropol.* **2**:271–310.

Fonaroff, A., 1968, Differential concepts of protein–calorie malnutrition in Jamaica: An exploratory study of information and beliefs, *J. Trop. Pediatr.* **14**(2).

Forbes, M. H. C., 1976, Farming and foraging in prehistoric Greece: The nutritional ecology of wild resource use, in: *Nutrition and Anthropology in Action* (T. Fitzgerald, ed.), pp. 46–51, Van Gorcum, Assen, The Netherlands.

Forster, E., and Forster, R. (eds.), 1975, *European Diet from Pre-Industrial to Modern Times*, Harper and Row, New York.

Fortes, M., and Fortes, S. L., 1936, Food in the domestic economy of the Tallensi, *Africa* **9:**237–276.

Fortune, R. F., 1963, *Sorcerers of Dobu*, Dutton, New York.

Foster, G., 1953, Relationships between Spanish and Spanish-American folk medicine, *J. Am. Folklore* **66:**201–217.

Foulks, E. F., 1972, The Arctic Hysterias of the North Alaskan Eskimo, Anthropological Studies No. 10, American Anthropological Association, Washington, D.C.

Franke, R. W., and Chasin, B. H., 1980, *Seeds of Famine: Ecological Destruction and the Development Dilemma in the West African Sahel*, Allanheld, Osmun, Montclair, New Jersey.

Frankenberg, R., and Leeson, J., 1976, Disease, illness, and sickness: Social aspects of the choice of healer in a Lusoka suburb, in: *Social Anthropology and Medicine* (J. B. Louden, ed.), A.S.A. Mongraph 13, pp. 223–258, Academic Press, New York.

Franklin, D., and Valdés, I. V., 1979, Desnutrición infantil y su relación con el tiempo y las habilidades de la madres, *Separata de Cuadernos de Economía* **16**(49):343–358.

Freedman, R. L., 1976, Nutritional anthropology: An overview, in: *Nutrition and Anthropology in Action* (T. K. Fitzgerald, ed.), pp. 1–23, Van Gorcum, Assen, The Netherlands.

Freedman, R. (compiler), 1981, *Human Food Uses: A Cross-Cultural, Comprehensive Annotated Bibliography*, Greenwood Press, Westport, Connecticut.

Gershoff, S. N., 1971, Childhood malnutrition in the United States, in: *Progress in Human Nutrition*, Vol. 1 (S. Margen and N. L. Wilson, eds.), pp. 68–74, Westport, Connecticut.

Gershoff, S. N., McGandy, R. B., Suttapreyasri, D., Choomnoom, D., Nondasuta, A., Uthai, P., Tantiwongse, P., and Viravaidhaya, V., 1977, Nutrition studies in Thailand. II. Effects of fortification of rice with lysine, threonine, thiamin, riboflavin, vitamin A, and iron on preschool children, *Am. J. Clin. Nutr.* **30:**1185–1195.

Gluckman, M., 1963, Rituals of rebellion in south-east Africa, in: *Order and Rebellion in Tribal Africa*, pp. 110–136, Free Press of Glencoe, New York.

Golpadas, T., Srinivasan, N., and Varadarasan, F., 1975, *Project Poshak: An Integrated Health Nutrition Macro/Pilot Study of Preschool Children in Rural Madhya Pradesh*, Vol. 1 and 2, CARE India, New Delhi.

Goody, J., 1958, The fission of domestic groups among the Lo Dagaba, in: *The Developmental Cycle in Domestic Groups* (J. Goody, ed.), pp. 53–91, Cambridge University Press, Cambridge.

Gopalan, C., 1978, Adaptation to low calorie and low protein intake: Does it exist?, in: *Progress in Human Nutrition*, Vol. II (J. Margen and R. Ogar, eds.), pp. 132–141, Avi, Westport, Connecticut.

Greenberg, J., 1981, *Santiago's Sword*, University of California Press, Berkeley and Los Angeles.

Greene, L. S., 1976, *Nutrition and Behavior in Highland Ecuador*, University Microfilms, Ann Arbor.

Greene, L. (ed.), 1977, *Malnutrition, Behavior, and Social Organization*, Academic Press, New York.

Grenier, T., 1975, *The Promotion of Bottle Feeding by Multinational Corporations: How Advertising and the Health Professions Have Contributed*, Cornell International Nutrition Monograph Series, No. 2.

Grenier, T., 1977a, *Regulation and Education: Strategies for Solving the Bottle-Feeding Problem*, Cornell International Monograph Series, No. 4.

Grenier, T., 1977b, *Infant Food Advertising and Malnutrition in St. Vincent*, Unpublished M.Sc. thesis, Cornell University, Ithaca, New York.

Grenier, T., Van Esterik, P., and Latham, M. C., 1981, Commentary on insufficient milk syndrome: A biocultural explanation, *Med. Anthropol.* **5:**233–247.

Gross, D. R., and Underwood, B. A., 1971, Technological change and caloric costs: Sisal agriculture in northeastern Brazil, *Am. Anthropol.* **7:**724–740.

Gross, D., Eiten, G., Flowers, N., Leoi, F., Ritter, M., and Werner, D., 1979, Ecology and acculturation among native peoples of Brazil, *Science* **206:**1043–1050.

Gussler, J. D., 1973, Social change, ecology and spirit possession among the South African Nguni, in: *Religion, Altered States of Consciousness, and Social Change* (E. Bourguignon, ed.), pp. 88–126, Ohio State University Press, Columbus.

Gussler, J. D., and Briesemeister, J., 1980, The insufficient milk syndrome: A biocultural explanation, *Med. Anthropol.* **4:**1–24.

Haas, J. D., and Harrison, G. G., 1977, Nutritional anthropology and biological adaptation, *Annu. Rev. Anthropol.* **8:**69–101.

Harner, M., 1977, The ecological basis for Aztec sacrifice, *Am. Ethnol.* **4:**117–135.

Harper, E. C., 1964, Ritual pollution as an integrator of caste and religion, in: *Religion in South Asia* (E. Harper, ed.), pp. 151–196, University of Washington Press, Seattle.

Harris, M., 1968, *The Rise of Anthropological Theory*, Thomas Y. Crowell, New York.

Harris, M., 1974, *Cows, Pigs, Wars, and Witches: The Riddles of Culture*, Random House, New York.

Harrison, G., 1975, Primary adult lactase deficiency: A problem in anthropological genetics, *Am. Anthropol.* **77:**812–835.

Harwood, A., 1971, The hot–cold theory of disease: Implications for treatment of Puerto Rican patients, *J. Am. Med. Assoc.* **216:**1153–1158.

Hayden, B., 1981, Subsistence and ecological adaptations of modern hunter–gatherers, in: *Omnivorous Primates* (R. S. O. Harding and G. Teleki, eds.), pp. 344–421, Columbia University Press, New York.

Helsing, E., 1976, Is the inevitable, inevitable?, *Ecol. Food Nutr.* **5:**115–117.

Herrera, M. G., Mora, J. O., Christiansen, N., Ortiz, N., Clement, J., Vuori, L., Waber, D., DeParedes, B., Wagner, M., 1980, Effects of nutritional supplementation and early education on physical and cognitive development, in: *Life-Span Developmental Psychology*, pp. 149–184, Academic Press, New York.

Hewitt de Alcantara, C., 1976, *Modernizing Mexican Agriculture: Socioeconomic Implications of Technological Change, 1940–1970*, United Nations Research Institute for Social Development, Geneva.

Holmberg, A. R., 1950, *Nomads of the Long Bow*, Smithsonian Institute Press, Washington, D.C.

Jerome, N. W., 1977, Taste experience and the development of a dietary preference for sweet in humans: Ethnic and cultural variations in early taste experience, in: *Taste and Development: The Genesis of Sweet Preference*, U.S. Department of Health, Education and Welfare Publication No. (NIH) 77-1068, Washington, D.C.

Jerome, N. W., 1979a, Diet and acculturation: The case of black American in-migrants, in: *Nutritional Anthropology* (N. W. Jerome, R. F. Kandel, and G. H. Pelto, eds.), pp. 275–325, Redgrave, Pleasantville, New York.

Jerome, N. W., 1979b, Medical anthropology and nutrition, *Med. Anthropol.* **3:**339–351.

Jerome, N. W., Kandel, R. F., and Pelto, G. H. (eds.), 1979, *Nutritional Anthropology*, Redgrave, Pleasantville, New York.

Jochim, M., 1981, *Strategies for Survival*, Academic Press, New York.

Johnson, O. R., 1980, The social context of intimacy and avoidance: A videotape study of Machiguenga meals, *Ethnology* **19:**353–366.

Kandel, R. F., and Pelto, G. F., 1979, The health food movement, in: *Nutritional Anthropology* (N. W. Jerome, R. F. Kandel, and G. H. Pelto, eds.), pp. 327–363, Redgrave, Pleasantville, New York.

Kardiner, A., 1945, The concept of basic personality structure as an operational tool in the social sciences, in: *The Science of Man* (R. Linton, ed.), pp. 107–122, Columbia University Press, New York.

Katona-Apte, J., 1976, Dietary aspects of acculturation: Meals, feasts, and fasts in a minority community in South Asia, in: *Gastronomy: The Anthropology of Food and Food Habits* (M. L. Arnott, ed.), pp. 315–326, World Anthropology Series, Aldine, Chicago.

Katz, S. H., Hediger, M. L., and Valleroy, L. A., 1974, Traditional maize processing techniques in the New World, *Science* **184:**765–773.

Kehoe, A. B., and Giletti, D. H., 1981, Women's preponderance in possession cults: The calcium deficiency hypothesis extended, *Am. Anthropol.* **83:**549–561.

Keith, M., and Armelagos, G., 1983, Naturally occurring dietary antibiotics and human health, in: *The Anthropology of Medicine, from Culture to Method* (L. Romanucci-Ross, D. E. Moerman, and L. R. Taneredi, eds.), pp. 221–230, Praeger, New York.

Kerr, M. A. D., Bogues, J. L., and Kerr, D. S., 1978, Psychological function of mothers of malnourished children, *Pediatrics* **62**:778–784.

Khare, R. S., 1970, *The Changing Brahmins: Associations and Elites among the Kanya Kubjas of North India*, University of Chicago Press, Chicago.

Khare, R. S., 1976, *The Hindu Hearth and Home*, Carolina Academic Press, Durham, North Carolina.

Kliks, M., 1975, Paleoepidemiological studies on Great Basin coprolites: Estimates of dietary intake and evaluation of the ingestion of anthelmintic plant substances, Archaeology Research Facility Department of Anthropology, University of California, Berkeley.

Kolasa, K., 1978, "I won't cook turnip greens if you won't cook kielbasa": Food behavior of Polonia and its health implications, in: *The Anthropology of Health* (E. E. Bauwens, ed.), pp. 130–140, C. V. Mosby, St. Louis.

Köster, E. P., 1981, Time and frequency analysis: A new approach to the measurement of some less-well-known aspects of food preferences, in: *Criteria of Food Acceptance: How Man Chooses What He Eats* (J. Solms and R. L. Hall, eds.), pp. 240–252, Forster, Zurich.

Kuhnlein, H., 1980, The trace element content of indigenous salts compared to commercially refined substitutes, *Ecol. Food Nutr.* **10**:113–121.

Kuipers, J., 1981, The vocabulary of taste in Sumbanese, Presented at the 80th Annual Meeting, American Anthropological Association, Los Angeles, December 2.

Kumar, S. K., 1977, *Role of Household Economy in Determining Child Nutrition at Low Income Levels: A Case Study in Kerala*, Occasional Paper No. 95, Department of Agricultural Economics, Cornell University, Ithaca, New York.

Lackey, C., 1978, Pica—a nutritional anthropology concern, in: *The Anthropology of Health* (E. E. Bauwens, ed.), pp. 121–129, C. V. Mosby, St. Louis.

Lappé, F. M., and Collins, J., 1977, *Food First: Beyond the Myth of Scarcity*, Houghton Mifflin, Boston.

Laughlin, C. D., and Brady, I. A., 1978, Introduction, in: *Extinction and Survival in Evolutionary Populations* (C. D. Laughlin and I. A. Brady, eds.), pp. 1–48, Columbia University Press, New York.

Leach, E., 1964, Anthropological aspects of language: Animal categories and verbal abuse, in: *New Directions in the Study of Language* (E. H. Lennenberg, ed.), pp. 23–63, M.I.T. Press, Cambridge, Massachusetts.

Leach, E., 1970, *Claude Lévi-Strauss*, Viking Press, New York.

Lee, R. B., 1968, What hunters do for a living, or, how to make out on scarce resources, in: *Man the Hunter* (R. Lee and I. DeVore, eds.), pp. 30–48, Aldine, Chicago.

Lee, R. B., 1969, !Kung bushman subsistence: An input–output analysis, in: *Environment and Cultural Behavior* (A. P. Vayda, ed.), pp. 47–76, Natural History Press, Garden City, New York.

Lee, R., and DeVore, I., 1968, *Man the Hunter*, Aldine, Chicago.

Lee, R., and DeVore, I., 1976, *Kalahari Hunter Gatherers*, Harvard University Press, Cambridge, Massachusetts.

Lévi-Strauss, C., 1963, *Totemism* (translated by R. Needham), Beacon Press, Boston.

Lévi-Strauss, C., 1966, *The Savage Mind*, University of Chicago Press.

Lévi-Strauss, C., 1969, *The Raw and the Cooked* (translated by J. and D. Weightman), Harper and Row, New York.

Lévi-Strauss, C., 1973, *From Honey to Ashes* (translated by J. and D. Weightman), Harper and Row, New York.

Lévi-Strauss, C., 1978, *The Origin of Table Manners* (translated by J. and D. Weightman), Harper and Row, New York.

Lewellyn, T. C., 1981, Aggression and hypoglycemia in the Andes: Another look at the evidence, *Curr. Anthropol.* **22**:347–361.

Lewin, K., 1943, Forces behind food habits and methods of change, in: *The Problem of Changing Food Habits* (Committee on Food Habits), pp. 35–65, Bulletin of the National Research Council No. 108, Washington, D.C.

Lindenbaum, S., 1976, The "last course": Nutrition and anthropology in Asia, in: *Nutrition and Anthropology in Action* (T. Fitzgerald, ed.), pp. 141–155, Van Gorcum, Assen, The Netherlands.

Logan, M., 1977, Anthropological research on the hot–cold theory of disease: Some methodological suggestions, *Med. Anthropol.* 1(4):87–112.

Malinowski, B., 1935, *Coral Gardens and Their Magic: A Study of the Methods of Tilling the Soil and of Agricultural Rites in the Trobriand Islands*, American Book Company, New York.

Malinowski, B., 1946, *The Dynamics of Culture Change: An Inquiry into Race Relations in Africa*, Yale University Press, New Haven, Connecticut.

Malinowski, B., 1948, Introduction, in: *Hunger and Work in a Savage Tribe* (by A. Richards), pp. ix–xvi, Free Press, Glencoe, Illinois.

Mamarbachi, D., Pellett, P. L., Basha, H. M., and Djani, F., 1980, Observations in nutritional marasmus in a newly rich nation, *Ecol. Food. Nutr.* 9:43–54.

Manderson, L., 1981, Traditional food classifications and humoral medical theory in peninsular Malaysia, *Ecol. Food Nutr.* 11:81–93.

Manderson, L., and Mathews, M., 1981, Vietnamese behavioral and dietary precautions during pregnancy, *Ecol. Food Nutr.* 11:81–93.

Marchione, T. J., 1977, Food and nutrition in self-reliant national development: The impact on child nutrition of Jamaican government policy, *Med. Anthropol.* 1(1):57–79.

Marquandt, H., Rufino, F., and Weisburger, J. H., 1977, Mutagenic activity of nitrite treated foods: Human stomach cancer may be related to dietary factors, *Science* 196:1000.

Marriott, M., 1964, Caste ranking and food transactions: A matrix analysis, in: *Structure and Change in Indian Society* (M. Singer and B. Cohn, eds.), Viking Fund Publications in Anthropology 47:133–171.

Marshall, L., 1976, *The !Kung of Nyae-Nyae*, Harvard University Press, Cambridge, Massachusetts.

Massara, E. B., 1980, Obesity and cultural weight valuations: A Puerto Rican case, *Appetite* 1:291–298.

Matossian, M. K., 1982, Ergot and the Salem witchcraft affair, *Am. Sci.*, 70:355–357.

Mauss, M., 1967, *The Gift* (translated by I. Cunnison), W. W. Norton, New York.

McArthur, M., 1974, Pigs for ancestors: A review article, *Oceania* 45:87–123.

McCullough, J. M., 1973, Human ecology, heat adaptation, and belief systems: The hot–cold system of Yucatan, *Southwest. J. Anthropol.* 29:32–39.

McElroy, A., and Townshend, P., 1979, *Medical Anthropology*, Duxbury Press, North Scituate, Massachusetts.

McKay, D. A., 1980, Food, illness, and folk medicine: Insights from Ulu Trengganu in: *Food, Ecology, and Culture* (J. Robson, ed.), pp. 61–66, Gordon and Breach, New York.

McKnight, D., 1973, Sexual symbolism of food among the Wik-Mungkan, *Man* 8:194–210.

Mead, M., 1962, The social psychology of food habits, in: *Malnutrition and Food Habits* (A. Burgess and R. F. A. Dean, eds.), p. 77, Macmillan, New York.

Mead, M., 1964, *Food Habits Research: Problems of the 1960s*, National Academy of Sciences–National Research Council Publication No. 1225, Washington, D.C.

Mead, M., 1977, Contemporary implications of the state of the art, in: *Malnutrition, Behavior, and Social Organization* (L. Greene, ed.), pp. 259–265, Academic Press, New York.

Messer, E., 1976a, The ecology of vegetarian diet in a modernizing Mexican community, in: *Nutrition and Anthropology in Action* (T. Fitzgerald, ed.), pp. 117–124, Van Gorcum, Assen, The Netherlands.

Messer, E., 1976b, The pragmatics of folk classification, *Mich. Discuss. Anthropol.* 2:91–106.

Messer, E., 1978, *Zapotec Plant Knowledge: Classifications, Uses, and Communications about Plants in Mitla, Oaxaca, Mexico*, Memoirs of the Museum of Anthropology, No. 10, University of Michigan, Ann Arbor.

Messer, E., 1979, Some like it sweet, Presented at the 78th Annual Meeting of the American Anthropological Association, November 1979, Cincinnati.

Messer, E., 1981a, Hot–cold classifications: Theoretical and practical implications of a Mexican study, *Soc. Sci. Med.* **15B:**133–145.

Messer, E., 1981b, Getting through three meals a day, Presented at the 80th Annual Meeting of the American Anthropological Association, Los Angeles.

Mintz, S., 1979–1980, Time, sugar and sweetness, *Marxist Perspect.* **2**(4):56–73.

Molony, C., 1975, Systematic valence coding of Mexican hot–cold food, *Ecol. Food. Nutr.* **4:**67–74.

Montgomery, E., and Bennett, J., 1979, Anthropological studies of food and nutrition: The 1940s and the 1970s, in: *The Uses of Anthropology* (W. Goldschmidt, ed.), pp. 124–143, Special Publication of the American Anthropological Association, Number 11, Washington, D.C.

Neel, J. V., 1962, Diabetes mellitus: A "thrifty" genotype rendered detrimental by "progress," *Am. J. Hum. Genet.* **14:**354–362.

Ogbu, J. M., 1973, Seasonal hunger in tropical Africa as a cultural phenomenon, *Africa* **13:**317–332.

O'Laughlin, B., 1974, Mediation of contradiction: Why Mbum women do not eat chicken, in: *Women, Culture, and Society* (M. Rosaldo and L. Lamphere, eds.), pp. 301–318, Stanford University Press, Stanford, California.

Omawale, 1980, Nutrition problem identification and development policy implications, *Ecol. Food Nutr.* **9:**113–122.

Ortiz de Montellano, B., 1982, Aztec sacrifice, Lecture given at Yale University, January 1982.

Pagezy, H., 1982, Seasonal hunger, as experienced by the Oto and the Twa of a Ntomba village in the Equatorial forest (Lake Tumba, Zaire), *Ecol. Food Nutr.* **12:**139–153.

Passin, H., and Bennett, J. W., 1943, Social process and dietary change, in: *The Problem of Changing Food Habits* (Committee on Food Habits), pp. 113–123, Bulletin of the National Research Council, No. 108, Washington, D.C.

Pelto, G., 1981, Perspectives on infant feeding: Decision-making and ecology, *Food Nutr. Bull.* **3**(3):16–29.

Pelto, G. H., and Jerome, N. W., 1979, Intracultural diversity and nutritional anthropology, in: *Health and the Human Condition: Perspectives on Medical Anthropology* (M. Logan and E. Hunt, eds.), pp. 322–328, Duxbury Press, North Scituate, Massachusetts.

Pelto, P. J., and Pelto, G. H., 1975, Intra-cultural diversity: Some theoretical issues, *Am. Ethnol.* **2:**1–18.

Pimentel, D., and Pimentel, M., 1979, *Food Energy and Society*, John Wiley, New York.

Pimentel, D., Hurd, L. E., Bellotti, A. C., Forster, M. J., Oka, I. N., Scholes, O. D., and Whitman, R. J., 1973, Food production and the energy crisis, *Science* **182:**443–449.

Pimentel, D., Dritschilo, W., Krummel, J., and Kutzman, J., 1975, Energy and land constraints in food production, *Science* **190:**754–761.

Pollitt, E., 1973, Behavior of infant in causation of nutritional marasmus, *Am. J. Clin. Nutr.* **26:**264–270.

Poorwo, S., 1962, Malnutrition in Indonesia: The malnourished mother, in: *Malnutrition and Food Habits* (A. Burgess and R. F. A. Dean, eds.), p. 48, Macmillan, New York.

Popkin, B. M., 1980, Time allocation of the mother and child nutrition, *Ecol. Food Nutr.* **9:**1–14.

Popkin, B., and Solon, F. S., 1976, Income, time and the working mother and nutriture, *J. Trop. Pediatr. Environ. Child Health* **6:**156–166.

Rappaport, R. A., 1968, *Pigs for the Ancestors*, Yale University Press, New Haven, Connecticut.

Rappaport, R. A., 1971, The flow of energy in an agricultural society, *Sci. Am.* **225:**117–132.

Reichel-Dolmatoff, G., 1968, *Amazonian Cosmos*, University of Chicago Press, Chicago.

Revelle, R., 1976, Energy use in rural India, *Science* **192:**969–975.

Richards, A. F., 1939, *Land, Labour, and Diet in Northern Rhodesia: An Economic Study of the Bemba Tribe*, G. Routledge, London.

Ritenbaugh, C., 1978, Human foodways: A window on evolution, in: *Anthropology and Health* (E. Bauwens, ed.), pp. 111–120, C. V. Mosby, St. Louis.

Rizvi, N., 1981, Socioeconomic and cultural factors of intrahousehold food distribution in rural Bangladesh, Presented at the 80th Annual Meeting of the American Anthropological Association, Los Angeles.

Robson, J. R. K., and Elias, J. N., 1978, *The Nutritional Value of Indigenous Wild Plants: An Annotated Bibliography*, Whitson, Troy, New York.

Rogers, D., 1965, Some botanical and ethnological considerations of *Manioc esculenta, Econ. Bot.* **19**:369–377.

Romero de Gwynn, E., and Sanjur, D., 1974, Nutritional anthropometry: Diet and health related correlates among preschool children in Bogota, Columbia, *Ecol. Food Nutr.* **3**:273–282.

Ross, E. B., 1978, Food taboos, diet and hunting strategy: The adaptation to animals in Amazon cultural ecology, *Curr. Anthropol.* **19**:1–36.

Rotenberg, R., 1981, The impact of industrialization on meal patterns in Vienna, Austria, *Ecol. Food Nutr.* **11**:25–35.

Rozin, E., 1973, *The Flavor Principle Cookbook*, Hawthorne, New York.

Rozin, P., and Fallon, A. E., 1981, The acquisition of likes and dislikes for foods, in: *Criteria of Food Acceptance: How Man Chooses What He Eats* (J. Solms and R. L. Hall, eds.), pp. 35–45, Forster, Zurich.

Rozin, P., and Schiller, D., 1980, The nature and acquisition of a preference for chili pepper by humans, *Motiv. Emotion* **4**:77–101.

Sahlins, M., 1972, *Stone Age Economics*, Aldine, Chicago.

Sanjur, D., and Romero, E., 1975, Conceptual levels of dietary indicators as predictors of nutritional status, *Nutrition* **2**:214–222.

Schorr, B., Sanjur, D., and Erickson, E., 1972, Teen-age food habits: A multidimensional analysis, *J. Am. Diet. Assoc.* **61**:4.

Schuftan, C., 1979, "Household purchasing-power deficit": A more operational indicator to express malnutrition, *Ecol. Food Nutr.* **8**:29–35.

Scrimshaw, N., and Wallerstein, M. (eds.), 1982, *Nutrition Policy and Implementation. Issues and Experiences*, Plenum Press, New York.

Seeger, A., 1981, *Nature and Society in Central Brazil: The Suya Indians of Mato Grosso*, Harvard University Press, Cambridge, Massachusetts.

Shack, D. N., 1969, Nutritional processes and personality development among the Gurage, *Ethnology* **8**:292–300.

Shack, W., 1971, Hunger, anxiety, and ritual: Deprivation and spirit possession among the Gurage of Ethiopia, *Man* **6**(1):30–43.

Sillitoe, P., 1981, The gender of crops in the Papua New Guinea highlands, *Ethnology* **20**:1–14.

Simoons, F. J., 1960, *Northwest Ethiopia: Peoples and Economy*, University of Wisconsin Press, Madison.

Simoons, F. J., 1973, The determinants of dairying and milk use in the Old World: Ecological, physiological and cultural, *Ecol. Food Nutr.* **2**:83–90.

Sims, L. S., 1978, Food-related value orientations, attitudes, and beliefs of vegetarians and non-vegetarians, *Ecol. Food Nutr.* **7**:23–35.

Solms, J., and Hall, R. L. (eds.), 1981, *Criteria of Food Acceptance: How Man Chooses What He Eats*, Forster, Zurich.

Sperber, D., 1975, *Rethinking Symbolism* (translated by A. L. Morton), Cambridge University Press, Cambridge.

Stanton, W. R., 1969, Some domesticated lower plants in Southeast Asian food technology, in: *The Domestication of Plants and Animals* (P. Ucko and G. W. Dimbleby, eds.), pp. 463–469, Aldine, Chicago.

Steiner, F., 1956, *Taboo*, Penguin, Middlesex.

Steinhart, J. S., and Steinhart, C. E., 1974, Energy use in the U.S. food system, *Science* **184**:307–316.

Steward, J., 1955, *Theory of Culture Change*, University of Illinois Press, Urbana.

Strum, S. C., 1981, Processes and products of change: Baboon predatory behavior at Gilgil, Kenya, in: *Omnivorous Primates: Gathering and Hunting in Human Evolution* (R. S. O. Harding, and G. Teleki, eds.), pp. 255–302, Columbia University Press, New York.

Suárez, M. M., 1974, Etiology, hunger, and folk diseases in the Venezuelan Andes, *J. Anthropol. Res.* **30**:41–54.

Taha, S. A., 1979, Ecological factors underlying protein–calorie malnutrition in an irrigated area of the Sudan, *Ecol. Food Nutr.* **7**:193–201.

Tambiah, S. H., 1968, Animals are good to think and good to prohibit, *Ethnology* **8**:42–59.

Tannahill, R., 1973, *Food in History*, Stein and Day, New York.

Theophano, J., and Curtis, K., 1981, Sisters, mothers and daughters: Food exchange and reciprocity in an Italian American community, Presented at the 80th Annual Meeting, American Anthropological Association, Los Angeles.

Thomas, R. B., 1973, Human adaptation to high Andean energy flow system, Occasional Paper in Anthropology No. 7, Department of Anthropology, Pennsylvania State University, University Park.

Tripp, R., 1981, Farmers and traders: Some economic determinants of nutritional status in Northern Ghana, *J. Trop. Pediatr.* **27**:15–22.

Turnbull, C., 1965, *Wayward Servants: The Two Worlds of the African Pygmies*, Natural History Press, Garden City, New York.

Turnbull, C., 1972, *The Mountain People*, Simon and Schuster, New York.

Turner, V., 1969, *The Ritual Process: Structure and Anti-Structure*, Aldine, Chicago.

Tuzin, D. F., 1972, Yam symbolism in Sepik: An interpretative account, *Southwest. J. Anthropol.* **28**:230–254.

Uddin Ahmed, R., Löwgren, M., Velarde, N., Abrahamsson, L., and Hambraeur, L., 1981, Study of home-prepared weaning foods for consumption in Bangladesh with special reference to protein quality, *Ecol. Food Nutr.* **11**:93–102.

U.S.A.I.D. Women and Food Information Network, *Newsletters*.

Van Veen, M. S., 1971, Some ecological considerations of nutritional problems on Java, *Ecol. Food Nutr.* **1**:25–38.

Vayda, A. P., and Rappaport, R. A., 1968, Ecology, cultural and noncultural, in: *Introduction to Cultural Anthropology* (J. A. Clifton, ed.), pp. 477–497, Houghton Mifflin, Boston.

Veith, I., 1949, *The Yellow Emperor's Classic of Internal Medicine*, University of California Press, Berkeley.

Viteri, F., 1982, Nutrition and work performance, in: *Nutrition Policy Implementation: Issues and Experience* (N. S. Scrimshaw and M. Wallerstein, eds.), pp. 3–13, Plenum Press, New York.

Waldmann, E., 1976, Early Bapedi growth in Sekhukuniland Transvaal, South Africa, *Ecol. Food Nutr.* **5**:133–141.

Waldmann, E., 1980, The ecology of the nutrition of the Bapedi, Sekhukuniland, in: *Food, Ecology, and Culture* (J. Robson, ed.), pp. 47–59, Gordon and Breach, New York.

Wallace, A. F. C., and Ackerman, R. E., 1960, An interdisciplinary approach to mental disorder among the polar Eskimos of Northwest Greenland, *Anthropologica N. Ser.* **2**:249–260.

Weiner, J. S., 1977, Nutritional ecology, in: *Human Biology*, 2nd ed. (G. A. Harrison and J. S. Weiner, eds.), pp. 400–423, Oxford University Press.

Weissner, P., 1981, Measuring the impact of social ties on nutritional status among the !Kung San, *Soc. Sci. Inform.* **415**:641–678.

Whiting, J. W. M., and Child, I. L., 1953, *Child Training and Personality: A Cross Cultural Study*, Yale University Press, New Haven, Connecticut.

Wilmsen, E. N., 1978, Seasonal effects of dietary intake on Kalahari San, *Fed. Proc. Fed. Am. Soc. Exp. Biol.* **37**:65–72.

Wilmsen, E. N., 1982, Studies in diet, nutrition, and fertility among a group of Kalahari bushmen in Botswana, *Soc. Sci. Inform.* **21**(1):95–125.

Wilson, C. S., 1973, Food habits: A selected annotated bibliography, *J. Nutr. Educ.* **5** (Suppl. 1 to No. 1):38–72.

Wilson, C. S., 1979, Food—custom and nurture: An annotated bibliography on sociocultural and biocultural aspects of nutrition, *J. Nutr. Educ.* **11**(4):Suppl. 1.

Wilson, N. L., and Wilson, R. H. L., 1971, Dysnutrition and boredom, in: *Progress in Human Nutrition*, Vol. 1 (S. Margen and N. L. Wilson, eds.), pp. 111–117, Avi, Westport, Connecticut.

Winikoff, B. (ed.), 1975, *Nutrition and National Policy*, M.I.T. Press, Cambridge, Massachusetts.

Winterhalder, B., and Smith, E. A. (eds.), 1981, *Hunter–Gatherer Foraging Strategies*, University of Chicago Press, Chicago.

Worsley, A., 1980, Thought for food: Investigations of cognitive aspects of food, *Ecol. Food Nutr.* **9:**65–80.

Worthington, E. B., 1936, On the food and nutrition of African natives, *Africa* **9**(2):150–165.

Yen, D. E., 1975, Indigenous food processing in Oceania, in: *Gastronomy: The Anthropology of Food and Food Habits* (M. Arnott, ed.), pp. 147–168, Mouton, The Hague.

Young, J. C., 1981, *Medical Choice in a Mexican Village*, Rutgers University Press, New Brunswick, New Jersey.

Young, M., 1971, *Fighting with Food*, Cambridge University Press, New York.

Glossary

Adenosine triphosphatase (Na$^+$-K$^+$-ATPase)   Membrane-bound enzyme needed for ATP utilization; low levels of Na$^+$-K$^+$-ATPase are thought to reflect low activity in the sodium–potassium pump and may contribute to obesity.

Adenosine triphosphate (ATP)   The "energy storehouse" of the body; needed to drive the sodium–potassium pump.

Affective disorder   Illness characterized by an alteration of the affect, or mood, including depression, mania, and anxiety states.

Albumin   A high-molecular weight plasma protein that binds tryptophan and nonesterfied free fatty acids.

Alliesthesia   A change in sensation of pleasantness depending on internal state.

Amenorrhea   Cessation of menstruation; may occur in 30–45% of patients with anorexia nervosa before weight loss.

Amino acids, large neutral   A class of amino acids (phenylalanine, tyrosine, valine, leucine, and isoleucine) that share a common carrier mechanism with tryptophan for transport into brain from blood.

Analysis of variance   A set of statistical procedures for determining whether three or more groups differ significantly in their mean scores on a variable.

Anaphylactic shock   A general systemic shock reaction caused by hypersensitivity to some substance.

Anorexia nervosa   A syndrome consisting of self-induced starvation, fear of fatness, and loss of menstrual periods or sexual drive and often associated with feelings that one's body is larger than it actually is.

Anthropometric measures   Those variables related to physical growth (e.g., height, weight, arm circumference).

Artifactual illness   Factitious or self-induced illness conditions that are self-inflicted and carried out in clear consciousness; the subject, however, is often unaware of why he is doing it, in contrast to malingering.

Attention-deficit disorder (ADD)   A common childhood disorder formerly referred to as hyperkinesis or minimal brain dysfunction (MBD).

Binge   The uncontrolled ingestion over a short time of large amounts of food.

Biopsychosocial model   The concept that illness often results from the interactional effects of psychological and social forces with individual vulnerabilities.

Body-weight set-point   Theoretical construct that suggests that body weight is biologically determined to be maintained at a stable level.

Bonding   The mutual attachment between the mother and her offspring that generally takes place very shortly after birth.

Bradycardia   Heart rate below 60.

Brown adipose tissue (BAT)   A specialized heat-producing organ that is thought to be involved in both cold-adaptation and diet-induced thermogenesis. (See also, **Diet-induced thermogenesis.**)

Bulimia   An eating binge that occurs after unsuccessful resistance; it is followed by a sense of guilt and loss of control and, often, self-induced vomiting or diuretic or laxative abuse.

"Cafeteria" or "junk-food" diet   Dietary paradigm that involves giving animals a choice of foods such as cookies, marshmallows, potato chips, cheese, and salami: the wide choice of presumably highly palatable foods results in overeating and weight gain.

Cognition   Refers to mental examination, comprehension, storage, and retrieval of information.

Cognitive–behavioral therapy   An integration of behavioral therapy with techniques to identify and change abnormal thought patterns.

Cognitive development   The sequential process of acquiring increasingly complex mental abilities.

Confounding   Variance of an uncontrolled factor with the experimental treatment, making it difficult to determine whether the observed changes are due to the experimental treatment.

Construct validity   The degree to which a measurement procedure is related to the theoretical construct it is meant to represent.

Covariation   Variance shared between two variables.

Criterion variable   The variable being measured that is hypothesized to be caused by the independent variable.

Cross-sectional analysis   A data collection–analysis in which individuals from different cohorts are sampled at the same point in time.

Cultural materialism   A scientific research strategy that seeks to understand the causes of differences and similarities among societies and cultures as responses to the practical (economic and ecological) problems of existence.

Curvilinear relationship   An empirical relationship between two variables that is best represented by a function that is curved in shape, for example, a parabolic function.

Cytotoxic test   A method of measuring allergic sensitivity from the effect that suspected substances have on living blood cells.

Delirium   A state of distorted consciousness, usually arising acutely in the setting of physical illness (e.g., infections, drug or alcohol abuse, or deficiency disease), in which awareness is clouded, perceptual abnormalities (i.e., illusions, hallucinations, or delusions) are common, and there is an affect of fear.

Dependent variable   Synonym for criterion variable.

Depression (endogenous or psychotic)   Autonomous depression that is largely independent of external factors and in which there are often delusions, frequently of a self-critical or nihilistic nature, motor changes, a varying degree of loss of insight and suicidal ideas; associated with biochemical changes.

Depression (reactive or neurotic)   Affective disorder that can often be attributed to identifiable external cause (divorce, illness, job loss) and that is less extensive and of shorter duration than endogenous depression; delusions and loss of contact with reality do not generally occur.

Dietary complexity   An index of the numbers of different food items consumed over a given period of time (meal, day, week) by a given social category (individual, household), which is used as a surrogate measure of nutritional status.

Dietary pattern   The arrangement of meals and snacks, principal and supplementary foods, over a given cycle of time (day, week, season, year) within a given culture.

Diet-induced thermogenesis (DIT)   An increase in metabolic heat production resulting from overnutrition.

Dopamine   A compound related to an amino acid named DOPA that is a precursor of norepinephrine, epinephrine, and melanin; it has important central nervous system neurotransmitter functions.

Ecology   The study of the interrelationships between living organisms and their environment, including both physical and biotic features and emphasizing intraspecies as well as interspecies relations, i.e., *cultural ecology*, the relationships among cultures and their environments; *human* *ecology*, the relationships among human populations and their environments; and *socioecology*, the relationships among human social groups and their environments.

Electroconvulsive therapy (ECT)   Induction of a seizure by means of passing a current through the head for purposes of treatment, usually in severe depression; the motor component of the convulsion is abolished by means of a muscle relaxant, and the procedure is normally carried out under general anesthesia.

Energy flow   How energy inflow (intake) matches energy outflow (expenditure).

Ethnoscientific–nutritional research   Anthropological research that elicits and analyzes native ("folk") classifications and evaluations of behaviors relative to environment, foods, health, and illness that, from the native perspective, affect appetite, food intake, and health.

Etiology   The assignment of cause to a physical or behavioral disorder.

Experimental design   A research design that involves the comparison of two or more groups to determine causal inferences with respect to an experimental treatment.

Externality theory of obesity   The theory that the eating behavior of obese individuals is triggered mainly by external food-related cues, rather than by internal physiological signals for hunger.

fa/fa(**Zucker**) **rat**   A genetic form of obesity in rats, also associated with a single recessive gene; obesity in both *ob/ob* and *fa/fa* animals has been hypothesized to involve defects in brown adipose tissue. (See also *ob/ob* **mouse**.)

Food acceptance   The probable consumption of an available food.

Food aversion   The nonconsumption of a food or dietary component developed through conditioning when the food or dietary component in a prior experience has been followed by real or imagined negative internal consequences.

Food basket   A list of food items relevant to a specific purpose.

Food chain   A sequential set of organisms that feed one on the other; an energy pathway.

Food code   Food classifications and ritual uses of particular foods or quantities of foods that are viewed as encoding sociocultural information about social categories, ritual concepts of time, and ritual or social events.

Food habits   The culturally standardized set of behaviors in regard to food that are manifested by individuals who have been raised within a given cultural tradition.

Food intake   The food ingestion or consumption process.

Food intolerance   Physical discomfort emanating from the gastrointestinal tract resulting from the ingestion of a specific food or dietary component.

Food meaning   A value that is attached to foods.

Food perception   Mental impressions or images of food formed as the result of visual, olfactory, gustatory, auditory, and tactile experiences.

Food preference   The ranking of foods according to the degree of liking or disliking.

Food-restricting anorexia   Severe weight loss produced by abstention from food rather than use of medications or vomiting.

Food selection   The choice of one food item over another based on the influence of various exogenous and endogenous factors.

Food sensitivity   An immunologically mediated adverse reaction to food.

Food system   The ecological, economic, sociocultural, and nutritional interrelationships characterizing production, classification, distribution, and consumption of food within a given sociocultural group.

Foodways   The behaviors that affect what people eat, including definitions of what is food and what is not, the methods of preparation and acceptable combinations of foods, and the rules for distributing particular foods.

Food web   A set of interrelated organisms that feed on each other; a set of energy pathways.

Foraging behavior   The activity of hungry animals characterized by searching the environment for food.

Gastric dysrhythmia   A term for abnormal gastic emptying (usually slowed); a common symptom in anorexia nervosa.

Gonadatropins   Pituitary hormones, including luteinizing hormone and follicle-stimulating hormore, that cause ovulation and lead to the production of estrogen by the ovary.

Hallucinations   Perceptual abnormalities in which the perception is false, without an objective stimulus; by contrast, an illusion is a perceptual misinterpretation of a real stimulus.

Health belief   The association between the health qualities of a given substance (e.g., food) and the belief system of the individual.

History effects   Changes in a measurement that are due not to the influence of the independent (presumed causal) variable, but rather to some event occurring during the course of the experiment or data collection that was outside the control of the investigator.

Holistic concept   A concept in Gestalt psychology that assumes that behavior is cohesive, organized, and nonreducible in nature and therefore must be examined and analyzed as an integrated whole.

Hot–cold   One of a number of possible cultural–symbolic (humoral) dimensions used to evaluate an intrinsic quality found in foods that affects the relative balance of that same quality within the human body.

5-Hydroxindole   A metabolic breakdown product of seratonin.

Hyperkinesis   A condition of childhood involving poor attention, restlessness, and impulsive behavior. (See also **Attention-deficit disorder**.)

Hyperphagia   Overeating (*hyper*, "over, more;" *phagia* "eating").

Hypochondriasis   Complaints arising from morbid self-scrutiny or body consciousness.

In vivo   Term denoting a measurement or procedure conducted in the living organism.

In vitro   Term denoting a measurement of procedure conducted outside the living organism, usually in laboratory conditions.

Independent variable   The variable in a research study that the investigator manipulates; hypothesized to be causally related to the outcome variable.

Instrumentation   Change in a measurement over time that is due to changes in the measuring instrument (e.g., a stopwatch "slowing down").

Insufficient-milk syndrome   Premature weaning or early termination of breast feeding due to a mother's perception that her infant is hungry, i.e., that she has "insufficient milk."

Interaction (Statistical interaction)   Results of an experiment showing that the effects of a particular independent variable on a particular dependent variable depend on (i.e., differ as a function of) the level of operation or functioning of some *other* independent variable.

Internal validity   The degree to which an investigator has confidence that an empirical relationship has been identified between an independent and a dependent variable and that this relationship is not a result of one or more confounding variables.

Intracultural variation   Differences in beliefs and behaviors within single societies and cultures.

Intricacy (Structuredness of cultural food systems)   A measure of the information it would take to describe a culture's food rules (meal components, feasting, food exchange).

Kwashiorkor   A form of severe childhood malnutrition resulting largely from protein deficiency; usually occurs postweaning.

Longitudinal study   A study that involves repeated measures of a given cohort over time.

"Luxusconsumption"   Heat production produced by overeating.

Marasmus   A form of severe malnutrition that results from a restriction of both protein and calorie intake in very young infants.

Maturation effects   Change in a measurement due not to the influence of the independent (presumed causal) variable, but rather to developmental change in the subjects.

Mental retardation or handicap   Mental impairment from physical, cultural, or genetic causes characterized by subnormality of intelligence and varying degrees of social and physical incompetence.

Metabolic obesity   Individual differences in energy utilization thought to contribute to differences in body weight and obesity.

Metabolic sodium–potassium pump   The active transport of sodium out of the cell and potassium into the cell; this membrane-bound pump is a major energy-using process in the body and may account for up to one half the basal metabolic rate.

Methylphenidate (Ritalin)   A stimulant drug frequently used in the treatment of hyperkinetic children.

Modified psychosomatic theory   A recent formulation that suggests that only diffuse, uncontrollable arousal leads to overeating, while labeled, controllable arousal does not.

Mother–infant interaction   The reciprocal interplay between a mother and her child, each influencing and being influenced by the other.

Motivated behavior   Behavior characterized by goal-seeking and consummatory phases, such as found in eating, drinking, sexual, and aggressive behavior; related to limbic-system activity.

Neurotransmitters   Compounds that are stored within nerve terminals, released on nerve-cell depolarization, and interact at target-cell receptors to mediate communication between nerve cells and other cells in the body.

Obese eating style   Controversial hypothesis that obese individuals' eating behavior is characterized by large meals and a very rapid rate of food intake.

Obesity   An excessive accumulation of body fat; not simply an increase in body weight.

ob/ob **mouse**   A genetic form of obesity in mice, associated with a single recessive gene. (See also *fa/fa* **rat**.)

Orienting response   A response to a novel stimulus indicative of alerting to the stimulus.

Perceptual distortion   The feeling that one's body is larger than it actually is; *not* a delusion, since patients know that medically they are too thin.

Personality change   Term denoting alterations of the persona or normal personal characteristics with a varying degree of impairment of striving, personal achievement, finer feelings, emotional expression, personal hygiene, self-respect, and behavior seen in a severe brain disturbance for any reason; the condition is usually chronic and frequently irreversible.

Physiological factors   The variety of endogenous stimuli affecting food selection and food intake; some physiological factors are: monoamines, e.g., serotonin; peptides, e.g., cholecystokinin; and others, such as blood glucose levels.

Pica   A disorder of childhood in which the individual has a disturbed appetite leading to the ingestion of foreign substances, dirt, or other matter; most common in children.

Placebo   An inert or inactive pill or substance used in drug studies to control for nonpharmacological effects of taking a pill.

Postprandial   Following a meal.

Power   The ability of a statistical test (i.e., in a given experimental situation) to correctly reject the null hypothesis (of no relationship between independent and dependent variables) when the null hypothesis is false.

Prospective analysis   A method of data analysis in which subjects are either selected for or placed into treatment and control conditions, or both, on the basis of their relative standings on an independent (presumed causal) variable, following which the groups are compared on a dependent variable.

Protective, hypometabolic effect   A general decrease in metabolism and endocrine function associated with starvation.

Psychometric test   Term denoting a measurement of the psyche, e.g., in standardized tests of various aspects of intelligence, personality, and other aspects of mental functioning.

Psychosis   Major mental illness often accompanied by delusions and hallucinations, cognitive impairment, and, occasionally, clouding of consciousness in which the grasp of reality is distorted and insight generally lost.

Psychosis, organic   Loss of contact with reality as a result of physical or drug or alcohol factors or deficiency diseases.

Psychosomatic theory of obesity   The theory that overeating is an attempt to reduce anxiety and other negative emotions; the inappropriate feeding response is thought to originate in childhood when indiscriminate feeding was used to alleviate stress.

Psychotropic drugs   Drugs influencing psychic function that are used in the course of treatment of psychiatric conditions.

Pursuit of thinness   The relentless decrease in body weight characteristic of anorexia nervosa patients.

Quasi-experimental design   Research design that does not permit the inference of causal relationships because of inadequate control groups.

RAST (radioallergosorbeat) test   A method of measuring the presence of antibodies with specific substances for determining allergic response.

Refeeding edema   Retention of fluid associated with feeding starved individuals; pathophysiology unknown.

Reliability   The consistency with which a test or assessment procedure measures a particular characteristic or trait.

Sampling error   Error variance that results from less than perfect agreement between a sample and the population it represents; a type of variability that reduces the capacity of the statistical test to identify a true relationship between independent and dependent variable (i.e., to detect real differences between the groups).

Satiety   A subjective experience resulting in a feeling of satisfaction or fullness on ingestion of food.

Schizophrenia   Mental illness in clear consciousness characterized by blunting or splintering of the personality with impairment of affect, striving, volition, thought processes and content, and finer feelings; insight is usually lost, and without treatment the condition generally becomes chronic.

Selection effect   Difference in mean scores between subjects from different experimental conditions (i.e., in a quasi-experimental design) that occur not because the groups differed on the independent or treatment variable, but because they differed on one or more confounding variables.

Sensory factors   The variety of stimuli affecting food selection, food intake, and food preferences; the five basic sensory stimuli are taste, sight, smell, sound, and touch.

Serotonin (5-Hydroxytryptamine)   A neurotransmitter compound used by neurons, the cell bodies of which are localized in the raphe nuclei of the brain; stimulates smooth-muscle contraction.

Social factors   The variety of exogenous stimuli affecting food selection and food intake; some social factors are price, convenience, and prestige.

Subject mortality   Loss of subjects from a condition in an experiment, resulting in changes in mean score in that condition due to change in group composition.

Taste   Oral sensation of pleasure or displeasure.

Total parenteral nutrition (TPN)   The introduction of large amounts of balanced nutrients through a large vein to rapidly increase weight.

Tryptophan   An essential amino acid that is involved in a number of biochemical pathways, including its precursor role important for synthesis of brain serotonin molecules; present in common protein.

Type I error   Rejection of the null hypothesis that there is no relationship between variables when in fact there is.

Validity   The degree to which the results of an experiment are generalizable (external validity); the degree to which a test measures what it claims to measure (internal validity).

Ventromedial hypothalamus (VMH)   The area of the brain involved in appetite regulation; destruction of this postulated satiety center leads to overeating and obesity.

Visual evoked response A measure of brain response to visually presented information derived from computer averaging of the time-locked EEG following a repeated signal.

Wechsler Intelligence Scale for Children (WISC) A standardized intelligence test for children ages 6–16.

Weight phobia Fear of normal weight, sometimes used as a synonym for anorexia nervosa.

Index